Monographs in Mathematics
Vol. 97

Managing Editors:
H. Amann
Universität Zürich, Switzerland
J.-P. Bourguignon
IHES, Bures-sur-Yvette, France
K. Grove
University of Maryland, College Park, USA
P.-L. Lions
Université de Paris-Dauphine, France

Associate Editors:
H. Araki, Kyoto University
F. Brezzi, Università di Pavia
K.C. Chang, Peking University
N. Hitchin, University of Warwick
H. Hofer, Courant Institute, New York
H. Knörrer, ETH Zürich
K. Masuda, University of Tokyo
D. Zagier, Max-Planck-Institut Bonn

Hans Triebel

The Structure
of Functions

Birkhäuser Verlag
Basel · Boston · Berlin

Author:
Hans Triebel
Mathematisches Institut
Friedrich-Schiller-Universität Jena
Ernst-Abbe-Platz 1–4
07743 Jena
Germany
e-mail: triebel@minet.uni-jena.de

2000 Mathematics Subject Classification 46E45, 28A80, 42C40; 35P, 47B06, 42B

A CIP catalogue record for this book is available from the Library of Congress, Washington D.C., USA

Deutsche Bibliothek Cataloging-in-Publication Data
Triebel, Hans:
The structure of functions / Hans Triebel. - Basel ; Boston ; Berlin : Birkhäuser, 2001
  (Monographs in mathematics ; Vol. 97)
  ISBN 3-7643-6546-3

ISBN 3-7643-6546-3 Birkhäuser Verlag, Basel – Boston – Berlin

This work is subject to copyright. All rights are reserved, whether the whole or part of the material is concerned, specifically the rights of translation, reprinting, re-use of illustrations, recitation, broadcasting, reproduction on microfilms or in other ways, and storage in data banks. For any kind of use the permission of the copyright owner must be obtained.

© 2001 Birkhäuser Verlag, P.O. Box 133, CH-4010 Basel, Switzerland
Member of the BertelsmannSpringer Publishing Group
Printed on acid-free paper produced from chlorine-free pulp. TCF ∞
Printed in Germany
ISBN 3-7643-6546-3

9 8 7 6 5 4 3 2 1   www.birkhauser.ch

Contents

Preface xi

I Decompositions of Functions

1. **Introduction, heuristics, and preliminaries** 1
   Classical function spaces; quarkonial representations; Weierstrassian approach

2. **Spaces on $\mathbb{R}^n$: the regular case** 10
   Sequence spaces b_{pq} and f_{pq}; β-quarks; $B^s_{pq}(\mathbb{R}^n)$ with $s > \sigma_p$ and $F^s_{pq}(\mathbb{R}^n)$ with $s > \sigma_{pq}$; unconditional convergence and optimal coefficients; Weierstrassian approach and Cauchy formula; generalized quarks

3. **Spaces on $\mathbb{R}^n$: the general case** 27
   β-quarks; $B^s_{pq}(\mathbb{R}^n)$ and $F^s_{pq}(\mathbb{R}^n)$ with $s \in \mathbb{R}$; optimal coefficients and Weierstrassian approach

4. **An application: the Fubini property** 34
   Definition; theorem

5. **Spaces on domains: localization and Hardy inequalities** 41
   Definition of $A^s_{pq}(\Omega)$, $\widetilde{A}^s_{pq}(\Omega)$, $\mathring{A}^s_{pq}(\Omega)$ with $A = B$ or $A = F$; relations between these spaces; density assertions; Hardy inequalities for F^s_{pq}-spaces; Hardy inequalities for B^s_{pq}-spaces; refined localization for $F^s_{pq}(\Omega)$ spaces with $s > \sigma_{pq}$; equivalent quasi-norms for F^s_{pq}-spaces; pointwise multipliers; approximate resolutions of unity; refined localization for other cases

6. **Spaces on domains: decompositions** 71
   Approximate resolutions of unity; β quarks; sequence spaces; quarkonial decompositions for F^s_{pq}-spaces on domains; unconditional convergence and optimal coefficients; alternative approach; references

7 **Spaces on manifolds** 81
Riemannian manifolds (M,g) with bounded geometry and positive injectivity radius; d-domains; resolution of unity; the spaces $F_{pq}^s(M,g^\varkappa)$; refined localizations; Laplace-Beltrami operator and related lifts; embeddings; families of approximate resolutions of unity; β-quarks; quarkonial decompositions of $F_{pq}^s(M,g^\varkappa)$; optimal coefficients

8 **Taylor expansions of distributions** 101
Weighted spaces and tempered distributions in $\mathbb{R}^n$; related β-quarks; quarkonial decompositions of weighted B_p^s-spaces on $\mathbb{R}^n$; quarkonial decompositions and Taylor expansions of tempered distributions on $\mathbb{R}^n$; tempered distributions in domains; related β-quarks; quarkonial decompositions and Taylor expansions of tempered distributions on domains

9 **Traces on sets, related function spaces and their decompositions** 120
Finite compactly supported Radon measures in $\mathbb{R}^n$ as distributions; trace operator and identification operator; local means; criteria for the existence of traces of $F_{pq}^s(\mathbb{R}^n)$ on compact sets; related potentials and maximal functions; trace spaces; related comments and references; ball condition; criterion for sets satisfying the ball condition; traces on sets satisfying the ball condition; approximate lattices and subordinated resolutions of unity on compact sets; β-quarks on sets; B_{pq}^s-spaces on sets and related quarkonial decompositions; trace spaces on sets; comments and references

II Sharp Inequalities

10 **Introduction: Outline of methods and results** 161
Preliminary versions of: rearrangements; three cases: sub-critical, critical, super-critical; growth and continuity envelopes and envelope functions; related Borel measures on the real line; related Hardy inequalities

11 **Classical inequalities** 167
Sharp embeddings of $B_{pq}^s(\mathbb{R}^n)$ and $F_{pq}^s(\mathbb{R}^n)$ in: $L_1^{loc}(\mathbb{R}^n)$, $L_r(\mathbb{R}^n)$, $C(\mathbb{R}^n)$, $C^1(\mathbb{R}^n)$; Lorentz-Zygmund spaces; sharp refined classical embeddings in limiting situations of $B_{pq}^s(\mathbb{R}^n)$ and $H_p^s(\mathbb{R}^n)$ in Lorentz spaces and Zygmund spaces and related inequalities; historical comments and references

12 Envelopes 181
Rearrangement and growth functions; related Borel measures on the real line; examples of admitted Borel measures; growth envelope functions; equivalence classes of growth envelope functions; growth envelopes; related inequalities; moduli of continuity; continuity envelope functions; equivalence classes of continuity envelope functions; continuity envelopes; inequalities between moduli of continuity and rearrangements

13 The critical case 202
Calculation of the growth envelopes $\mathfrak{E}_G B_{pq}^{\frac{n}{p}}$ and $\mathfrak{E}_G F_{pq}^{\frac{n}{p}}$; related inequalities; extremal functions; historical comments and references; spaces on domains and related growth envelopes; bmo and its growth envelope

14 The super-critical case 218
Calculation of the continuity envelopes $\mathfrak{E}_C B_{pq}^{1+\frac{n}{p}}$ and $\mathfrak{E}_C F_{pq}^{1+\frac{n}{p}}$; extremal functions; related inequalities; borderline cases; envelope functions and non-compactness; historical comments and references

15 The sub-critical case 229
Calculation of the growth envelopes $\mathfrak{E}_G B_{pq}^s$ and $\mathfrak{E}_G F_{pq}^s$ in the sub-critical case; related inequalities; optimal embeddings in Lorentz spaces and Zygmund spaces; references

16 Hardy inequalities 235
Hardy inequalities as consequences of sharp embeddings: critical case, sub-critical case; references; Hardy inequalities related to half-space, domains, d-sets, Borel measures

17 Complements 243
Green's functions as envelope functions; further limiting embeddings; moduli of continuity; generalized B-spaces and Lip-spaces; logarithmic spaces; compact embeddings; references

III Fractal Elliptic Operators

18 Introduction 251
Preliminary versions of: d-sets and (d,Ψ)-sets, types of fractal elliptic operators, related spectral problems

19 Spectral theory for the fractal Laplacian 253
Physical and mathematical background; the Dirichlet Laplacian and its mapping properties; the fractal elliptic operators $B = (-\Delta)^{-1} \circ tr^\Gamma$ and $(-\Delta)^{-1} \circ \mu$; density of $D(\mathbb{R}^n\backslash\Gamma)$ in some

function spaces: spectral synthesis; related references; spectral theory of the operator B related to d-sets: eigenvalues and eigenfunctions; the case $n = 1$; Weyl measures; strongly diffuse measures; entropy numbers and approximation numbers; fractal drums in the plane; the music of the ferns: Weylian or alien; degenerate case

20 The fractal Dirichlet problem 294

Smooth case, single layer potentials; duality of H^1-spaces in arbitrary domains; H^s-spaces on d-sets; the operator B and single layer potentials on d-sets; mapping properties; the fractal Dirichlet problem; references; Dirichlet problem in domains with fractal boundary; the smooth case; L_2-L_p-theory; classical solutions, Wiener criterion

21 Spectral theory on manifolds 310

Continuation of Sect. 7: Riemannian manifolds M, function spaces $F_{pq}^s(M, g^\varkappa)$, compact embeddings, related entropy numbers; the operator $H_\beta = -\Delta_g + \varrho id - \beta g^{-\varkappa}$; calculation of the negative spectrum of H_β; local singularities and hydrogen-like operators, related entropy numbers and negative spectra; the opinion of the physicists in such hyperbolic worlds

22 Isotropic fractals and related function spaces 329

Analysis versus (fractal) geometry; (d, Ψ)-sets and properties of admissible functions Ψ; related measures are Weyl measures; pseudo self-similar sets and ψIFS; function spaces of type $B_{pq}^{(s,\Psi)}(\mathbb{R}^n)$ and $F_{pq}^{(s,\Psi)}(\mathbb{R}^n)$; related atoms, β-quarks, and quarkonial decompositions; tailored spaces for (d, Ψ)-sets; spaces on (d, Ψ)-sets and related entropy numbers

23 Isotropic fractal drums 348

Extension of the spectral theory in Sect. 19 from d-sets to (d, Ψ)-sets; Weyl measures

IV Truncations and Semi-linear Equations

24 Introduction 355

Preliminary versions: truncation property, truncation couple, Q-operator, semi-linear equations

25 Truncations 357

Real spaces $\mathbb{B}_{pq}^s(\mathbb{R}^n)$, $\mathbb{F}_{pq}^s(\mathbb{R}^n)$, $\mathbb{H}_p^s(\mathbb{R}^n)$; truncation operators T, T^+; examples; Fatou property; truncation property, uniform continuity; truncation couples; calculation of all truncation couples; truncation property for $\mathbb{F}^s_{pq}(\mathbb{R}^n)$; non-linear interpolation; references; truncation property for $\mathbb{H}^s_p(\mathbb{R}^n)$ and $bmo(\mathbb{R}^n)$; Lipschitz-continuity; Hölder-continuity

26 The Q-operator 385
Definition of the Q-operator; properties: boundedness and Lipschitz-continuity

27 Semi-linear equations; the Q-method 389
Semi-linear integral equations and related regularity theory: the Q-method; semi-linear differential equations and related regularity theory; local singularities and related regularity theory; the Q-method revisited

References 403

Symbols 419

Index 423

Preface

This book deals with the symbiotic relationship between

    I Quarkonial decompositions of functions,

on the one hand, and

    II Sharp inequalities and embeddings in function spaces,
    III Fractal elliptic operators,
    IV Regularity theory for some semi-linear equations,

on the other hand.

Accordingly, the book has four chapters. In Chapter I we present the Weierstrassian approach to the theory of function spaces, which can be roughly described as follows. Let ψ be a non-negative C^∞ function in $\mathbb{R}^n$ with compact support such that $\{\psi(\cdot - m) : m \in \mathbb{Z}^n\}$ is a resolution of unity in $\mathbb{R}^n$. Let $\psi^\beta(x) = x^\beta \psi(x)$ where $x \in \mathbb{R}^n$ and $\beta \in \mathbb{N}_0^n$. One may ask under which circumstances functions and distributions f in $\mathbb{R}^n$ admit expansions

$$f(x) = \sum_{\beta \in \mathbb{N}_0^n} \sum_{j=0}^{\infty} \sum_{m \in \mathbb{Z}^n} \lambda_{jm}^\beta \psi^\beta(2^j x - m), \quad x \in \mathbb{R}^n, \qquad (0.1)$$

with the coefficients $\lambda_{jm}^\beta \in \mathbb{C}$. This resembles, at least formally, the Weierstrassian approach to holomorphic functions (in the complex plane), combined with the wavelet philosophy: translations $x \mapsto x - m$ where $m \in \mathbb{Z}^n$ and dyadic dilations $x \mapsto 2^j x$ where $j \in \mathbb{N}_0$ in $\mathbb{R}^n$. Such representations pave the way to constructive definitions of function spaces. We are mainly interested in the two scales B_{pq}^s and F_{pq}^s with $s \in \mathbb{R}$, $0 < p \le \infty$, $0 < q \le \infty$, which cover many well-known classical spaces, such as (fractional) Sobolev spaces, Hölder-Zygmund spaces, Besov spaces and (inhomogeneous) Hardy spaces. The theory of those spaces has been developed systematically by many mathematicians since the early 1960s. The first chapter in [Triγ] is a historically-minded survey of this subject with many references covering the period up to 1990. In Chapter I of the present book we offer the indicated fresh constructive approach to these spaces on $\mathbb{R}^n$, domains, fractals and some manifolds. Chapters II, III, and IV deal with various applications of the results of Chapter I. In Chapter II we contribute to one of the main topics in the theory of function spaces: embeddings and inequalities. We are mostly interested in delicate limiting situations, asking for necessary and sufficient conditions. Chapter III deals with elliptic operators, preferably the Laplacian, in diverse fractal settings, such as fractal boundaries of underlying domains, measure-valued coefficients or potentials, etc. We wish to demonstrate the symbiotic relationship between some basic notation of fractal geometry and spectral theory. This chapter might be considered as the continuation of the Chapters IV and V in [Triδ]. Finally, in Chapter IV of the present book we study truncations in function spaces and we use these results, combined with the quarkonial decompositions indicated above, to develop a new regularity theory for some semi-linear integral and differential equations. Each chapter begins with a separate introduction, where we outline in somewhat greater detail what can be expected.

This book is mainly based on the results of the author and his co-workers obtained in the last few years. We tried to present the material in such a way that the main ideas can be understood independently of the existing literature. On the other hand, after proving in Chapter I that the function spaces introduced via quarkonial decompositions coincide with the well-established spaces B^s_{pq} and F^s_{pq} we feel free to use known results about these spaces, especially when we have nothing new to say about the assertions used. A reader who is mostly interested in the material presented in one of the Chapters II, III, or IV, which are largely independent of each other, may skip Chapter I, at the first glance. But most of the related proofs in these chapters depend substantially on the theory developed in the first chapter.

It is a pleasure to acknowledge the great help I have received from my collaborators in Jena, in particular Dorothee Haroske and Winfried Sickel, who made valuable suggestions which have been incorporated in the text. I am especially indebted to Dorothee Haroske for producing all the figures in this book. Last, but not least, I wish to thank my friend and colleague David Edmunds in Brighton who looked through the whole manuscript and offered many comments.

Jena, Spring 2001   Hans Triebel

Chapter I
Decompositions of Functions

1 Introduction, heuristics, and preliminaries

1.1 Introduction

The function spaces B^s_{pq} and F^s_{pq} on $\mathbb{R}^n$ and on domains with respect to the full range of the parameters

$$s \in \mathbb{R}, \quad 0 < p \le \infty, \quad 0 < q \le \infty, \tag{1.1}$$

were introduced between 1959 and 1975. They cover many well-known classical concrete function spaces having their own history. In 1.2 we give a corresponding short list. These two scales of spaces and their special cases attracted a lot of attention and have been treated systematically, with numerous applications given. We mention in particular the following books, reflecting also the development of this theory: [Sob50], [Nik77] (first edition 1969), [Ste70], [BIN75], [Pee76], [Triα] (1978), [Triβ] (1983) and [Triγ] (1992). Special aspects but related to our intentions have been studied in [Maz85] (Sobolev spaces), [Zie89] (Sobolev spaces), and [ST87] (periodic spaces, anisotropic spaces and spaces with dominating mixed derivatives). The two surveys [BKLN88] and [KuN88] cover in particular the Russian literature. More recent developments of the spaces B^s_{pq} and F^s_{pq} may be found in [ET96], [RuS96], [AdH96], and [Triδ] (1997). We refer the interested reader to Chapter 1 in [Triγ] : in this survey we tried to describe the historical roots and how the diverse ideas developed over the years (up to 1990). There one finds also many references to the original papers.

In the present book we wish to shed new light on functions, distributions and function spaces of the above type. We ask for quarkonial decompositions as described in the Preface and Taylor expansions of functions and distributions; and we extract from the related coefficients all the information we wish to have, for example, to which spaces B_{pq}^s or F_{pq}^s the considered function or distribution belongs. Or, as the other side of the same coin, one may ask for characterizations of these function spaces in terms of quarkonial decompositions and Taylor expansions. It will be our point of view to define function spaces via such expansions and to prove that they coincide with the spaces considered so far. This approach may be considered as a modification of so-called atomic decompositions and (to a lesser extent) wavelet characterizations of function spaces. We add a few references. At least as far as the spaces B_{pq}^s and F_{pq}^s are concerned atomic decompositions go back to [FrJ85] and [FrJ90]. The historical roots may be found in [Triγ], 1.9. Descriptions are also given in [FJW91], [Tor91], [Triγ], [ET96], and [Triδ], Sect. 13. This technique has been widely used. In addition to the papers quoted we refer to [AdH96] and [RuS96]. Wavelet descriptions of function spaces may be found in [Mey92]. Quarkonial decompositions were introduced in [Triδ], Sect. 14. They provided a tool for handling entropy numbers of compact embeddings between function spaces in connection with fractals, which resulted in a related spectral theory of partial differential operators. Other instruments available in the literature seemed to be rather inadequate for this purpose. But now it is quite clear that this approach is more basic than originally thought. It is the aim of Chapter I to give a systematic treatment of this subject.

We outline the contents of this chapter. In the remaining subsections of Section 1 we give in 1.2 a brief description of some classical function spaces. In 1.3, 1.4 and 1.5 we try to provide at a heuristic level an understanding of quarkonial decompositions and Taylor expansions of functions and distributions. For this purpose we look at the Weierstrassian approach to holomorphic functions on the complex plane. Sections 2, 3, 4 deal systematically with quarkonial decompositions for functions and distributions and related function spaces of type B_{pq}^s and F_{pq}^s on $\mathbb{R}^n$, including an application to study the Fubini property. Corresponding considerations for functions and spaces on domains in $\mathbb{R}^n$, some manifolds, and fractals are given in Sections 5, 6, 7, 9. Section 8 deals with Taylor expansions of distributions on domains and on $\mathbb{R}^n$.

1.2 Concrete spaces

The systematic study in this book begins with Section 2. Then we collect the notation needed in the sequel in detail. On this somewhat preliminary basis we list a few special cases of the spaces $B_{pq}^s(\mathbb{R}^n)$ and $F_{pq}^s(\mathbb{R}^n)$ without further comments. The aim is twofold. First we wish to substantiate what has been said in 1.1. Secondly, as far as classical function spaces are concerned, we fix some notation. Of course, $\mathbb{R}^n$ is euclidean n-space and $L_p(\mathbb{R}^n)$ is the usual (complex-valued) Lebesgue space with respect to Lebesgue measure. Otherwise we use standard notation. In case of doubt one might consult the beginning of Section 2 or the list of symbols at the end of the book. We follow here closely [Triδ], 10.5. More details may be found in [Triβ], especially 2.2.2, p. 35, and [Triγ], especially Chapter 1.

(i) Let $1 < p < \infty$. Then

$$F_{p2}^0(\mathbb{R}^n) = L_p(\mathbb{R}^n). \tag{1.2}$$

This is a Paley-Littlewood theorem, see [Triβ], 2.5.6, p. 87.

(ii) Let $1 < p < \infty$ and $s \in \mathbb{N}_0$. Then

$$F_{p2}^s(\mathbb{R}^n) = W_p^s(\mathbb{R}^n) \tag{1.3}$$

are the *classical Sobolev spaces*, usually normed by

$$\|f \mid W_p^s(\mathbb{R}^n)\| = \left(\sum_{|\alpha|\leq s} \|D^\alpha f \mid L_p(\mathbb{R}^n)\|^p\right)^{\frac{1}{p}}. \tag{1.4}$$

This generalizes assertion (i), see again [Triβ], 2.5.6, p. 87.

(iii) Let $\sigma \in \mathbb{R}$. Then

$$I_\sigma : \quad f \mapsto (\langle\xi\rangle^\sigma \widehat{f})^\vee, \tag{1.5}$$

with $\langle\xi\rangle = (1+|\xi|^2)^{\frac{1}{2}}$, is a one-to-one map of the Schwartz space $S(\mathbb{R}^n)$ onto itself and of the space of tempered distributions $S'(\mathbb{R}^n)$ onto itself. We use standard notation,

$$f \mapsto \widehat{f} \quad \text{and} \quad f \mapsto f^\vee \tag{1.6}$$

for the Fourier transform on $\mathbb{R}^n$ and its inverse, respectively. As for the spaces $B_{pq}^s(\mathbb{R}^n)$ and $F_{pq}^s(\mathbb{R}^n)$ with $s \in \mathbb{R}$, $0 < p \leq \infty$ ($p < \infty$ for the F-scale), $0 < q \leq \infty$, the transform I_σ acts as a lift (equivalent quasi-norms):

$$I_\sigma B_{pq}^s(\mathbb{R}^n) = B_{pq}^{s-\sigma}(\mathbb{R}^n) \quad \text{and} \quad I_\sigma F_{pq}^s(\mathbb{R}^n) = F_{pq}^{s-\sigma}(\mathbb{R}^n), \tag{1.7}$$

see [Triβ], 2.3.8, p. 58. In particular, let

$$H_p^s(\mathbb{R}^n) = I_{-s} L_p(\mathbb{R}^n), \quad s \in \mathbb{R}, \quad 1 < p < \infty, \tag{1.8}$$

be the *Sobolev spaces*. Then (1.2), (1.3), and (1.7) yield

$$H_p^s(\mathbb{R}^n) = F_{p2}^s(\mathbb{R}^n), \quad s \in \mathbb{R}, \quad 1 < p < \infty, \tag{1.9}$$

and
$$H_p^s(\mathbb{R}^n) = W_p^s(\mathbb{R}^n) \quad \text{if} \quad s \in \mathbb{N}_0 \quad \text{and} \quad 1 < p < \infty. \tag{1.10}$$

As for notation we call H_p^s *Sobolev spaces* (sometimes also denoted in the literature as fractional Sobolev spaces) and its special cases (1.10) with (1.4) *classical Sobolev spaces*.

(iv)   We denote
$$\mathcal{C}^s(\mathbb{R}^n) = B_{\infty\infty}^s(\mathbb{R}^n), \quad s \in \mathbb{R}, \tag{1.11}$$

as *Hölder-Zygmund spaces*. Let
$$(\Delta_h^1 f)(x) = f(x+h) - f(x), \quad (\Delta_h^{l+1} f)(x) = \Delta_h^1(\Delta_h^l f)(x), \tag{1.12}$$

where $x \in \mathbb{R}^n$, $h \in \mathbb{R}^n$, $l \in \mathbb{N}$, be the iterated differences in $\mathbb{R}^n$. As usual, $\Delta_h^1 = \Delta_h$. Let $0 < s < m \in \mathbb{N}$. Then
$$\|f \mid \mathcal{C}^s(\mathbb{R}^n)\| = \sup_{x \in \mathbb{R}^n} |f(x)| + \sup |h|^{-s} |\Delta_h^m f(x)|, \tag{1.13}$$

where the second supremum is taken over all $x \in \mathbb{R}^n$ and $h \in \mathbb{R}^n$ with $0 < |h| \leq 1$, are equivalent norms in $\mathcal{C}^s(\mathbb{R}^n)$. For more details we refer again to [Triβ], 2.2.2, 2.5.12. We extended the notation of Hölder-Zygmund spaces to $s \leq 0$.

(v)   Assertion (iv) can be generalized as follows. Once more let $0 < s < m \in \mathbb{N}$ and $1 \leq p \leq \infty$, $1 \leq q \leq \infty$. Then
$$\|f \mid L_p(\mathbb{R}^n)\| + \left(\int_{|h| \leq 1} |h|^{-sq} \|\Delta_h^m f \mid L_p(\mathbb{R}^n)\|^q \frac{dh}{|h|^n} \right)^{\frac{1}{q}} \tag{1.14}$$

(with the usual modification if $q = \infty$) are equivalent norms on $B_{pq}^s(\mathbb{R}^n)$. See again [Triβ], 2.2.2, 2.5.12, for more details. These are the *classical Besov spaces*.

There are further concrete spaces such as the inhomogeneous Hardy spaces $h_p(\mathbb{R}^n)$ with $0 < p < \infty$ and the spaces $bmo(\mathbb{R}^n)$ of all functions with bounded mean oscillations which can be incorporated in the above scales B_{pq}^s and F_{pq}^s. Furthermore there are numerous equivalent norms and quasi-norms. We refer to the literature quoted above, especially [Triα], [Triβ], and [Triγ]. In the present book our main aim is slightly different. We define the spaces $B_{pq}^s(\mathbb{R}^n)$ and $F_{pq}^s(\mathbb{R}^n)$ via quarkonial decompositions and prove afterwards that they coincide with the spaces usually denoted in this way.

1.3 Holomorphic functions and heuristics

Holomorphic functions Let $f(x)$, where $x = (x_1, x_2)$, be a holomorphic function with respect to the complex variable $z = x_1 + ix_2$ in a domain (connected open set) Ω in $\mathbb{R}^2$. Let K_j be open circles centred at $x^j \in \Omega$ and of radius r_j with $0 < r_j < 1$ such that for a suitable number $N \in \mathbb{N}$ at most N of these circles have a non-empty intersection,

$$\Omega = \bigcup_{l=1}^{\infty} K_l \quad \text{and} \quad dist(K_j, \partial\Omega) > r_j \quad \text{where} \quad j \in \mathbb{N}. \tag{1.15}$$

We assume in addition that the circles K_j are chosen in such a way that there is a resolution of unity by non-negative C^∞ functions $\psi_j(x)$ with

$$supp\,\psi_j \subset K_j, \quad 1 = \sum_{j=1}^{\infty} \psi_j(x) \quad \text{if} \quad x \in \Omega, \tag{1.16}$$

and

$$|D^\gamma \psi_j(x)| \leq c_\gamma r_j^{-|\gamma|} \quad \text{where} \quad \gamma \in \mathbb{N}_0^2, \quad j \in \mathbb{N}. \tag{1.17}$$

Here $c_\gamma > 0$ are suitable constants which are independent of $j \in \mathbb{N}$. With $z = x_1 + ix_2$ and $z^j = x_1^j + ix_2^j$ it follows from the classical Taylor series for holomorphic functions,

$$\begin{aligned} f(x) &= \sum_{j=1}^{\infty} \sum_{k=0}^{\infty} \varrho_j^k (z - z^j)^k \psi_j(x) \\ &= \sum_{j=1}^{\infty} \sum_{\beta \in \mathbb{N}_0^2} \lambda_j^\beta (x - x^j)^\beta \psi_j(x), \quad x \in \Omega. \end{aligned} \tag{1.18}$$

Obviously, the terms with ψ_j are extended outside of K_j by zero. (Again we use in this introductory section standard symbols. In case of doubt one may consult the beginning of Section 2 or the list of symbols at the end of the book.) Here $\varrho_j^k \in \mathbb{C}$ are the Taylor coefficients with respect to $z^j \in \Omega$ which may be calculated, for example, via *Cauchy's formula*. Since by (1.15) for some $c_j \geq 0$,

$$|\varrho_j^k| \leq c_j (2r_j)^{-k} \quad \text{if} \quad j \in \mathbb{N} \quad \text{and} \quad k \in \mathbb{N}_0, \tag{1.19}$$

$$(z - z^j)^k = (x_1 - x_1^j + i(x_2 - x_2^j))^k = \sum_{|\beta|=k} a_\beta (x - x^j)^\beta \tag{1.20}$$

with $\sum_{|\beta|=k} |a_\beta| = 2^k$, and

$$\lambda_j^\beta = \varrho_j^k a_\beta \quad \text{where} \quad |\beta| = k, \tag{1.21}$$

both the complex and the real representation (1.18) converge absolutely at any point $x \in \Omega$. This is the *Weierstrassian approach* to holomorphic functions in the complex plane modified in such a way that it makes sense to ask whether there are counterparts for non-smooth functions or distributions on $\mathbb{R}^n$ and how do they look.

Functions on $\mathbb{R}^n$ Let $f(x)$ be a function on $\mathbb{R}^n$. One might think about an element belonging to one of the concrete function spaces described above, maybe the classical Besov spaces in 1.2(v). Asking for real representations of type (1.18) one point is clear from the very beginning. In contrast to holomorphic functions, arbitrary functions do not admit something like Taylor expansions in suitable balls which are clipped together afterwards as indicated. As a suitable substitute one might try to replace one ball by a sequence of balls with radii tending to zero. The simplest choice is to work with the (dyadic) sequence of lattices $2^{-\nu}\mathbb{Z}^n$ in $\mathbb{R}^n$, where $\nu \in \mathbb{N}_0$, suitable resolutions of unity, and to take over otherwise the typical structure of the real representation in (1.18). In other words, we arrive at (0.1) in the Preface. More precisely, let ψ be a non-negative C^∞ function in $\mathbb{R}^n$ with compact support such that

$$\sum_{m \in \mathbb{Z}^n} \psi(x - m) = 1 \quad \text{if} \quad x \in \mathbb{R}^n. \tag{1.22}$$

This is called a resolution of unity. Let $\psi^\beta(x) = x^\beta \psi(x)$ where $x \in \mathbb{R}^n$ and $\beta \in \mathbb{N}_0^n$ (recall $x^\beta = x_1^{\beta_1} \cdots x_n^{\beta_n}$ if $x = (x_1, \ldots, x_n) \in \mathbb{R}^n$ and $\beta = (\beta_1, \ldots, \beta_n) \in \mathbb{N}_0^n$). Then (1.18) together with the indicated modifications results in the question of what is meant by

$$f(x) = \sum_{\beta \in \mathbb{N}_0^n} \sum_{\nu=0}^\infty \sum_{m \in \mathbb{Z}^n} \lambda_{\nu m}^\beta \psi^\beta(2^\nu x - m), \quad x \in \mathbb{R}^n. \tag{1.23}$$

First one must ask for what conditions are the coefficients $\lambda_{\nu m}^\beta \in \mathbb{C}$ and in what sense the right-hand side of (1.23) converges. Furthermore it must be discussed which functions f can be represented in this way and whether for given f there are distinguished coefficients $\lambda_{\nu m}^\beta$ which may serve as the counterpart of the Taylor coefficients for the holomorphic functions. To provide an understanding of what follows in Section 2 we discuss in 1.4 the convergence of (1.23) on a preliminary basis.

Other cases Quarkonial representations of type (1.23) for regular distributions and their relations to function spaces are the heart of the matter. At least in principle, all other problems of interest in this connection can be reduced to this case: Singular distributions, functions and function spaces on domains, manifolds and fractals. This will be done step by step in this chapter.

1.4 Convergence

Before the systematic treatment starts in Section 2 it is reasonable to discuss one crucial point. Whereas one does not have problems with the convergence of the (complex or real) expansion (1.18) of holomorphic functions, the situation for representations of functions or distributions via (1.23) is not so clear. We have to deal later on with the following typical situation. (Again for notation we refer to the beginning of Section 2 or to the list of symbols at the end of the book.) Let $0 < p \leq \infty$ and

$$\|\lambda^\beta \,|\, \ell_p\| = \left(\sum_{\nu=0}^\infty \sum_{m \in \mathbb{Z}^n} |\lambda_{\nu m}^\beta|^p \right)^{\frac{1}{p}}, \quad \beta \in \mathbb{N}_0^n, \tag{1.24}$$

and

$$\|\lambda \,|\, \ell_p\|_\varrho = \sup_{\beta \in \mathbb{N}_0^n} 2^{\varrho |\beta|} \|\lambda^\beta \,|\, \ell_p\|, \tag{1.25}$$

where $\lambda_{\nu m}^\beta \in \mathbb{C}$ and $\varrho > 0$ (obviously modified if $p = \infty$ in (1.24)). It is our aim to introduce function spaces via modified expansions (1.23) and sequence spaces of type (1.24). We discuss the typical situation which is related, at the end, to the spaces $B_{pp}^s(\mathbb{R}^n)$, where

$$0 < p \leq \infty, \quad s > \sigma_p = n \left(\frac{1}{p} - 1 \right)_+, \tag{1.26}$$

(here $a_+ = \max(a, 0)$ where $a \in \mathbb{R}$). We adapt (1.23) to this special case by

$$f(x) = \sum_{\beta \in \mathbb{N}_0^n} \sum_{\nu=0}^\infty \sum_{m \in \mathbb{Z}^n} \lambda_{\nu m}^\beta 2^{-\nu(s-\frac{n}{p})} \psi^\beta(2^\nu x - m). \tag{1.27}$$

This modification can be justified by the observation that

$$2^{-\nu(s-\frac{n}{p})} \psi^\beta(2^\nu x - m)$$

are normalized building blocks in $B_{pp}^s(\mathbb{R}^n)$: there are positive constants c_1^β and c_2^β, which are independent of $\nu \in \mathbb{N}_0$ and $m \in \mathbb{Z}^n$ such that

$$c_1^\beta \leq 2^{-\nu(s-\frac{n}{p})} \|\psi^\beta(2^\nu \cdot - m) \,|\, B_{pp}^s(\mathbb{R}^n)\| \leq c_2^\beta. \tag{1.28}$$

The coefficients $\lambda_{\nu m}^\beta$ belong to ℓ_p according to (1.24) and the influence of β will be compensated by sufficiently large ϱ in (1.25). As for the convergence of (1.27) we first remark that for some $r > 0$,

$$|x^\beta \psi(x)| \leq 2^{r|\beta|}, \quad \beta \in \mathbb{N}_0^n. \tag{1.29}$$

Then we have by (1.27),

$$|f(x)| \leq \sum_{\beta \in \mathbb{N}_0^n} \sum_{\nu=0}^\infty \sum_{m \in \mathbb{Z}^n} 2^{r|\beta|} |\lambda_{\nu m}^\beta| 2^{-\nu(s-\frac{n}{p})} \eta_{\nu m}(x), \tag{1.30}$$

where $\eta_{\nu m}(x)$ is the characteristic function of a ball centred at $2^{-\nu}m$ with radius $2^{-\nu+r}$. Of course, for fixed $\beta \in \mathbb{N}_0^n$, $\nu \in \mathbb{N}_0$, and $x \in \mathbb{R}^n$ the sum over $\mathbb{Z}^n$ in (1.30) reduces to finitely many terms (independently of β, ν, and x). Let first $p = \infty$. Then $s > 0$. With $\varrho > r$ we have

$$|f(x)| \leq c \, \|\lambda \,|\ell_\infty\|_\varrho, \quad x \in \mathbb{R}^n. \tag{1.31}$$

In particular, (1.27) converges pointwise, uniformly and absolutely and $f(x)$ is a uniformly continuous and bounded function in $\mathbb{R}^n$. Let $1 \leq p < \infty$. Then again $s > 0$. For any $\varepsilon > 0$ there is a constant $c_\varepsilon > 0$ such that

$$\begin{aligned} |f(x)|^p &\leq c_\varepsilon \left(\sup_{\beta,\nu,m} 2^{(r+\varepsilon)|\beta|} |\lambda_{\nu m}^\beta| 2^{-\nu(s-\frac{n}{p}-\varepsilon)} \eta_{\nu m}(x) \right)^p \\ &\leq c_\varepsilon \sum_{\beta,\nu,m} 2^{(r+\varepsilon)|\beta|p} |\lambda_{\nu m}^\beta|^p 2^{-\nu(s-\varepsilon)p+n\nu} \eta_{\nu m}(x). \end{aligned} \tag{1.32}$$

If we choose $0 < \varepsilon < s$ and $\varrho > r + \varepsilon$, then integration of (1.32) results in

$$\|f \,|L_p(\mathbb{R}^n)\| \leq c \, \|\lambda \,|\ell_p\|_\varrho. \tag{1.33}$$

In particular, if $\|\lambda \,|\ell_p\|_\varrho < \infty$ then (1.27) converges unconditionally in $L_p(\mathbb{R}^n)$ (that means, any rearrangement of the terms involved results in the same function). Let $0 < p < 1$. Then integration of (1.30), $s > \frac{n}{p} - n$ by (1.26), and the monotonicity of the sequence spaces ℓ_p result in

$$\begin{aligned} \|f \,|L_1(\mathbb{R}^n)\| &\leq c \sum_{\beta,\nu,m} 2^{r|\beta|} |\lambda_{\nu m}^\beta| 2^{-\nu(s-\frac{n}{p}+n)} \\ &\leq c' \, \|\lambda \,|\ell_1\|_\varrho \leq c'' \, \|\lambda \,|\ell_p\|_\varrho. \end{aligned} \tag{1.34}$$

Hence, if the right-hand side of (1.34) is finite, then (1.27) converges unconditionally in $L_1(\mathbb{R}^n)$. In particular, in any case the right-hand side of (1.27) converges unconditionally in $S'(\mathbb{R}^n)$ if $\|\lambda\,|\ell_p|\|_\varrho < \infty$ for sufficiently large ϱ. The last observation remains valid if one replaces in (1.27)

$$\psi^\beta(2^\nu x - m) \quad \text{by} \quad \Delta^k\left(\psi^\beta(2^\nu x - m)\right), \tag{1.35}$$

where

$$\Delta^k = \left(\sum_{l=1}^n \frac{\partial^2}{\partial x_l^2}\right)^k \quad \text{with} \quad k \in \mathbb{N}, \tag{1.36}$$

is the iterated Laplacian. This follows from the well-known fact that Δ^k maps $S'(\mathbb{R}^n)$ continuously into itself.

1.5 The Weierstrassian approach

We return briefly to holomorphic functions as discussed at the beginning of 1.3. Again let $f(z)$ be a complex-valued function in the domain Ω in the complex plane $\mathbb{C}$. There are two versions of conditions under which $f(z)$ is called holomorphic:

(i) The Riemannian (or qualitative) approach: $f(z)$ is holomorphic in Ω if it is complex differentiable at any point $z \in \Omega$.

(ii) The Weierstrassian (or quantitative) approach: $f(z)$ is holomorphic in Ω if it can be expanded locally in a Taylor series at any point $z \in \Omega$. Then one has (1.18) and the (uniquely determined) Taylor coefficients ϱ_j^k can be calculated by *Cauchy integral formulas*.

Function spaces are usually defined in the Riemannian spirit: qualitative requirements for (distributional) derivatives, differences, or other means. The descriptions of the concrete spaces in 1.2 may serve as examples. In this Chapter we complement this standard way of introducing function spaces by a *Weierstrassian approach*. We expand functions and distributions as indicated in (1.27), maybe with the modification (1.35). Then we introduce function spaces by the behaviour of the (not uniquely determined) coefficients $\lambda_{\nu m}^\beta$, expressed in terms of sequence spaces, where (1.24), (1.25) may serve as prototypes. Furthermore, one can ask whether for given f there are *optimal* coefficients $\lambda_{\nu m}^\beta$ depending linearly on f,

$$\lambda_{\nu m}^\beta = \left(f, \Psi_{\nu m}^\beta\right), \quad f \in S'(\mathbb{R}^n), \quad \Psi_{\nu m}^\beta \in S(\mathbb{R}^n), \tag{1.37}$$

(dual pairing). In analogy to holomorphic functions one might call (1.37) *Cauchy formulas*.

2 Spaces on $\mathbb{R}^n$: the regular case

2.1 Basic notation

Let $\mathbb{N}$ be the collection of all natural numbers and $\mathbb{N}_0 = \mathbb{N} \cup \{0\}$. Let $\mathbb{R}^n$ be euclidean n-space, where $n \in \mathbb{N}$; put $\mathbb{R} = \mathbb{R}^1$, whereas $\mathbb{C}$ is the complex plane. Let $S(\mathbb{R}^n)$ be the Schwartz space of all complex-valued, rapidly decreasing, infinitely differentiable functions on $\mathbb{R}^n$. By $S'(\mathbb{R}^n)$ we denote its topological dual, the space of all tempered distributions on $\mathbb{R}^n$. Furthermore, $L_p(\mathbb{R}^n)$ with $0 < p \leq \infty$, is the standard quasi-Banach space with respect to Lebesgue measure, quasi-normed by

$$\|f \,|L_p(\mathbb{R}^n)\| = \left(\int_{\mathbb{R}^n} |f(x)|^p \, dx\right)^{\frac{1}{p}} \qquad (2.1)$$

with the obvious modification if $p = \infty$.

As usual, $\mathbb{Z}$ is the collection of all integers; and $\mathbb{Z}^n$ where $n \in \mathbb{N}$, denotes the lattice of all points $m = (m_1, \ldots, m_n) \in \mathbb{R}^n$ with $m_j \in \mathbb{Z}$. Let $\mathbb{N}_0^n$, where $n \in \mathbb{N}$, be the set of all multi-indices

$$\alpha = (\alpha_1, \ldots, \alpha_n) \quad \text{with} \quad \alpha_j \in \mathbb{N}_0 \quad \text{and} \quad |\alpha| = \sum_{j=1}^n \alpha_j. \qquad (2.2)$$

If $x = (x_1, \ldots, x_n) \in \mathbb{R}^n$ and $\beta = (\beta_1, \ldots, \beta_n) \in \mathbb{N}_0^n$ then we put

$$x^\beta = x_1^{\beta_1} \cdots x_n^{\beta_n} \quad \text{(monomial)}. \qquad (2.3)$$

2.2 Sequence spaces

We collect some more specific notation in connection with sequence spaces. Let $Q_{\nu m}$ be a cube in $\mathbb{R}^n$ with sides parallel to the axes of coordinates, centred at $2^{-\nu}m$, and with side length $2^{-\nu}$, where $m \in \mathbb{Z}^n$ and $\nu \in \mathbb{N}_0$. If Q is a cube in $\mathbb{R}^n$ and $r > 0$ then rQ is the cube in $\mathbb{R}^n$ concentric with Q and with side length r times the side length of Q. We denote by $\chi_{\nu m}^{(p)}$ the p-normalized characteristic function of the cube $Q_{\nu m}$, which means

$$\chi_{\nu m}^{(p)}(x) = 2^{\frac{\nu n}{p}} \quad \text{if} \quad x \in Q_{\nu m} \quad \text{and} \quad \chi_{\nu m}^{(p)}(x) = 0 \quad \text{if} \quad x \notin Q_{\nu m}, \qquad (2.4)$$

where $\nu \in \mathbb{N}_0$, $m \in \mathbb{Z}^n$, and $0 < p \leq \infty$. Of course

$$\|\chi_{\nu m}^{(p)} \,|L_p(\mathbb{R}^n)\| = 1, \quad \text{where} \quad 0 < p \leq \infty. \qquad (2.5)$$

2. Spaces on $\mathbb{R}^n$: the regular case

Let $0 < p \le \infty$, $0 < q \le \infty$, and

$$\lambda = \{\lambda_{\nu m} \in \mathbb{C}: \quad \nu \in \mathbb{N}_0, \, m \in \mathbb{Z}^n\}. \tag{2.6}$$

Then we introduce the sequence spaces

$$b_{pq} = \left\{ \lambda: \quad \|\lambda\,|b_{pq}\| = \left(\sum_{\nu=0}^{\infty} \left(\sum_{m \in \mathbb{Z}^n} |\lambda_{\nu m}|^p \right)^{\frac{q}{p}} \right)^{\frac{1}{q}} < \infty \right\} \tag{2.7}$$

and

$$f_{pq} = \left\{ \lambda: \quad \|\lambda\,|f_{pq}\| = \left\| \left(\sum_{\nu=0}^{\infty} \sum_{m \in \mathbb{Z}^n} |\lambda_{\nu m}\, \chi_{\nu m}^{(p)}(\cdot)|^q \right)^{\frac{1}{q}} \,|L_p(\mathbb{R}^n) \right\| < \infty \right\} \tag{2.8}$$

with the usual modification if $p = \infty$ and/or $q = \infty$. By (2.5) we have

$$b_{pp} = f_{pp} \quad \text{and} \quad \|\lambda\,|b_{pp}\| = \left(\sum_{\nu=0}^{\infty} \sum_{m \in \mathbb{Z}^n} |\lambda_{\nu m}|^p \right)^{\frac{1}{p}}, \tag{2.9}$$

where $0 < p \le \infty$ (with the obvious modification if $p = \infty$). In case of doubt we write also $b_{p,q}$ or $f_{p,q}$ in place of b_{pq} or f_{pq}, respectively.

2.3 Proposition

Let $0 < p \le \infty$ and $0 < q \le \infty$. Then b_{pq} and f_{pq} are quasi-Banach spaces. Furthermore,

$$b_{p,\min(p,q)} \subset f_{pq} \subset b_{p,\max(p,q)}, \tag{2.10}$$

and, in particular, $b_{pp} = f_{pp}$.

Proof The proof that b_{pq} and f_{pq} are quasi-Banach spaces is standard. With

$$\lambda_\nu(x) = \sum_{m \in \mathbb{Z}^n} \lambda_{\nu m}\, \chi_{\nu m}^{(p)}(x), \quad x \in \mathbb{R}^n, \tag{2.11}$$

we have by (2.5),

$$\|\lambda\,|b_{pq}\| = \left(\sum_{\nu=0}^{\infty} \|\lambda_\nu\,|L_p(\mathbb{R}^n)\|^q \right)^{\frac{1}{q}} \tag{2.12}$$

and

$$\|\lambda \,|f_{pq}\| = \left\| \left(\sum_{\nu=0}^{\infty} |\lambda_\nu(\cdot)|^q \right)^{\frac{1}{q}} |L_p(\mathbb{R}^n) \right\|. \qquad (2.13)$$

Then (2.10) follows from the triangle inequality for Banach spaces combined with (2.9) and the monotonicity of f_{pq} with respect to q (for fixed p).

2.4 Definition

Let ψ be a non-negative C^∞ function in $\mathbb{R}^n$ with

$$\operatorname{supp} \psi \subset \{y \in \mathbb{R}^n \,:\, |y| < 2^r\} \qquad (2.14)$$

for some $r \geq 0$ and

$$\sum_{m \in \mathbb{Z}^n} \psi(x - m) = 1 \quad \text{if} \quad x \in \mathbb{R}^n. \qquad (2.15)$$

Let $s \in \mathbb{R}$, $0 < p \leq \infty$, $\beta \in \mathbb{N}_0^n$ and $\psi^\beta(x) = x^\beta \psi(x)$, where x^β is the monomial (2.3). Then

$$(\beta qu)_{\nu m}(x) = 2^{-\nu(s-\frac{n}{p})} \psi^\beta(2^\nu x - m), \quad x \in \mathbb{R}^n, \qquad (2.16)$$

is called an (s,p)-β-quark related to $Q_{\nu m}$. Here $\nu \in \mathbb{N}_0$ and $m \in \mathbb{Z}^n$.

2.5 Discussion

We have for some $d > 0$,

$$\operatorname{supp}(\beta qu)_{\nu m} \subset dQ_{\nu m} \quad \text{where} \quad \nu \in \mathbb{N}_0, \quad m \in \mathbb{Z}^n, \qquad (2.17)$$

$$|\psi^\beta(x)| \leq 2^{r|\beta|}, \quad \beta \in \mathbb{N}_0^n, \qquad (2.18)$$

where r has the same meaning as in (2.14) and hence

$$|(\beta qu)_{\nu m}(x)| \leq 2^{-\nu(s-\frac{n}{p})+r|\beta|} \quad \text{with} \quad \beta \in \mathbb{N}_0^n. \qquad (2.19)$$

As for the normalizing factors $2^{-\nu(s-\frac{n}{p})}$ in (2.16) we refer to the heuristical discussion in 1.4, especially to (1.28) with (1.26). We now complement (1.26). Let $0 < p \leq \infty$ and $0 < q \leq \infty$. Then we put

$$\sigma_p = n\left(\frac{1}{p} - 1\right)_+ \quad \text{and} \quad \sigma_{pq} = n\left(\frac{1}{\min(p,q)} - 1\right)_+. \qquad (2.20)$$

Here, as usual, $a_+ = \max(a, 0)$ where $a \in \mathbb{R}$.

2.6 Definition

(i) Let

$$0 < p \le \infty, \quad 0 < q \le \infty \quad \text{and} \quad s > \sigma_p. \tag{2.21}$$

Let $(\beta qu)_{\nu m}$ be (s,p)-β-quarks according to Definition 2.4 with respect to the fixed function ψ. We put

$$\lambda = \{\lambda^\beta : \beta \in \mathbb{N}_0^n\} \quad \text{with} \quad \lambda^\beta = \{\lambda^\beta_{\nu m} \in \mathbb{C} : \nu \in \mathbb{N}_0, m \in \mathbb{Z}^n\}. \tag{2.22}$$

Let $\varrho > r$, where r has the same meaning as in (2.14), $\lambda^\beta \in b_{pq}$ and

$$\|\lambda \,|b_{pq}\|_\varrho = \sup_{\beta \in \mathbb{N}_0^n} 2^{\varrho|\beta|} \|\lambda^\beta \,|b_{pq}\| < \infty. \tag{2.23}$$

Then $B^s_{pq}(\mathbb{R}^n)$ is the collection of all $f \in S'(\mathbb{R}^n)$ which can be represented as

$$f = \sum_{\beta \in \mathbb{N}_0^n} \sum_{\nu=0}^{\infty} \sum_{m \in \mathbb{Z}^n} \lambda^\beta_{\nu m} (\beta qu)_{\nu m}(x) \tag{2.24}$$

with (2.23). Furthermore,

$$\|f\,|B^s_{pq}(\mathbb{R}^n)\|_{\psi,\varrho} = \inf \|\lambda\,|b_{pq}\|_\varrho, \tag{2.25}$$

where the infimum is taken over all admissible representations.

(ii) Let

$$0 < p < \infty, \quad 0 < q \le \infty \quad \text{and} \quad s > \sigma_{pq}. \tag{2.26}$$

Let again $(\beta qu)_{\nu m}$ be (s,p)-β-quarks, $\lambda^\beta \in f_{pq}$ and

$$\|\lambda\,|f_{pq}\|_\varrho = \sup_{\beta \in \mathbb{N}_0^n} 2^{\varrho|\beta|} \|\lambda^\beta\,|f_{pq}\| < \infty, \tag{2.27}$$

where again $\varrho > r$ and λ^β and λ are the sequences according to (2.22). Then $F^s_{pq}(\mathbb{R}^n)$ is the collection of all $f \in S'(\mathbb{R}^n)$ which can be represented by (2.24) with (2.27). Furthermore,

$$\|f\,|F^s_{pq}(\mathbb{R}^n)\|_{\psi,\varrho} = \inf \|\lambda\,|f_{pq}\|_\varrho, \tag{2.28}$$

where the infimum is taken over all admissible representations.

2.7 Discussion

As above, let the function ψ with (2.14) and (2.15) and the numbers r and ϱ be fixed. Let p, s, q be given by (2.21) and let $\varepsilon > 0$ such that $s - \varepsilon > \sigma_p$. Then $2^{\nu\varepsilon} (\beta qu)_{\nu m}$ with (2.16) are $(s - \varepsilon, p)$-β-quarks and (2.24) can be re-written as

$$f = \sum_{\beta \in \mathbb{N}_0^n} \sum_{\nu=0}^{\infty} \sum_{m \in \mathbb{Z}^n} (2^{-\nu\varepsilon} \lambda_{\nu m}^{\beta}) (\beta qu)_{\nu m}(x) \, 2^{\nu\varepsilon}. \qquad (2.29)$$

By (2.7) and (2.22)

$$\left(\sum_{\nu=0}^{\infty} \sum_{m \in \mathbb{Z}^n} |2^{-\nu\varepsilon} \lambda_{\nu m}^{\beta}|^p \right)^{\frac{1}{p}} \leq c_{\varepsilon} \, \|\lambda^{\beta} \, |b_{pq}\| \qquad (2.30)$$

(usual modification if $p = \infty$), where c_{ε} depends only on ε. By (2.23) we can apply the arguments from 1.4 with $2^{-\nu\varepsilon} \lambda_{\nu m}^{\beta}$ in place of $\lambda_{\nu m}^{\beta}$. Hence, (2.24) converges unconditionally in $L_{\bar{p}}(\mathbb{R}^n)$ with $\bar{p} = \max(p, 1)$. In particular, $f \in S'(\mathbb{R}^n)$ is a regular distribution and we can simplify (2.24) by

$$f = \sum_{\beta, \nu, m} \lambda_{\nu m}^{\beta} (\beta qu)_{\nu m}(x). \qquad (2.31)$$

By (2.10) this argument applies also to part (ii) of the above definition. There arise two questions. First one has to prove that the spaces $B_{pq}^s(\mathbb{R}^n)$ and $F_{pq}^s(\mathbb{R}^n)$ are independent of the given function ψ used in Definition 2.4 and of the fixed number ϱ with $\varrho > r$: we need equivalent quasi-norms in (2.25) and (2.28), respectively, for different choices of ψ and ϱ. Secondly one wishes to know whether the spaces $B_{pq}^s(\mathbb{R}^n)$ and $F_{pq}^s(\mathbb{R}^n)$ introduced in this way coincide with the well-established spaces denoted in the same way. We shall give affirmative answers simultaneously.

2.8 Some more notation

It is the main aim of Section 2 to prove that the spaces $B_{pq}^s(\mathbb{R}^n)$ and $F_{pq}^s(\mathbb{R}^n)$ introduced in Definition 2.6 coincide with those spaces usually denoted in this way. For this purpose we recall some standard notation. Of course, the Schwartz space $S(\mathbb{R}^n)$ and the space $S'(\mathbb{R}^n)$ of tempered distributions have the same meaning as in 2.1. If $\varphi \in S(\mathbb{R}^n)$ then

$$\widehat{\varphi}(\xi) = (F\varphi)(\xi) = (2\pi)^{-\frac{n}{2}} \int_{\mathbb{R}^n} e^{-ix\xi} \varphi(x) \, dx, \quad \xi \in \mathbb{R}^n, \qquad (2.32)$$

2. Spaces on $\mathbb{R}^n$: the regular case

denotes the Fourier transform of φ. As usual, $F^{-1}\varphi$ or $\varphi^\vee$, stands for the inverse Fourier transform, given by the right-hand side of (2.32) with i in place of $-i$. Here $x\xi$ denotes the scalar product in $\mathbb{R}^n$. Both F and F^{-1} are extended to $S'(\mathbb{R}^n)$ in the standard way. Let $\varphi \in S(\mathbb{R}^n)$ with

$$\varphi(x) = 1 \quad \text{if} \quad |x| \le 1 \quad \text{and} \quad \varphi(x) = 0 \quad \text{if} \quad |x| \ge \frac{3}{2}. \tag{2.33}$$

We put $\varphi_0 = \varphi$, $\varphi_1(x) = \varphi(\frac{x}{2}) - \varphi(x)$, and

$$\varphi_k(x) = \varphi_1(2^{-k+1}x), \quad x \in \mathbb{R}^n, \quad k \in \mathbb{N}. \tag{2.34}$$

Then, since

$$1 = \sum_{k=0}^{\infty} \varphi_k(x) \quad \text{for all} \quad x \in \mathbb{R}^n, \tag{2.35}$$

the φ_k form a dyadic resolution of unity in $\mathbb{R}^n$. Recall that $(\varphi_k \widehat{f})^\vee$ is an entire analytic function on $\mathbb{R}^n$ for any $f \in S'(\mathbb{R}^n)$. In particular, $(\varphi_k \widehat{f})^\vee(x)$ makes sense pointwise. Let

$$s \in \mathbb{R}, \quad 0 < p \le \infty, \quad 0 < q \le \infty. \tag{2.36}$$

Then we put for $f \in S'(\mathbb{R}^n)$,

$$\|f\,|B^s_{pq}(\mathbb{R}^n)\|_\varphi = \left(\sum_{j=0}^{\infty} 2^{jsq} \left\| (\varphi_j \widehat{f})^\vee \,|L_p(\mathbb{R}^n) \right\|^q \right)^{\frac{1}{q}} \tag{2.37}$$

and

$$\|f\,|F^s_{pq}(\mathbb{R}^n)\|_\varphi = \left\| \left(\sum_{j=0}^{\infty} 2^{jsq} |(\varphi_j \widehat{f})^\vee(\cdot)|^q \right)^{\frac{1}{q}} \,|L_p(\mathbb{R}^n) \right\| \tag{2.38}$$

with the usual modification if $q = \infty$. The right-hand sides of (2.37) and (2.38) may be finite or infinite.

2.9 Theorem

(i) Let

$$0 < p \le \infty, \quad 0 < q \le \infty \quad \text{and} \quad s > \sigma_p. \tag{2.39}$$

Then $B^s_{pq}(\mathbb{R}^n)$, introduced in Definition 2.6(i), is a quasi-Banach space. Furthermore

$$B^s_{pq}(\mathbb{R}^n) = \{ f \in S'(\mathbb{R}^n) : \quad \|f\,|B^s_{pq}(\mathbb{R}^n)\|_\varphi < \infty \}, \qquad (2.40)$$

and $\|f\,|B^s_{pq}(\mathbb{R}^n)\|_\varphi$, given by (2.37), is an equivalent quasi-norm. In particular, all the quasi-norms in (2.25) and (2.37) for admitted functions ψ, numbers ϱ, and functions φ, are equivalent to each other.

(ii) Let

$$0 < p < \infty, \quad 0 < q \leq \infty, \quad \text{and} \quad s > \sigma_{pq}. \qquad (2.41)$$

Then $F^s_{pq}(\mathbb{R}^n)$, introduced in Definition 2.6(ii), is a quasi-Banach space. Furthermore,

$$F^s_{pq}(\mathbb{R}^n) = \{ f \in S'(\mathbb{R}^n) : \quad \|f\,|F^s_{pq}(\mathbb{R}^n)\|_\varphi < \infty \}, \qquad (2.42)$$

and $\|f\,|F^s_{pq}(\mathbb{R}^n)\|_\varphi$, given by (2.38), is an equivalent quasi-norm. In particular, all the quasi-norms in (2.28) and (2.38) for admitted functions ψ, numbers ϱ, and functions φ, are equivalent to each other.

Proof *Step 1* Let $f \in F^s_{pq}(\mathbb{R}^n)$ according to Definition 2.6(ii). We decompose the representation (2.24) by

$$f = \sum_{\beta \in \mathbb{N}_0^n} f^\beta \qquad (2.43)$$

with

$$f^\beta = \sum_{\nu=0}^\infty \sum_{m \in \mathbb{Z}^n} \lambda^\beta_{\nu m} (\beta qu)_{\nu m}(x). \qquad (2.44)$$

By the discussions in 2.7 there are no problems about the convergence of the series involved. Let $K \in \mathbb{N}$ with $K > s$. By [Triδ], Theorem 13.8 on p. 75, one might consider (2.44) as an atomic decomposition. By Step 2 of the proof of this theorem it follows that

$$\|f^\beta\,|F^s_{pq}(\mathbb{R}^n)\|_\varphi \leq c \sup_{x \in \mathbb{R}^n} \sum_{|\alpha| \leq K} |D^\alpha \psi^\beta(x)|\,\|\lambda^\beta\,|f_{pq}\|. \qquad (2.45)$$

Here we used that (2.38) is the standard Fourier-analytical quasi-norm in $F^s_{pq}(\mathbb{R}^n)$ and that this quasi-norm is equivalent to the quasi-norm on which the proof in [Triδ], pp. 76–79, relies, based on so-called local means. By $\psi^\beta(x) = x^\beta \psi(x)$ and the discussion 2.5, especially (2.18), we have for any $\varepsilon > 0$,

$$\sup_{x \in \mathbb{R}^n} \sum_{|\alpha| \leq K} |D^\alpha x^\beta \psi(x)| \leq c_1 (1+|\beta|)^K 2^{r|\beta|} \leq c_2 2^{(r+\varepsilon)|\beta|}, \qquad (2.46)$$

where c_2 is independent of β (but depends on ψ, K and ε). We insert (2.46) in (2.45) and choose $\varepsilon > 0$ so small that $r + \varepsilon < \varrho$. Let $\eta = \min(1, p, q)$. Then we have

$$\|f^\beta \,|F^s_{pq}(\mathbb{R}^n)\|_\varphi \le c\, 2^{-(\varrho-r-\varepsilon)|\beta|}\, \|\lambda\,|f_{pq}\|_\varrho \tag{2.47}$$

and by the triangle inequality for F^s_{pq}-quasi-norms

$$\|f\,|F^s_{pq}(\mathbb{R}^n)\|^\eta_\varphi \le \sum_{\beta \in \mathbb{N}_0^n} \|f^\beta\,|F^s_{pq}(\mathbb{R}^n)\|^\eta_\varphi \le c\, \Big(\sum_{\beta \in \mathbb{N}_0^n} 2^{-(\varrho-r-\varepsilon)\eta|\beta|}\Big) \|\lambda\,|f_{pq}\|^\eta_\varrho, \tag{2.48}$$

where c is independent of ϱ. It follows that

$$\|f\,|F^s_{pq}(\mathbb{R}^n)\|_\varphi \le c\, \|\lambda\,|f_{pq}\|_\varrho. \tag{2.49}$$

If, in addition, $\varrho \ge r + 1$, then c in (2.49) is independent of ϱ. Taking the infimum over all admitted representations we obtain

$$\|f\,|F^s_{pq}(\mathbb{R}^n)\|_\psi \le c\, \|f\,|F^s_{pq}(\mathbb{R}^n)\|_{\psi,\varrho}, \tag{2.50}$$

where c is independent of f. If, in addition, $\varrho \ge r+1$ then c is also independent of ϱ. If $f \in B^s_{pq}(\mathbb{R}^n)$ according to Definition 2.6(i), then it follows similarly that

$$\|f\,|B^s_{pq}(\mathbb{R}^n)\|_\varphi \le c\, \|f\,|B^s_{pq}(\mathbb{R}^n)\|_{\psi,\varrho}, \tag{2.51}$$

where c is independent of f, and for $\varrho \ge r+1$ also independent of ϱ.

Step 2 To prove the converse inequalities to (2.50) and (2.51) we modify the arguments in [Triδ], 14.15, pp. 101–104. Let f be an element of the right-hand side of (2.42). By (2.35) we have

$$\widehat{f}(\xi) = \sum_{k=0}^\infty \varphi_k(\xi)\, \widehat{f}(\xi), \quad \xi \in \mathbb{R}^n, \tag{2.52}$$

(convergence in $S'(\mathbb{R}^n)$). Let Q_k be a cube in $\mathbb{R}^n$ centred at the origin and with side-length, say, $2\pi\, 2^k$. In particular we have $\mathrm{supp}\, \varphi_k \subset Q_k$. We interpret $\varphi_k \widehat{f}$ as a periodic distribution and expand it in Q_k by

$$(\varphi_k \widehat{f})(\xi) = \sum_{m \in \mathbb{Z}^n} b_{km} \exp(-i2^{-k}\, m\xi), \quad \xi \in Q_k, \tag{2.53}$$

with

$$b_{km} = c\, 2^{-kn} \int_{Q_k} \exp(i2^{-k}\, m\xi)\, (\varphi_k \widehat{f})(\xi)\, d\xi = c'\, 2^{-kn}\, (\varphi_k \widehat{f})^\vee(2^{-k} m). \tag{2.54}$$

18   I. Decompositions of functions

Let
$$\Lambda = \{\Lambda_{km} : \quad k \in \mathbb{N}_0, \ m \in \mathbb{Z}^n\} \tag{2.55}$$
with
$$\Lambda_{km} = 2^{k(s-\frac{n}{p})} (\varphi_k \widehat{f})^\vee (2^{-k} m). \tag{2.56}$$
We may assume
$$\|f \,|\, F_{pq}^s(\mathbb{R}^n)\|_\varphi \sim \|\Lambda \,|\, f_{pq}\|, \tag{2.57}$$
(equivalent quasi-norms), where the right-hand side is given by (2.8), (2.6) with Λ in place of λ. We comment on (2.57) in Remark 2.10 below, where we also give references. We take (2.57) for granted. Let $\varkappa \in S(\mathbb{R}^n)$,
$$\varkappa_k(\xi) = \varkappa(2^{-k}\xi), \quad \varkappa_k(\xi) = 1 \quad \text{if} \quad \xi \in \operatorname{supp} \varphi_k, \quad \operatorname{supp} \varkappa_k \subset Q_k, \tag{2.58}$$
where $k \in \mathbb{N}_0$. We multiply (2.53) with $\varkappa_k$ and extend it by zero from Q_k to $\mathbb{R}^n$. Then we have
$$\begin{aligned}
(\varphi_k \widehat{f})^\vee(x) &= \sum_{m \in \mathbb{Z}^n} b_{km} \, \varkappa_k^\vee(x - 2^{-k} m) \\
&= 2^{kn} \sum_{m \in \mathbb{Z}^n} b_{km} \, \varkappa^\vee(2^k x - m) \\
&= c \sum_{m \in \mathbb{Z}^n} \Lambda_{km} \, 2^{-k(s-\frac{n}{p})} \varkappa^\vee(2^k x - m), \quad x \in \mathbb{R}^n,
\end{aligned} \tag{2.59}$$
where we used (2.54) and (2.56). The entire analytic function $\varkappa^\vee(x) \in S(\mathbb{R}^n)$ can be extended from $\mathbb{R}^n$ to $\mathbb{C}^n$. By the Paley-Wiener-Schwartz theorem we have for some $c > 0$, any $b > 0$ and an appropriate number c_b,
$$|\varkappa^\vee(x + iy)| \le c_b \, e^{c|y|} \langle x \rangle^{-b}, \tag{2.60}$$
where $x \in \mathbb{R}^n$ and $y \in \mathbb{R}^n$, with $\langle x \rangle = (1 + |x|^2)^{\frac{1}{2}}$, see e. g., [Tri$\beta$], 1.2.1, p. 13. Iterative application of Cauchy's representation theorem in the complex plane yields
$$\varkappa^\vee(z_1, \ldots, z_n)$$
$$= (2\pi i)^{-n} \int_{|\zeta_1 - z_1|=1} \cdots \int_{|\zeta_n - z_n|=1} \frac{\varkappa^\vee(\zeta_1, \ldots, \zeta_n)}{(\zeta_1 - z_1) \cdots (\zeta_n - z_n)} \, d\zeta_1 \cdots d\zeta_n, \tag{2.61}$$
where $z_k \in \mathbb{C}$. By (2.60) we obtain
$$|D^\alpha \varkappa^\vee(x)| \le c_b' \, \alpha! \langle x \rangle^{-b}, \quad x \in \mathbb{R}^n, \tag{2.62}$$

where c'_b is independent of $x \in \mathbb{R}^n$ and $\alpha \in \mathbb{N}_0^n$. Let ψ and ψ^β with $\beta \in \mathbb{N}_0^n$ be the functions introduced in Definition 2.4. We expand the function $\varkappa^\vee(2^k x - m)$ in (2.59) at the point $2^{-k-\varrho} l$ where $l \in \mathbb{Z}^n$ and $\varrho \in \mathbb{N}$ is fixed. (It is our intention to prove the converse of (2.50) for these numbers ϱ, which is sufficient.) We have

$$\psi(2^{k+\varrho} x - l)\, \varkappa^\vee(2^k x - m)$$
$$= \sum_{\beta \in \mathbb{N}_0^n} \frac{2^{k|\beta|}}{\beta!} (D^\beta \varkappa^\vee)(2^{-\varrho} l - m)(x - 2^{-k-\varrho} l)^\beta \, \psi(2^{k+\varrho} x - l)$$
$$= \sum_{\beta \in \mathbb{N}_0^n} \frac{(D^\beta \varkappa^\vee)(2^{-\varrho} l - m)}{\beta!} \, \psi^\beta(2^{k+\varrho} x - l)\, 2^{-\varrho|\beta|}. \tag{2.63}$$

By (2.15) we have

$$1 = \sum_{l_0} \sum_{l \in \mathbb{Z}^n} \psi(2^{k+\varrho} x - 2^\varrho l - l_0) = \sum_{l \in \mathbb{Z}^n} \psi(2^{k+\varrho} x - 2^\varrho l) + \cdots, \tag{2.64}$$

where the finite sum with respect to l_0 is taken over all lattice points $l_0 = (l_{0,1}, \ldots, l_{0,n}) \in \mathbb{Z}^n$ with $0 \le l_{0,j} < 2^\varrho$, where $j = 1, \ldots, n$. As indicated in (2.64) we concentrate on the term with $l_0 = 0$, where $+\cdots$ stands for the remaining terms which can be treated in the same way. Then we obtain by (2.59)

$$(\varphi_k \widehat{f})^\vee(x) = c\, 2^{-k(s - \frac{n}{p})} \sum_{m \in \mathbb{Z}^n} \Lambda_{km} \sum_{l \in \mathbb{Z}^n} \psi(2^{k+\varrho} x - 2^\varrho l)\, \varkappa^\vee(2^k x - m) + \cdots \tag{2.65}$$

and hence by (2.63)

$$(\varphi_k \widehat{f})^\vee(x)$$
$$= c\, 2^{-k(s - \frac{n}{p})} \sum_{\beta \in \mathbb{N}_0^n} \sum_{l \in \mathbb{Z}^n} \psi^\beta(2^{k+\varrho} x - 2^\varrho l) \sum_{m \in \mathbb{Z}^n} \frac{D^\beta \varkappa^\vee(l - m)}{\beta!} \Lambda_{km} 2^{-\varrho|\beta|} + \cdots$$
$$= c' \sum_{\beta \in \mathbb{N}_0^n} \sum_{l \in \mathbb{Z}^n} \lambda^\beta_{k+\varrho, 2^\varrho l}\, (\beta qu)_{k+\varrho, 2^\varrho l}(x)\, 2^{\varrho(s - \frac{n}{p})} + \cdots, \tag{2.66}$$

where $(\beta qu)_{k+\varrho, 2^\varrho l}$ are (s,p)-β-quarks according to Definition 2.4 and

$$\lambda^\beta_{k+\varrho, 2^\varrho l} = 2^{-\varrho|\beta|} \sum_{m \in \mathbb{Z}^n} \frac{D^\beta \varkappa^\vee(l - m)}{\beta!} \Lambda_{km}. \tag{2.67}$$

As in (2.22) we denote the collection of all these coefficients by λ^β. We wish to prove that there is a number $c_\varrho > 0$ such that for all $\beta \in \mathbb{N}_0^n$ and, of course, all admitted f,

$$\|\lambda^\beta \,|f_{pq}\| \leq c_\varrho \, 2^{-\varrho|\beta|} \, \|f \,|F_{pq}^s(\mathbb{R}^n)\|_\varphi \,. \tag{2.68}$$

By (2.62) we have

$$|\lambda^\beta_{k+\varrho, 2^\varrho l}| \leq c_1 2^{-\varrho|\beta|} \sum_{m \in \mathbb{Z}^n} \frac{\Lambda_{km}}{1 + |l - m|^b}$$

$$\leq c_2 2^{-\varrho|\beta|} \sum_{m \in \mathbb{Z}^n} \frac{\Lambda_{k, m+l}}{1 + |m|^b} \,. \tag{2.69}$$

If one replaces $2^\varrho l$ on the left-hand side by $2^\varrho l + l_0$ according to (2.64) then one gets (2.69) with the same right-hand side for these $2^{\varrho n}$ terms. The cube related to the coefficients on the left-hand side of (2.69) is $Q_{k+\varrho, 2^\varrho l}$ centred at

$$2^{-k-\varrho} 2^\varrho l = 2^{-k} l \quad \text{with the side-length } 2^{-k-\varrho} \,. \tag{2.70}$$

Hence it is the cube $2^{-\varrho} Q_{k,l}$ which is a sub-cube of $Q_{k,l}$. This fits in the definition (2.8) of f_{pq}. Let $\eta = \min(1, p, q)$. Then the η-triangle inequality for f_{pq} results in

$$\|\lambda^\beta \,|f_{pq}\|^\eta \leq c \, 2^{\varrho n} \, 2^{-\varrho|\beta|\eta} \sum_{m \in \mathbb{Z}^n} \left(\frac{1 + |m|^a}{1 + |m|^b} \right)^\eta \|\Lambda \,|f_{pq}\|^\eta , \tag{2.71}$$

where c is independent of ϱ and where Λ is given by (2.55). The additional factor $1 + |m|^a$ comes from the influence of the index-shifting $l \mapsto l + m$ in the f_{pq}-quasi-norms. Some details will be given below in 2.15, especially (2.103). If b is large then the sum in (2.71) converges. Now (2.57) proves (2.68) with $c_\varrho = c \, 2^{\frac{\varrho n}{\eta}}$. Combining this estimate with (2.28) and (2.27) we get the converse of (2.50),

$$\|f \,|F_{pq}^s(\mathbb{R}^n)\|_{\psi, \varrho} \leq \|\lambda \,|f_{pq}\|_\varrho \leq c \, 2^{\frac{\varrho n}{\eta}} \|f \,|F_{pq}^s(\mathbb{R}^n)\|_\varphi , \tag{2.72}$$

where c is independent of f and ϱ. The proof of part (ii) of the theorem is complete.

Step 3 The proof of part (i) is similar, but technically simpler. The converse of (2.51) is essentially covered by [Triδ], 14.15, p. 101–104.

2.10 Remark

We used (2.57) without restriction of generality. This means that, if necessary, $\varphi(x)$ must be replaced by $\varphi(cx)$ for some $c > 0$. This follows in a somewhat implicit way from the technique of maximal functions in [Triγ], 2.4.1, in particular Corollary 2 on pp. 108/109, and may be found more explicitly in [FrJ90], [FJW91], Ch. 7, and [Din95], [Far00] (where the two latter papers deal also with the anisotropic case). Some basic ideas of the above proof for the simpler case of the spaces $B_{pq}^s(\mathbb{R}^n)$ with (2.39) may be found in [Triδ], 14.15, pp. 101–104. For the spaces $F_{pq}^s(\mathbb{R}^n)$ with (2.41) there are additional technical complications which forced us to break down the resolution of unity in (2.64) into $2^{\varrho n}$ partial sums and which resulted in the factor $2^{\frac{\varrho n}{\eta}}$ in (2.72) and in the additional shifting factor $(1+|m|)^a$ in (2.71) with a reference to (2.103) in 2.15 below. A full proof of the anisotropic generalization of Theorem 2.9 may be found in [Far00]. To overcome the indicated technical problems connected with the F-case, W. Farkas used more directly the technique of maximal functions than we did here (but it is behind all the technical assertions mentioned in 2.15 below).

2.11 Unconditional convergence, estimates of constants, and optimal coefficients

By Theorem 2.9 the spaces $B_{pq}^s(\mathbb{R}^n)$ with (2.39) and the spaces $F_{pq}^s(\mathbb{R}^n)$ with (2.41) coincide with those spaces usually denoted in this way. The main point of the approach described here is the representation (2.24) with (2.23) and (2.27), respectively. By the discussion in 2.7 the series in (2.24) converges absolutely and unconditionally at least in $L_r(\mathbb{R}^n)$ for some r, $1 \le r \le \infty$. This justifies writing (2.31),

$$f = \sum_{\beta,\nu,m} \lambda_{\nu m}^\beta \, (\beta qu)_{\nu m}(x), \quad x \in \mathbb{R}^n, \qquad (2.73)$$

with the abbreviation

$$\sum_{\beta,\nu,m} = \sum_{\beta \in \mathbb{N}_0^n} \sum_{\nu=0}^{\infty} \sum_{m \in \mathbb{Z}^n}, \qquad (2.74)$$

which will be used in the sequel. Here $(\beta qu)_{\nu m}(x)$ are the (s,p)-β-quarks, which, of course, depend on s and p. At least in the above context, s and p are considered to be fixed. This may justify the omission of s and p in the notation of β-quarks. Looking at the interplay between individual elements $f \in S'(\mathbb{R}^n)$ and the above function spaces, one may adopt two points of view, which are the two sides of the same coin:

(i) Characterize all $f \in S'(\mathbb{R}^n)$ which belong to a given space, say, $B^s_{pq}(\mathbb{R}^n)$ or $F^s_{pq}(\mathbb{R}^n)$.

(ii) Characterize all spaces, say, $B^s_{pq}(\mathbb{R}^n)$ and $F^s_{pq}(\mathbb{R}^n)$, to which a given distribution $f \in S'(\mathbb{R}^n)$ belongs.

Our preference, so far, is the point of view (i). On the other hand, Step 2 of the proof of Theorem 2.9 is essentially in the spirit of the point of view (ii). To make clear what is meant we introduce the notion of *optimal coefficients*. Let $f \in S'(\mathbb{R}^n)$ be given. Then one asks for a *constructive procedure*

$$f \mapsto \lambda^\beta_{\nu m}(f), \quad \beta \in \mathbb{N}^n_0, \ \nu \in \mathbb{N}_0, \ m \in \mathbb{Z}^n, \qquad (2.75)$$

which allows us to decide whether f can be represented by (2.24) with (2.23) or (2.27). Optimality means that in addition there is a number $c > 0$ such that for all f in question,

$$\|\lambda(f) | b_{pq}\|_\varrho \leq c \, \|f | B^s_{pq}(\mathbb{R}^n)\| \qquad (2.76)$$

or

$$\|\lambda(f) | f_{pq}\|_\varrho \leq c \, \|f | F^s_{pq}(\mathbb{R}^n)\|. \qquad (2.77)$$

Here we used the notation introduced in Definition 2.6, now indicating the dependence of the coefficients on f. Since by definition, the converse of (2.76) and (2.77) is obvious, these inequalities can be re-written as equivalences,

$$\|\lambda(f) | b_{pq}\|_\varrho \sim \|f | B^s_{pq}(\mathbb{R}^n)\| \qquad (2.78)$$

and

$$\|\lambda(f) | f_{pq}\|_\varrho \sim \|f | F^s_{pq}(\mathbb{R}^n)\|, \qquad (2.79)$$

where the related equivalence constants are independent of f. Furthermore, we used in writing (2.76)–(2.79) the fact that all the quasi-norms in Definition 2.6 and Theorem 2.9 are equivalent to each other and that there is no need to distinguish between them in the sequel. The dependence of the optimal coefficients $\lambda^\beta_{\nu m}(f)$ constructed in Step 2 of the proof of Theorem 2.9 on given s, p, ϱ reduces, as we shall see, to the above normalizing factors and an additional mild influence of ϱ. For our later purposes it is useful to fix some of these dependencies of the above constants on ϱ. We assume that ψ and hence r with (2.14), (2.15), and φ from 2.8 are fixed. In particular we take the quasi-norms in $F^s_{pq}(\mathbb{R}^n)$ and $B^s_{pq}(\mathbb{R}^n)$ with respect to such a fixed φ. Let again $\lambda^\beta_{\nu m}(f)$, $\lambda^\beta(f)$, $\lambda(f)$ be the optimal coefficients and their sequences constructed in Step 2 of the proof of Theorem 2.9. We claim that

$$\|f | F^s_{pq}(\mathbb{R}^n)\| \leq c_1 \, \|\lambda(f) | f_{pq}\|_\varrho \leq c_2 \, 2^{\frac{\varrho n}{\eta}} \, \|f | F^s_{pq}(\mathbb{R}^n)\|, \qquad (2.80)$$

where c_1 and c_2 are independent of ϱ with $\varrho \geq r+1$. Recall $\eta = \min(1, p, q)$. The left-hand side follows from (2.50) (here one needs $\varrho \geq r+1$), whereas the right-hand side is covered by (2.72). Finally we need later on the interplay of ϱ, say, with $\varrho > r+\varepsilon$ for some $\varepsilon > 0$, and $\beta \in \mathbb{N}_0^n$. Let f be given by (2.43) with (2.44) where we assume that $\lambda_{\nu m}^\beta = \lambda_{\nu m}^\beta(f)$ are the above optimal coefficients. Then we have by (2.47) and (2.80),

$$\|f^\beta \,|F_{pq}^s(\mathbb{R}^n)\| \leq c\, 2^{\frac{\varrho n}{\eta}} \, 2^{-(\varrho-r-\varepsilon)|\beta|} \,\|f\,|F_{pq}^s(\mathbb{R}^n)\|, \qquad (2.81)$$

where c is independent of ϱ and $\beta \in \mathbb{N}_0^n$. Hence for large ϱ's one pays for a strong decay of the quasi-norms of f^β with bad equivalence constants. Of course one has a counterpart of (2.80) and (2.81) for the B-spaces.

2.12 Corollary

Under the hypotheses of Theorem 2.9 there are optimal coefficients according to 2.11 which can be represented as dual pairings in $S(\mathbb{R}^n)$ and $S'(\mathbb{R}^n)$ given by

$$\lambda_{\nu m}^\beta = 2^{\nu(s-\frac{n}{p})} 2^{-\varrho|\beta|} \left(f, \Psi_{\nu m}^{\beta,\varrho}\right) \qquad (2.82)$$

with

$$\Psi_{\nu m}^{\beta,\varrho} \in S(\mathbb{R}^n), \quad \beta \in \mathbb{N}_0^n, \ \nu \in \mathbb{N}_0, \ m \in \mathbb{Z}^n. \qquad (2.83)$$

Proof We follow the construction in Step 2 of the proof of Theorem 2.9. Let $f \in S'(\mathbb{R}^n)$. Appropriately interpreted, (2.56) can be written as

$$\Lambda_{\nu l} = c\, 2^{\nu(s-\frac{n}{p})} \int_{\mathbb{R}^n} f(y)\, \varphi_\nu^\vee(2^{-\nu}l - y)\, dy, \quad \nu \in \mathbb{N}_0, \ l \in \mathbb{Z}^n. \qquad (2.84)$$

Inserting (2.84) in (2.67) we get
$\lambda_{\nu+\varrho,m}^\beta$

$$= c\, 2^{-\varrho|\beta|}\, 2^{\nu(s-\frac{n}{p})} \sum_{l \in \mathbb{Z}^n} \frac{D^\beta \varkappa^\vee(2^{-\varrho}m-l)}{\beta!} \int_{\mathbb{R}^n} f(y)\, \varphi_\nu^\vee(2^{-\nu}l-y)\,dy$$

$$= c_\varrho\, 2^{-\varrho|\beta|}\, 2^{(\nu+\varrho)(s-\frac{n}{p})} \left(f, \Psi_{\nu+\varrho,m}^{\beta,\varrho}\right) \qquad (2.85)$$

with

$$\Psi_{\nu+\varrho,m}^{\beta,\varrho}(y) = \sum_{l \in \mathbb{Z}^n} \frac{D^\beta \varkappa^\vee(2^{-\varrho}m-l)}{\beta!} \varphi_\nu^\vee(2^{-\nu}l - y). \qquad (2.86)$$

We prove that these functions belong to $S(\mathbb{R}^n)$. Let $K \geq 0$. By (2.62) we have

$$|y|^K \left|\Psi^{\beta,\varrho}_{\nu+\varrho,m}(y)\right| \leq c \sum_{l \in \mathbb{Z}^n} \frac{|y|^K}{(1+|l|^b)(1+|2^{-\nu}l - y|^d)} \qquad (2.87)$$

where $b > 0$ and $d > 0$ are at our disposal. Here c may depend on ϱ, m, ν, b, and d. If one chooses d and b sufficiently large, for example $d = K$ and $b > K + n$, then one obtains

$$|y|^K \left|\Psi^{\beta,\varrho}_{\nu+\varrho,m}(y)\right| \leq c < \infty, \quad y \in \mathbb{R}^n. \qquad (2.88)$$

One gets a similar estimate with $D^\alpha \Psi^{\beta,\varrho}_{\nu+\varrho,m}$ in place of $\Psi^{\beta,\varrho}_{\nu+\varrho,m}$. Hence $\Psi^{\beta,\varrho}_{\nu+\varrho,m} \in S(\mathbb{R}^n)$.

2.13 The Weierstrassian approach and the Cauchy formula

The factors $2^{\nu(s-\frac{n}{p})}$ and $2^{-\varrho|\beta|}$ in (2.82) compensate the normalizing factors in (2.16) and (2.23), respectively. On the other hand, the dependence of $\Psi^{\beta,\varrho}_{\nu m}(x)$ on ϱ can be estimated more carefully by (2.86). It follows that c in (2.88) can be replaced by $c' \, 2^{L\varrho}$ where $L > 0$ is sufficiently large and $c' > 0$ is independent of ϱ. Then (2.78) and (2.79) can be strengthened by

$$c_1 \, 2^{-L\varrho} \|f \, |B^s_{pq}(\mathbb{R}^n)\| \leq \|\lambda(f)\,|b_{pq}\|_\varrho \leq c_2 \, 2^{L\varrho} \|f \, |B^s_{pq}(\mathbb{R}^n)\| \qquad (2.89)$$

and

$$c_1 \, 2^{-L\varrho} \|f \, |F^s_{pq}(\mathbb{R}^n)\| \leq \|\lambda(f)\,|f_{pq}\|_\varrho \leq c_2 \, 2^{L\varrho} \|f \, |F^s_{pq}(\mathbb{R}^n)\|, \qquad (2.90)$$

respectively, where $c_1 > 0$ and $c_2 > 0$ are independent of ϱ.

Summarizing the main observations in Section 2, especially Theorem 2.9 and Corollary 2.12, we arrived at the goal outlined in 1.5: The constructive *Weierstrassian approach* (2.73),

$$f = \sum_{\beta,\nu,m} \lambda^\beta_{\nu m}(f)\,(\beta q u)_{\nu m}(x), \quad x \in \mathbb{R}^n, \qquad (2.91)$$

to the function spaces $B^s_{pq}(\mathbb{R}^n)$ and $F^s_{pq}(\mathbb{R}^n)$ with the optimal *Cauchy coefficients* (2.82),

$$\lambda^\beta_{\nu m}(f) = 2^{\nu(s-\frac{n}{p})} \, 2^{-\varrho|\beta|} \, (f, \Psi^{\beta,\varrho}_{\nu m}) \qquad (2.92)$$

where the $S(\mathbb{R}^n)$-functions $\Psi^{\beta,\varrho}_{\nu m}$ are given by (2.86).

2.14 Generalized quarks

We return to Step 2 of the proof of Theorem 2.9. In the reformulation (2.65) of (2.59) we used (2.15) (resolution of unity) and no other specific properties. Then we obtained the β-quarks $(\beta qu)_{k+\varrho,l}$ in (2.66). The estimates of the coefficients which followed afterwards and which resulted finally in (2.72) have not very much to do with these constructions. In other words, without substantial changes we can modify this part of the proof of Theorem 2.9 as follows. Let $k \in \mathbb{N}_0$ and let

$$\{x^{k,m} : m \in \mathbb{Z}^n\} \subset \mathbb{R}^n \qquad (2.93)$$

be a sequence of *approximate lattices* such that there are two positive numbers c_1 and c_2 with

$$|x^{k,m_1} - x^{k,m_2}| \geq c_1 \, 2^{-k}, \quad k \in \mathbb{N}_0, \; m_1 \neq m_2, \qquad (2.94)$$

and

$$\mathbb{R}^n = \bigcup_{m \in \mathbb{Z}^n} B(x^{k,m}, c_2 \, 2^{-k}), \quad k \in \mathbb{N}_0. \qquad (2.95)$$

Here

$$B(x, c) = \{y \in \mathbb{R}^n : |x - y| < c\} \qquad (2.96)$$

is a ball centred at x and of radius $c > 0$. Let $\{\psi^{k,m} : m \in \mathbb{Z}^n\}$ be subordinated resolutions of unity with

$$\sum_{m \in \mathbb{Z}^n} \psi^{k,m}(x) = 1, \quad x \in \mathbb{R}^n, \; k \in \mathbb{N}_0, \qquad (2.97)$$

and

$$\operatorname{supp} \psi^{k,m} \subset B(x^{k,m}, c_2 \, 2^{-k+1}), \quad k \in \mathbb{N}_0, \; m \in \mathbb{Z}^n. \qquad (2.98)$$

Then we can replace $\sum_{l \in \mathbb{Z}^n} \psi(2^{k+\varrho} x - l) = 1$ in (2.65) by (2.97) (where the index-shifting caused by ϱ is immaterial). This results in (2.66) in a replacement of $(\beta qu)_{kl}(x)$ by

$$2^{-k(\sigma - \frac{n}{p})} 2^{k|\beta|} (x - x^{k,l})^\beta \, \psi^{k,l}(x), \quad k \in \mathbb{N}_0, \; l \in \mathbb{Z}^n. \qquad (2.99)$$

Ignoring the technical adaptions in the arguments after (2.66) we obtain (2.72), now based on the generalized β-quarks given by (2.99). We check Step 1 of the proof of Theorem 2.9 where we now replace $(\beta qu)_{\nu m}(x)$ in (2.44) by the notationally adapted generalized β-quarks in (2.99) (with ν in place of k). We apply again the atomic decompositions from [Triδ], Theorem 13.8 on p. 75. Again let $K > s$. We need the following additional assumption for the above functions $\psi^{k,m}$: there is a constant $c_3 > 0$ such that

$$\left|D^\alpha \psi^{k,m}(x)\right| \leq c_3\, 2^{k|\alpha|}, \quad x \in \mathbb{R}^n,\; k \in \mathbb{N}_0,\; |\alpha| \leq K. \tag{2.100}$$

Finally with $2^r = 2c_2$ and $\varrho > r$ one can follow the arguments in Step 1 of the proof of Theorem 2.9 and obtain the counterpart of (2.50). Hence:

Let $\psi^{k,m}$ be K times differentiable functions in $\mathbb{R}^n$ with (2.97), (2.98), (2.100). Then one can replace the β-quarks $(\beta qu)_{\nu m}$ both in Definition 2.6 and in Theorem 2.9 by the generalized β-quarks in (2.99).

We add further comments in 2.16 below.

2.15 Modified sequence spaces

Let $Q_{\nu m}$ and $c Q_{\nu m}$ be the cubes in $\mathbb{R}^n$ introduced at the beginning of 2.2, where $\nu \in \mathbb{N}_0$, $m \in \mathbb{Z}^n$, $c > 0$. Let $\chi_{\nu m}^{(p),c}(x)$ be given by (2.4) with $c Q_{\nu m}$ in place of $Q_{\nu m}$. Let f_{pq}^c be the sequence space given by (2.8) with $\chi_{\nu m}^{(p),c}$ in place of $\chi_{\nu m}^{(p)}$. Of course $f_{pq}^1 = f_{pq}$. Then

$$f_{pq}^c = f_{pq}, \quad 0 < p < \infty,\; 0 < q \leq \infty,\; c > 0. \tag{2.101}$$

For fixed p, q the quasi-norms $\|\lambda\,|f_{pq}\|$ in (2.8) and, in obvious notation, $\|\lambda\,|f_{pq}^c\|$, are equivalent. At first glance this is a little bit surprising, especially if $0 < c < 1$. But it is essentially a consequence of the vector-valued Hardy-Littlewood maximal inequality. Details of this technique may be found in [Triδ], p. 79. This observation is due to Frazier and Jawerth, [FrJ90], Proposition 2.7, where also a short direct proof is given. By the same argument one can also replace $c Q_{\nu m}$ with $0 < c < 1$ in (2.101) by arbitrary cubes $E_{\nu m}$ (or balls $B_{\nu m}$) with

$$E_{\nu m} \subset Q_{\nu m} \quad (\text{or} \quad B_{\nu m} \subset Q_{\nu m}) \tag{2.102}$$

with side-length (or radius) $c\, 2^{-\nu}$, where again $0 < c < 1$ may be arbitrarily small (but independent of ν and m). This type of argument also covers the following assertion. Let λ be given by (2.6), let $k \in \mathbb{Z}^n$, and let $\|\lambda\,|f_{pq}\|^{(k)}$ be given by the quasi-norm in (2.8) with $\lambda_{\nu,m+k}$ in place of $\lambda_{\nu m}$ (index-shifting). Then there are constants $a > 0$ and $c > 0$ such that for all $\lambda \in f_{pq}$ and all $k \in \mathbb{Z}^n$,

$$\|\lambda\,|f_{pq}\|^{(k)} \leq c\,(1 + |k|)^a\, \|\lambda\,|f_{pq}\|. \tag{2.103}$$

We used this technical observation before, for example in (2.71). The number a depends on p, q (and n).

2.16 Summary

In 2.13 we summarized the (so we hope) elegant way of looking at all the spaces $B^s_{pq}(\mathbb{R}^n)$ and $F^s_{pq}(\mathbb{R}^n)$ covered by Definition 2.6 starting from one single function ψ with (2.15). We have (2.91), (2.92), and the quality of a given function f is judged via the sequence spaces b_{pq} and f_{pq} introduced in 2.2. The considerations in 2.14 and 2.15 relax this otherwise rather rigid setting. It becomes clear that one needs only a dyadic sequence of approximate lattices and that the generalized β-quarks (2.99) may serve as building blocks. Again there are *optimal Cauchy coefficients* analogous to (2.92). Also the higher flexibility of the sequence spaces f^c_{pq} as expressed by (2.101) or (2.102) will be of great service. This not only reveals the nature of the matter but will also be useful later on when spaces related to domains, manifolds or fractal sets are considered. Then it is convenient and natural to adapt the (approximate) lattices and the (generalized) quarks to the geometry of the domains, manifolds or fractals.

3 Spaces on $\mathbb{R}^n$: the general case

3.1 Preliminaries

By Theorem 2.9, the spaces $B^s_{pq}(\mathbb{R}^n)$ and $F^s_{pq}(\mathbb{R}^n)$ introduced in Definition 2.6 coincide with the well-established spaces usually denoted in this way. In particular, we have the following lifting property: Let $\sigma \in \mathbb{R}$ and

$$I_\sigma : \quad f \mapsto (id - \Delta)^{\frac{\sigma}{2}} f, \quad \Delta = \sum_{j=1}^n \frac{\partial^2}{\partial x_j^2} ; \tag{3.1}$$

then I_σ is not only an isomorphic map from $S(\mathbb{R}^n)$ onto itself, and from $S'(\mathbb{R}^n)$ onto itself, but also, in obvious notation,

$$I_\sigma B^s_{pq}(\mathbb{R}^n) = B^{s-\sigma}_{pq}(\mathbb{R}^n) \quad \text{and} \quad I_\sigma F^s_{pq}(\mathbb{R}^n) = F^{s-\sigma}_{pq}(\mathbb{R}^n), \tag{3.2}$$

at least, so far as all the spaces involved fit in Definition 2.6. We refer to [Triβ], 2.3.8. Of course, one could use (3.2) to introduce the spaces $B^s_{pq}(\mathbb{R}^n)$ and $F^s_{pq}(\mathbb{R}^n)$ also for those values of s which are not covered so far. But in order to be consistent we prefer a definition extending 2.6. For this purpose one has first to adapt Definition 2.4 to this more general situation.

3.2 Definition

Let ψ be a non-negative C^∞ function in $\mathbb{R}^n$ with

$$\operatorname{supp} \psi \subset \{y \in \mathbb{R}^n \,:\, |y| < 2^r\} \tag{3.3}$$

for some $r \geq 0$ and

$$\sum_{m \in \mathbb{Z}^n} \psi(x - m) = 1 \quad \text{if} \quad x \in \mathbb{R}^n. \tag{3.4}$$

Let $\psi^\beta(x) = x^\beta \psi(x)$, where x^β are the monomials (2.3). Let

$$s \in \mathbb{R}, \quad 0 < p \leq \infty, \quad \beta \in \mathbb{N}_0^n, \quad \text{and} \quad \frac{L+1}{2} \in \mathbb{N}_0. \tag{3.5}$$

Then

$$(\beta qu)_{\nu m}^L(x) = 2^{-\nu(s - \frac{n}{p})} \left((-\Delta)^{\frac{L+1}{2}} \psi^\beta \right) (2^\nu x - m), \quad x \in \mathbb{R}^n, \tag{3.6}$$

is called an $(s, p)^L$-β-quark related to $Q_{\nu m}$. Here $\nu \in \mathbb{N}_0$ and $m \in \mathbb{Z}^n$.

3.3 Remark

If $L = -1$, then the above definition coincides with Definition 2.4. In particular, $(\beta qu)_{\nu m}^{-1} = (\beta qu)_{\nu m}$ are called as before (s, p)-β-quarks. If $L > -1$ then moment conditions up to order L,

$$\int_{\mathbb{R}^n} x^\alpha \, (\beta qu)_{\nu m}^L(x) \, dx = 0, \quad |\alpha| \leq L, \tag{3.7}$$

are ensured. This explains also the above notation which is consistent with a corresponding notation for atoms, we refer for example to [Triδ], 13.3, p. 73. As in (2.17) we have for some $d > 0$,

$$\mathrm{supp}\, (\beta qu)_{\nu m}^L \subset d\, Q_{\nu m} \quad \text{where} \quad \nu \in \mathbb{N}_0, \, m \in \mathbb{Z}^n. \tag{3.8}$$

Let r be given by (3.3). Then there is a number $c > 0$, which may depend on L, such that

$$\left| (\beta qu)_{\nu m}^L(x) \right| \leq c\, 2^{-\nu(s - \frac{n}{p}) + r|\beta|} \quad \text{where} \quad \beta \in \mathbb{N}_0^n. \tag{3.9}$$

This generalizes (2.19). Our aim is to extend Definition 2.6 to all $s \in \mathbb{R}$. Let again σ_p and σ_{pq} be given by (2.20). We complement (2.22) by

$$\eta = \{ \eta^\beta : \beta \in \mathbb{N}_0^n \} \quad \text{with} \quad \eta^\beta = \{ \eta^\beta_{\nu m} \in \mathbb{C} : \nu \in \mathbb{N}_0, \, m \in \mathbb{Z}^n \} \tag{3.10}$$

and use the quasi-norms (2.23) and (2.27) and their η-counterparts. We use the abbreviation (2.74).

3.4 Definition

(i) Let
$$0 < p \le \infty, \quad 0 < q \le \infty \quad \text{and} \quad s \in \mathbb{R}. \tag{3.11}$$

Let $\sigma > \max(\sigma_p, s)$ and $\frac{L+1}{2} \in \mathbb{N}_0$ with $L \ge \max(-1, [\sigma_p - s])$ be fixed. Let $(\beta qu)_{\nu m}$ be (σ, p)-β-quarks and $(\beta qu)_{\nu m}^L$ be $(s, p)^L$-β-quarks with respect to a fixed function ψ. Let $\varrho > r$, where r has the same meaning as in (3.3) and (3.9). Then $B_{pq}^s(\mathbb{R}^n)$ is the collection of all $f \in S'(\mathbb{R}^n)$ which can be represented as

$$f = \sum_{\beta,\nu,m} \left(\eta_{\nu m}^\beta (\beta qu)_{\nu m} + \lambda_{\nu m}^\beta (\beta qu)_{\nu m}^L \right) \tag{3.12}$$

with
$$\|\eta \,|b_{pq}\|_\varrho + \|\lambda \,|b_{pq}\|_\varrho < \infty. \tag{3.13}$$

Furthermore,
$$\|f \,|B_{pq}^s(\mathbb{R}^n)\| = \inf \left(\|\eta \,|b_{pq}\|_\varrho + \|\lambda \,|b_{pq}\|_\varrho \right), \tag{3.14}$$

where the infimum is taken over all admissible representations.

(ii) Let
$$0 < p < \infty, \quad 0 < q \le \infty, \quad \text{and} \quad s \in \mathbb{R}. \tag{3.15}$$

Let $\sigma > \max(\sigma_{pq}, s)$ and $\frac{L+1}{2} \in \mathbb{N}_0$ with $L \ge \max(-1, [\sigma_{pq} - s])$ be fixed. Let again $(\beta qu)_{\nu m}$ be (σ, p) β quarks and $(\beta qu)_{\nu m}^L$ be $(s, p)^L$-β-quarks, and let $\varrho > r$ as above. Then $F_{pq}^s(\mathbb{R}^n)$ is the collection of all $f \in S'(\mathbb{R}^n)$ which can be represented as (3.12) with

$$\|\eta \,|f_{pq}\|_\varrho + \|\lambda \,|f_{pq}\|_\varrho < \infty. \tag{3.16}$$

Furthermore,
$$\|f \,|F_{pq}^s(\mathbb{R}^n)\| = \inf \left(\|\eta \,|f_{pq}\|_\varrho + \|\lambda \,|f_{pq}\|_\varrho \right), \tag{3.17}$$

where the infimum is taken over all admissible representations.

3.5 Discussion

If $s > \sigma_p$ then one may choose $L = -1$ in part (i) of the above definition and we have essentially part (i) of Definition 2.6. In this case the η-terms in (3.12) and (3.14) are not necessary and $\eta = 0$ is the best possible choice in (3.14).

Similarly in part (ii) if $s > \sigma_{pq}$. Of course, the quasi-norms in (3.14) depend on the fixed function ψ and the fixed numbers σ, L, ϱ. By Theorem 3.6 below, all these quasi-norms are pairwise equivalent and $B_{pq}^s(\mathbb{R}^n)$ does not depend on these parameters. Similarly for $F_{pq}^s(\mathbb{R}^n)$ and (3.17). We now justify the use of the abbreviation (2.74),

$$\sum_{\beta,\nu,m} = \sum_{\beta \in \mathbb{N}_0^n} \sum_{\nu=0}^{\infty} \sum_{m \in \mathbb{Z}^n} . \tag{3.18}$$

By the discussion in 2.7 the η-term in (3.12) converges unconditionally in $L_{\bar{p}}(\mathbb{R}^n)$ with $\bar{p} = \max(p, 1)$ and hence it converges also unconditionally in $S'(\mathbb{R}^n)$. As for the λ-term we first remark that the $(s,p)^L$-β-quark in (3.6) can be written as

$$(\beta qu)_{\nu m}^L(x) = (-\Delta)^{\frac{L+1}{2}} \left\{ 2^{-\nu(s+L+1-\frac{n}{p})} \psi^\beta(2^\nu x - m) \right\}$$
$$= (-\Delta)^{\frac{L+1}{2}} (\beta qu)_{\nu m}^+(x), \tag{3.19}$$

where $(\beta qu)_{\nu m}^+(x)$ are $(s+L+1, p)$-β-quarks according to Definition 2.4. Since in particular,

$$s + L + 1 > \sigma_p, \quad \text{resp.} \quad s + L + 1 > \sigma_{pq}, \tag{3.20}$$

we can apply the discussion in 2.7. It follows that

$$g = \sum_{\beta,\nu,m} \lambda_{\nu m}^\beta (\beta qu)_{\nu m}^+(x) \tag{3.21}$$

converges in $L_{\bar{p}}(\mathbb{R}^n)$ unconditionally. By (3.19), the λ-term in (3.12) coincides with $(-\Delta)^{\frac{L+1}{2}} g$. Hence, also the λ-term in (3.12) converges unconditionally in $S'(\mathbb{R}^n)$. This justifies the use of (3.18). Now we have an almost obvious extension of Theorem 2.9. We use again the notation introduced in (2.37) and (2.38).

3.6 Theorem

(i) Let

$$0 < p \leq \infty, \quad 0 < q \leq \infty, \quad \text{and} \quad s \in \mathbb{R}. \tag{3.22}$$

Then $B_{pq}^s(\mathbb{R}^n)$, introduced in Definition 3.4(i), is a quasi-Banach space. Furthermore,

$$B_{pq}^s(\mathbb{R}^n) = \{ f \in S'(\mathbb{R}^n) : \| f \,|\, B_{pq}^s(\mathbb{R}^n) \|_\varphi < \infty \}, \tag{3.23}$$

and $\|f\,|B_{pq}^s(\mathbb{R}^n)\|_\varphi$, given by (2.37), is an equivalent quasi-norm. In particular, all the quasi-norms in (3.14) and (2.37) for admitted functions ψ, numbers σ, L, ϱ, and functions φ, are equivalent to each other.

(ii) Let

$$0<p<\infty,\quad 0<q\le\infty,\quad\text{and}\quad s\in\mathbb{R}.\tag{3.24}$$

Then $F_{pq}^s(\mathbb{R}^n)$, introduced in Definition 3.4(ii), is a quasi-Banach space. Furthermore,

$$F_{pq}^s(\mathbb{R}^n)=\left\{f\in S'(\mathbb{R}^n)\,:\,\|f\,|F_{pq}^s(\mathbb{R}^n)\|_\varphi<\infty\right\},\tag{3.25}$$

and $\|f\,|F_{pq}^s(\mathbb{R}^n)\|_\varphi$, given by (2.38), is an equivalent quasi-norm. In particular, all the quasi-norms in (3.17) and (2.38) for admitted functions ψ, numbers σ, L, ϱ, and functions φ, are equivalent to each other.

Proof *Step 1* The proof of

$$\|f\,|F_{pq}^s(\mathbb{R}^n)\|_\varphi\le c\,\|f\,|F_{pq}^s(\mathbb{R}^n)\|,\tag{3.26}$$

where the right-hand side is given by (3.17) for fixed functions ψ and numbers σ, L, ϱ and of its B-counterpart is essentially the same as in Step 1 of the proof of Theorem 2.9. The η-part of (3.12) fits in the scheme of Step 1 of the proof of Theorem 2.9 (complemented by the monotonicity of $F_{pq}^s(\mathbb{R}^n)$ for fixed p, q with respect to s). The obvious counterpart of (2.44) for the λ-part of (3.12) is given by

$$f^{\beta,L}=\sum_{\nu=0}^\infty\sum_{m\in\mathbb{Z}^n}\lambda_{\nu m}^\beta\,(\beta qu)_{\nu m}^L(x).\tag{3.27}$$

Again by [Triδ], Theorem 13.8 on p.75, this might be considered as an atomic decomposition in $F_{pq}^s(\mathbb{R}^n)$, where the moment conditions needed are covered by (3.7). Otherwise one can follow the arguments in Step 1 of the proof of Theorem 2.9.

Step 2 We prove the converse of inequality (3.26). Let f be an element of the right-hand side of (3.25). Let σ and L be fixed according to Definition 3.4(ii) and let $\frac{M+1}{2}\in\mathbb{N}_0$ such that

$$s+M+1\ge\sigma\quad\text{and}\quad M\ge L.\tag{3.28}$$

In modification of the lifting properties (3.1), (3.?) we have

$$f=\left(id+(-\Delta)^{\frac{M+1}{2}}\right)g\tag{3.29}$$

with

$$\|g\,|F_{pq}^{s+M+1}(\mathbb{R}^n)\|_\varphi \sim \|f\,|F_{pq}^s(\mathbb{R}^n)\|_\varphi \qquad (3.30)$$

(isomorphic mapping). This follows from [Triβ], 2.3.8, and obvious modifications. Hence by (3.28) and again by [Triβ], 2.3.8, we have

$$f = g_1 + (-\Delta)^{\frac{L+1}{2}} g_2 \quad \text{with} \quad g_1 = g \quad \text{and} \quad g_2 = (-\Delta)^{\frac{M-L}{2}} g, \qquad (3.31)$$

where

$$\|g_1\,|F_{pq}^\sigma(\mathbb{R}^n)\|_\varphi + \|g_2\,|F_{pq}^{s+L+1}(\mathbb{R}^n)\|_\varphi \le c\,\|g\,|F_{pq}^{s+M+1}(\mathbb{R}^n)\|_\varphi. \qquad (3.32)$$

By $\sigma > \sigma_{pq}$ and (3.20) we can apply Theorem 2.9(ii) both to $F_{pq}^\sigma(\mathbb{R}^n)$ and $F_{pq}^{s+L+1}(\mathbb{R}^n)$. Hence there are (optimal) decompositions

$$g_1 = \sum_{\beta,\nu,m} \eta_{\nu m}^\beta\,(\beta qu)_{\nu m}(x) \qquad (3.33)$$

and

$$g_2 = \sum_{\beta,\nu,m} \lambda_{\nu m}^\beta\,(\beta qu)_{\nu m}^+(x), \qquad (3.34)$$

where $(\beta qu)_{\nu m}$ and $(\beta qu)_{\nu m}^+$ are (σ,p)-β-quarks and $(s+L+1,p)$-β-quarks, respectively, and

$$\|\eta\,|f_{pq}\|_\varrho \sim \|g_1\,|F_{pq}^\sigma(\mathbb{R}^n)\|_\varphi \qquad (3.35)$$

and

$$\|\lambda\,|f_{pq}\|_\varrho \sim \|g_2\,|F_{pq}^{s+L+1}(\mathbb{R}^n)\|_\varphi. \qquad (3.36)$$

By (3.31) and (3.19) we have (3.12). Using (3.32) and (3.30) we obtain

$$\|\eta\,|f_{pq}\|_\varrho + \|\lambda\,|f_{pq}\|_\varrho \le c\,\|f\,|F_{pq}^s(\mathbb{R}^n)\|_\varphi, \qquad (3.37)$$

where c is independent of f. This is the converse of (3.26). The proof for the B-spaces is similar.

3.7 Optimal coefficients and the Weierstrassian approach

There are counterparts of our discussions in 2.11, 2.13 and of Corollary 2.12. Let $\Psi_{\nu+\varrho,m}^{\beta,\varrho} \in S(\mathbb{R}^n)$ be the function given by (2.86). Then we may assume that the optimal coefficients $\eta^\beta_{\nu m}$ and $\lambda^\beta_{\nu m}$ in (3.33) and (3.34) with (3.35) and (3.36), respectively, are calculated according to Corollary 2.12,

$$\eta^\beta_{\nu m} = 2^{-\nu(\sigma-\frac{n}{p})} 2^{-\varrho|\beta|} \left(g_1, \Psi^{\beta,\varrho}_{\nu m}\right) \tag{3.38}$$

and

$$\lambda^\beta_{\nu m} = 2^{-\nu(s+L+1-\frac{n}{p})} 2^{-\varrho|\beta|} \left(g_2, \Psi^{\beta,\varrho}_{\nu m}\right). \tag{3.39}$$

By $g_1 = g$ and (3.29) we have

$$\eta^\beta_{\nu m} = 2^{-\nu(\sigma-\frac{n}{p})} 2^{-\varrho|\beta|} \left(f, (id + (-\Delta)^{\frac{M+1}{2}})^{-1} \Psi^{\beta,\varrho}_{\nu m}\right). \tag{3.40}$$

By (3.31) we have a similar formula for $\lambda^\beta_{\nu m}$. In other words, also in this general case we have *Cauchy coefficients* much as in (2.92) and we have a constructive *Weierstrassian approach* (3.12) in analogy to (2.91). Also the discussion about the dependence of some constants on ϱ can be carried over from the regular case treated in Section 2 to the general case considered now. We need later on the extended version of (2.81). Let f be given by (3.12) and let f^β be the obvious counterpart of (2.44). We have again (2.47). Let $v = \min(1, p, q)$. Then we obtain again the right-hand side of (2.80) for the optimal coefficients according to Step 2 of the proof of Theorem 3.6. Hence we have

$$\|f^\beta \,|F^s_{pq}(\mathbb{R}^n)\| \le c\, 2^{\frac{\varrho n}{v}} 2^{-(\varrho-r-\varepsilon)|\beta|} \|f \,|F^s_{pq}(\mathbb{R}^n)\|, \tag{3.41}$$

where c is independent of ϱ with $\varrho > r + \varepsilon$ and $\beta \in \mathbb{N}_0^n$.

3.8 Generalizations

By (2.101) or (2.102) one can replace f_{pq} in connection with Definition 3.4 and Theorem 3.6 by f^c_{pq} with $c > 0$. Furthermore, the arguments in Step 2 of the proof of Theorem 3.6 are based on liftings which reduce the problem to spaces covered by Section 2. Then it is quite clear that also in this context the β-quarks $(\beta qu)_{k,l}(x)$ can be replaced by the generalized quarks in (2.99). The arguments in Step 1 of the proof of Theorem 3.6 apply without any changes to this more general situation. Then one gets an obvious analogue of the end of 2.14.

3.9 Discussion

We introduced the general spaces $B^s_{pq}(\mathbb{R}^n)$ and $F^s_{pq}(\mathbb{R}^n)$ in Definition 3.4 and Theorem 3.6 via the representations (3.12) with the respective conditions for the coefficients. It is well known that moment conditions of type (3.7) cannot be avoided for general values of s. The simplest way to incorporate them is to apply powers of the Laplacian as has been done in (3.6). Hence, in general, the λ-terms in (3.12) are indispensable. What about the η-terms in (3.12) ? Of course they cannot be totally omitted. Otherwise one would have moment conditions of type (3.7) for the whole function f. This is not the case. But it comes out that one needs only the η-terms with $\nu = 0$. This will be discussed in a somewhat different context in Section 8. We refer in particular to Theorem 8.7. From the point of view of the above spaces we deal there only with special spaces. But the arguments given there can be extended to all of the above spaces. From an aesthetical point of view, the outcome, for example (8.43), looks better than (3.12). But (3.12) is more stable and hence in many applications more useful: If one multiplies f in (3.12) with functions in connection with pointwise multiplier problems, or if one wishes to use such representations as a starting point to study function spaces in domains, on manifolds, or on fractals, then the full η-sum in (3.12) is very helpful. This is also the case in connection with the question to what extent the support of f is reflected in these representations. We refer to [Triδ], Corollary 14.11, p. 99. For this reason we prefer here the above version. But we refer to Section 8 for (so we hope) elegant complements.

4 An application: the Fubini property

4.1 The Fubini property

Let $1 < p < \infty$ and $s = k \in \mathbb{N}_0$. Then $W_p^k(\mathbb{R}^n)$ are the classical Sobolev spaces mentioned in (1.4). Let $n \geq 2$,

$$x = (x_1, \ldots, x_n) \in \mathbb{R}^n \quad \text{and} \quad x^j = (x_1, \ldots, x_{j-1}, x_{j+1}, \ldots, x_n) \in \mathbb{R}^{n-1}.$$

Let $f \in W_p^k(\mathbb{R}^n)$ and

$$x_j \mapsto f^{x^j}(x_j) = f(x), \quad x \in \mathbb{R}^n, \tag{4.1}$$

considered as a function on $\mathbb{R}$ for any fixed $x^j \in \mathbb{R}^{n-1}$. It is well known that $W_p^k(\mathbb{R}^n)$ has the so-called *Fubini property*

$$\|f\,|W_p^k(\mathbb{R}^n)\| \sim \sum_{j=1}^n \left\| \|f^{x^j}\,|W_p^k(\mathbb{R})\| \,|\, L_p(\mathbb{R}^{n-1}) \right\|, \tag{4.2}$$

where (in the usual measure-theoretical a.e. interpretation) the inner norm is taken with respect to x_j and the $L_p(\mathbb{R}^{n-1})$-norms with respect to x^j. This is essentially a Fourier multiplier assertion, [Triβ], 2.5.13, p. 114, and Theorem

2.5.6, p. 88. Of course, if $k = 0$, then (4.2) is essentially the classical Fubini theorem, from where the name comes. The extension of this assertion to the Sobolev spaces $H_p^s(\mathbb{R}^n)$ with $1 < p < \infty$, $s > 0$, according to (1.8), (1.9), and to the special Besov spaces $B_{pp}^s(\mathbb{R}^n)$ with $1 \le p \le \infty$, $s > 0$ is due to Strichartz, [Str67] and [Str68], respectively. An extension of this assertion to the special Besov spaces $B_{pp}^s(\mathbb{R}^n)$ with

$$0 < p \le \infty, \quad s > \sigma_p = n(\frac{1}{p} - 1)_+, \tag{4.3}$$

may be found in [Triβ], 2.5.13, p. 115. The Fubini property for the spaces $F_{pq}^s(\mathbb{R}^n)$ with

$$1 < p < \infty, \quad 1 < q < \infty, \quad s > 0, \tag{4.4}$$

has been proved by Kaljabin, [Kal80]. We refer also to [RuS96], p. 70. Some assertions concerning the spaces $F_{pq}^s(\mathbb{R}^n)$ with $p \le 1$ may be found again in [Triβ], 2.5.13, p. 115. Finally we proved in [Tri99a], Theorem 2.1.12, p. 696, the Fubini property for the spaces $F_{pq}^s(\mathbb{R}^n)$ under the natural restrictions

$$0 < p < \infty, \quad 0 < q \le \infty, \quad s > \sigma_{pq} = n\left(\frac{1}{\min(p,q)} - 1\right)_+. \tag{4.5}$$

The treatment given here is a more detailed and extended version of the related parts in [Tri99a].

4.2 Definition

Let $n \ge 2$,

$$0 < p \le \infty, \quad 0 < q \le \infty, \quad s > \sigma_p. \tag{4.6}$$

Let $A_{pq}^s(\mathbb{R}^n)$ be either $B_{pq}^s(\mathbb{R}^n)$ or $F_{pq}^s(\mathbb{R}^n)$ (with $p < \infty$ in the F-case). Then $A_{pq}^s(\mathbb{R}^n)$ is said to have the *Fubini property* if for all $f \in A_{pq}^s(\mathbb{R}^n)$,

$$\|f\,|A_{pq}^s(\mathbb{R}^n)\| \sim \sum_{j=1}^n \left\|\,\|f^{x^j}\,|A_{pq}^s(\mathbb{R})\|\,|L_p(\mathbb{R}^{n-1})\right\| \tag{4.7}$$

(equivalent quasi-norms).

4.3 Remark

We used again the notation (4.1), and the explanation of (4.7) is similar as in connection with (4.2). Recall that the spaces $B_{pq}^s(\mathbb{R}^n)$ with (4.3) consist entirely of regular distributions. This follows from Definition 2.6 and the discussion in 2.7. But this applies also to $F^s_{pq}(\mathbb{R}^n)$ with (4.6) and $p < \infty$, since

$$B^s_{p,\min(p,q)}(\mathbb{R}^n) \subset F^s_{pq}(\mathbb{R}^n) \subset B^s_{p,\max(p,q)}(\mathbb{R}^n), \qquad (4.8)$$

in analogy to (2.10), [Triβ], Proposition 2, p. 47. The final clarification under which circumstances $B^s_{pq}(\mathbb{R}^n)$ and $F^s_{pq}(\mathbb{R}^n)$ are subspaces of $L^{loc}_1(\mathbb{R}^n)$ (and hence, consist entirely of regular distributions) may be found in [RuS96], pp. 33/34, with a reference to [SiT95], and in Theorem 11.2 below. As for the technical side, to check (4.7) we may assume that f is smooth: If $p < \infty$, $q < \infty$, then $S(\mathbb{R}^n)$ is dense both in $B^s_{pq}(\mathbb{R}^n)$ and $F^s_{pq}(\mathbb{R}^n)$, and the rest is a matter of completion. But even if $p = \infty$ and/or $q = \infty$ then one can rely on Fatou arguments, see [ET96], p. 48, [RuS96], p. 15, or 25.1 below. Recall that σ_p and σ_{pq} are given by (4.3) and (4.5), respectively.

4.4 Theorem

Let $n \geq 2$.
(i) The spaces $F^s_{pq}(\mathbb{R}^n)$ with (4.5) have the Fubini property.
(ii) The spaces $B^s_{pq}(\mathbb{R}^n)$ with (4.6) have the Fubini property if, and only if, $p = q$.

Proof *Step 1* Let $x = (x_1, x') \in \mathbb{R}^n$ where $x' = (x_2, \ldots, x_n) \in \mathbb{R}^{n-1}$. Let p, q, s be restricted by (4.5) and $f \in F^s_{pq}(\mathbb{R}^n)$ be smooth. By (4.1) it is clear what is meant by $f^{x'}(x_1)$. We wish to prove that there is a constant $c > 0$ such that for all such f,

$$\left\| \|f^{x'}|F^s_{pq}(\mathbb{R})\| \,|L_p(\mathbb{R}^{n-1})\right\| \leq c \|f|F^s_{pq}(\mathbb{R}^n)\|, \qquad (4.9)$$

where the inner quasi-norm is taken with respect to x_1 and $L_p(\mathbb{R}^{n-1})$ refers to $x' \subset \mathbb{R}^{n-1}$. Let f be given by (2.24) with

$$\|\lambda|f_{pq}\|_\varrho \sim \|f|F^s_{pq}(\mathbb{R}^n)\|. \qquad (4.10)$$

Decomposing f by (2.43), (2.44), it is sufficient to prove (4.9) with f^β in place of f. This follows from (2.81). Hence we may assume without restriction of generality that f coincides with f^β where $\beta = 0$. We put $\lambda_{\nu m} = \lambda^\beta_{\nu m}$ in (2.24) with $\beta = 0$ and assume that the function ψ in (2.15) has a product structure $\psi(x) = \psi(x_1)\psi(x')$. Then we have

$$f = \sum_{\nu=0}^\infty \sum_{m_1 \in \mathbb{Z}} \left[\sum_{m' \in \mathbb{Z}^{n-1}} \lambda_{\nu,(m_1,m')} 2^{\nu\frac{n-1}{p}} \psi(2^\nu x' - m') \right] 2^{-\nu(s-\frac{1}{p})} \psi(2^\nu x_1 - m_1). \qquad (4.11)$$

4. An application: the Fubini property

This is for fixed $x' \in \mathbb{R}^{n-1}$ a quarkonial decomposition of $f^{x'}(x_1)$ with the indicated coefficients in the brackets $[\cdots]$. Let $cQ_{\nu m'}$ be the cubes in $\mathbb{R}^{n-1}$ according to the beginning of 2.2, where $c > 0$, and let $\chi_{\nu m'}(x')$ be the corresponding characteristic functions. If c is suitably chosen then we may assume $\psi(2^\nu x' - m') \le \chi_{\nu m'}(x')$. This applies also to the x_1-variable. By the one-dimensional version of (2.28) with (2.8), modified by (2.101), we have

$$\|f^{x'} \,|\, F^s_{pq}(\mathbb{R})\|$$

$$\le c_1 \left\| \left(\sum_{m_1 \in \mathbb{Z}} \sum_{\nu=0}^\infty \left| \sum_{m' \in \mathbb{Z}^{n-1}} \lambda_{\nu,(m_1,m')} \psi(2^\nu x' - m') \right|^q 2^{\nu \frac{nq}{p}} \chi_{\nu m_1}(\cdot) \right)^{\frac{1}{q}} \,|\, L_p(\mathbb{R}) \right\|$$

$$\le c_2 \left\| \left(\sum_{m_1 \in \mathbb{Z}} \sum_{\nu=0}^\infty \sum_{m' \in \mathbb{Z}^{n-1}} |\lambda_{\nu,(m_1,m')}|^q \, 2^{\nu \frac{n}{p} q} \chi_{\nu m'}(x') \chi_{\nu m_1}(\cdot) \right)^{\frac{1}{q}} \,|\, L_p(\mathbb{R}) \right\|$$
(4.12)

(with the usual modification if $q = \infty$). The last sum is just the same as in the n-dimensional version of f_{pq} in (2.8) with the modification (2.101). We apply $L_p(\mathbb{R}^{n-1})$ with respect to the x'-variables to (4.12). Since the coefficients $\lambda_{\nu m}$ are chosen optimally, (4.10), we obtain (4.9). If f is optimally decomposed by (2.43), (2.44), each term f^β can be estimated in this way and we get first (4.9) with f^β, and afterwards by (2.81) in the general case. Together with the remarks in 4.3 we have in the notation of (4.7),

$$\sum_{j=1}^n \left\| \|f^{x^j} \,|\, F^s_{pq}(\mathbb{R})\| \,|\, L_p(\mathbb{R}^{n-1}) \right\| \le c \|f \,|\, F^s_{pq}(\mathbb{R}^n)\| \tag{4.13}$$

for some $c > 0$ and all $f \in F^s_{pq}(\mathbb{R}^n)$.

Step 2 We prove the converse of (4.13), again under the restriction (4.5). Let

$$0 < u < \min(1, p, q), \quad s < N \in \mathbb{N}.$$

Then

$$d^N_{t,u} f(x) = \left(t^{-n} \int_{|h| \le t} |\Delta^N_h f(x)|^u \, dh \right)^{\frac{1}{u}}, \quad t > 0, \tag{4.14}$$

are ball means where $\Delta^N_h f(x)$ are the usual (n-dimensional) differences given by (1.12). By [Triγ], 3.5.2, 3.5.3, pp. 193–4,

$$\|f \,|\, L_p(\mathbb{R}^n)\| + \left\| \left(\int_0^1 t^{-sq} d^N_{t,u} f(\cdot)^q \frac{dt}{t} \right)^{\frac{1}{q}} \,|\, L_p(\mathbb{R}^n) \right\| \tag{4.15}$$

(usual modification if $q = \infty$) is an equivalent quasi-norm in $F_{pq}^s(\mathbb{R}^n)$. Let $\Delta_{\tau,j}^N f(x)$ with $j = 1, \ldots, n$ and $x \in \mathbb{R}^n$, $\tau \in \mathbb{R}$, be the one-dimensional differences in $\mathbb{R}^n$ with respect to the jth direction of the coordinates. Replacing N in Δ_h^N by nN we wish to estimate

$$\left|\Delta_h^{nN} f(x)\right|^u \leq c \sum_{k=1}^K \sum_{j=1}^n \left|\Delta_{h_j,j}^N f(x + h^k)\right|^u \tag{4.16}$$

with

$$h^k = (b_1^k h_1, \ldots, b_n^k h_n) \quad \text{for} \quad h = (h_1, \ldots, h_n) \in \mathbb{R}^n, \tag{4.17}$$

where c, K, $b_1^k, \ldots, b_n^k$ are suitable numbers. To prove (4.16) we first recall

$$(\Delta_h^{nN} f)(x) = \left((e^{ih\xi} - 1)^{nN} \widehat{f}(\xi)\right)^\vee (x), \quad x \in \mathbb{R}^n. \tag{4.18}$$

Inserting the (nN)th power of

$$e^{ih\xi} - 1 = e^{ih_1\xi_1} - 1 + e^{ih_1\xi_1}(e^{ih_2\xi_2} - 1) + \cdots + e^{ih_1\xi_1 + \cdots + ih_{n-1}\xi_{n-1}}(e^{ih_n\xi_n} - 1) \tag{4.19}$$

in (4.18) we obtain (4.16) with (4.17). Let $(Mf)(x)$ be the Hardy-Littlewood maximal function in $\mathbb{R}^n$. Then by (4.14) with nN in place of N and (4.16) it follows a.e.

$d_{t,u}^{nN} f(x)$

$$\leq c \sum_{k=1}^K \sum_{j=1}^n \left(M \left[\frac{1}{t} \int_{-t}^t |\Delta_{\tau,j}^N f(\cdots, \cdot + c_k \tau, \cdots)|^u \, d\tau \right](x) \right)^{\frac{1}{u}} + + \tag{4.20}$$

for some real constants c_k, where $++$ indicates terms with partial maximal functions (with respect to $m < n$ variables). The interior integrals are the one-dimensional versions of (4.14) but with shifted arguments, which come from (4.17). Checking the arguments in Step 1 of the proof of Theorem 3.5.3 in [Triγ], pp. 194–5, it follows that (4.15) (here the one-dimensional version) remains an equivalent quasi-norm also with the indicated shiftings in (4.20). Furthermore one can discretize (4.15) by $t \to 2^{-l}$ with $l \in \mathbb{N}_0$ and ℓ_q in place of the t-integral. We insert (4.20) in the discretized version of (4.15). We use the vector-valued maximal inequality by Fefferman and Stein with respect to $\frac{q}{u} > 1$ and $\frac{p}{u} > 1$. A formulation and references may be found in [Triγ], 2.2.2, p. 89. Then it follows again by the one-dimensional version of [Triγ], 3.5.3,

p. 194 (with the indicated modification), that the right-hand side of (4.13) can be estimated from above by the left-hand side of (4.13). But this is just what we want. The proof of part (i) is complete.

Step 3 Since $B^s_{pp}(\mathbb{R}^n) = F^s_{pp}(\mathbb{R}^n)$, $p < \infty$, these spaces with $s > \sigma_p$ have the Fubini property. In this case the above arguments can be extended to $p = \infty$. But the Fubini property for $B^s_{pp}(\mathbb{R}^n)$ with $0 < p \leq \infty$, $s > \sigma_p$, is known, [Triβ], 2.5.13, p. 115.

Step 4 It remains to disprove that the spaces $B^s_{pq}(\mathbb{R}^n)$ with (4.6) and $p \neq q$ have the Fubini property. Let

$$\psi(x) = \psi(x_1) \cdots \psi(x_n) = \psi(x_1) \psi(x')$$

be a non-trivial C^∞-function in $\mathbb{R}^n$ with a compact support near the origin and with the indicated product structure. Let

$$f(x) = \sum_{j=1}^{\infty} a_j \, 2^{-j(s-\frac{n}{p})} \, \psi\left(2^j(x - m^j)\right), \quad (4.21)$$

where $a_j \in \mathbb{C}$ and $m^j = (j, \ldots, j)$. We have by Definition 2.6(i),

$$\|f \,|\, B^s_{pq}(\mathbb{R}^n)\| \leq c \left(\sum_{j=1}^{\infty} |a_j|^q \right)^{\frac{1}{q}} \quad (4.22)$$

(usual modification if $q = \infty$) for some $c > 0$ which is independent of a_j. Although it is almost clear that the quarkonial decomposition (4.21) of f is optimal and that (4.22) is an equivalence, we give a detailed proof. Let again $\Delta^M_h f(x)$ be the usual differences according to (1.12). Let $M > s$. Let

$$K_j = \left\{ h \in \mathbb{R}^n : 2^{-j-1} \leq |h| \leq 2^{-j} \right\} \quad \text{where} \quad j \in \mathbb{N}_0.$$

Then

$$\|f \,|\, B^s_{pq}(\mathbb{R}^n)\| \sim \|f \,|\, L_p(\mathbb{R}^n)\| + \left(\sum_{j=1}^{\infty} 2^{jsq} \sup_{h \in K_j} \|\Delta^M_h f \,|\, L_p(\mathbb{R}^n)\|^q \right)^{\frac{1}{q}} \quad (4.23)$$

is an equivalent quasi-norm. This is the discretized version of Theorem 2.5.12 on p. 110 in [Triβ]. Let $h \in K_j$. We may assume that the support of ψ is near the origin. Then we have by the supports of the functions involved

$$\left\| \Delta^M_h \sum_{k=1}^{\infty} a_k \, 2^{-k(s-\frac{n}{p})} \, \psi(2^k(\cdot - m^k)) \,|\, L_p(\mathbb{R}^n) \right\|^p \geq c \, 2^{-jsp} |a_j|^p \quad (4.24)$$

for some $c > 0$ (again obviously modified if $p = \infty$). Inserting this estimate in (4.23) we obtain the converse of (4.22). Hence,

$$\|f\,|B^s_{pq}(\mathbb{R}^n)\| \sim \left(\sum_{j=1}^{\infty}|a_j|^q\right)^{\frac{1}{q}}, \tag{4.25}$$

where the equivalent constants are independent of a_j (they may depend on ψ). Now we assume that $B^s_{pq}(\mathbb{R}^n)$ has the Fubini property, hence by Definition 4.2,

$$\|f\,|B^s_{pq}(\mathbb{R}^n)\| \sim \sum_{j=1}^{n}\left\|\,\|f^{x_j}\,|B^s_{pq}(\mathbb{R})\|\,|L_p(\mathbb{R}^{n-1})\right\|. \tag{4.26}$$

By the assumed product structure $\psi(x) = \psi(x_1)\,\psi(x')$ we can rewrite (4.21)

$$f^{x'}(x_1) = f(x) = \sum_{j=1}^{\infty} a_j\, 2^{j\frac{n-1}{p}}\, \psi\left(2^j(x' - m'^j)\right)\, 2^{-j(s-\frac{1}{p})}\, \psi\left(2^j(x_1 - j)\right). \tag{4.27}$$

The counterpart of (4.25) is given by

$$\|f^{x'}(x_1)\,|B^s_{pq}(\mathbb{R})\| \sim \left(\sum_{j=1}^{\infty}|a_j|^q\, 2^{j\frac{n-1}{p}q}\, \psi^q\left(2^j(x' - m'^j)\right)\right)^{\frac{1}{q}}, \tag{4.28}$$

where we assume that for fixed x' at most one term on the right-hand side of (4.28) is different from zero. But then it is clear that application of $L_p(\mathbb{R}^{n-1})$ with respect to x' results in

$$\left\|\,\|f^{x'}(x_1)\,|B^s_{pq}(\mathbb{R})\|\,|L_p(\mathbb{R}^{n-1})\right\| \sim \left(\sum_{j=1}^{\infty}|a_j|^p\right)^{\frac{1}{p}}. \tag{4.29}$$

Of course this applies to any of the terms on the right-hand side of (4.26). Together with (4.25) we get

$$\left(\sum_{j=1}^{\infty}|a_j|^p\right)^{\frac{1}{p}} \sim \left(\sum_{j=1}^{\infty}|a_j|^q\right)^{\frac{1}{q}} \tag{4.30}$$

(usual modification if $p = \infty$ and/or $q = \infty$) where the equivalent constants are independent of a_j. This proves $p = q$.

5 Spaces on domains: localization and Hardy inequalities

5.1 Introduction to Sections 5 and 6

Let Ω be a domain in $\mathbb{R}^n$. Then the spaces $B^s_{pq}(\Omega)$ and $F^s_{pq}(\Omega)$ can be introduced by restriction of $B^s_{pq}(\mathbb{R}^n)$ and $F^s_{pq}(\mathbb{R}^n)$ on Ω, respectively. This applies in particular to the special cases considered in 1.2: Sobolev spaces (classical and fractional), Hölder-Zygmund spaces, classical Besov spaces etc. These concrete spaces, but also the general scales $B^s_{pq}(\Omega)$ and $F^s_{pq}(\Omega)$, have been studied in great detail. We refer to the books listed in 1.1. It is the aim of this section and the following one to contribute to this theory in the spirit of Sections 2 and 3, where we discussed corresponding spaces on $\mathbb{R}^n$.

We describe our intentions by looking first at a simple case, where all assertions which are of interest for us are known. Let Ω be a bounded C^∞ domain in $\mathbb{R}^n$. Let $1 < p < \infty$ and $s \in \mathbb{N}_0$. Then

$$W^s_p(\Omega) = \left\{ f \in D'(\Omega) \,:\, \text{there is a } g \in W^s_p(\mathbb{R}^n) \text{ with } g|\Omega = f \right\}, \tag{5.1}$$

$$\|f \,|\, W^s_p(\Omega)\| = \inf \|g \,|\, W^s_p(\mathbb{R}^n)\|, \tag{5.2}$$

where the infimum is taken over all $g \in W^s_p(\mathbb{R}^n)$ such that its restriction $g|\Omega$ to Ω coincides in $D'(\Omega)$ with f. Furthermore, $\overset{\circ}{W}{}^s_p(\Omega)$ is the completion of $D(\Omega) = C_0^\infty(\Omega)$ in $W^s_p(\Omega)$, and

$$\widetilde{W}^s_p(\Omega) = \left\{ f \in W^s_p(\mathbb{R}^n) \,:\, \mathrm{supp}\, f \subset \overline{\Omega} \right\}, \tag{5.3}$$

$$\|f \,|\, \widetilde{W}^s_p(\Omega)\| = \|f \,|\, W^s_p(\mathbb{R}^n)\|, \tag{5.4}$$

considered as subspaces of $D'(\Omega)$. It is well known that

$$\|f \,|\, W^s_p(\Omega)\| \sim \left(\sum_{|\alpha| \le s} \|D^\alpha f \,|\, L_p(\Omega)\|^p \right)^{\frac{1}{p}}, \quad f \in W^s_p(\Omega), \tag{5.5}$$

(equivalent norms) and

$$\overset{\circ}{W}{}^s_p(\Omega) = \widetilde{W}^s_p(\Omega) \tag{5.6}$$

(equivalent norms). Let

$$d(x) = \mathrm{dist}(x, \partial\Omega), \quad x \in \mathbb{R}^n, \tag{5.7}$$

be the distance of a point $x \in \mathbb{R}^n$ to the boundary $\partial\Omega$ of Ω. We have the well-known *Hardy inequality*: there is a $c > 0$ such that

$$\int_\Omega d^{-sp}(x) |f(x)|^p\, dx \le c \,\|f \mid W_p^s(\Omega)\|^p \quad \text{for all} \quad f \in \overset{\circ}{W}{}_p^s(\Omega). \tag{5.8}$$

As a consequence one obtains easily the following *localization* assertion: Let K_j be balls of radius $r_j > 0$ in $\mathbb{R}^n$ with

$$\bigcup_{j=1}^\infty K_j = \Omega, \quad dist(K_j, \partial\Omega) \sim r_j, \tag{5.9}$$

such that at most N of them have a non-empty intersection, where $N \in \mathbb{N}$ is a suitable number. Assume that there is a subordinated resolution of unity $\varphi_j \in D(\Omega)$ with

$$\sum_{j=1}^\infty \varphi_j(x) = 1 \quad \text{if } x \in \Omega; \quad supp\, \varphi_k \subset K_k \quad \text{if } k \in \mathbb{N}, \tag{5.10}$$

and

$$|D^\gamma \varphi_j(x)| \le c_\gamma\, r_j^{-|\gamma|}, \quad \gamma \in \mathbb{N}_0^n, \, j \in \mathbb{N}, \tag{5.11}$$

for some $c_\gamma > 0$. Then

$$\|f \mid W_p^s(\Omega)\|^p \sim \sum_{j=1}^\infty \|\varphi_j f \mid W_p^s(\mathbb{R}^n)\|^p, \quad f \in \widetilde{W}_p^s(\Omega). \tag{5.12}$$

One can introduce corresponding spaces, say, $F_{pq}^s(\Omega)$, $\widetilde{F}_{pq}^s(\Omega)$, $\overset{\circ}{F}{}_{pq}^s(\Omega)$ and their B-counterparts. It is the aim of this section to study their interrelations, where we are especially interested in counterparts of (5.12). But there is a significant difference between the F-spaces and the B-spaces. Let $\psi \in D(\mathbb{R}^n)$ be the same function as in Definition 2.4 with the resolution of unity (2.15). Put $\psi_m(x) = \psi(x-m)$, where $m \in \mathbb{Z}^n$. Let $0 < p < \infty$, $0 < q \le \infty$ and $s \in \mathbb{R}$. Then we have

$$\|f \mid F_{pq}^s(\mathbb{R}^n)\| \sim \left(\sum_{m \in \mathbb{Z}^n} \|\psi_m f \mid F_{pq}^s(\mathbb{R}^n)\|^p \right)^{\frac{1}{p}}, \quad f \in F_{pq}^s(\mathbb{R}^n), \tag{5.13}$$

(equivalent quasi-norms). We refer to [Triγ], Theorem 2.4.7, p. 124. By the same theorem it follows that (5.13) with B_{pq}^s in place of F_{pq}^s holds if, and only if, $p = q$ (recall $B_{pp}^s = F_{pp}^s$), including $p = q = \infty$ (we denoted $\mathcal{C}^s = B_{\infty\infty}^s$ as Hölder-Zygmund spaces). In other words, counterparts of (5.12), and of the underlying inequality (5.8), can only be expected for the F-spaces. This explains why we concentrate in this section, and also in the following one, almost exclusively on the F-scale. We are mostly interested in the regular case, which often results in the restriction

$$0 < p < \infty, \quad 0 < q \leq \infty, \quad s > \sigma_{pq} = n \left(\frac{1}{\min(p,q)} - 1 \right)_+, \tag{5.14}$$

of the parameters involved for the spaces $F^s_{pq}(\Omega)$. But we describe what is known in some other cases. Section 6 is based on the combination of the F^s_{pq}-counterpart of (5.12), say, with (5.14), and the quarkonial description of $F^s_{pq}(\mathbb{R}^n)$ as given by Definition 2.6. This will result in a quarkonial characterization of $\widetilde{F}^s_{pq}(\Omega)$. Later on we use these results in Section 7 in a quarkonial approach to function spaces on some hyperbolic manifolds, which, in turn, is the basis for a spectral theory of related elliptic operators in Section 21. The present section, and also the following one, is mainly based on [Tri99a].

5.2 Some notation

Let Ω be a domain (i. e., open set) in $\mathbb{R}^n$. Its boundary is denoted by $\partial \Omega$. Let

$$d(x) = dist\,(x, \partial \Omega) = \inf |x - y|, \quad x \in \mathbb{R}^n, \tag{5.15}$$

be the distance of a point $x \in \mathbb{R}^n$ to $\partial \Omega$, where the infimum is taken over all $y \in \partial \Omega$.

Let $0 < p \leq \infty$ and $\sigma \in \mathbb{R}$. Then $L_p(\Omega, d^\sigma)$ is the quasi-Banach space of all complex Lebesgue measurable functions in Ω such that

$$\| f \,|\, L_p(\Omega, d^\sigma) \| = \left(\int_\Omega d^{\sigma p}(x) \, |f(x)|^p \, dx \right)^{\frac{1}{p}} < \infty \tag{5.16}$$

(with the usual modification if $p = \infty$).

As usual, $D(\Omega) = C_0^\infty(\Omega)$ stands for the collection of all complex infinitely differentiable functions in $\mathbb{R}^n$ with compact support in Ω. Let $D'(\Omega)$ be the dual space of distributions on Ω.

Let $g \in S'(\mathbb{R}^n)$. Then we denote by $g|\Omega$ its restriction to Ω,

$$g|\Omega \in D'(\Omega) : \quad (g|\Omega)(\varphi) = g(\varphi) \quad \text{for} \quad \varphi \in D(\Omega). \tag{5.17}$$

As said in 5.1 we are mainly interested in the spaces F^s_{pq}. Then we have $0 < p < \infty$, $0 < q \leq \infty$. This includes the spaces $B^s_{pp} = F^s_{pp}$ with $0 < p < \infty$. Occasionally it will be convenient for us to extend the F-scale notationally to $p = q = \infty$,

$$\mathcal{C}^s = B^s_{\infty\infty} = F^s_{\infty\infty}, \quad s \in \mathbb{R}. \tag{5.18}$$

5.3 Definition

Let Ω be a bounded domain in $\mathbb{R}^n$. Let $0 < p \leq \infty$, $0 < q \leq \infty$, $s \in \mathbb{R}$. Then A^s_{pq} stands either for B^s_{pq} or F^s_{pq} (with $p < \infty$ in the F-case).

(i)
$$A^s_{pq}(\Omega) = \{f \in D'(\Omega) : \text{ there is a } g \in A^s_{pq}(\mathbb{R}^n) \text{ with } g|\Omega = f\}, \quad (5.19)$$

$$\|f \mid A^s_{pq}(\Omega)\| = \inf \|g \mid A^s_{pq}(\mathbb{R}^n)\|, \quad (5.20)$$

where the infimum is taken over all $g \in A^s_{pq}(\mathbb{R}^n)$ such that its restriction $g|\Omega$ to Ω coincides in $D'(\Omega)$ with f.

(ii) Furthermore, $\mathring{A}^s_{pq}(\Omega)$ is the completion of $D(\Omega)$ in $A^s_{pq}(\Omega)$.

(iii) Finally,
$$\widetilde{A}^s_{pq}(\Omega) = \{f \in D'(\Omega) : \quad (5.21)$$
$$\text{there is a } g \in A^s_{pq}(\mathbb{R}^n) \text{ with } g|\Omega = f \text{ and } \operatorname{supp} g \subset \overline{\Omega}\},$$

$$\|f \mid \widetilde{A}^s_{pq}(\Omega)\| = \inf \|g \mid A^s_{pq}(\mathbb{R}^n)\|, \quad (5.22)$$

where the infimum is taken over all admitted g in (5.21).

5.4 Remark

As said, we are mainly interested in the F-scale, complemented by (5.18). For this reason we restrict the formulations below to the F-scale, although some of the assertions have more or less obvious B-counterparts (but not, as indicated in 5.1, our main results in this section: the counterpart of (5.12)). It is of interest to compare the subspace

$$\widetilde{F}^s_{pq}(\overline{\Omega}) = \{f \in F^s_{pq}(\mathbb{R}^n) : \operatorname{supp} f \subset \overline{\Omega}\} \quad (5.23)$$

of $F^s_{pq}(\mathbb{R}^n)$ with the space $\widetilde{F}^s_{pq}(\Omega)$ given by (5.21), which is a subspace of $D'(\Omega)$. We have

$$\widetilde{F}^s_{pq}(\Omega) = \widetilde{F}^s_{pq}(\overline{\Omega}) / \{h \in F^s_{pq}(\mathbb{R}^n) : \operatorname{supp} h \subset \partial\Omega\} \quad (5.24)$$

as a factor space (in the usual interpretation). We are interested in those cases where the spaces in (5.23) and (5.24) coincide.

5.5 Proposition

Let Ω be a bounded C^∞ domain in $\mathbb{R}^n$. Let

$$0 < p \leq \infty, \quad 0 < q \leq \infty \quad (\text{with } q = \infty \text{ if } p = \infty), \tag{5.25}$$

and

$$\max\left(\frac{1}{p} - 1, n(\frac{1}{p} - 1)\right) < s < \infty. \tag{5.26}$$

Then

$$\widetilde{F}^s_{pq}(\Omega) = \widetilde{F}^s_{pq}(\overline{\Omega}) \tag{5.27}$$

(in the usual interpretation).

Proof *Step 1* By (5.24) we must prove

$$\{h \in F^s_{pq}(\mathbb{R}^n) : \operatorname{supp} h \subset \partial\Omega\} = \{0\}. \tag{5.28}$$

If $s > \sigma_p = n(\frac{1}{p} - 1)_+$, then $F^s_{pq}(\mathbb{R}^n)$ consists entirely of regular distributions. This is well known and was discussed again in 4.3. We refer also to Theorem 11.2 below. Hence we have (5.28), since $|\partial\Omega| = 0$, where $|\partial\Omega|$ is the Lebesgue measure of $\partial\Omega$. The remaining cases

$$1 < p \leq \infty, \quad 0 < q \leq \infty \ (q = \infty \text{ if } p = \infty), \quad \frac{1}{p} - 1 < s \leq 0 \tag{5.29}$$

can be reduced to

$$1 < p = q < \infty, \quad \frac{1}{p} - 1 < s < 0, \tag{5.30}$$

since $\partial\Omega$ is compact and the spaces on the left-hand side of (5.28) are separately monotone with respect to each of the three indices s, p, q. Hence we may assume (5.30). By [Triβ], 2.11.2, p. 178, we have the duality

$$\left(F^s_{pp}(\mathbb{R}^n)\right)' = F^{-s}_{p'p'}(\mathbb{R}^n), \quad \frac{1}{p} + \frac{1}{p'} = 1, \tag{5.31}$$

according to the dual pairing $(S(\mathbb{R}^n), S'(\mathbb{R}^n))$. Of course, $B^\sigma_{pp} = F^\sigma_{pp}$. We claim that under the above conditions $1 < p' < \infty$, $0 < -s < \frac{1}{p'}$,

$$\{\varphi \in S(\mathbb{R}^n) : \varphi(x) = 0 \text{ near } \partial\Omega\} \tag{5.32}$$

is dense in $F_{p'p'}^{-s}(\mathbb{R}^n)$. We shift the proof of this assertion to Step 2 and take it temporarily for granted. Let now $h \in F_{pq}^s(\mathbb{R}^n)$ with $\operatorname{supp} h \subset \partial\Omega$. Then we have by (5.31)

$$\|h\,|F_{pp}^s(\mathbb{R}^n)\| = \sup |h(\varphi)| = 0, \qquad (5.33)$$

where the supremum is taken over all φ with (5.32) and $\|\varphi\,|F_{p'p'}^{-s}(\mathbb{R}^n)\| = 1$. This proves (5.28).

Step 2 It remains to prove the following assertion: Let $1 < u < \infty$ and $0 < \sigma < \frac{1}{u}$, then (5.32) is dense in $F_{uu}^\sigma(\mathbb{R}^n) = B_{uu}^\sigma(\mathbb{R}^n)$. Since $S(\mathbb{R}^n)$ is dense in $F_{uu}^\sigma(\mathbb{R}^n)$ it is sufficient to approximate a function $\chi \in S(\mathbb{R}^n)$ in $F_{uu}^\sigma(\mathbb{R}^n)$ by functions belonging to (5.32). Let ψ be the same function as in Definition 2.4 with the resolution of unity (2.15). Let $\nu \in \mathbb{N}_0$. Then we decompose

$$\chi(x) = \chi_1(x) + \chi_2(x) = \chi_1(x) + {\sum_m}' \psi(2^\nu x - m)\,\chi(x) \qquad (5.34)$$

where the latter sum is taken over those lattice points in $\mathbb{Z}^n$ such that the cubes $d\,Q_{\nu m}$ in (2.17) have a non-empty intersection with $\partial\Omega$. Then χ_1 belongs to the set in (5.32) and χ_2 can be written as the atomic decomposition

$$\chi_2(x) = {\sum_m}' 2^{\nu(\sigma - \frac{n}{u})}\, 2^{-\nu(\sigma - \frac{n}{u})}\,\psi(2^\nu x - m)\,\chi(x) \qquad (5.35)$$

in $B_{uu}^\sigma(\mathbb{R}^n)$. The number of cubes involved in (5.35) can be estimated by $c\,2^{\nu(n-1)}$, where $c > 0$ is independent of $\nu \in \mathbb{N}_0$. Hence we have by [Triδ], Theorem 13.8, p. 75,

$$\|\chi_2\,|B_{uu}^\sigma(\mathbb{R}^n)\| \leq c\,2^{\nu(\sigma - \frac{n}{u})} \left({\sum_m}' 1\right)^{\frac{1}{u}} \leq c'\,2^{\nu(\sigma - \frac{1}{u})}. \qquad (5.36)$$

However the right-hand side tends to zero if $\nu \to \infty$, since $\sigma < \frac{1}{u}$. But this is just what we wanted to prove.

5.6 Proposition

Let Ω be a bounded C^∞ domain in $\mathbb{R}^n$.

(i) Let $0 < p \leq \infty$ and $s \in \mathbb{R}$, and let $F_{p\infty}^s(\Omega)$ be the spaces introduced in Definition 5.3(i), complemented by (5.18). Let $C^\infty(\overline{\Omega})$ be the restriction of $S(\mathbb{R}^n)$ on Ω. Then $C^\infty(\overline{\Omega})$ is not dense in $F_{p\infty}^s(\Omega)$.

(ii) Let $F_{pq}^s(\Omega)$ and $\overset{\circ}{F}{}_{pq}^s(\Omega)$ be the spaces introduced in Definition 5.3. Then

$$F_{pq}^s(\Omega) = \overset{\circ}{F}{}_{pq}^s(\Omega) \qquad (5.37)$$

5. Spaces on domains: localization and Hardy inequalities

if, and only if, one of the following two conditions is satisfied:

(a) $\quad 0 < p < \infty, \quad -\infty < s < \dfrac{1}{p}, \quad 0 < q < \infty;$ (5.38)

(b) $\quad 1 < p < \infty, \quad s = \dfrac{1}{p}, \quad 0 < q < \infty.$ (5.39)

Proof *Step 1* We prove part (i). Let $p < \infty$. Then by (5.13) it is sufficient to prove that $S(\mathbb{R}^n)$ is not dense in $F_{p\infty}^s(\mathbb{R}^n)$. We may assume $s = 0$. Let $\varrho \in S(\mathbb{R}^n)$ be a non-trivial function such that $\hat\varrho$ has compact support near the origin. Let

$$f = \sum_{k=0}^{\infty} e^{i 2^k x_1} \varrho(x) \quad \text{where} \quad x \in \mathbb{R}^n. \tag{5.40}$$

We assume that

$$(\varphi_j \hat{f})^\vee(x) = e^{i 2^j x_1} \varrho(x) \quad \text{with} \quad x \in \mathbb{R}^n, \tag{5.41}$$

where $\{\varphi_j\}$ are the same functions as used in (2.38) and in Theorems 2.9 and 3.6. Let $\psi \in S(\mathbb{R}^n)$. Then it follows from (2.38) and (5.41) that

$$\|f - \psi \,|F_{p\infty}^0(\mathbb{R}^n)\| \geq \|\varrho \,|L_p(\mathbb{R}^n)\| > 0. \tag{5.42}$$

This proves that $S(\mathbb{R}^n)$ is not dense in $F_{p\infty}^s(\mathbb{R}^n)$. In case of $p = q = \infty$, $0 < s < 1$, and $\Omega = (-1, 1)$, the function

$$f(t) = |t|^s, \quad t \in (-1, 1), \tag{5.43}$$

belongs to $\mathcal{C}^s(\Omega) = F_{\infty\infty}^s(\Omega)$, but it cannot be approximated in $\mathcal{C}^s(\Omega)$ by smooth functions. There is an n-dimensional counterpart. By lifting it follows that $C^\infty(\overline\Omega)$ is not dense in $\mathcal{C}^s(\Omega)$ for any $s \in \mathbb{R}$.

Step 2 We prove part (ii). By Step 1 we may assume that $0 < p < \infty$ and $0 < q < \infty$. We begin with a few preparations. Recall that both $S(\mathbb{R}^n)$ and $D(\mathbb{R}^n) = C_0^\infty(\mathbb{R}^n)$ are dense in $F_{pq}^s(\mathbb{R}^n)$. First we claim that (5.37) is equivalent to the assertion that

$$D_\Omega(\mathbb{R}^n) = \{g \in D(\mathbb{R}^n) : g(x) = 0 \text{ near } \partial\Omega\} \tag{5.44}$$

is dense in $F_{pq}^s(\mathbb{R}^n)$. One direction is obvious: If $D_\Omega(\mathbb{R}^n)$ is dense in $F_{pq}^s(\mathbb{R}^n)$ then it follows from Definition 5.3 that $D(\Omega)$ is dense in $F_{pq}^s(\Omega)$, and we get (5.37). To prove the converse we assume that we have (5.37). It is sufficient to approximate a function $f \in D(\mathbb{R}^n)$ in $F_{pq}^s(\mathbb{R}^n)$ by functions belonging to (5.44).

Let $f|\Omega \in C^\infty(\overline{\Omega})$ be the restriction of $f \in D(\mathbb{R}^n)$ on Ω and let $ext\,(f|\Omega)$ be the extension of $f|\Omega$ according to the linear extension procedure described in [Triγ], Theorem 5.1.3, p. 239. We have $ext\,(f|\Omega) \in F^s_{pq}(\mathbb{R}^n)$, but we may also assume $ext\,(f|\Omega) \in C^N(\mathbb{R}^n)$ for any given $N \in \mathbb{N}$ and that $ext\,(f|\Omega)$ vanishes outside of a suitable ball. Let $g \in D(\Omega)$. Then

$$\|ext\,(f|\Omega) - ext\,g\,|F^s_{pq}(\mathbb{R}^n)\| \le c\,\|f|\Omega - g\,|F^s_{pq}(\Omega)\|, \qquad (5.45)$$

and, by construction, $ext\,g \in C^N(\mathbb{R}^n)$ vanishes near $\partial\Omega$ and outside a ball. Approximating $ext\,g$, it follows that for any $\varepsilon > 0$ there is a function $h \in D_\Omega(\mathbb{R}^n)$ with

$$\|ext\,(f|\Omega) - h\,|F^s_{pq}(\mathbb{R}^n)\| \le \varepsilon. \qquad (5.46)$$

By construction, $f - ext\,(f|\Omega) \in C^N(\mathbb{R}^n)$ vanishes identically in $\overline{\Omega}$ and outside a large ball. By standard arguments one can approximate such a function in $F^s_{pq}(\mathbb{R}^n)$ by functions belonging to $D_\Omega(\mathbb{R}^n)$ in (5.44). Hence, f can be approximated in the desired way.

Step 3 We need a second preparation. Let

$$s \in \mathbb{R}, \quad 0 < p < \infty, \quad 0 < q_1 < \infty, \quad 0 < q_2 < \infty. \qquad (5.47)$$

Then we claim that (5.37) holds for $F^s_{pq_1}(\Omega)$ if, and only if, (5.37) holds for $F^s_{pq_2}(\Omega)$ (independence from the third index). By Step 2 it is sufficient to ask for an approximation of $f \in D(\mathbb{R}^n)$ by functions belonging to $D_\Omega(\mathbb{R}^n)$ in the respective spaces on $\mathbb{R}^n$. Let $f \in D(\mathbb{R}^n)$ and $g \in D_\Omega(\mathbb{R}^n)$. We use the quarkonial decomposition (3.12) for $f - g$ in place of f where we may assume that the coefficients $\eta^\beta_{\nu m}$ and $\lambda^\beta_{\nu m}$, given by (3.38)–(3.40), depend on s and p, but not on q_1 and q_2,

$$\begin{aligned} f - g &= \sum_{\beta,\nu,m} \left(\eta^\beta_{\nu m} (\beta qu)_{\nu m} + \lambda^\beta_{\nu m} (\beta qu)^L_{\nu m} \right) \chi \\ &= (h_1 + h_2 + h_3)\chi \end{aligned} \qquad (5.48)$$

where $\chi \in D(\mathbb{R}^n)$ is a cut-off function, which is identically 1 on the supports of f and g,

$$h_1 = \sum_{|\beta|>B} \sum_{\nu=0}^\infty \sum_{m\in\mathbb{Z}^n} \cdots + \sum_{|\beta|\le B} \sum_{\nu=N}^\infty \sum_{m\in\mathbb{Z}^n} \cdots, \qquad (5.49)$$

$$h_2 = \sum_{|\beta|\le B} \sum_{\nu=0}^{N-1} {\sum_m}' \cdots, \qquad (5.50)$$

$$h_3 = \sum_{|\beta|\le B} \sum_{\nu=0}^{N-1} {\sum_m}'' \cdots, \qquad (5.51)$$

5. Spaces on domains: localization and Hardy inequalities   49

for some $B \in \mathbb{N}$ and $N \in \mathbb{N}$. Here $\sum_m'$ covers (in dependence on ν) those lattice points on $\mathbb{Z}^n$ for which $d\,Q_{\nu m}$ with (2.17) has a non-empty intersection with $\partial\Omega$, and $\sum_m''$ collects the remaining lattice points. Let $0 < q_1 < q_2 < \infty$. If $\varepsilon > 0$ is given and if B and N are chosen sufficiently large, then we claim that

$$\|\chi\, h_1\,|F^s_{pq_1}(\mathbb{R}^n)\| \le \varepsilon. \tag{5.52}$$

So far as the sum over $|\beta| > B$ in (5.49) is concerned we refer to (3.41). As for the second sum we have standard measure-theoretical arguments in (3.17) based on (2.8), (2.27); while for h_2 we may assume $h_2 = \chi\, h_2$. Furthermore

$$\|h_2\,|F^s_{pq_1}(\mathbb{R}^n)\| \sim \|h_2\,|F^s_{pq_2}(\mathbb{R}^n)\|, \tag{5.53}$$

where the equivalence constants are independent of B, N, and h_2, as far as it has the special structure (5.50). This independence of q (equivalent quasi-norms) of $\|h_2\,|F^s_{pq}(\mathbb{R}^n)\|$ for functions of this special structure is one of the major technical discoveries in this context and is due to Frazier and Jawerth. We refer to [FrJ90], §11, for a detailed argument on an atomic level, which applies (independently) to any fixed $\beta \in \mathbb{N}_0^n$. Then the extension to the quarkonial level follows immediately from (2.27). Finally, $\chi h_3 \in D_\Omega(\mathbb{R}^n)$. It follows by (5.48), (5.52), (5.53) and $q_2 > q_1$ that

$$\begin{aligned}\|f - g - \chi h_3\,|F^s_{pq_1}(\mathbb{R}^n)\| &\le c\varepsilon + \|h_2\,|F^s_{pq_1}(\mathbb{R}^n)\| \\ &\sim \varepsilon + \|h_2\,|F^s_{pq_2}(\mathbb{R}^n)\| \\ &\le c'\varepsilon + c'\,\|f - g - \chi h_3\,|F^s_{pq_2}(\mathbb{R}^n)\|,\end{aligned} \tag{5.54}$$

where c and c' are independent of f, g, and ε. Now the independence of (5.37) on q follows from (5.54). Assume we have (5.37) for $F^s_{pq_2}(\Omega)$. For given f, one calculates $\chi h_3 \in D_\Omega(\mathbb{R}^n)$ and chooses $g \in D_\Omega(\mathbb{R}^n)$ such that the right-hand side becomes small. Then we obtain also an approximation in $F^s_{pq_1}(\mathbb{R}^n)$. The converse assertion follows from the monotonicity of the spaces $F^s_{pq}(\mathbb{R}^n)$ with respect to q.

Step 4   We prove (5.37) under the assumption (5.38). Let, in addition,

$$n\left(\frac{1}{p} - 1\right)_+ = \sigma_p < s < \frac{1}{p}. \tag{5.55}$$

We may assume $q = p$. (This follows from the monotonicity of the spaces F^s_{pq} with respect to s and elementary embeddings. The more sophisticated argument from Step 3 is needed only in limiting situations.) Then we are in the same situation as in Step 2 of the proof of Proposition 5.5 and we get the desired result. For small p the restriction (5.55) might not be available. Then one can modify the above argument as follows. Any $f \in D(\mathbb{R}^n)$ can be represented as

$$f(x) = (-\Delta)^M g(x), \quad g \in D(\mathbb{R}^n), \quad x \text{ near } \partial\Omega. \tag{5.56}$$

This follows from mapping properties of the operator $(-\Delta)^M$, where $M \in \mathbb{N}$, the related regularity theory and the usual cut-off arguments. Hence, for our purpose we assume that f is given by (5.56). We decompose g much as in (5.34) and apply afterwards $(-\Delta)^M$. If M is chosen sufficiently large, then one gets a counterpart of (5.35) where now the atoms involved have the needed moment conditions according to [Triδ], Theorem 13.8, p. 75. Now the counterpart of (5.36) gives the desired result.

Step 5 We prove the proposition under the restriction $p \leq 1$. By Step 4 we must disprove (5.37) if $s \geq \frac{1}{p}$ and $0 < q < \infty$. By Step 3 we may assume $p = q$. According to [Triγ], 4.4.3, p. 220, and the usual standard arguments we have the trace property

$$tr_{\partial\Omega} B_{pp}^{\frac{1}{p}}(\mathbb{R}^n) = L_p(\partial\Omega), \quad 0 < p \leq 1. \tag{5.57}$$

(If $n = 1$, then $f \in B_{pp}^{\frac{1}{p}}(\mathbb{R}^n)$ is continuous.) This disproves (5.37) if $s = \frac{1}{p}$, and, by monotonicity, also if $s \geq \frac{1}{p}$ and $p \leq 1$.

Step 6 We prove the proposition for the remaining case $1 < p < \infty$ and $s \geq \frac{1}{p}$. If $s > \frac{1}{p}$, then we have the well-known trace assertion which may be found in [Triβ], 2.7.2, p. 132, or [Triγ], 4.4.1, 4.4.2, pp. 212–213. This disproves (5.37). Finally we prove (5.37) under the restriction (5.39). By Step 3 we may assume $q = 2$. According to (1.9), $F_{p,2}^{\frac{1}{p}}(\mathbb{R}^n) = H_p^{\frac{1}{p}}(\mathbb{R}^n)$ are the well-known Sobolev spaces. In this case, (5.37) is known. We refer to [Triα], Theorem 2.9.3(d), p. 220. The proof is complete.

5.7 Proposition

Let Ω be a bounded C^∞ domain in $\mathbb{R}^n$ and let $L_p(\Omega, d^\sigma)$ be the spaces introduced in 5.2. Let

$$0 < p \leq \infty, \quad 0 < q \leq \infty, \quad \text{and} \quad s > \sigma_p = n\left(\frac{1}{p} - 1\right)_+, \tag{5.58}$$

(with $q = \infty$ if $p = \infty$). Let $\widetilde{F}_{pq}^s(\Omega)$ be the spaces introduced in Definition 5.3(iii) and notationally complemented by (5.18). Then there is a positive num-

5. Spaces on domains: localization and Hardy inequalities

ber c such that

$$\|f\,|L_p(\Omega, d^{-s})\| \le c\, \|f\,|\widetilde{F}^s_{pq}(\Omega)\| \tag{5.59}$$

for all $f \in \widetilde{F}^s_{pq}(\Omega)$.

Proof By Proposition 5.5, the quasi-norm on the right-hand side of (5.59) can be replaced by the $F^s_{pq}(\mathbb{R}^n)$-quasi-norm. Since the spaces $F^s_{pq}(\mathbb{R}^n)$ are monotone for fixed s and p with respect to q we may assume $q \ge p$. Hence, by the usual localization technique it is sufficient to prove

$$\int_{\mathbb{R}^n_+} x_n^{-sp} |f(x)|^p \, dx \le c \, \|f\,|F^s_{pq}(\mathbb{R}^n)\|^p \tag{5.60}$$

for all $f \in F^s_{pq}(\mathbb{R}^n)$ with

$$\mathrm{supp}\, f \subset \{y \in \mathbb{R}^n \,:\, |y| < 1,\, y_n \ge 0\}, \tag{5.61}$$

(usual modification if $p = \infty$). Of course, $y = (y_1, \ldots, y_n)$. Recall

$$\mathbb{R}^n_+ = \{x \in \mathbb{R}^n \,:\, x = (x_1, \ldots, x_n),\, x_n > 0\}.$$

Since $q \ge p$ we can rely on the equivalent quasi-norm (4.15) with (4.14). We refer again to [Triγ], Theorem 3.5.3, p. 194, which also covers the case $p = q = \infty$. Let $x \in \mathbb{R}^n_+$ with $0 < x_n < \varepsilon$, and let t in (4.14) be restricted by

$$c_1\, x_n < t < c_2\, x_n, \tag{5.62}$$

where $0 < c_1 < c_2$ are large. Let $h \in \mathbb{R}^n$ in (4.14) with $|h| \le t$ be, in addition, an element of the cone

$$K^- = \{h = (h', h_n) \in \mathbb{R}^n \,:\, h_n < 0,\, |h'| < |h_n|\}. \tag{5.63}$$

Then we have for $h \in K^-$,

$$\Delta^N_h f(x) = f(x) \quad \text{if, additionally,} \quad |h_n| \sim t. \tag{5.64}$$

We insert (5.64) in (4.14) and obtain

$$|f(x)| \le c\, d^N_{t,u} f(x) \tag{5.65}$$

for t restricted by (5.62). Hence,

$$\begin{aligned} x_n^{-s}|f(x)| &\le c \left(\int_{c_1 x_n}^{c_2 x_n} t^{-sq} d^N_{t,u} f(x)^q \, \frac{dt}{t} \right)^{\frac{1}{q}} \\ &\le c \left(\int_0^1 t^{-sq} d^N_{t,u} f(x)^q \, \frac{dt}{t} \right)^{\frac{1}{q}}, \end{aligned} \tag{5.66}$$

(with the usual modification if $q = \infty$). Now (5.60) follows from (5.66) and the equivalent quasi-norm (4.15).

5.8 Hardy inequalities: the F-scale

Let again Ω be a bounded C^∞ domain in $\mathbb{R}^n$. We collect those special cases of (5.59) which are known so far, and which can be obtained by other means. By (1.10) and Definition 5.3,

$$W_p^s(\Omega) = F_{p,2}^s(\Omega), \quad 1 < p < \infty, \quad s \in \mathbb{N}, \tag{5.67}$$

are the classical Sobolev spaces, which can be normed as in (5.5). By (5.59) and (5.6) we have

$$\int_\Omega d^{-sp}(x)\,|f(x)|^p\,dx \le c\,\|f\,|W_p^s(\Omega)\|^p \sim \sum_{|\alpha|\le s} \int_\Omega |D^\alpha f(x)|^p\,dx \tag{5.68}$$

for all $f \in \overset{\circ}{W}{}_p^s(\Omega)$. This is a well-known immediate consequence of the classical Hardy inequality and coincides with (5.8). By (1.9) and Definition 5.3,

$$H_p^s(\Omega) = F_{p,2}^s(\Omega), \quad 1 < p < \infty, \quad s \in \mathbb{R}, \tag{5.69}$$

are the Sobolev spaces with the classical Sobolev spaces in (5.67) as special cases. One can extend the Hardy inequality (5.68) by real and complex interpolation of $\overset{\circ}{W}{}_p^k(\Omega)$ with $1 < p < \infty$, $k \in \mathbb{N}_0$, to

$$\|f\,|L_p(\Omega, d^{-s})\| \le c\,\|f\,|\widetilde{H}_p^s(\Omega)\|, \quad 1 < p < \infty,\ s > 0, \tag{5.70}$$

and

$$\|f\,|L_p(\Omega, d^{-s})\| \le c\,\|f\,|\widetilde{F}_{pp}^s(\Omega)\|, \quad 1 < p < \infty,\ s > 0. \tag{5.71}$$

As for the interpolation needed we refer to [Tria], 4.3.2, Theorem 2 on p. 318 (in the 1995 edition). In [Tria], 3.2.6, p. 259, one finds also weighted versions of (5.68) and (5.71). Recall that $B_{pp}^s = F_{pp}^s$. The natural counterpart of (5.59) for the spaces $\widetilde{B}_{pq}^s(\Omega)$ looks somewhat different and will be described briefly in 5.9 below. We shift also the discussion under what circumstances the spaces $\widetilde{F}_{pq}^s(\Omega)$ with (5.58) coincide with $\overset{\circ}{F}{}_{pq}^s(\Omega)$ to 5.21–5.24. This is the case if, in addition, $p < \infty$, $q < \infty$ and $s - \frac{1}{p} \notin \mathbb{N}_0$. But more details will be given later on. Finally we mention the interesting connection between the question of whether the characteristic function on Ω is a pointwise multiplier in $F_{pq}^s(\mathbb{R}^n)$ and inequalities of type (5.59). Let

$$0 < p < \infty, \quad 0 < q < \infty, \quad \sigma_{pq} < s < \min\left(1, \frac{1}{p}\right), \tag{5.72}$$

where σ_{pq} is given by (2.20). Then, on the one hand, by [RuS96], 4.6.3, p. 208, the characteristic function of Ω is a pointwise multiplier in $F^s_{pq}(\mathbb{R}^n)$ and on the other hand one can apply [Triβ], 2.8.6, Proposition 1 and Remark 1, on p. 155, with the outcome

$$\|f\,|L_p(\Omega, d^{-s})\| \le c\,\|f\,|F^s_{pq}(\Omega)\| \tag{5.73}$$

for all $f \in F^s_{pq}(\Omega)$. Since the characteristic function of Ω is a pointwise multiplier in $F^s_{pq}(\mathbb{R}^n)$ with (5.72), the spaces $F^s_{pq}(\Omega)$ and $\widetilde{F}^s_{pq}(\Omega)$ coincide, and we have again (5.59). In other words, the proposition extends some known special cases of Hardy inequalities in a rather definitive way. We continue this discussion in 5.11.

5.9 Hardy inequalities: the B-scale

Let again Ω be a bounded C^∞ domain in $\mathbb{R}^n$ and let $\widetilde{B}^s_{pq}(\Omega)$ be the spaces introduced in Definition 5.3. Asking for counterparts of (5.59) we have the natural restriction (5.58) for the parameters involved. Then $B^s_{pq}(\mathbb{R}^n)$ consists entirely of regular distributions and we have an immediate counterpart of (5.27),

$$\widetilde{B}^s_{pq}(\Omega) = \{f \in B^s_{pq}(\mathbb{R}^n)\ :\ \operatorname{supp} f \subset \overline{\Omega}\}\,, \tag{5.74}$$

where $\|f\,|B^s_{pq}(\mathbb{R}^n)\|$ is an equivalent quasi-norm compared with (5.22). In this section we are mostly interested in the F-spaces. So we restrict ourselves to a description of the B-counterpart of (5.59). Let

$$\Omega^t = \{x \in \Omega\ :\ d(x) < t\} \quad \text{where } t > 0\,. \tag{5.75}$$

Let

$$0 < p < \infty,\quad 0 < q < \infty,\quad s > \sigma_p = n\left(\frac{1}{p} - 1\right)_+\,. \tag{5.76}$$

Then there is a constant $c > 0$ such that

$$\int_0^\infty t^{-sq} \left(\int_{\Omega^t} |f(x)|^p\,dx\right)^{\frac{q}{p}} \frac{dt}{t} \le c\,\|f\,|\widetilde{B}^s_{pq}(\Omega)\|^q \tag{5.77}$$

for all $f \in \widetilde{B}^s_{pq}(\Omega)$. A proof is given in [Tri99b]. The basic idea is to start with (5.59) and to use real interpolation both for the spaces $\widetilde{F}^s_{pq}(\Omega)$ and the weighted L_p-spaces involved. One checks easily that (5.59) and (5.77) coincide if $p = q$ (as it should be). As in the F-case the spaces $B^s_{pq}(\Omega)$ and $\overset{\circ}{B}{}^s_{pq}(\Omega)$ coincide if p, q, ε are given by (5.76) and, in addition, $s - \frac{1}{p} \notin \mathbb{N}_0$. We refer to [Tri99b]. Restricted to $1 < p < \infty$ and $1 \le q \le \infty$ inequality (5.77) may be found in [Triα], p. 319 (usual modification if $q = \infty$). We continue this discussion in 5.12.

5.10 Theorem

Let Ω be a bounded C^∞ domain in $\mathbb{R}^n$ and let $L_p(\Omega, d^\sigma)$ be the spaces introduced in 5.2. Let

$$0 < p \le \infty, \quad 0 < q \le \infty, \quad \text{and} \quad s > \sigma_{pq} = n\left(\frac{1}{\min(p,q)} - 1\right)_+, \tag{5.78}$$

(with $q = \infty$ if $p = \infty$). Let $\widetilde{F}^s_{pq}(\Omega)$ be the spaces introduced in Definition 5.3(iii) and notationally complemented by (5.18). Then

$$\widetilde{F}^s_{pq}(\Omega) = F^s_{pq}(\Omega) \cap L_p(\Omega, d^{-s}) \tag{5.79}$$

(equivalent quasi-norms).

Proof *Step 1* By Definition 5.3 and Proposition 5.7 we have

$$\|f \mid F^s_{pq}(\Omega)\| + \|f \mid L_p(\Omega, d^{-s})\| \le c \, \|f \mid \widetilde{F}^s_{pq}(\Omega)\| \tag{5.80}$$

for any $f \in \widetilde{F}^s_{pq}(\Omega)$.

Step 2 We prove the converse inequality and begin with some preparation. By localization we may assume (in obvious notation)

$$f \in F^s_{pq}(\mathbb{R}^n_+), \quad x_n^{-s} f \in L_p(\mathbb{R}^n_+), \tag{5.81}$$

and

$$f(x) = 0 \quad \text{if} \quad x \in \mathbb{R}^n_+ \quad \text{and} \quad |x| > 1. \tag{5.82}$$

Let

$$ext\, f(x) = f(x) \text{ if } x \in \mathbb{R}^n_+ \quad \text{and} \quad ext\, f(x) = 0 \text{ if } x \in \mathbb{R}^n_-, \tag{5.83}$$

be the extension of f from $\mathbb{R}^n_+$ by zero to $\mathbb{R}^n$. Of course,

$$\mathbb{R}^n_- = \{x \in \mathbb{R}^n : x = (x_1, \ldots, x_n),\ x_n < 0\}. \tag{5.84}$$

By Proposition 5.5 one must prove $g = ext\, f \in F^s_{pq}(\mathbb{R}^n)$. We need a modification of the quasi-norms in $F^s_{pq}(\mathbb{R}^n)$ described by (4.14), (4.15). Let, in analogy to (5.63),

$$K^+_t = \{h = (h', h_n) \in \mathbb{R}^n : 0 < h_n < t,\ |h'| < h_n\} \tag{5.85}$$

be truncated cones. Then we modify the ball means $d^N_{t,u}$ in (4.14) by the cone means

$$D^N_{t,u} g(x) = \left(t^{-n} \int_{h \in K^+_t} |\Delta^N_h g(x)|^u \, dh\right)^{\frac{1}{u}}. \tag{5.86}$$

5. Spaces on domains: localization and Hardy inequalities

As before $0 < u < \min(1, p, q)$ and $s < N \in \mathbb{N}$, where $\Delta_h^N g(x)$ are the usual (n-dimensional) differences given by (1.12). We claim that also after this modification

$$\|g\,|L_p(\mathbb{R}^n)\| + \left\|\left(\int_0^1 t^{-sq} D_{t,u}^N g(\cdot)^q \frac{dt}{t}\right)^{\frac{1}{q}} |L_p(\mathbb{R}^n)\right\| \tag{5.87}$$

is an equivalent quasi-norm in $F_{pq}^s(\mathbb{R}^n)$. In case of the ball means $d_{t,u}^N g$ this is covered by [Triγ], 3.5.2, 3.5.3, p. 194. The proof given there applies also to the cone means (5.86). The crucial estimates in Step 2 of the proof on p. 195 in [Triγ] rely on formula (4) on p. 181 in [Triγ]. But this applies equally to balls and cones. Hence, (5.87) is an equivalent quasi-norm in $F_{pq}^s(\mathbb{R}^n)$. By the localization technique in [Triγ], 5.2.2, p. 245, based on the means in [Triγ], 3.5.2, p. 193, we have a similar assertion for spaces F_{pq}^s on domains Ω and on $\mathbb{R}_+^n$. In particular, with f given by (5.81), (5.82) and $g = \operatorname{ext} f$ defined by (5.83) it is sufficient to prove

$$\|g\,|L_p(\mathbb{R}^n)\| + \left\|\left(\int_0^1 t^{-sq} D_{t,u}^N g(\cdot)^q \frac{dt}{t}\right)^{\frac{1}{q}} |L_p(\mathbb{R}^n)\right\|$$

$$\leq c\,\|x_n^{-s} f(x)\,|L_p(\mathbb{R}_+^n)\| + c\left\|\left(\int_0^1 t^{-sq} D_{t,u}^N f(\cdot)^q \frac{dt}{t}\right)^{\frac{1}{q}} |L_p(\mathbb{R}_+^n)\right\| \tag{5.88}$$

for some $c > 0$. If $x_n > 0$ then $g(x) = f(x)$ and $D_{t,u}^N g(x) = D_{t,u}^N f(x)$ and (5.88) with $\mathbb{R}_+^n$ in place of $\mathbb{R}^n$ on the left-hand side is obvious. As for $x_n < 0$ we may assume $-1 < x_n < 0$. Then

$$D_{t,u}^N f(x) = 0 \quad \text{if} \quad 0 < t \leq c_1\,|x_n| \tag{5.89}$$

for some $c_1 > 0$. Hence we may assume $t > c_1\,|x_n|$. There is a number $c_2 > 0$ such that

$$h_n \sim |h| \sim x_n + h_n \quad \text{if} \quad h \in K_t^+ \quad \text{and} \quad |h| \geq c_2\,|x_n|, \tag{5.90}$$

which means that all these numbers can be pairwise estimated from above and from below by positive constants which are independent of $x = (x', x_n)$, where $x' \in \mathbb{R}^{n-1}$ and $h = (h', h_n)$, where $h' \in \mathbb{R}^{n-1}$ with the above restrictions. The two cases

$$|h| < c_2\,|x_n| \quad \text{and} \quad |h| \geq c_2\,|x_n| \tag{5.91}$$

are treated separately in the two following steps. We wish to prove that the related parts of the left-hand side of (5.88) can be estimated by the first term on the right-hand side of (5.88).

Step 3 Let $|h| < c_2 |x_n|$ according to Step 2. Let $\widetilde{x} = (x', -x_n) \in \mathbb{R}_n^+$. Since $t > c_1 |x_n|$ we have

$$\left(t^{-n} \int_{h \in K_t^+, |h| < c_2 |x_n|} |\Delta_h^N g(x)|^u \, dh \right)^{\frac{1}{u}} \leq c' \left(M_{c|x_n|} |f|^u \right)(\widetilde{x})^{\frac{1}{u}} \qquad (5.92)$$

with the truncated Hardy-Littlewood maximal function M_a, where the related mean values are restricted to balls of radius at most $a > 0$. Then we have

$$\left(\int_{c|x_n|}^1 t^{-sq} \left(M_{c|x_n|} |f|^u \right)^{\frac{q}{u}} (\widetilde{x}) \, \frac{dt}{t} \right)^{\frac{1}{q}}$$

$$\leq c' |x_n|^{-s} \left(M_{c|x_n|} |f|^u \right)^{\frac{1}{u}} (\widetilde{x})$$
$$\leq c'' \left(M_{c|x_n|} |y_n^{-s} f(y)|^u \right)^{\frac{1}{u}} (\widetilde{x})$$
$$\leq c'' \left(M |y_n^{-s} f(y)|^u \right)^{\frac{1}{u}} (\widetilde{x}), \qquad (5.93)$$

where M stands for the usual maximal function. Recall $u < p$. We apply the $L_p(\mathbb{R}_-^n)$-quasi-norm to (5.93). Since $\frac{p}{u} > 1$ we obtain by the Hardy-Littlewood maximal inequality

$$\left\| \left(\int_{c|x_n|}^1 t^{-sq} \left(M_{c|x_n|} |f|^u \right)^{\frac{q}{u}} \frac{dt}{t} \right)^{\frac{1}{q}} | L_p(\mathbb{R}_-^n) \right\|$$

$$\leq c' \| x_n^{-s} f(x) | L_p(\mathbb{R}_+^n) \|. \qquad (5.94)$$

By (5.92) and (5.89) one gets the desired estimate for the respective parts of the left-hand side of (5.88).

Step 4 Let $|h| \geq c_2 |x_n|$ according to (5.91). Then we have (5.90). Let

$$\chi(t,h) = 1 \text{ if } |h| \leq t \quad \text{and} \quad \chi(t,h) = 0 \text{ if } |h| > t. \qquad (5.95)$$

We estimate the counterpart of the left-hand side of (5.92) with $|h| \geq c_2 |x_n|$ in place of $|h| < c_2 |x_n|$. It is sufficient to deal with integrands of type $|f(x+h)|^u$. Furthermore we replace K_t^+ by $K^+ = K_1^+$ and $|f(x+h)|^u$ by $|f(x+h)|^u \chi(t,h)$.

5. Spaces on domains: localization and Hardy inequalities   57

Since $u < q$ we obtain by the triangle inequality for norms

$$\left[\int_{c|x_n|}^{1} t^{-sq}\left(t^{-n}\int_{h\in K^+, |h|\geq c'|x_n|} |f(x+h)|^u \chi(t,h)\,dh\right)^{\frac{q}{u}} \frac{dt}{t}\right]^{\frac{u}{q}\cdot\frac{1}{u}}$$

$$\leq \left[\int_{h\in K^+, |h|\geq c'|x_n|} |f(x+h)|^u \left(\int_{c|x_n|}^{1} t^{-sq-n\frac{q}{u}}\chi(t,h)\frac{dt}{t}\right)^{\frac{u}{q}} dh\right]^{\frac{1}{u}}. \quad (5.96)$$

By (5.95) and $|h| \geq c_2|x_n|$ the inner integral can be estimated from above by $c|h|^{-sq-n\frac{q}{u}}$, and hence (5.96) can be estimated from above by

$$c\left[\int_{h\subset K^+, |h|\geq c|x_n|} |f(x+h)|^u |h|^{-su} \frac{dh}{|h|^n}\right]^{\frac{1}{u}} \quad (5.97)$$

and thus, after $h \to |x_n|h$, by

$$c|x_n|^{-s}\left[\int_{h\in K^+_\infty, |h|\geq c} |f(x+|x_n|h)|^u |h|^{-su} \frac{dh}{|h|^n}\right]^{\frac{1}{u}}, \quad (5.98)$$

where K^+_∞ is given by (5.85) with $t = \infty$. We apply the L_p-quasi-norm to (5.96) and hence to (5.98) with respect to $\mathbb{R}^{n-1}\times[-1,0]$. We transform $x \to (x', -x_n)$. Since $u < p$ we see by the triangle inequality for $L_{\frac{p}{u}}(\mathbb{R}^{n-1}\times[0,1])$ that the indicated L_p-quasi-norm of (5.96) can be estimated from above by

$$c\left[\int_{h\in K^+_\infty, |h|\geq c} |h|^{-su} \frac{dh}{|h|^n} \left(\int_{\mathbb{R}^{n-1}\times[0,1]} |f(x+x_nh)|^p x_n^{-sp}\,dx\right)^{\frac{u}{p}}\right]^{\frac{1}{u}}. \quad (5.99)$$

In the inner integral we use the transformation $y = x + x_nh$. The respective Jacobian is proportional to $|h|^{-1}$. Using (5.90) we estimate (5.99) from above by

$$c\,\|y_n^{-s}f(y)\,|L_p(\mathbb{R}^n_+)\|\left[\int_{h\in K^+_\infty, |h|\geq c'} |h|^{-su} h_n^{su} \frac{dh}{|h|^{n+\frac{u}{p}}}\right]^{\frac{1}{u}}$$

$$\leq c''\,\|y_n^{-s}f(y)\,|L_p(\mathbb{R}^n_+)\|. \quad (5.100)$$

Now the proof of (5.88) follows from (5.92), (5.94) on the one hand and the just-derived estimate in (5.100) on the other hand. Hence $g \in F^s_{pq}(\mathbb{R}^n)$ and we have finally the converse of (5.80).

5.11 Discussion: the F-scale

We continue our discussion from 5.8. Let again Ω be a bounded C^∞ domain in $\mathbb{R}^n$. For the classical Sobolev spaces $W^s_p(\Omega)$ one can strengthen (5.68) by

$$\mathring{W}^s_p(\Omega) = W^s_p(\Omega) \cap L_p(\Omega, d^{-s}), \quad s \in \mathbb{N}, \ 1 < p < \infty. \tag{5.101}$$

This follows from (5.79) and (5.6). Later on, in 5.21–5.24, we will discuss the circumstances under which, in more general cases, the spaces $\widetilde{F}^s_{pq}(\Omega)$ and $\mathring{F}^s_{pq}(\Omega)$ coincide. At the moment we restrict ourselves to a few remarks directly connected with Proposition 5.7 and Theorem 5.10. The additional restriction $s > \sigma_{pq}$ in (5.78), compared with $s > \sigma_p$ in (5.58), comes from the use of the equivalent quasi-norm (5.87) in $F^s_{pq}(\mathbb{R}^n)$, which is available (so far) only for $s > \sigma_{pq}$. In other words, it is unclear whether (5.79) holds under the weaker restriction (5.58). We describe an example of (5.79) which is not covered by (5.78). First we recall:

The characteristic function of the bounded C^∞ domain Ω is a pointwise multiplier in $F^s_{pq}(\mathbb{R}^n)$ if

$$0 < p \leq \infty, \quad 0 < q \leq \infty, \quad \max\left(\frac{1}{p} - 1, n(\frac{1}{p} - 1)\right) < s < \frac{1}{p}, \tag{5.102}$$

(with $q = \infty$ if $p = \infty$).
We refer to [RuS96], 4.6.3, p. 208, for the final version, which in turn is based on [Triβ], [Kal85], [Fra86], and the forerunners mentioned there. As a consequence we have in these cases

$$F^s_{pq}(\Omega) = \widetilde{F}^s_{pq}(\Omega). \tag{5.103}$$

Combining this observation with Proposition 5.7 we obtain the following assertion: *Let*

$$0 < p \leq \infty, \quad 0 < q \leq \infty, \quad \sigma_p < s < \frac{1}{p}, \tag{5.104}$$

(with $q = \infty$ if $p = \infty$). Then we have

$$\|f \,|L_p(\Omega, d^{-s})\| \leq c \,\|f \,|F^s_{pq}(\Omega)\| \tag{5.105}$$

5. Spaces on domains: localization and Hardy inequalities

for all $f \in F^s_{pq}(\Omega)$ and (obviously) (5.79). Maybe it is worth mentioning that the above pointwise multiplier assertion combined with Proposition 5.6 gives the following result: *Let*

$$0 < p < \infty, \quad 0 < q < \infty, \quad \max\left(\frac{1}{p} - 1, n\left(\frac{1}{p} - 1\right)\right) < s < \frac{1}{p}, \quad (5.106)$$

then

$$\overset{\circ}{F}{}^s_{pq}(\Omega) = F^s_{pq}(\Omega) = \widetilde{F}^s_{pq}(\Omega). \quad (5.107)$$

5.12 Discussion: the B-scale

We continue our discussion from 5.9. Again let Ω be a bounded C^∞ domain in $\mathbb{R}^n$. One may ask whether there is a counterpart of (5.79) for the B-spaces. The natural counterpart of the quasi-norm of $L_p(\Omega, d^{-s})$ is now the left-hand side of (5.77). By [Triα], 4.3.2, especially p. 319, we have

$$\|f\,|\widetilde{B}^s_{pq}(\Omega)\| \sim \|f\,|B^s_{pq}(\Omega)\| + \left(\int_0^\infty t^{-sq} \left(\int_{\Omega^t} |f(x)|^p\, dx\right)^{\frac{q}{p}} \frac{dt}{t}\right)^{\frac{1}{q}} \quad (5.108)$$

if

$$1 < p < \infty, \quad 1 \le q \le \infty, \quad s > 0. \quad (5.109)$$

One can expect that this assertion can be extended to the parameters p, q, s with (5.76). But this is not covered by [Tri99b].

5.13 Localization: the set-up

In Section 5 we prove two central assertions: The first one is Theorem 5.10 and the related Hardy inequalities in Proposition 5.7. The second one is the extension of (5.13) to bounded domains. This will be done in 5.14. We need some preparation. Let Ω be a bounded C^∞ domain in $\mathbb{R}^n$. It is not difficult to see that there exist open balls

$$\{K_{jr} : j \in \mathbb{N}_0;\ r = 1, \ldots, N_j\} \quad (5.110)$$

and subordinated C^∞ functions

$$\{\varphi_{jr}(x) : j \in \mathbb{N}_0;\ r = 1, \ldots, N_j\} \quad (5.111)$$

with the following properties:

(i) There exist a natural number $M \in \mathbb{N}$ and positive numbers $c_1, \ldots, c_5$ such that at most M of all these balls have a non-empty intersection, $c_1 \, 2^{-j}$ is the radius of K_{jr},

$$c_2 \leq N_j \, 2^{-j(n-1)} \leq c_3 \quad \text{where} \quad j \in \mathbb{N}_0, \tag{5.112}$$

$$\bigcup_{j \in \mathbb{N}_0} \bigcup_{r=1}^{N_j} K_{jr} = \Omega, \tag{5.113}$$

and

$$c_4 \, 2^{-j} \leq \operatorname{dist}(\overline{K_{jr}}, \partial\Omega) \leq c_5 \, 2^{-j}. \tag{5.114}$$

(ii) There are positive numbers c_γ such that

$$|D^\gamma \varphi_{jr}(x)| \leq c_\gamma \, 2^{j|\gamma|} \quad \text{where} \quad \gamma \in \mathbb{N}_0^n, \tag{5.115}$$

$$\operatorname{supp} \varphi_{jr} \subset K_{jr}, \tag{5.116}$$

and

$$\sum_{j=0}^{\infty} \sum_{r=1}^{N_j} \varphi_{jr}(x) = 1 \quad \text{if} \quad x \in \Omega. \tag{5.117}$$

To prove the existence of the balls K_{jr} and related functions $\varphi_{jr}(x)$ one can first cover

$$\Omega^j = \{x \in \Omega : \ 2^{-j-1} \leq \operatorname{dist}(x, \partial\Omega) \leq 2^{-j+1}\} \tag{5.118}$$

with cubes $d\,Q_{jm}$ as in 2.2, where $d > 0$ is suitably chosen. Afterwards one can find a resolution of unity (5.117) with (5.115), (5.116).

5.14 Theorem

Let Ω be a bounded C^∞ domain in $\mathbb{R}^n$ and let

$$\{\varphi_{jr}(x) : \ j \in \mathbb{N}_0; \ r = 1, \ldots, N_j\}$$

be the above resolution of unity. Let

$$0 < p \leq \infty, \quad 0 < q \leq \infty, \quad s > \sigma_{pq} = n\left(\frac{1}{\min(p,q)} - 1\right)_+, \tag{5.119}$$

5. Spaces on domains: localization and Hardy inequalities   61

(with $q = \infty$ if $p = \infty$). Let $\widetilde{F}^s_{pq}(\Omega)$ be the spaces introduced in Definition 5.3(iii) and notationally complemented by (5.18). Then $f \in L_1(\Omega)$ belongs to $\widetilde{F}^s_{pq}(\Omega)$ if, and only if,

$$\left(\sum_{j=0}^{\infty} \sum_{r=1}^{N_j} \| \varphi_{jr} f \, | F^s_{pq}(\mathbb{R}^n) \|^p \right)^{\frac{1}{p}} < \infty, \qquad (5.120)$$

(usual modification if $p = q = \infty$). Furthermore, (5.120) is an equivalent quasi-norm.

Proof   *Step 1*   Let $f \in \widetilde{F}^s_{pq}(\Omega)$ with $p < \infty$ (if $p = \infty$, and hence also $q = \infty$, one has to modify the arguments appropriately). By Theorem 5.10 we have
$\| f \, | \widetilde{F}^s_{pq}(\Omega) \|^p$

$$\leq c \| f \, | F^s_{pq}(\Omega) \|^p + c \int_{\Omega} d^{-sp}(x) \left| \sum_{j=0}^{\infty} \sum_{r=1}^{N_j} \varphi_{jr}(x) f(x) \right|^p dx, \qquad (5.121)$$

where $\{\varphi_{jr}\}$ is the resolution of unity introduced in 5.13. We wish to estimate the expression from above by (5.120). Since everything is local, we may again replace Ω by $\mathbb{R}^n_+$ and assume that $f(x)$ is restricted by (5.82). Let $D^N_{t,u} f(x)$ be the means given by (5.86) with f in place of g and $x \in \mathbb{R}^n_+$. In analogy to (5.87),

$$\| f \, | L_p(\mathbb{R}^n_+) \| + \left\| \left(\int_0^1 t^{-sq} D^N_{t,u} f(\cdot)^q \frac{dt}{t} \right)^{\frac{1}{q}} \, | L_p(\mathbb{R}^n_+) \right\| \qquad (5.122)$$

is an equivalent quasi-norm in $F^s_{pq}(\mathbb{R}^n_+)$. This follows from [Triγ], 5.2.2, p. 245, and 3.5.2, p. 193, and the remarks in connection with (5.87). As there, $0 < s < N \in \mathbb{N}$ and $0 < u < \min(1, p, q)$. We fix $x \in \mathbb{R}^n_+$ with $x_n \sim 2^{-j}$. Let either $0 < t \leq c\, 2^{-j}$ or $t \sim 2^{-k}$ with $k \in \mathbb{N}_0$ and $k < j$. The typical term in $D^N_{t,u} f(x)$, which we have to estimate, is given by

$$D^N_{t,u} f(x) \leq c \left(t^{-n} \int_{h \in K_t^+} |\Delta^N_h \varphi_{j\, r_j} f(x)|^u \, dh \right)^{\frac{1}{u}}$$

$$+ c \left(2^{-kn} \sum_{l=k}^{j} \int_{h \in K_t^+} |(\varphi_{l\, r_l} f)(x+h)|^u \, dh \right)^{\frac{1}{u}} + +, \qquad (5.123)$$

where ++ indicates both replacements of h by mh with $m = 1, \ldots, N$, and also the replacement of φ_{lr_l} by neighbouring φ_{lr}. But in any case by the construction of the resolution of unity $\{\varphi_{lr}\}$ and the restriction of h to the upward cone K_t^+ it follows that there are at most L terms of this type, where L is independent of x, j, and k. Furthermore in case of $0 < t \leq 2^{-j}$ one has only the first term on the right-hand side of (5.123) (and neighbouring terms of the same type). Inserting (5.123) in (5.122) the first terms on the right-hand side of (5.123) result in what we wish to have. As for the remaining terms we are much in the same situation as in (5.96)–(5.100). For this purpose one must replace $|f(x+h)|^u$ on the left-hand side of (5.96) by

$$\sum_{l=k}^{j} |(\varphi_{lr_l} f)(x+h)|^u \sim \left(\sum_{l=k}^{j} |(\varphi_{lr_l} f)(x+h)|^p \right)^{\frac{u}{p}} \tag{5.124}$$

where we used the support properties of φ_{lr_l}. The sum on the right-hand side of (5.124) can be extended to the whole resolution of unity. After this replacement one can follow the arguments from (5.96)–(5.100) and one obtains an obvious counterpart of the right-hand side of (5.100). Clipping together all these estimates we get

$$\|f \,|F_{pq}^s(\Omega)\|^p \leq c \sum_{j=0}^{\infty} \sum_{r=1}^{N_j} \|\varphi_{jr} f \,|F_{pq}^s(\mathbb{R}^n)\|^p$$

$$+ c \sum_{j=0}^{\infty} \sum_{r=1}^{N_j} \int_{\Omega} d^{-sp}(x) \, |\varphi_{jr}(x) \, f(x)|^p \, dx \,, \tag{5.125}$$

where we switched back from $\mathbb{R}_+^n$ in (5.122) to Ω. In Step 3 below we prove the following assertion: *Let p, q, s, u and N be as above. There is a number $c > 0$ such that for all $0 < \lambda \leq 1$ and all*

$$f \in F_{pq}^s(\mathbb{R}^n), \quad \operatorname{supp} f \subset B_\lambda = \{ y \in \mathbb{R}^n : |y| < \lambda \}, \tag{5.126}$$

it holds that

$$\|f \,|L_p(\mathbb{R}^n)\| \leq c \lambda^s \left\| \left(\int_0^\lambda t^{-sq} d_{t,u}^N f(\cdot)^q \frac{dt}{t} \right)^{\frac{1}{q}} \,|L_p(\mathbb{R}^n) \right\|, \tag{5.127}$$

where the ball means from (4.14) can be replaced by the cone means according to (5.86). In particular the second terms on the right-hand side of (5.125), and

5. Spaces on domains: localization and Hardy inequalities

also of (5.121), can be estimated from above by the first terms on the right-hand side of (5.125). Hence,

$$\|f\,|\widetilde{F}^s_{pq}(\Omega)\|^p \le c \sum_{j=0}^{\infty} \sum_{r=1}^{N_j} \|\varphi_{jr} f\,|F^s_{pq}(\mathbb{R}^n)\|^p. \tag{5.128}$$

If $p = q = \infty$ then one has an obvious counterpart of (5.128). If now $f \in L_1(\Omega)$ with (5.120), then one can follow the above arguments with the outcome that the quasi-norms in (5.122), (5.125), and (5.128) are finite. By the characterization in [Triγ], 5.2.2, p. 245, and (5.79) it follows that $f \in \widetilde{F}^s_{pq}(\Omega)$.

Step 2 We prove the reverse inequality to (5.128). By the usual localization argument it is sufficient to prove the converse of (5.128) for functions $f \in F^s_{pq}(\mathbb{R}^n)$ with

$$\operatorname{supp} f \subset \mathbb{R}^n_+ \cap \{y \,:\, |y| < 1\}. \tag{5.129}$$

We use the equivalent quasi-norm (4.15) with ball means (4.14) applied to $\varphi_{jr} f$. Of course, we suppose again $u < \min(1, p, q)$ and $s < N \in \mathbb{N}$. We wish to prove that for a sufficiently small number $a > 0$,

$$\int_{\mathbb{R}^n} \sum_{j=0}^{\infty} \sum_{r=1}^{N_j} \left(\int_0^{a2^{-j}} t^{-sq} d^N_{t,u} \varphi_{jr} f(x)^q \frac{dt}{t} \right)^{\frac{p}{q}} dx \le c\,\|f\,|F^s_{pq}(\mathbb{R}^n)\|^p \tag{5.130}$$

and

$$\sum_{j=0}^{\infty} \sum_{r=1}^{N_j} \left\| \left(\int_{a2^{-j}}^1 t^{-sq} d^N_{t,u} \varphi_{jr} f(\cdot)^q \frac{dt}{t} \right)^{\frac{1}{q}} |L_p(\mathbb{R}^n) \right\|^p \le c\,\|x_n^{-s} f(x)\,|L_p(\mathbb{R}^n_+)\|^p, \tag{5.131}$$

where, by (4.14),

$$d^N_{t,u} \varphi_{jr} f(x) = \left(t^{-n} \int_{|h|\le t} |\Delta^N_h \varphi_{jr} f(x)|^u \, dh \right)^{\frac{1}{u}}. \tag{5.132}$$

Then the converse of (5.128) follows from (5.130), (5.131), the equivalent quasi-norm (4.15) and (5.79). First we prove (5.130). By mathematical induction we have for some constants c_m,

$$(\Delta^N_h \varphi_{jr} f)(x) = \sum_{m=0}^{N} c_m (\Delta^{N-m}_h f)(x) (\Delta^m_h \varphi_{jr})(x + (N-m)h). \tag{5.133}$$

We choose $K \in \mathbb{N}$ and $N \in \mathbb{N}$ such that $K > s$ and $N - K > s$. Let $b\,K_{jr}$ be a ball concentric with the ball K_{jr} from 5.13 and with radius b times the radius of K_{jr}. If $a > 0$ is small, then there is a b near 1 such that (5.133) with $|h| < t < a\,2^{-j}$ is zero if $x \in \mathbb{R}^n \setminus b\,K_{jr}$. In particular, the supports of the respective terms in (5.130) are in $b\,K_{jr}$. We have

$$|\Delta_h^N \varphi_{jr} f(x)| \le c \sum_{l=0}^{K} |\Delta_h^{N-l} f(x)| + c \sum_{l=K+1}^{N} \sum_{r=0}^{N} |f(x+rh)|\, 2^{jl}\, t^l, \qquad (5.134)$$

where $|h| < t < a\,2^{-j}$ and $x \in b\,K_{jr}$. We insert (5.134) in (5.132) and obtain $d_{t,u}^N \varphi_{jr} f(x)$

$$\le c \sum_{l=0}^{K} d_{t,u}^{N-l} f(x) + c\, 2^{jK}\, t^K\, 2^{-js} \left(M\,|y_n^{-s} f(y)|^u \right)^{\frac{1}{u}}(x), \qquad (5.135)$$

where M is the Hardy-Littlewood maximal function and $x \in b\,K_{jr}$. Hence,

$$\left(\int_0^{a2^{-j}} t^{-sq}\, d_{t,u}^N \varphi_{jr} f(x)^q\, \frac{dt}{t} \right)^{\frac{1}{q}} \le c \left(\sum_{l=0}^{K} \int_0^{a2^{-j}} t^{-sq}\, d_{t,u}^{N-l} f(x)^q\, \frac{dt}{t} \right)^{\frac{1}{q}}$$

$$+ c\, \left(M\,|y_n^{-s} f(y)|^u \right)^{\frac{1}{u}}(x)\, 2^{j(K-s)} \left(\int_0^{a2^{-j}} t^{(K-s)q}\, \frac{dt}{t} \right)^{\frac{1}{q}}, \qquad (5.136)$$

where $x \in b\,K_{jr}$. Since $K > s$ we have

$$\left(\int_0^{a2^{-j}} t^{-sq}\, d_{t,u}^N \varphi_{jr} f(x)^q\, \frac{dt}{t} \right)^{\frac{1}{q}}$$

$$\le c \left(\sum_{l=0}^{K} \int_0^{a2^{-j}} t^{-sq}\, d_{t,u}^{N-l} f(x)^q\, \frac{dt}{t} \right)^{\frac{1}{q}} + c\, \left(M\,|y_n^{-s} f(y)|^u \right)^{\frac{1}{u}}(x), \qquad (5.137)$$

where $x \in b\,K_{jr}$. Recall $N - K > s$ and that the left-hand side of (5.137) is zero if $x \notin b\,K_{jr}$. Hence, the sum over the pth power of (5.137) is the integrand on the left-hand side of (5.130). By the equivalent quasi-norm (4.15) the left-hand side of (5.130) can be estimated from above by

$$c\,\|f\,|F_{pq}^s(\mathbb{R}^n)\|^p + \left\| \left(M\,|y_n^{-s} f(y)|^u \right)^{\frac{1}{u}}(\cdot)\,|L_p(\mathbb{R}_+^n) \right\|^p. \qquad (5.138)$$

Since $p > u$ we can apply the Hardy-Littlewood maximal inequality. Hence the second term in (5.138) can be estimated from above by

$$\left\| y_n^{-s} f(y)\,|L_p(\mathbb{R}_+^n) \right\|^p.$$

5. Spaces on domains: localization and Hardy inequalities

Together with (5.79) we obtain (5.130). We prove (5.131). From (5.132) and the above maximal function follows

$$\left(\int_{a2^{-j}}^{1} t^{-sq} d_{t,u}^{N} \varphi_{jr} f(x)^q \frac{dt}{t} \right)^{\frac{1}{q}}$$

$$\leq c \, (M|\varphi_{jr} f|^u)^{\frac{1}{u}} (x) \left(\int_{a2^{-j}}^{1} t^{-sq} \frac{dt}{t} \right)^{\frac{1}{q}}$$

$$\leq c' \, 2^{sj} \, (M|\varphi_{jr} f|^u)^{\frac{1}{u}} (x)$$

$$\leq c'' \, \left(M|y_n^{-s}(\varphi_{jr} f)(y)|^u \right)^{\frac{1}{u}} (x). \tag{5.139}$$

Again by the maximal inequality one can estimate the left-hand side of (5.131) term by term and obtain that

$$c \sum_{j=0}^{\infty} \sum_{r=1}^{N_j} \|x_n^{-s} (\varphi_{jr} f)(x) \,|L_p(\mathbb{R}_+^n)\|^p \leq c' \, \|x_n^{-s} f(x) \,|L_p(\mathbb{R}_+^n)\|^p \tag{5.140}$$

as an estimate from above. By the above remarks the proof is complete.

Step 3 We prove (5.126), (5.127). There is a number $c > 0$ such that

$$\|g \,|L_p(\mathbb{R}^n)\| \leq c \left\| \left(\int_0^1 t^{-sq} d_{t,u}^{N} g(\cdot)^q \frac{dt}{t} \right)^{\frac{1}{q}} |L_p(\mathbb{R}^n) \right\| \tag{5.141}$$

for all

$$g \in F_{pq}^s(\mathbb{R}^n), \quad \text{supp } g \subset B_1 = \{y \in \mathbb{R}^n : |y| < 1\}. \tag{5.142}$$

This follows by standard arguments from the compact embedding of, say, $F_{pq}^s(B_2)$, where B_2 has the same meaning as in (5.126), in all spaces $L_r(B_2)$ with $0 < r \leq p*$ for some $1 < p* \leq \infty$: If (5.141) does not hold for a suitable $c > 0$, then it would follow by the usual compactness argument that there is a non-trivial function $g \in L_{p*}(\mathbb{R}^n)$ with compact support in B_1 and $(\Delta_h^N g)(x) = 0$ for almost all $x \in \mathbb{R}^n$ and $|h| \leq 1$. But this is a contradiction. Now let f be given by (5.126). We apply (5.141) to $g(x) = f(\lambda x)$. For the ball means we have

$$[d_{t,u}^N f(\lambda \cdot)](x) = \left(t^{-n} \int_{|h| \leq t} |\Delta_{\lambda h}^N f(\lambda x)|^u \, dh \right)^{\frac{1}{u}} = (d_{\lambda t, u}^N f)(\lambda x). \tag{5.143}$$

We insert (5.143) in (5.141) and obtain (5.127).

5.15 Corollary

Let Ω be a bounded C^∞ domain in $\mathbb{R}^n$ and let $d(x)$ be the distance of $x \in \Omega$ to $\partial\Omega$ as in (5.15). Let p, q, s be given by (5.119) (with $q = \infty$ if $p = \infty$). Let $d_{t,u}^N f(x)$ be the ball means introduced in (4.14) with $u < \min(1, p, q)$ and $s < N \in \mathbb{N}$. Then $f \in L_1(\Omega)$ belongs to $\widetilde{F}_{pq}^s(\Omega)$ if, and only if,

$$\left\| \left(\int_0^{cd(\cdot)} t^{-sq} d_{t,u}^N f(\cdot)^q \frac{dt}{t} \right)^{\frac{1}{q}} \,|L_p(\Omega)\right\| + \left\| d^{-s} f \,|L_p(\Omega)\right\| < \infty \quad (5.144)$$

(usual modification if $q = \infty$), where c may be chosen arbitrarily small and $0 < c < c_0$ for some c_0 (intrinsic equivalent quasi-norms).

Proof This assertion is covered by Step 2 of the proof of Theorem 5.14, especially (5.139), (5.140).

5.16 Corollary

Let B_λ be the balls centred at the origin and of radius λ, where $0 < \lambda \le 1$, (5.126). Let p, q, s be given by (5.119) (with $q = \infty$ if $p = \infty$). Let $d_{t,u}^N f(x)$ be the ball means introduced in (4.14) with $u < \min(1, p, q)$ and $s < N \in \mathbb{N}$. Let

$$d_q^{s,\lambda} f(x) = \left(\int_0^\lambda t^{-sq} d_{t,u}^N f(x)^q \frac{dt}{t} \right)^{\frac{1}{q}}. \quad (5.145)$$

Then

$$\| f \,|\widetilde{F}_{pq}^s(B_\lambda)\| \sim \| d_q^{s,\lambda} f \,|L_p(\mathbb{R}^n)\| \quad (5.146)$$

and

$$\| f(\lambda \cdot) \,|\widetilde{F}_{pq}^s(B_1)\| \sim \lambda^{s - \frac{n}{p}} \| f \,|\widetilde{F}_{pq}^s(B_\lambda)\| \quad (5.147)$$

where the equivalence constants both in (5.146) and (5.147) are independent of $f \in \widetilde{F}_{pq}^s(B_\lambda)$ and of λ with $0 < \lambda \le 1$.

Proof By the arguments of Step 2 of the proof of Theorem 5.14, in particular by (5.139), (5.140) with $\lambda \sim 2^{-j}$, it follows that

$$\| f \,|\widetilde{F}_{pq}^s(B_\lambda)\| \sim \| d_q^{s,\lambda} f \,|L_p(\mathbb{R}^n)\| + \lambda^{-s} \| f \,|L_p(\mathbb{R}^n)\|, \quad (5.148)$$

where the equivalence constants are independent of λ. Then (5.146) follows from (5.148) and (5.127). Finally (5.147) is a consequence of (5.146), (5.143), and (5.145).

5.17 Pointwise multipliers

A given function φ is called a *pointwise multiplier* in, say, $F^s_{pq}(\mathbb{R}^n)$ or $B^s_{pq}(\mathbb{R}^n)$ if

$$f \mapsto \varphi f \tag{5.149}$$

maps the considered space into itself. More details and explanations may be found in [Triγ], 4.2, pp. 201–206, and the literature mentioned there. An extensive treatment has been given in [RuS96], Chapter 4. Our aim here is rather modest. We describe a direct consequence of the above corollary. Again let B_λ be the balls introduced in (5.126) and let p, q, s be given by (5.119) (with $q = \infty$ if $p = \infty$). Let φ be a function having classical derivatives in, say, $B_{2\lambda}$ up to order $1 + [s]$ with

$$|D^\gamma \varphi(x)| \le a\,\lambda^{-|\gamma|}, \quad |\gamma| \le 1 + [s], \quad x \in B_{2\lambda}. \tag{5.150}$$

Then

$$\|\varphi f \,|\widetilde{F}^s_{pq}(B_\lambda)\| \le c\,\|f\,|\widetilde{F}^s_{pq}(B_\lambda)\|, \tag{5.151}$$

where c is independent of $f \in \widetilde{F}^s_{pq}(B_\lambda)$ and of λ with $0 < \lambda \le 1$ (but depends on a in (5.150)): By [Triγ], Corollary 4.2.2 on p. 205, the function $\varphi(\lambda \cdot)$ is a pointwise multiplier in $\widetilde{F}^s_{pq}(B_1)$. Then (5.151) is a consequence of (5.147).

5.18 Approximate resolution of unity

We return to 5.13 and assume that the domain Ω, the balls K_{jr} and the functions $\varphi_{jr}(x)$ have the same meaning as there with the exception of (5.117), generalized now by

$$a \le \sum_{j=0}^{\infty} \sum_{r=1}^{N_j} \varphi_{jr}(x) \le b \quad \text{if} \quad x \in \Omega, \tag{5.152}$$

where a and b are two positive numbers (independent of $x \in \Omega$). This is called an *approximate resolution of unity*.

5.19 Corollary

Theorem 5.14 remains valid if the resolution of unity

$$\{\varphi_{jr}(x) : \; j \in \mathbb{N}_0\,,\; r = 1,\ldots,N_j\}$$

used there is replaced by the approximate resolution of unity according to 5.18.

Proof This is an immediate consequence of Theorem 5.14 and the pointwise multiplier property described in 5.17.

5.20 Refined localization: other cases

The localization property (5.13) holds for all spaces $F^s_{pq}(\mathbb{R}^n)$ with $0 < p \le \infty$, $0 < q \le \infty$, ($q = \infty$ if $p = \infty$), and $s \in \mathbb{R}$. On the other hand we have the *refined localization property* (5.120) for the spaces $\widetilde{F}^s_{pq}(\Omega)$ so far only for the parameters restricted by (5.119) (with $q = \infty$ if $p = \infty$). Hence it is natural to ask whether Theorem 5.14 and Corollary 5.19 can be extended to other values of p, q, s. We have no final answer. For our later purposes, Theorem 5.14 and Corollary 5.19 are sufficient. So we restrict ourselves to a description of what is known, referring for details to [Tri99a], especially Theorem 2.3.3, p. 701, and Corollary 2.3.6, p. 703. Let Ω be again a bounded C^∞ domain in $\mathbb{R}^n$ and let

$$\{\varphi_{jr}(x) : \quad j \in \mathbb{N}_0, \, r = 1, \ldots, N_j\}$$

be the resolution of unity according to 5.13. Let either

$$1 < p \le \infty, \quad 1 < q \le \infty, \quad -\infty < s < \frac{1}{p}, \tag{5.153}$$

($q = \infty$ if $p = \infty$), or

$$\frac{n-1}{n} < p \le 1, \quad p \le q \le \infty, \quad -\infty < s < \frac{1}{p}. \tag{5.154}$$

Then $f \in D'(\Omega)$ belongs to $F^s_{pq}(\Omega)$ if, and only if, it can be represented as

$$f = \sum_{j=0}^{\infty} \sum_{r=1}^{N_j} \varphi_{jr}(x) f \quad \text{(convergence in } D'(\Omega)\text{)} \tag{5.155}$$

such that

$$\left(\sum_{j=0}^{\infty} \sum_{r=1}^{N_j} \|\varphi_{jr} f \,|\, F^s_{pq}(\mathbb{R}^n)\|^p \right)^{\frac{1}{p}} < \infty \tag{5.156}$$

(usual modification if $p = q = \infty$). Furthermore, (5.156) is an equivalent quasi-norm. As for proofs we refer to [Tri99a]. We do not go into detail. But to provide an understanding we mention that in case of (5.153) we combine Theorem 5.14 with the duality assertion

$$\left(\widetilde{F}^{-s}_{p'q'}(\Omega) \right)' = F^s_{pq}(\Omega), \tag{5.157}$$

where p, q, s are given by (5.153) and $s < 0$. Afterwards one uses some complex interpolation. This makes clear that the natural replacement of $\widetilde{F}^s_{pq}(\Omega)$ from

5. Spaces on domains: localization and Hardy inequalities

Theorem 5.14 with (5.119) is now $F^s_{pq}(\Omega)$ (and not $\widetilde{F}^s_{pq}(\Omega)$) with (5.153) or (5.154). By duality one has also the counterpart of (5.147), which means

$$\|f(\lambda \cdot)\,|F^s_{pq}(B_1)\| \sim \lambda^{s-\frac{n}{p}}\,\|f\,|F^s_{pq}(B_\lambda)\|, \qquad (5.158)$$

where p, q, s are restricted by (5.153). Here $0 < \lambda \le 1$, and the equivalence constants in (5.158) are independent of $f \in F^s_{pq}(B_\lambda)$ and λ. One also has an obvious counterpart of the pointwise multiplier assertion from 5.17 for these spaces. Then (5.156) is still an equivalent quasi-norm in the respective spaces if $\{\varphi_{jr}\}$ is an approximate resolution of unity in 5.18. Details may be found in [Tri99a], 3.9, pp. 726–727.

5.21 The spaces $\overset{\circ}{F}{}^s_{pq}(\Omega)$: preliminaries

Let Ω be a bounded C^∞ domain in $\mathbb{R}^n$. The spaces $\overset{\circ}{F}{}^s_{pq}(\Omega)$ and $\overset{\circ}{B}{}^s_{pq}(\Omega)$ were introduced in Definition 5.3 as the completion of $D(\Omega)$ in the respective spaces $F^s_{pq}(\Omega)$ and $B^s_{pq}(\Omega)$. In Proposition 5.6 we clarified under what conditions $\overset{\circ}{F}{}^s_{pq}(\Omega)$ coincides with $F^s_{pq}(\Omega)$. If $s > \frac{1}{p}$, then the adequate substitute for (5.37) is the question under what conditions

$$\overset{\circ}{F}{}^s_{pq}(\Omega) = \widetilde{F}^s_{pq}(\Omega) \quad \text{and/or} \quad \overset{\circ}{B}{}^s_{pq}(\Omega) = \widetilde{B}^s_{pq}(\Omega) \qquad (5.159)$$

hold. Since the 1960s problems of this type have attracted a lot of attention. In [Triα], 4.3.2, p. 320, one finds the necessary references to related earlier results by J. L. Lions, Magenes, Grisvard and others with respect to classical spaces. In [Triα], 4.3.2, pp. 317–320, we have given a more or less definitive discussion of this problem as far as the spaces $H^s_p = F^s_{p,2}$ and B^s_{pq} with $1 < p < \infty$, $1 \le q < \infty$, are concerned. We continued this study in [Tri99a], Section 2.4, pp. 703–706. We will not need these results later on. This may justify that we restrict ourselves to a description of the main results obtained there. For more details and also a related discussion of the interesting spaces $\mathcal{C}^s = B^s_{\infty\infty}$ we refer to this paper.

5.22 The problem (5.159)

Let Ω be a bounded C^∞ domain in $\mathbb{R}^n$. Let $0 < p < \infty$, $0 < q < \infty$,

$$s > \sigma_p = n\left(\frac{1}{p} - 1\right)_+ \quad \text{and} \quad s - \frac{1}{p} \notin \mathbb{N}_0. \qquad (5.160)$$

Then

$$\overset{\circ}{F}{}^s_{pq}(\Omega) = \widetilde{F}^s_{pq}(\Omega) \quad \text{and} \quad \overset{\circ}{B}{}^s_{pq}(\Omega) = \widetilde{B}^s_{pq}(\Omega). \qquad (5.161)$$

5.23 Remark

This coincides with [Tri99a], 2.4.2, p. 703, as far as the F-spaces are concerned. As for the B-spaces we refer to [Tri99b]. Although this assertion has not the final character of Proposition 5.6, it makes clear that

$$s - \frac{1}{p} \in \mathbb{N}_0 \tag{5.162}$$

are delicate limiting cases. For the classical spaces $H_p^s(\Omega)$ and $B_{pq}^s(\Omega)$ with $1 < p < \infty$, $1 < q < \infty$, sufficiently definitive answers may be found in [Triα], Section 2.4, pp. 703–706. Otherwise one can prove

$$\overset{\circ}{F}{}_{pq}^s(\Omega) \neq \widetilde{F}_{pq}^s(\Omega) \quad \text{if } 1 < p < \infty,\; 0 < q \leq \infty,\; s - \frac{1}{p} \in \mathbb{N}_0. \tag{5.163}$$

We refer to [Tri99a], pp. 704–705, where one finds also further discussions. Furthermore we return later on in Corollary 16.7 and Remark 16.8 to problems of this type. Closely related to the above questions is the problem of characterizing $\overset{\circ}{F}{}_{pq}^s(\Omega)$ by vanishing boundary data.

5.24 Vanishing boundary data

Let Ω be a bounded C^∞ domain in $\mathbb{R}^n$. Let $0 < p < \infty$, $0 < q < \infty$, and

$$0 < s - \frac{1}{p} \notin \mathbb{N}. \tag{5.164}$$

(i) Then

$$\overset{\circ}{F}{}_{pq}^s(\Omega) = \left\{ f \in F_{pq}^s(\Omega) \,:\, D^\gamma f|\partial\Omega = 0 \;\text{ if }\; |\gamma| < s - \frac{1}{p} \right\}. \tag{5.165}$$

(ii) Let, in addition, $s > \sigma_p$. Then the characteristic function of Ω is a pointwise multiplier for the subspace

$$\left\{ f \in F_{pq}^s(\mathbb{R}^n) \,:\, D^\gamma f|\partial\Omega = 0 \;\text{if}\; |\gamma| < s - \frac{1}{p} \right\} \tag{5.166}$$

of $F_{pq}^s(\mathbb{R}^n)$. We refer again to [Tri99a], 2.4.5–2.4.8, pp. 705–706, where one finds also further information of this type.

5.25 Irregular domains

Definition 5.3 applies to arbitrary bounded domains Ω in $\mathbb{R}^n$. But in all assertions of Section 5 we assumed in addition that Ω is a C^∞ domain. The question arises to what extent some properties remain valid, or how they must be modified, if $\partial\Omega$ is not C^∞, but Lipschitz or even fractal. It turns out that the upper Minkowski dimension of $\partial\Omega$ is an appropriate notion in this context. We refer to [Cae99] for details.

6 Spaces on domains: decompositions

6.1 Introduction

Let Ω be a bounded C^∞ domain in $\mathbb{R}^n$, let $\{\varphi_{jr}\}$ be the resolution of unity according to 5.13, and let p, q, s be restricted by (5.119). Then by Theorem 5.14 any $f \in \widetilde{F}^s_{pq}(\Omega)$ can be (almost obviously) represented as

$$f = \sum_{j,r} \varphi_{jr}(x) f \quad \text{(convergence in } L_1(\Omega)\text{)}, \tag{6.1}$$

and (not so obviously) (5.120) is an equivalent quasi-norm. It is the main aim of this section to combine this refined localization assertion with quarkonial decompositions, now applied to $\varphi_{jr} f$, according to Definition 2.6 and Theorem 2.9. Although clear in principle, the details cause some technical problems. Recall that according to 5.22, with exception of some singular cases, the above spaces $\widetilde{F}^s_{pq}(\Omega)$ coincide with $\overset{\circ}{F}{}^s_{pq}(\Omega)$, and can be described by (5.166).

6.2 Some notation

Let Ω be a bounded C^∞ domain in $\mathbb{R}^n$. First we combine the idea of approximate lattices in $\mathbb{R}^n$ from 2.14 with the localization set-up in 5.13. There are positive numbers c_l (l = 1, ... ,8) and c_α ($\alpha \in \mathbb{N}_0^n$), (irregular) lattices

$$\{x^{j,m} : m = 1, \ldots, M_j\} \subset \Omega, \quad \text{where } j \in \mathbb{N}_0, \tag{6.2}$$

balls B_{jm}, and subordinated approximate resolutions of unity

$$\{\varphi_{jm} : \quad m = 1, \ldots, M_j\}$$

with the following properties:

(i)

$$c_1 \le M_j\, 2^{-jn} \le c_2 \quad \text{where } j \in \mathbb{N}_0, \tag{6.3}$$

$$\left|x^{j,m_1} - x^{j,m_2}\right| \geq c_3\, 2^{-j} \quad \text{where } j \in \mathbb{N}_0 \text{ and } m_1 \neq m_2\,, \tag{6.4}$$

$$\operatorname{dist}(B_{jm}, \partial\Omega) \geq c_4\, 2^{-j} \quad \text{where } j \in \mathbb{N}_0\,,\ 1 \leq m \leq M_j\,, \tag{6.5}$$

with the balls

$$B_{jm} = \left\{ y : |y - x^{j,m}| < c_5\, 2^{-j} \right\} \subset \Omega\,. \tag{6.6}$$

(ii) $\varphi_{jm}(x)$ are real C^∞ functions with

$$\operatorname{supp} \varphi_{jm} \subset B_{jm} \quad \text{where } j \in \mathbb{N}_0\,,\ 1 \leq m \leq M_j\,, \tag{6.7}$$

$$|D^\alpha \varphi_{jm}(x)| \leq c_\alpha\, 2^{j|\alpha|} \quad \text{where } \alpha \in \mathbb{N}_0^n\,,\ j \in \mathbb{N}_0\,,\ 1 \leq m \leq M_j\,, \tag{6.8}$$

and

$$c_6 \leq \sum_{m=1}^{M_j} \varphi_{jm}(x) \leq c_7 \quad \text{if } x \in \Omega \text{ and } \operatorname{dist}(x, \partial\Omega) \geq c_8\, 2^{-j} \tag{6.9}$$

for all $j \in \mathbb{N}_0$. For short, we call $\{\varphi_{jm}\}$ a *family of approximate resolutions of unity in Ω* if one finds lattices $\{x^{j,m}\}$ with all the properties (6.2)–(6.9). After these preparations we can introduce the counterpart of the β-quarks from (2.16) in a version of (2.99).

6.3 Definition

Let Ω be a bounded C^∞ domain in $\mathbb{R}^n$. Let $s \in \mathbb{R}$, $0 < p \leq \infty$ and $\beta \in \mathbb{N}_0^n$. Let $\{\varphi_{jm}\}$ be a family of approximate resolutions of unity in Ω according to 6.2. Then

$$(\beta qu)_{jm}(x) = 2^{-j(s-\frac{n}{p})+j|\beta|}\, (x - x^{j,m})^\beta\, \varphi_{jm}(x)\,, \quad x \in \Omega\,, \tag{6.10}$$

with $j \in \mathbb{N}_0$ and $m = 1, \ldots, M_j$ is called an $(s,p) - \beta$-quark.

6.4 Remark

As said previously, this is the direct counterpart of (2.99). As there, the factor $2^{-j(s-\frac{n}{p})}$ reflects the normalization of these elementary building blocks with respect to the different j-levels. Now we call $(\beta qu)_{jm}$ a β-quark and not a generalized β-quark. As for the uniform normalization for the diverse $\beta \in \mathbb{N}_0^n$ one needs the counterpart of the number r in (2.14). In our case one can take

$$c_5 = 2^r\,, \quad r \geq 0\,, \tag{6.11}$$

where c_5 has the same meaning as in (6.6), (6.7). Then

$$2^{-r|\beta|}\, (\beta qu)_{jm}(x) \tag{6.12}$$

are normalized building blocks with respect to j, m, β.

6.5 Sequence spaces

We need the counterparts of the sequence spaces f_{pq} introduced in (2.8). Let $\chi_{jm}(x)$ be the characteristic function of the ball B_{jm} given by (6.6). We put, in analogy to (2.4),

$$\chi_{jm}^{(p)}(x) = 2^{\frac{jn}{p}} \chi_{jm}(x) \quad \text{where } x \in \mathbb{R}^n \text{ and } 0 < p \leq \infty. \tag{6.13}$$

Let $0 < p \leq \infty$, $0 < q \leq \infty$, and

$$\lambda = \{\lambda_{jm} \in \mathbb{C} : \ j \in \mathbb{N}_0, \ m = 1, \ldots, M_j\}. \tag{6.14}$$

Then

$$f_{pq}^\Omega = \left\{ \lambda : \ \|\lambda \,|\, f_{pq}^\Omega\| = \left\| \left(\sum_{j=0}^\infty \sum_{m=1}^{M_j} \left|\lambda_{jm} \chi_{jm}^{(p)}(\cdot)\right|^q \right)^{\frac{1}{q}} \,|\, L_p(\Omega) \right\| < \infty \right\} \tag{6.15}$$

are the counterparts of f_{pq} in (2.8). Finally we have to adapt the sequence spaces from Definition 2.6 to our situation. Let $\varrho > r$ where r is the same number as in (6.11). Then we put

$$\lambda = \{\lambda^\beta : \ \beta \in \mathbb{N}_0^n\} \tag{6.16}$$

with

$$\lambda^\beta = \{\lambda_{jm}^\beta \in \mathbb{C} : \ j \in \mathbb{N}_0, \ m = 1, \ldots, M_j\} \tag{6.17}$$

and

$$\|\lambda \,|\, f_{pq}^\Omega\|_\varrho = \sup_{\beta \in \mathbb{N}_0^n} 2^{\varrho|\beta|} \, \|\lambda^\beta \,|\, f_{pq}^\Omega\|. \tag{6.18}$$

After these preparations we are now in a similar position as we were in Definition 2.6(ii) and Theorem 2.9(ii), now with Ω in place of $\mathbb{R}^n$.

6.6 Theorem

Let Ω be a bounded C^∞ domain in $\mathbb{R}^n$. There is a family of approximate resolutions of unity according to 6.2 with the following property. Let

$$0 < p \leq \infty, \quad 0 < q \leq \infty, \quad s > \sigma_{pq} = n \left(\frac{1}{\min(p,q)} - 1\right)_+, \tag{6.19}$$

(with $q = \infty$ if $p = \infty$). Let $\widetilde{F}^s_{pq}(\Omega)$ be the spaces introduced in Definition 5.3(iii) and notationally complemented by (5.18). Let $\varrho > r$, where r is given by (6.11). Then $f \in L_1(\Omega)$ belongs to $\widetilde{F}^s_{pq}(\Omega)$ if, and only if, it can be represented as

$$f(x) = \sum_{\beta \in \mathbb{N}_0^n} \sum_{j=0}^{\infty} \sum_{m=1}^{M_j} \lambda^\beta_{jm} (\beta qu)_{jm}(x) \tag{6.20}$$

with

$$\|\lambda \,|\, f^\Omega_{pq}\|_\varrho < \infty \tag{6.21}$$

(absolute and unconditional convergence in $L_1(\Omega)$), where $(\beta qu)_{jm}(x)$ and (6.21) are given by (6.10) and (6.18), respectively. Furthermore, the infimum in (6.21) over all admissible representations (6.20) is an equivalent quasi-norm in $\widetilde{F}^s_{pq}(\Omega)$.

Proof *Step 1* We begin with some preparation. Let again

$$B_\Lambda = \{y : |y| < \Lambda\}, \qquad \Lambda > 0, \tag{6.22}$$

be the ball in $\mathbb{R}^n$ centred at the origin and of radius Λ. Let

$$f \in F^s_{pq}(\mathbb{R}^n) \quad \text{with} \quad \operatorname{supp} f \subset B_1, \tag{6.23}$$

where p, q, s are restricted by (6.19). Let $\psi^{k,m}(x)$ be the resolution of unity (2.97) based on the approximate lattices described in 2.14. We denote temporarily the related β-quarks in (2.99) by $(\beta Qu)_{kl}(x)$. Then we have by 2.14 the corresponding quarkonial representation according to Definition 2.6 and Theorem 2.9. Let $\Phi(x)$ be a C^∞ function in $\mathbb{R}^n$ with

$$\operatorname{supp} \Phi \subset B_4 \quad \text{and} \quad \Phi(x) = 1 \quad \text{if} \quad x \in B_3, \tag{6.24}$$

where B_3 and B_4 are given by (6.22) with $\Lambda = 3$ and $\Lambda = 4$, respectively. Then we have

$$f(x) = f(x)\Phi(x) = \sum_{\beta \in \mathbb{N}_0^n} \sum_{k=0}^{\infty} \sum_{l \in \mathbb{Z}^n} \lambda^\beta_{kl} (\beta Qu)_{kl}(x) \Phi(x) \tag{6.25}$$

(absolute and unconditional convergence in $L_1(\mathbb{R}^n)$). Assume that the supports of $\psi^{k,l}$ and, hence, of $(\beta Qu)_{kl}$, are restricted by (2.98) with $c_2 = 1$. Then we have $\Phi(x) = 1$ if $x \in B(x^{k,l}, 2^{-k+1})$ (in the notation of (2.98)) and

$$B\left(x^{k,l}, 2^{-k+1}\right) \cap B_1 \neq \emptyset. \tag{6.26}$$

In other words, for these terms the β-quarks $(\beta Qu)_{kl}(x)$ remain unchanged whereas for the other terms with supports in B_4 one has now the additional terms $(\beta Qu)_{kl}(x)\,\Phi(x)$. Of course we may assume $\psi^{k,l}(x) \geq 0$.

Step 2   Assume now

$$f \in F^s_{pq}(\mathbb{R}^n) \quad \text{with} \quad supp\, f \subset B_\Lambda \quad \text{with, say,} \quad \Lambda = 2^{-K}, \tag{6.27}$$

for some $K \in \mathbb{N}_0$. We use the homogeneity property (5.147) from Corollary 5.16 and apply (6.25) to $f(2^{-K}x)$. Recall that the $\widetilde{F}^s_{pq}(B_1)$-quasi-norm is the $F^s_{pq}(\mathbb{R}^n)$-quasi-norm. By (2.99) and the factor $2^{-K(s-\frac{n}{p})}$ in (5.147) this causes an index shifting $k \to k+K$, and hence we have the quarkonial decomposition

$$f(x) = \sum_{\beta \in \mathbb{N}_0^n} \sum_{k=K}^{\infty} \sum_{l \in \mathbb{Z}^n} \lambda^\beta_{kl}\,(\beta Qu)_{kl}(x)\,\Phi(2^K x), \tag{6.28}$$

which starts at the level K. Again the (irregular) approximate lattices $x^{k,l}$ in (2.99) are at our disposal.

Step 3   Now we prove the theorem by constructing step by step an appropriate family of approximate resolutions of unity and related β-quarks (6.10) with the properties claimed. To avoid confusion with the notation used in 6.2 we denote the resolution of unity in Theorem 5.14 by

$$\{\Phi_{jr}(x) \,:\, j \in \mathbb{N}_0\,,\; r = 1,\ldots,N_j\}\,. \tag{6.29}$$

Then we have by Theorem 5.14 for $f \in \widetilde{F}^s_{pq}(\Omega)$,

$$f = \sum_{j=0}^{\infty} \sum_{r=1}^{N_j} \Phi_{jr}(x)\,f \tag{6.30}$$

(absolute and unconditional convergence in $L_1(\Omega)$) and

$$\left\|f\,|\widetilde{F}^s_{pq}(\Omega)\right\| \sim \left(\sum_{j=0}^{\infty}\sum_{r=1}^{N_j} \left\|\Phi_{jr}\,f\,|F^s_{pq}(\mathbb{R}^n)\right\|^p\right)^{\frac{1}{p}} \tag{6.31}$$

(equivalent quasi-norms). We start, say, with $\Phi_{0,1}\,f$ and expand this function optimally according to (6.28) with $\Phi_{0,1}\,f$ in place of f, ignoring possible translations. This results in some sequences of (irregular) lattice points $\{x^{k,l}_{0,1}\}$ near $supp\,\Phi_{0,1}$, where, say, $k = 0, 1, 2, \ldots$, with the required properties, in particular (6.4). Then we do the same with, say, $\Phi_{0,2}\,f$. We get new lattice points $\{x^{k,l}_{0,2}\}$.

They might interfere with $\{x_{0,1}^{k,l}\}$. But the new lattice points $\{x_{0,2}^{k,l}\}$ can be chosen in such a way that $\{x_{0,1}^{k,l}\} \cup \{x_{0,2}^{k,l}\}$ are again approximate lattices according to 6.2, especially with the counterpart of (6.4), near $supp\,\Phi_{0,1} \cup supp\,\Phi_{0,2}$. Now one can continue this procedure step by step. If $K \in \mathbb{N}_0$ is fixed, then only those lattices $\{x_{j,r}^{K,l}\}$ with $j \leq K$ of level K contribute to the combined lattice $\{x^{K,l}\}$. This follows from (6.28) applied to $\Phi_{jr} f$ (ignoring translations). But now it is clear that one gets lattices (6.2) with the properties (6.3)–(6.6). Furthermore we get β-quarks (6.10) where the functions $\varphi_{jm}(x)$ originate from the constructions in Step 1 and Step 2. Then we have (6.7), (6.8), and

$$1 \leq \sum_{m=1}^{M_j} \varphi_{jm}(x) \leq c, \quad x \in \Omega, \quad dist\,(x, \partial\Omega) \geq c'\, 2^{-j}. \qquad (6.32)$$

The right-hand side of (6.32) is clear. As for the left-hand side we fix $x \in \Omega$ with $dist\,(x, \partial\Omega) \geq c'\, 2^{-j}$. Then $x \in supp\,\Phi_{k,r}$ with $k \leq j$. But at that point all non-vanishing terms from (2.97) (after a suitable translation) are incorporated in $\{\varphi_{jm}(x)\}$ by the specific construction from Step 1 and Step 2. Recall that we have always $\varphi_{jm}(x) \geq 0$. There might be additional non-negative terms. This proves the left-hand side of (6.32) and hence (6.9). All other properties, in particular (6.21), are covered by the construction, (6.31), and the optimally chosen coefficients in (6.28) according to Definition 2.6 and Theorem 2.9.

6.7 Unconditional convergence and optimal coefficients

We have a full counterpart of 2.11. As mentioned in the above proof, (6.20) converges absolutely and unconditionally in $L_1(\Omega)$. Hence, in analogy to (2.73) and (2.74) one can write

$$f(x) = \sum_{\beta, j, m} \lambda_{jm}^{\beta}\, (\beta qu)_{jm}(x) \qquad (6.33)$$

now with

$$\sum_{\beta, j, m} = \sum_{\beta \in \mathbb{N}_0^n} \sum_{j=0}^{\infty} \sum_{m=1}^{M_j}. \qquad (6.34)$$

Furthermore, one can ask for optimal coefficients $\lambda_{jm}^{\beta}(f)$ in analogy to (2.75), (2.77), and (2.79). The above proof is constructive and the resulting optimal coefficients can be reduced to Corollary 2.12 and to (2.82) with $\Phi_{jr} f$ in place of f, where $\{\Phi_{jr}\}$ is the above resolution of unity according to (6.30). Then one ends up with the following assertion: *There are optimal coefficients*

$$\lambda^\beta_{jm}(f) = 2^{j(s-\frac{n}{p})} 2^{-\varrho|\beta|} \int_\Omega f(x)\, \Psi^{\beta,\varrho}_{jm}(x)\, dx, \tag{6.35}$$

where

$$\Psi^{\beta,\varrho}_{jm}(x) \in D(\Omega) \quad \text{with} \quad \beta \in \mathbb{N}_0^n,\ j \in \mathbb{N}_0,\ m = 1,\ldots,M_j. \tag{6.36}$$

These functions even can be constructed explicitly. One has to start with (2.86) and to follow the arguments in the above proof. But the outcome is rather involved.

6.8 Quarkonial decompositions: other cases

The proof of Theorem 6.6 was based on the refined localization (6.30), (6.31) on the one hand and quarkonial representations of the spaces $F^s_{pq}(\mathbb{R}^n)$ with (6.19) in the version of 2.14 on the other hand. Both ingredients are available also in some other cases. Let again Ω be a C^∞ domain in $\mathbb{R}^n$. As in (6.29) again we denote the resolution of unity according to 5.13 and 5.14 by

$$\{\Phi_{jr}(x) : j \in \mathbb{N}_0,\ r = 1,\ldots,N_j\}. \tag{6.37}$$

Let either

$$1 < p \leq \infty,\quad 1 < q \leq \infty,\quad -\infty < s < \frac{1}{p}, \tag{6.38}$$

($q = \infty$ if $p = \infty$), or

$$\frac{n-1}{n} < p \leq 1,\quad p \leq q \leq \infty,\quad -\infty < s < \frac{1}{p}. \tag{6.39}$$

As mentioned in 5.20 with a reference to [Tri99a], an element $f \in D'(\Omega)$ belongs to $F^s_{pq}(\Omega)$ if, and only if, it can be represented as

$$f = \sum_{j=0}^\infty \sum_{r=1}^{N_j} \Phi_{jr}(x)\, f \quad \text{(convergence in } D'(\Omega)\text{)} \tag{6.40}$$

such that (usual modification if $p = q = \infty$)

$$\left(\sum_{j=0}^\infty \sum_{r=1}^{N_j} \|\Phi_{jr}\, f\, |F^s_{pq}(\mathbb{R}^n)\|^p \right)^{\frac{1}{p}} < \infty \tag{6.41}$$

(equivalent quasi-norms). Next one can try to follow the arguments of the proof of Theorem 6.6 and to replace the quarkonial decomposition in 2.14 by Definition 3.4(ii) and Theorem 3.6, always notationally complemented by (5.18). This causes some trouble since the multiplication by $\Phi(x)$ in (6.25) destroys the moment conditions which are needed in connection with the β-quarks according to Definition 3.2. We outline how to circumvent this difficulty. Let B_Λ be the balls given by (6.22) and again let

$$f \in F_{pq}^s(\mathbb{R}^n) \quad \text{with} \quad supp \, f \subset B_1, \tag{6.42}$$

where we may assume, temporarily,

$$0 < p \leq \infty, \quad 0 < q \leq \infty, \quad s \in \mathbb{R}, \tag{6.43}$$

($q = \infty$ if $p = \infty$). Then f can be represented by

$$f = g_1 + (-\Delta)^M g_2 \quad \text{with} \quad g_j \in F_{pq}^{s+2M}(\mathbb{R}^n) \tag{6.44}$$

with

$$supp \, g_j \subset B_2, \quad (j = 1, 2). \tag{6.45}$$

Here $(-\Delta)^M$ is the Mth power of the Laplacian $-\Delta$. This follows from lifting and some additional modifications which may be found in [Triδ], p. 83. Let

$$0 < p \leq \infty, \quad 0 < q \leq \infty, \quad s + 2M > \sigma_{pq} = n\left(\frac{1}{\min(p,q)} - 1\right)_+. \tag{6.46}$$

Then one can apply the arguments from Step 1 of the proof of Theorem 6.6 to g_1 and g_2 as elements of $F_{pq}^{s+2M}(\mathbb{R}^n)$, and we have the counterparts of (6.25). By (6.44) one gets a quarkonial decomposition of f in $F_{pq}^s(\mathbb{R}^n)$ which fits in our scheme. In Step 2 of the proof of Theorem 6.6 we used decisively the homogeneity property (5.147). At least for the spaces $F_{pq}^s(\mathbb{R}^n)$ with (6.38) we have the counterpart

$$\|f(\Lambda \cdot) \,|\, F_{pq}^s(B_1)\| \sim \Lambda^{s-\frac{n}{p}} \|f \,|\, F_{pq}^s(B_\Lambda)\|, \tag{6.47}$$

where the equivalence constants are independent of $f \in F_{pq}^s(B_\Lambda)$ and Λ with $0 < \Lambda \leq 1$. We refer to 5.20 and for details to [Tri99a], (3.88), p. 726. Afterwards one can follow the arguments from the proof of Theorem 6.6 without substantial changes. Then one obtains the following assertion: *Let Ω, the family of approximate resolutions of unity, and $\varrho > r$, be as in Theorem 6.6, and let p, q, s be restricted by (6.38). Let $\frac{L+1}{2} = M \in \mathbb{N}$ with (6.46). Let, in analogy to (6.10) and (3.6),*

$$(\beta qu)_{jm}(x) = 2^{-j(s+2M-\frac{n}{p})+j|\beta|} (x - x^{j,m})^\beta \varphi_{jm}(x) \tag{6.48}$$

be $(s+2M,p) - \beta$-quarks and

$$(\beta qu)_{jm}^L(x) = (-\Delta)^M (\beta qu)_{jm}(x) \qquad (6.49)$$

be $(s,p)^L - \beta$-quarks. Then $f \in D'(\Omega)$ belongs to $F_{pq}^s(\Omega)$ if, and only if, it can be represented (in the notation of (6.34)) as

$$f = \sum_{\beta,j,m} \left(\eta_{jm}^\beta (\beta qu)_{jm} + \lambda_{jm}^\beta (\beta qu)_{jm}^L \right) \qquad (6.50)$$

(unconditional convergence in $D'(\Omega)$) with

$$\|\eta \,|\, f_{pq}^\Omega\|_\varrho + \|\lambda \,|\, f_{pq}^\Omega\|_\varrho < \infty . \qquad (6.51)$$

Furthermore, the infimum in (6.51) over all admissible representations (6.50) is an equivalent quasi-norm in $F_{pq}^s(\Omega)$.

This is the counterpart of Definition 3.4 and Theorem 3.6 adapted to bounded C^∞ domains and with the restriction (6.38). But there is hardly any doubt that the latter restriction comes from the technique used and that the above assertion should be true for all p, q, s with (6.43) ($q = \infty$ if $p = \infty$). In (6.51) we used the notation introduced in (6.16)–(6.18). As for optimal coefficients we are now in the same situation as in (6.35), (6.36), which must be interpreted as the dual pairing in $(D(\Omega), D'(\Omega))$. This explains why we can admit any $f \in D'(\Omega)$ in the above assertion. We mention that also the splitting (6.44) with reference to [Triδ], p. 83, is constructive. Hence one gets an intrinsic description of the above spaces $F_{pq}^s(\Omega)$ in terms of (6.50), (6.51), with constructive optimal coefficients η_{jm}^β and λ_{jm}^β which depend linearly on f.

6.9 An alternative approach

We describe a second approach to prove decompositions of type (6.50) with an even simpler outcome since the η-terms are not needed. Again let Ω be a bounded C^∞ domain in $\mathbb{R}^n$. Let $M \in \mathbb{N}$ and let

$$0 < p < \infty, \quad 0 < q < \infty, \quad s > \max\left(\frac{1}{p} - 1, n(\frac{1}{p} - 1)\right) - M . \qquad (6.52)$$

Then J. Franke and Th. Runst proved in [FrR95], Theorem 13, p. 144, and in [RuS96], Theorem 1 on p. 133, that $(-\Delta)^M$ is an isomorphic map from

$$\{ f \in F_{pq}^{s+2M}(\Omega) : \quad D^\gamma f | \partial\Omega = 0 \text{ if } |\gamma| \leq M - 1 \} \qquad (6.53)$$

onto $F_{pq}^s(\Omega)$. If, in addition,

$$\max\left(\frac{1}{p}-1,\, n(\frac{1}{p}-1)\right) - M < s < \frac{1}{p} - M, \qquad (6.54)$$

then the space in (6.53) coincides with $\widetilde{F}_{pq}^s(\Omega)$. This follows from (5.161) and (5.165). Hence, if p, q, s are given by (6.52), (6.54), then

$$(-\Delta)^M : \widetilde{F}_{pq}^{s+2M}(\Omega) \mapsto F_{pq}^s(\Omega) \qquad (6.55)$$

is an isomorphic map. Now one can apply first Theorem 6.6 to $\widetilde{F}_{pq}^{s+2M}(\Omega)$ and then one can apply $(-\Delta)^M$. One gets in $F_{pq}^s(\Omega)$ the quarkonial representation

$$f = \sum_{\beta,j,m} \lambda_{jm}^\beta\, (\beta qu)_{jm}^L \qquad (6.56)$$

with $\frac{L+1}{2} = M$ and the same assertions as there. This covers in particular all cases

$$1 \le p < \infty, \quad 1 \le q < \infty, \quad \frac{1}{p} - 1 - M < s < \frac{1}{p} - M \quad \text{with } M \in \mathbb{N}. \quad (6.57)$$

But, of course, we used the indicated deep mapping theorem which, in turn, has a long history. Detailed references may be found in [FrR95] and [RuS96], Chapter 3. In connection with Taylor expansions of distributions we return in 8.15 to this approach.

6.10 Atoms and quarks in $\mathbb{R}^n$ and in domains: references and historical comments

We described in [Triγ], 1.9, the historical roots of atoms in some function spaces, preferably Hardy spaces on $\mathbb{R}^n$. Atomic decompositions of spaces of type $B_{pq}^s(\mathbb{R}^n)$ and $F_{pq}^s(\mathbb{R}^n)$ go back to M. Frazier and B. Jawerth. We refer to [FrJ85], [FrJ90], [FJW91], [Tor91]. As so often in the history of mathematics, especially in the theory of function spaces, independent and parallel developments in the East (in the Russian literature) remained unnoticed in the West. In connection with atoms in spaces of type $B_{pq}^s(\mathbb{R}^n)$ and $F_{pq}^s(\mathbb{R}^n)$ we refer in particular to the significant work of Yu. V. Netrusov, [Net87a], [Net87b], [Net89b]. Further references may be found in [Triγ], 1.9.2, Remark 2, p. 63. A new approach and new complete proofs of atomic decompositions may be found in [Triδ], Sect. 13. Quarkonial (or subatomic) decompositions for the spaces $B_{pq}^s(\mathbb{R}^n)$ and $F_{pq}^s(\mathbb{R}^n)$ go back to [Triδ], Sect. 14. The main motivation at that time came from the interplay between function spaces on $\mathbb{R}^n$ and fractal geometry, especially fractal partial differential operators and their spectral theory. Sections 2 and 3 of the present book might be considered as a more systematic treatment, now partly for its own sake, but also to serve as a basis for later applications.

As for atoms in domains there are only a few papers in the literature. Let Ω be a bounded, maybe irregular, domain in $\mathbb{R}^n$. Then we developed in [TrW96] a theory of intrinsic atomic descriptions of the spaces $B^s_{pq}(\Omega)$ and $F^s_{pq}(\Omega)$. A brief summary of these results may also be found in [ET96], 2.5, pp. 57–65. In [Miy90], [CKS92], [CKS93], [CDS99] the authors study several types of Hardy spaces in domains in $\mathbb{R}^n$ (general, C^∞, Lipschitz). In the notation of Definition 5.3 and Remark 5.4, these spaces coincide with $F^0_{p,2}(\Omega)$, $\widetilde{F}^0_{p,2}(\Omega)$ and subspaces of $\widetilde{F}^0_{p,2}(\overline{\Omega})$, where always $0 < p \leq 1$. There are intrinsic atomic descriptions and, at least in case of the Hardy spaces $F^0_{p,2}(\Omega)$ there are two types of atoms, say, *boundary atoms*, and *interior atoms*, where for the boundary atoms no, or only a limited number of, moment conditions must be fulfilled. But this is very similar to [TrW96]. On an atomic level there is also a connection with the procedure outlined in 6.9, but the intentions are quite different. We used the results about mapping properties of elliptic operators in [FrR95] to get intrinsic (sub)atomic decompositions, whereas in [CKS93] and [CDS99] the authors constructed first atomic decompositions and applied them afterwards to study the (Dirichlet and Neumann) Laplacian. Quarkonial representations of the spaces of type $F^s_{pq}(\Omega)$ on domains, as presented here, go back to [Tri99a]. Sections 5 and 6 above are modified versions of the respective parts of [Tri99a].

7 Spaces on manifolds

7.1 Introduction

Let M be a (n-dimensional non-compact) Riemannian manifold with bounded geometry and positive injectivity radius. In [Triγ], Chapter 7, we introduced spaces $F^s_{pq}(M)$ with

$$0 < p \leq \infty, \quad 0 < q \leq \infty, \quad s \in \mathbb{R}, \tag{7.1}$$

($q = \infty$ if $p = \infty$) via local charts. The motivation comes from $M = \mathbb{R}^n$ (euclidean metric) where one might interpret (5.13) as a characterization in terms of local charts in $\mathbb{R}^n$. In other words, in [Triγ], Chapter 7, we took (5.13) as a starting point, extended it naturally, and introduced in this way spaces F^s_{pq} on M. The outcome is satisfactory. One has not only intrinsic descriptions in geometric terms, but also

$$F^s_{p,2}(M) = H^s_p(M), \quad 1 < p < \infty, \quad s \in \mathbb{R}, \tag{7.2}$$

and
$$F^s_{p,2}(M) = W^s_p(M), \quad 1 < p < \infty, \quad s \in \mathbb{N}_0, \tag{7.3}$$

where $W^s_p(M)$ are the usual Sobolev spaces on M defined by covariant derivatives, and $H^s_p(M)$ are the (fractional) Sobolev spaces obtained from $L_p(M)$ by lifting via fractional powers of the Laplace-Beltrami operator. Corresponding spaces $B^s_{pq}(M)$ can be obtained afterwards by real interpolation. Details and references, in particular to the substantial history of Sobolev spaces on manifolds, may be found in [Triγ], 1.11, and Chapter 7. Our aim here is slightly different and, as far as the Riemannian background is concerned, much simpler. We specify M and assume that this manifold can be represented as a bounded domain Ω in $\mathbb{R}^n$, equipped with the Riemannian metric

$$(d\sigma)^2 = g^2(x)\,(dx)^2, \quad x = (x_1, \ldots, x_n) \in \Omega, \tag{7.4}$$

where g is a positive C^∞ function in Ω with

$$g(x) \sim \text{dist}\,(x, \partial\Omega)^{-1}, \quad x \in \Omega. \tag{7.5}$$

The outcome is an infinite (non-compact) complete Riemannian manifold with bounded geometry and positive injectivity radius (tacitly assuming that (7.5) can be complemented by estimates from above of $|D^\gamma g(x)|$ by the right-hand side of (7.5) with $-|\gamma| - 1$ in place of -1). The Poincaré ball

$$\Omega = B = \{x \in \mathbb{R}^n : |x| < 1\} \tag{7.6}$$

with

$$(d\sigma)^2 = \frac{(dx)^2}{(1 - |x|^2)^2}, \quad |x| < 1, \tag{7.7}$$

may serve as a prototype. We used this type of interplay between weighted spaces in a euclidean setting and (so far unweighted) spaces in a Riemannian setting in [Tri88] to study weighted pseudo-differential operators in domains. Now we follow [Tri99c]. In rough terms, instead of the localization (5.13) in $\mathbb{R}^n$ we use the refined localization in Theorem 5.14 as a motivation to introduce corresponding spaces now in a more general setting. Naturally, fractional powers $g^\varkappa(x)$ of $g(x)$ in (7.4) come in, as well as related weighted spaces $F^s_{pq}(M, g^\varkappa)$. As for weights $w(x)$ in connection with F^s_{pq}-spaces on $\mathbb{R}^n$ one has two possibilities. Either one replaces, say, $L_p(\mathbb{R}^n)$ in (2.38) by the weighted counterpart $L_p(\mathbb{R}^n, w)$ or one says that f belongs to $F^s_{pq}(\mathbb{R}^n, w)$ if, by definition, wf belongs to $F^s_{pq}(\mathbb{R}^n)$. Fortunately enough, for a large class of weights these two possibilities result in the same spaces. Some details are given in 7.3

below, whereas 7.2 contains some more information concerning the geometrical background, although not very much is needed in the sequel. We use it more as an adequate language. The aim of this section is twofold. First we extend the theory developed so far in the previous sections, especially in Sections 5 and 6, to spaces on manifolds of the above type. Secondly we prepare for our later considerations about the spectrum of elliptic operators in these infinite hyperbolic worlds.

7.2 Riemannian manifolds

As said previously, it is not our aim to repeat what is meant by an (abstract) Riemannian manifold M, equipped with the metric g, with bounded geometry and positive injectivity radius, in general. All that one needs in connection with related spaces $F_{pq}^s(M)$ may be found in [Triγ], 7.2, pp. 281–308. We restrict ourselves to the special case which is of interest for us later on. We follow [Tri88] and [Tri99c]. Let Ω be a bounded domain in $\mathbb{R}^n$. Let again

$$d(x) = dist\,(x, \partial\Omega), \quad x \in \Omega, \tag{7.8}$$

be the (euclidean) distance of $x \in \Omega$ to the boundary $\partial\Omega$. It is well known that there is a positive C^∞ function $g(x)$ in Ω with

$$g(x) \sim d(x)^{-1}, \quad |D^\gamma g(x)| \leq c_\gamma\, g^{1+|\gamma|}(x), \ x \in \Omega, \ \gamma \in \mathbb{N}_0^n, \tag{7.9}$$

for some positive constants c_γ. Recall that we use "$\sim$" for two positive functions $a(x)$ and $b(x)$ or for two sequences of positive numbers a_k and b_k (say, $k \in \mathbb{N}$), if there are two positive numbers c and C such that

$$c\,a(x) \leq b(x) \leq C\,a(x) \quad \text{or} \quad c\,a_k \leq b_k \leq C\,a_k \tag{7.10}$$

for all admitted variables x or k. A short proof of the existence of a function g with (7.9) may be found in [Triα], 3.2.3, p. 250–251. Now we equip Ω with the Riemannian metric

$$(d\sigma)^2 = g^2(x)\,(dx)^2, \quad x \in \Omega. \tag{7.11}$$

This means that at a given point $x \in \Omega$ the Riemannian distance $d\sigma$ can be compared locally with the euclidean distance dx multiplied with the dilation factor $g(x)$. To be more precise and to explain what is meant in *our context* by *bounded geometry* and *positive injectivity radius* we look at a sequence of balls B^j,

$$B^j = \left\{ x \in \mathbb{R}^n : \ |x - x^j| < c_1\, 2^{-j} \right\} \subset \Omega, \quad j \in \mathbb{N}_0, \tag{7.12}$$

centred at $x^j \in \Omega$ with $d(x^j) \sim 2^{-j}$ such that

$$dist\,(B^j,\, \partial\Omega) \geq c_2\, 2^{-j}, \quad j \in \mathbb{N}_0, \qquad (7.13)$$

where c_1 and c_2 are suitably chosen positive numbers (which are independent of $j \in \mathbb{N}_0$). Then $g(x) \sim 2^j$ in B^j, and B^j are uniformly (with respect to $j \in \mathbb{N}_0$) comparable with Riemannian balls, centred at x^j, and of radius 1. Let K^j be the image of B^j under the affine transform in $\mathbb{R}^n$,

$$x \mapsto y = 2^j(x - x^j), \quad j \in \mathbb{N}_0. \qquad (7.14)$$

Then each K^j is comparable with the unit ball in $\mathbb{R}^n$ centred at the origin. The Riemannian metric g in B^j, expressed by $g(x)$, transforms locally in the Riemannian metric G_j, expressed by

$$G_j(y) = 2^{-j}\, g(2^{-j} y + x^j), \quad j \in \mathbb{N}_0. \qquad (7.15)$$

By (7.9) one has now uniform (with respect to $j \in \mathbb{N}_0$) estimates with respect to $G_j(y)$. All relevant geometrical quantities as geodesics, curvature tensor etc. can be calculated via the local charts of type K^j equipped with the metric G_j. By the G-counterpart of (7.9) this can be done uniformly with respect to j. The outcome is what is called a manifold with *bounded geometry*. In our case *positive injectivity radius* means simply that all local charts of type K^j contain uniformly (with respect to $j \in \mathbb{N}_0$) a ball of some positive radius in $\mathbb{R}^n$ centred at the origin. We refer for more details to [Triγ], 7.2.1, pp. 281–285, where we gave a precise description of this geometric background in standard terms of differential geometry. There one finds also references to the underlying literature. Hence:

In this section M or (M, g) stands for the n-dimensional complete Riemannian manifold with bounded geometry and positive injectivity radius represented by a bounded domain Ω in $\mathbb{R}^n$ furnished with the Riemannian metric (7.9), (7.11).

Complete means that (M, g), considered as a metric space, is complete. This follows from the Hopf-Rinow theorem and the fact that all geodesics are infinitely extendible with respect to their arc length. Finally we mention that in our special case the Laplace-Beltrami operator, denoted by Δ_g, is given by

$$\Delta_g = g^{-n} \sum_{j=1}^{n} \frac{\partial}{\partial x_j} \left(g^{n-2} \frac{\partial}{\partial x_j} \right). \qquad (7.16)$$

We refer to [Triγ], 7.2.5, pp. 298–305, and 7.4.3, p. 316, where we described the Laplace-Beltrami operator for general Riemannian manifolds with bounded geometry and positive injectivity radius and its mapping properties in the context of the spaces $F_{pq}^s(M)$.

7.3 Weights

For the above manifold M we are not only interested in the spaces $F_{pq}^s(M)$ but also in the weighted spaces $F_{pq}^s(M, g^\varkappa)$, where g comes from the corresponding metric and $\varkappa \in \mathbb{R}$. This extension is rather natural, even unavoidable. To justify our approach we have a brief look at the corresponding situation in $\mathbb{R}^n$. We follow [ET96], 4.2.1, 4.2.2, pp. 153–160, which in turn is based on [HaT94a] and [HaT94b]. Let w be a positive C^∞ function in $\mathbb{R}^n$ with the following properties:

(i) For all multi-indices β there is a positive constant c_β with

$$|D^\beta w(x)| \leq c_\beta w(x) \quad \text{for all} \quad x \in \mathbb{R}^n. \tag{7.17}$$

(ii) There exist two constants $c > 0$ and $\alpha \geq 0$ such that

$$0 < w(x) \leq c\, w(y) \left(1 + |x-y|^2\right)^{\frac{\alpha}{2}} \quad \text{for all} \quad x \in \mathbb{R}^n,\ y \in \mathbb{R}^n. \tag{7.18}$$

Let $\{\varphi_k : k \in \mathbb{N}_0\}$ be the resolution of unity introduced in 2.8. Then the weighted spaces $F_{pq}^s(\mathbb{R}^n, w)$ with (7.1) (where $q = \infty$ if $p = \infty$) are defined, in modification of (2.38), as the collection of all $f \in S'(\mathbb{R}^n)$ such that

$$\left\| \left(\sum_{j=0}^\infty 2^{jsq} |(\varphi_j \widehat{f})^\vee(\cdot)|^q \right)^{\frac{1}{q}} w(\cdot) \,|\, L_p(\mathbb{R}^n) \right\| \tag{7.19}$$

is finite. By [ET96], Theorem 4.2.2, p. 156, one has the remarkable assertion that

$$F_{pq}^s(\mathbb{R}^n, w) = \{f \in S'(\mathbb{R}^n) : \quad wf \in F_{pq}^s(\mathbb{R}^n)\} \tag{7.20}$$

(equivalent quasi-norms). Hence, at least under the assumptions (7.17), (7.18) it does not matter whether one puts the weight w to the distribution or to the underlying measure. The above Riemannian situation is very much simpler. Taking the local charts K^j from 7.2 with the transformed metric, expressed by the functions G_j from (7.15), as the starting point then the transformed weights $w(x) = g^\varkappa(x)$, $\varkappa \in \mathbb{R}$, behave in K^j like (7.17). This would justify introducing weighted spaces $F_{pq}^s(M, g^\varkappa)$ as the collection of all f such that $g^\varkappa f \subset F_{pq}^s(M)$.

7.4 A preparatory remark

In [Triγ], Chapter 7, we developed the full theory of the spaces $F_{pq}^s(M)$ with (7.1) ($q = \infty$ if $p = \infty$) for arbitrary Riemannian manifolds M with bounded geometry and positive injectivity radius. This applies also to [Tri88], where the underlying manifold was specified as in 7.2. Our aim here is slightly different. We wish to rely on the results and the techniques developed in Chapters 5 and 6. In particular, homogeneity assertions of type (5.147) and the resulting pointwise multiplier properties as described in 5.17 are of great service for us. They allow us essentially to argue directly in the domain Ω, converted in the manifold (M,g), and to avoid shifting everything to local charts. The price to pay is some restrictions for the parameters involved where we concentrate on (5.119). But other possible cases will be mentioned. For these spaces we wish to get quarkonial decompositions much as in Theorem 6.6. This is sufficient for our later purposes: A spectral theory for elliptic operators in infinite hyperbolic worlds. This will be done in Section 21.

7.5 The set-up, d-domains

Let Ω be a bounded domain in $\mathbb{R}^n$ and let $d(x)$ be given by (7.8). Let

$$\Omega_j = \{x \in \Omega : \quad 2^{-j-1} < d(x) \le 2^{-j}\}, \quad j \in \mathbb{N}_0. \tag{7.21}$$

We assume, without restriction of generality, that $\Omega_j \ne \emptyset$ for all $j \in \mathbb{N}_0$ and that $d(x) < 1$ for any $x \in \Omega$. For a fixed small positive number c we cover $\overline{\Omega_j}$ by balls B_{jm} of radius $c\, 2^{-j}$ centred in $\overline{\Omega_j}$ such that

$$B_{jm} \subset \Omega_{j-1} \cup \Omega_j \cup \Omega_{j+1}, \quad j \in \mathbb{N}_0, \tag{7.22}$$

(with $\Omega_{-1} = \emptyset$) and $m = 1, \ldots, M_j$. The covering

$$\Omega = \bigcup_{j=0}^{\infty} \bigcup_{m=1}^{M_j} B_{jm} \tag{7.23}$$

is assumed to be locally finite: there is a natural number N such that at most N balls involved in (7.23) have a non-empty intersection. Furthermore we assume that there is a number λ with $0 < \lambda < 1$ such that all balls λB_{jm} are disjoint. Here λB_{jm} stands for the ball with the same centre as B_{jm} and of radius λ times the radius of B_{jm}. We equip this domain Ω with the Riemannian metric (7.11), (7.9) and denote the resulting Riemannian manifold likewise by M or (M,g). We described in 7.2 the Riemannian background.

Later on in 21.2 we specify Ω in the following way. Let $n-1 \le d < n$. Then Ω, or likewise M or (M,g) is called a d-domain if

$$M_j \sim 2^{jd}, \quad j \in \mathbb{N}_0. \tag{7.24}$$

7.6 Discussion and examples

First we assume that Ω is a Lipschitz domain. This means that there is a finite covering of $\partial\Omega$ by, say, balls in $\mathbb{R}^n$, such that in any of these neighbourhoods of $\partial\Omega$ the respective part of $\partial\Omega$ can be represented (after translation, rotation and dilation) by

$$x_n = H(x'), \quad x' = (x_1, \ldots, x_{n-1}), \quad |x'| < 1, \tag{7.25}$$

where H is a Lipschitz function,

$$|H(x') - H(y')| \leq c\,|x' - y'|, \quad |x'| < 1, \ |y'| < 1. \tag{7.26}$$

Covering first the unit ball in $\mathbb{R}^{n-1}$ by balls of radius $c\,2^{-j}$ and shifting this afterwards to the graph of H according to (7.25), then it follows easily that $d = n - 1$ in (7.24).

Next we assume that Ω is an arbitrary bounded domain Ω in $\mathbb{R}^n$ such that, without restriction of generality,

$$\{x = (x', 0) : \quad x' \in \mathbb{R}^{n-1}, \ |x'| < 1\} \subset \Omega. \tag{7.27}$$

Starting from the 'footpoint' $(x', 0)$ and climbing up or down to $(x', x_n) \in \Omega_j$, it is quite clear that one needs at least $c\,2^{j(n-1)}$ balls B_{jm} with (7.22), (7.23), where $c > 0$ is a suitable number. Hence the restriction $d \geq n - 1$ in connection with (7.24) comes in naturally. It is the *interior Minkowski dimension*, we refer to [Fal97], pp. 226–227. If $n - 1 < d < n$, then $\partial\Omega$ is fractal. The general fractal background may be found in [Fal85], [Fal90], [Mat95], including the diverse notation for what is called a (fractal) dimension of a set, in our case of $\partial\Omega$. But at least in connection with the spectral theory of the (euclidean) Laplacian $-\Delta$ in bounded domains with fractal boundary the notation *interior Minkowski dimensions, interior Minkowski contents, Minkowski sausages* seems to be adequate. We refer to [Lap91], [HeL97], [EvH93], [Ber98]. We follow this line and prove in Section 21 spectral assertions for some elliptic operators in d-domains. In the present section we develop the necessary tools for these later applications. But the main assertions in this section apply to arbitrary bounded domains and the specifications indicated will not be needed here.

Finally we describe a few examples of d-domains. Obviously, the classical prototype of an $(n-1)$-domain is the Poincaré ball (7.6), (7.7). Furthermore, for any d with $n - 1 < d < n$ there are *thorny star-like d-domains*. We give a brief description. Let Q be the unit cube in $\mathbb{R}^{n-1}$ and let $0 < s < 1$. In [Triδ], Sect. 16, pp. 119–123, we constructed a non-negative function $x_n = f(x')$

where $x' = (x_1, \ldots, x_{n-1}) \in \mathbb{R}^{n-1}$, with compact support in Q such that $f \in \mathcal{C}^s(\mathbb{R}^{n-1})$ (Hölder spaces) and

$$\{(x', f(x')) : \quad x' \in \mathbb{R}^{n-1} \text{ with } f(x') > 0\} \tag{7.28}$$

is an $(n-s)$-set in $\mathbb{R}^n$ (in the notation used there). As indicated in [Triδ], p. 123, with obvious modifications one can replace Q by the unit sphere $\{x \in \mathbb{R}^n : |x| = 1\}$ and x_n by the radial direction. Then one obtains a thorny star-like (with respect to the origin) simply connected d-domain with $d = n - s$.

7.7 The set-up: resolutions of unity

Let Ω or likewise M or (M, g) as in 7.5 be a bounded domain in $\mathbb{R}^n$. The covering (7.23) can be constructed in such a way that there is a related resolution of unity,

$$\sum_{j=0}^{\infty} \sum_{m=1}^{M_j} \varphi_{jm}(x) = 1 \quad \text{if} \quad x \in \Omega, \tag{7.29}$$

of C^∞ functions φ_{jm} with

$$\operatorname{supp} \varphi_{jm} \subset B_{jm} \tag{7.30}$$

and, for suitable $c_\gamma > 0$,

$$|D^\gamma \varphi_{jm}(x)| \leq c_\gamma \, 2^{j|\gamma|}, \quad \text{where} \quad \gamma \in \mathbb{N}_0^n. \tag{7.31}$$

Let x^{jm} be the centre of B_{jm}. Recall that $D'(\Omega)$ stands for the (complex-valued) distributions in Ω.

7.8 Definition

Let Ω, or likewise M or (M, g), be a bounded domain in $\mathbb{R}^n$ equipped with the covering $\{B_{jm}\}$ and the resolution of unity $\{\varphi_{jm}\}$ according to 7.5 and 7.7. Let

$$0 < p \leq \infty, \quad 0 < q \leq \infty, \quad s \in \mathbb{R}, \tag{7.32}$$

(with $q = \infty$ if $p = \infty$).

(i)   Then $F_{pq}^s(M)$ is the collection of all $f \in D'(\Omega)$ such that

$$\|f \mid F_{pq}^s(M)\| = \left(\sum_{j=0}^{\infty} \sum_{m=1}^{M_j} \|(\varphi_{jm} f)(x^{jm} + 2^{-j} \cdot) \mid F_{pq}^s(\mathbb{R}^n)\|^p \right)^{\frac{1}{p}} < \infty \tag{7.33}$$

(usual modification if $p = q = \infty$).

(ii) Let, in addition, $\varkappa \in \mathbb{R}$. Then

$$F^s_{pq}(M, g^\varkappa) = \{ f \in D'(\Omega) : \quad g^\varkappa f \in F^s_{pq}(M) \} \tag{7.34}$$

with

$$\| f \, | F^s_{pq}(M, g^\varkappa) \| = \| g^\varkappa f \, | F^s_{pq}(M) \|. \tag{7.35}$$

7.9 Comments

Recall that we introduced the spaces $F^s_{pq}(\mathbb{R}^n)$ in Definitions 2.6 and 3.4, or, likewise, in Theorems 2.9 and 3.6, using the notational complement (5.18). Part (i) of the definition is the specification of the Definition in [Triγ], 7.2.2, pp. 285–286, to our situation, and coincides essentially with [Tri88], 2.4.1, p. 42. In particular, $F^s_{pq}(M)$ is independent of the admitted coverings $\{B_{jm}\}$ and the related resolutions of unity $\{\varphi_{jm}\}$. In particular, all the quasi-norms in (7.33) are equivalent to each other. This may justify our omission of $\{B_{jm}\}$ and $\{\varphi_{jm}\}$ on the left hand side of (7.33). For further details and proofs we refer to [Triγ], Chapter 7. However for the cases of interest for us one gets the claimed independence as an immediate by-product of the considerations below. Otherwise we gave the necessary explanations about the Riemannian background in 7.2. In 7.3 we justified the way in which we incorporated the weights $g^\varkappa$ in (7.34). Recall our usual notation

$$\sigma_{pq} = n \left(\frac{1}{\min(p,q)} - 1 \right)_+, \quad \text{where} \quad 0 < p \leq \infty, \quad 0 < q \leq \infty. \tag{7.36}$$

7.10 Theorem

Let Ω be a bounded domain in $\mathbb{R}^n$ and let

$$0 < p \leq \infty, \quad 0 < q \leq \infty, \quad \text{and} \quad s > \sigma_{pq} \tag{7.37}$$

(with $q = \infty$ if $p = \infty$). Let $\varkappa \in \mathbb{R}$ and let $F^s_{pq}(M, g^\varkappa)$ be the spaces introduced in Definition 7.8. Then $f \in D'(\Omega)$ belongs to $F^s_{pq}(M, g^\varkappa)$ if, and only if,

$$\left(\sum_{j=0}^{\infty} \sum_{m=1}^{M_j} 2^{j(\varkappa - s + \frac{n}{p})p} \| \varphi_{jm} f \, | F^s_{pq}(\mathbb{R}^n) \|^p \right)^{\frac{1}{p}} < \infty \tag{7.38}$$

(with the usual modification if $p = q = \infty$). Furthermore, (7.38) is an equivalent quasi-norm.

Proof *Step 1* Let $\varkappa = 0$. By the comments in 7.9 it remains to prove that (7.38) is an equivalent quasi-norm. We use Corollary 5.16 and obtain

$$\left\|(\varphi_{jm}f)(x^{jm} + 2^{-j}\cdot)\,|F^s_{pq}(\mathbb{R}^n)\right\| \sim 2^{-j(s-\frac{n}{p})} \left\|\varphi_{jm}f\,|F^s_{pq}(\mathbb{R}^n)\right\|, \qquad (7.39)$$

where the equivalence constants are independent of $f \in D'(\Omega)$ and $j \in \mathbb{N}_0$ and m. We used Definition 5.3, Proposition 5.5, and (5.23).

Step 2 Let $\varkappa \in \mathbb{R}$. We apply the pointwise multiplier assertion from 5.17 to $\varphi_{jm}f$ in place of f and to $2^{-j\varkappa}g^{\varkappa}$, respectively $2^{j\varkappa}g^{-\varkappa}$ in place of φ and obtain

$$\left\|g^{\varkappa}\varphi_{jm}f\,|F^s_{pq}(\mathbb{R}^n)\right\| \sim 2^{j\varkappa} \left\|\varphi_{jm}f\,|F^s_{pq}(\mathbb{R}^n)\right\|. \qquad (7.40)$$

We used (7.9) and again 5.3, 5.5, and (7.38). The theorem follows now from (7.35) and Step 1.

7.11 Other cases, comments

As mentioned in 7.9, under the restriction (7.37) the independence of $F^s_{pq}(M)$ from the underlying covering $\{B_{jm}\}$ and the resolution of unity $\{\varphi_{jm}\}$ is an immediate by-product of the proof. This also makes clear that one can extend the theorem to other cases for which one has appropriate counterparts of the two decisive ingredients 5.16 and 5.17 of the above proof.

Let

$$1 < p \leq \infty, \quad 1 < q \leq \infty, \quad \text{and} \quad s \in \mathbb{R} \qquad (7.41)$$

($q = \infty$ if $p = \infty$). Let $\varkappa \in \mathbb{R}$. Then $f \in D'(\Omega)$ belongs to $F^s_{pq}(M, g^{\varkappa})$ if, and only if, (7.38) holds (equivalent norms).

As for the proof we may assume $s \leq 0$ (the case $s > 0$ is covered by the above theorem). One must extend (7.39) to these new cases. The equivalence (5.158) with (5.153) is near to what one needs, but (7.39) is not an immediate consequence. But the corresponding proof in [Tri99a], p. 726, formulas (3.85)–(3.88), also covers (7.39) with (7.41) and $s < 0$. Finally, $s = 0$ is a matter of complex interpolation. This applies also to the pointwise multiplier assertions from 5.17 extended to (7.41). Then one gets the assertion of the theorem also for p, q, s with (7.41).

Maybe the most interesting cases are the weighted Sobolev spaces

$$H^s_p(M, g^{\varkappa}) = F^s_{p,2}(M, g^{\varkappa}), \quad 1 < p < \infty,\, s \in \mathbb{R},\, \varkappa \in \mathbb{R}, \qquad (7.42)$$

which are covered by (7.41). As for the notation we refer to (7.2), (7.3).

Later on we are mainly interested in arbitrary d-domains according to 7.5. However if Ω is a bounded C^∞ domain in $\mathbb{R}^n$ then we have $d = n - 1$ and one can compare the spaces $F^s_{pq}(M, g^\varkappa)$ from Definition 7.8 and Theorem 7.10 with the spaces $\widetilde{F}^s_{pq}(\Omega)$ considered in Theorem 5.14. The set-up in 5.13 can be identified in this case with the set-up in 7.5 and 7.7.

7.12 Proposition

Let Ω be a bounded C^∞ domain in $\mathbb{R}^n$ and let M, or likewise (M, g), be the related Riemannian manifold according to 7.5. Let

$$0 < p \leq \infty, \quad 0 < q \leq \infty, \quad \text{and} \quad s > \sigma_{pq}, \tag{7.43}$$

(with $q = \infty$ if $p = \infty$). Let $\widetilde{F}^s_{pq}(\Omega)$ and $F^s_{pq}(M, g^\varkappa)$ with $\varkappa \in \mathbb{R}$ be the spaces introduced in Definitions 5.3 and 7.8, respectively. Then

$$F^s_{pq}(M, g^{s-\frac{n}{p}}) = \widetilde{F}^s_{pq}(\Omega), \tag{7.44}$$

(equivalent quasi-norms).

Proof This is an immediate consequence of Theorems 5.14 and 7.10.

7.13 Comments, other cases

Hence, by (7.44) there is a close connection between F^s_{pq} spaces defined in a euclidean setting and their Riemannian counterparts, where weights come in naturally. We add a few comments. By Definition 5.3 the completion of $D(\Omega)$ in $F^s_{pq}(\Omega)$ is denoted by $\overset{\circ}{F}{}^s_{pq}(\Omega)$.

Let Ω and M be as in Proposition 7.12 and let

$$0 < p < \infty, \quad 0 < q < \infty, \quad s > \sigma_{pq}, \quad s - \frac{1}{p} \notin \mathbb{N}_0. \tag{7.45}$$

Then

$$F^s_{pq}(M, g^{s-\frac{n}{p}}) = \overset{\circ}{F}{}^s_{pq}(\Omega). \tag{7.46}$$

This follows from Proposition 7.12 and 5.22. Furthermore, *let Ω and M as in Proposition 7.12 and let*

$$1 < p \leq \infty, \quad 1 < q \leq \infty, \quad -\infty < s < \frac{1}{p}, \tag{7.47}$$

($q = \infty$ if $p = \infty$). Then

$$F^s_{pq}(M, g^{s-\frac{n}{p}}) = F^s_{pq}(\Omega). \tag{7.48}$$

We prove this assertion. By 7.11, especially the assertion in connection with (7.41), a distribution $f \in D'(\Omega)$ belongs to the space on the left-hand side of (7.48) if, and only if,

$$\left(\sum_{j=0}^{\infty} \sum_{m=1}^{M_j} \|\varphi_{jm} f \,|\, F^s_{pq}(\mathbb{R}^n)\|^p \right)^{\frac{1}{p}} < \infty \tag{7.49}$$

(equivalent norm). Then (7.48) follows from (5.156), (5.153).

By (7.34) any space $F^s_{pq}(M, g^{\varkappa})$ can be mapped by

$$f \mapsto g^{\varkappa - (s - \frac{n}{p})} f \tag{7.50}$$

isomorphically onto $F^s_{pq}(M, g^{s-\frac{n}{p}})$. If again Ω is a bounded C^∞ domain in $\mathbb{R}^n$, then we have (7.44) with (7.43) and (7.48) with (7.47). Hence in all these cases, the weighted F^s_{pq} spaces on the Riemannian manifold M can be described by the (euclidean) F^s_{pq} spaces on Ω, where, for example, one has a lot of equivalent quasi-norms. Again of particuliar interest might be the Sobolev spaces (7.42).

7.14 Lifts

Let

$$0 < p \leq \infty, \quad 0 < q \leq \infty, \quad s \in \mathbb{R}, \tag{7.51}$$

(with $q = \infty$ if $p = \infty$). Recall that we put

$$F^s_{\infty\infty} = B^s_{\infty\infty}. \tag{7.52}$$

It is well known that for any $\varrho > 0$,

$$-\Delta + \varrho\, id : \quad F^{s+2}_{pq}(\mathbb{R}^n) \mapsto F^s_{pq}(\mathbb{R}^n) \tag{7.53}$$

is an isomorphic map. Of course, $-\Delta$ is the (euclidean) Laplacian. As discussed in 7.2, the Riemannian counterpart of $-\Delta$ in the above manifold M is given by the Laplace-Beltrami operator

$$-\Delta_g = -g^{-n} \sum_{j=1}^{n} \frac{\partial}{\partial x_j} \left(g^{n-2} \frac{\partial}{\partial x_j} \right). \tag{7.54}$$

We extended in [Triγ], Theorem 7.4.3, p. 316, the euclidean lifting assertion (7.53) to arbitrary Riemannian manifolds with bounded geometry and positive injectivity radius, where $-\Delta$ must be interpreted as the respective Laplace-Beltrami operator. This applies in particular to our case and can be generalized to the above weighted spaces. Let

$$G_\mu : \quad f \mapsto g^\mu f, \quad \mu \in \mathbb{R}, \quad f \in D'(\Omega), \tag{7.55}$$

where g has the previous meaning (7.9).

7.15 Theorem

Let Ω be a bounded domain in $\mathbb{R}^n$ and let

$$0 < p \leq \infty, \quad 0 < q \leq \infty, \quad s \in \mathbb{R}, \tag{7.56}$$

($q = \infty$ if $p = \infty$, (7.52)), and $\varkappa \in \mathbb{R}$. Let $F_{pq}^s(M, g^\varkappa)$ be the spaces introduced in Definition 7.8.

(i)   Let $-\Delta_g$ be the Laplace-Beltrami operator (7.54). If $\varrho \in \mathbb{R}$ is sufficiently large, then

$$-\Delta_g + \varrho\, id : \quad F_{pq}^{s+2}(M, g^\varkappa) \mapsto F_{pq}^s(M, g^\varkappa) \tag{7.57}$$

is an isomorphic mapping.

(ii)   Let $\mu \in \mathbb{R}$ and G_μ be given by (7.55). Then

$$G_\mu : \quad F_{pq}^s(M, g^\varkappa) \mapsto F_{pq}^s(M, g^{\varkappa - \mu}) \tag{7.58}$$

is an isomorphic mapping.

Proof   We proved (7.57) with $\varkappa = 0$ in [Triγ], 7.4.3, p. 316, for arbitrary Riemannian manifolds with bounded geometry and positive injectivity radius. This covers in particular our situation. This proof can be extended to $\varkappa \in \mathbb{R}$. Part (ii) follows from Definition 7.8.

7.16 Embeddings

Let Ω be a bounded domain in $\mathbb{R}^n$. In Definition 5.3 we introduced the spaces $F_{pq}^s(\Omega)$ with (7.56). We collect a few well-known embedding assertions. Let

$$-\infty < s_2 < s_1 < \infty, \quad 0 < p_1 \leq p_2 \leq \infty, \quad 0 < q_1 \leq \infty, \quad 0 < q_2 \leq \infty, \tag{7.59}$$

($q_1 = \infty$ if $p_1 = \infty$ and $q_2 = \infty$ if $p_2 = \infty$, (7.52)). Let

$$\delta = \left(s_1 - \frac{n}{p_1}\right) - \left(s_2 - \frac{n}{p_2}\right). \tag{7.60}$$

Then (continuous embedding)

$$F_{p_1 q_1}^{s_1}(\Omega) \subset F_{p_2 q_2}^{s_2}(\Omega) \tag{7.61}$$

if, and only if, $\delta \geq 0$. Furthermore, the embedding (7.61) is compact if, and only if, $\delta > 0$. As for the continuous embedding (7.61) we refer to [Triβ], 2.7.1, 3.3.1, pp. 129, 196–197. Compact embeddings of type (7.61) and their B-counterparts have been studied in detail in [ET96], 3.3. The main point is to calculate the behaviour of the entropy numbers of compact embeddings of type (7.61), which, in turn, paves the way to a spectral theory of related elliptic differential operators. It is our aim to do the same with the spaces $F_{pq}^s(M, g^\varkappa)$ introduced in Definition 7.8 in case of d-domains. We shift this task to Theorem 21.3 where it is used afterwards to develop a spectral theory for elliptic operators in those manifolds (M, g) which are d-domains. The rest of this section might be considered as preparation for this aim: We clarify continuity and compactness of embeddings between the spaces $F_{pq}^s(\Omega, g^\varkappa)$ and establish some quarkonial decompositions, again for arbitrary bounded domains.

7.17 Proposition

Let Ω be a bounded domain in $\mathbb{R}^n$. Let M and $F_{pq}^s(M, g^\varkappa)$ be as in Definition 7.8. Let s_1, s_2, p_1, p_2, q_1, q_2 be as in (7.59). Let $\varkappa_1 \in \mathbb{R}$ and $\varkappa_2 \in \mathbb{R}$. Let δ be given by (7.60). Then

$$F_{p_1 q_1}^{s_1}(M, g^{\varkappa_1}) \subset F_{p_2 q_2}^{s_2}(M, g^{\varkappa_2}) \tag{7.62}$$

(continuous embedding) if, and only if, $\delta \geq 0$, and $\varkappa_1 \geq \varkappa_2$. Furthermore, (7.62) is compact if, and only if, $\delta > 0$, and $\varkappa_1 > \varkappa_2$.

Proof *Step 1* By iterative application of the lifts in Theorem 7.15 we may assume, without restriction of generality,

$$s_1 > \sigma_{p_1 q_1} \quad \text{and} \quad s_2 > \sigma_{p_2 q_2}. \tag{7.63}$$

Then we can apply Theorem 7.10. Let $\delta \geq 0$ and $\varkappa_1 \geq \varkappa_2$. Now the continuous embedding (7.62) follows from (7.38) and (7.39), (7.61) with $\delta \geq 0$, together with the monotonicity $\ell_{p_1} \subset \ell_{p_2}$ of the ℓ_p-spaces. Next we prove that the embedding (7.62) is compact if $\delta > 0$ and $\varkappa_1 > \varkappa_2$. Let $\{\varphi_{jm}\}$ be the resolution of unity described in 7.7 and let

$$\varphi_J(x) = \sum_{j=0}^{J} \sum_{m=1}^{M_j} \varphi_{jm}(x), \quad J \in \mathbb{N}, \tag{7.64}$$

be a cut-off function. By the pointwise multiplier property 5.17, as used in Step 2 of the proof of Theorem 7.10, it follows that

$$\left\| \varphi_J f \, | F_{p_1 q_1}^{s_1}(M, g^{\varkappa_1}) \right\| \leq c \left\| f \, | F_{p_1 q_1}^{s_1}(M, g^{\varkappa_1}) \right\| \tag{7.65}$$

and

$$\left\| (1-\varphi_J) f \, | F_{p_2 q_2}^{s_2}(M, g^{\varkappa_2}) \right\|$$

$$\begin{aligned}
&\leq c_1 \left\| (1-\varphi_J) f \, | F_{p_1 q_1}^{s_1}(M, g^{\varkappa_2}) \right\| \\
&\leq c_2 \, 2^{-J(\varkappa_1 - \varkappa_2)} \left\| (1-\varphi_J) f \, | F_{p_1 q_1}^{s_1}(M, g^{\varkappa_1}) \right\| \\
&\leq c_3 \, 2^{-J(\varkappa_1 - \varkappa_2)} \left\| f \, | F_{p_1 q_1}^{s_1}(M, g^{\varkappa_1}) \right\|,
\end{aligned} \tag{7.66}$$

where c, c_1, c_2, c_3 are independent of J. By (7.65), $\delta > 0$, and the compactness of the embedding (7.61) it follows that the map

$$f \mapsto \varphi_J f : \quad F_{p_1 q_1}^{s_1}(M, g^{\varkappa_1}) \mapsto F_{p_2 q_2}^{s_2}(M, g^{\varkappa_2}) \tag{7.67}$$

is compact. Together with $\varkappa_1 > \varkappa_2$ and (7.66) one gets the compactness of the embedding (7.62).

Step 2 We prove the converse assertion. Let B be an (open) ball with $\overline{B} \subset \Omega$. If we have the continuous (or compact) embedding (7.62), denoted with id, then it follows by

$$id \left(F_{p_1 q_1}^{s_1}(B) \to F_{p_2 q_2}^{s_2}(B) \right) = re \circ id \circ ext \tag{7.68}$$

that also the embedding

$$F_{p_1 q_1}^{s_1}(B) \subset F_{p_2 q_2}^{s_2}(B) \tag{7.69}$$

is continuous (or compact). Here ext is a (linear and bounded) extension operator from $F_{p_1 q_1}^{s_1}(B)$ into $F_{p_1 q_1}^{s_1}(M, g^{\varkappa_1})$ and re is the restriction operator from $F_{p_2 q_2}^{s_2}(M, g^{\varkappa_2})$ onto $F_{p_2 q_2}^{s_2}(B)$. Then $\delta \geq 0$ in the continuous case and $\delta > 0$ in the compact case follow from 7.16. Next we prove $\varkappa_1 \geq \varkappa_2$ if the embedding (7.62) is continuous and $\varkappa_1 > \varkappa_2$ if this embedding is compact. Let

$$B^j = \left\{ x \in \mathbb{R}^n : \ |x - x^j| < c\, 2^{-j} \right\} \subset \Omega_j, \quad j \in \mathbb{N}_0, \tag{7.70}$$

where Ω_j is given by (7.21), and $c > 0$ is small. Let φ be a non-trivial C^∞ function with compact support near the origin. Let

$$f = \sum_{j=0}^{\infty} a_j \varphi\left(2^j (x - x^j)\right), \quad a_j \in \mathbb{C}. \tag{7.71}$$

We may assume that the balls B^j and the function φ are chosen such that we can apply (7.38) as follows,

$\|f \,|F^{s_1}_{p_1 q_1}(M, g^{\varkappa_1})\|$

$$\sim \left(\sum_{j=0}^{\infty} 2^{j(\varkappa_1 - s_1 + \frac{n}{p_1})p_1} \|\varphi(2^j \cdot) \,|F^{s_1}_{p_1 q_1}(\mathbb{R}^n)\|^{p_1} |a_j|^{p_1} \right)^{\frac{1}{p_1}}$$

$$\sim \left(\sum_{j=0}^{\infty} 2^{j \varkappa_1 p_1} |a_j|^{p_1} \right)^{\frac{1}{p_1}}, \tag{7.72}$$

where in the latter equivalence we used (5.147) with $\lambda = 2^{-j}$ in the same way as in connection with (7.39). We have the same equivalence with the index 2 in place of 1. Hence, if the embedding (7.62) is continuous then

$$\left(\sum_{j=0}^{\infty} 2^{j \varkappa_2 p_2} |a_j|^{p_2} \right)^{\frac{1}{p_2}} \le c \left(\sum_{j=0}^{\infty} 2^{j \varkappa_1 p_1} |a_j|^{p_1} \right)^{\frac{1}{p_1}} \tag{7.73}$$

for all $a_j \in \mathbb{C}$ such that the right-hand side of (7.73) is finite. It follows that $\varkappa_2 \le \varkappa_1$. If the embedding (7.62) is compact, then we have $\varkappa_2 < \varkappa_1$ since the embedding (7.73) with $\varkappa_1 = \varkappa_2$ is not compact.

7.18 Remark

This proposition is the direct counterpart of a corresponding assertion for the spaces $F^s_{pq}(\mathbb{R}^n, w)$ briefly mentioned in 7.3. We refer to [HaT94a] or [ET96], 4.2.3, p. 160. As said above our main concern later on in Theorem 21.3 is to measure the degree of compactness of the embedding (7.62) in terms of entropy numbers. For that purpose we need quarkonial decompositions of these spaces which we are going to describe now.

7.19 Some notation

We wish to study quarkonial decompositions of the spaces $F^s_{pq}(M, g^{\varkappa})$ from Theorem 7.10. We are very much in the same situation as in Theorem 6.6. We adapt the notation and the definitions in 6.2–6.5 to our situation now.

7. Spaces on manifolds

Let Ω be a bounded domain in $\mathbb{R}^n$. First we introduce *families of approximate resolutions of unity in Ω* in analogy to 6.2, avoiding letters used in Sect. 7 for other purposes. There are positive numbers c_k ($k = 1, \ldots, 8$) and c_α ($\alpha \in \mathbb{N}_0^n$), (irregular) lattices

$$\{y^{jl} : l = 1, \ldots, L_j\} \subset \Omega, \quad \text{where} \quad j \in \mathbb{N}_0, \tag{7.74}$$

balls K_{jl}, and subordinated approximate resolutions of unity

$$\{\psi_{jl} : l = 1, \ldots, L_j\}$$

with the following properties:

(i)
$$c_1 \leq L_j 2^{-jn} \leq c_2 \quad \text{where} \quad j \in \mathbb{N}_0, \tag{7.75}$$

$$|y^{jl_1} - y^{jl_2}| \geq c_3 2^{-j} \quad \text{where} \quad j \in \mathbb{N}_0 \quad \text{and} \quad l_1 \neq l_2, \tag{7.76}$$

$$\operatorname{dist}(K_{jl}, \partial\Omega) \geq c_4 2^{-j} \quad \text{where} \quad j \in \mathbb{N}_0, \quad 1 \leq l \leq L_j, \tag{7.77}$$

with the balls

$$K_{jl} = \{y : |y - y^{jl}| < c_5 2^{-j}\} \subset \Omega. \tag{7.78}$$

(ii) $\psi_{jl}(x)$ are real C^∞ functions with

$$\operatorname{supp} \psi_{jl} \subset K_{jl} \quad \text{where} \quad j \in \mathbb{N}_0, \quad 1 \leq l \leq L_j, \tag{7.79}$$

$$|D^\alpha \psi_{jl}(x)| \leq c_\alpha 2^{j|\alpha|} \quad \text{where} \quad \alpha \in \mathbb{N}_0^n, \ j \in \mathbb{N}_0, \ 1 \leq l \leq L_j, \tag{7.80}$$

and

$$c_6 \leq \sum_{l=1}^{L_j} \psi_{jl}(x) \leq c_7 \quad \text{if} \quad x \in \Omega \quad \text{and} \quad \operatorname{dist}(x, \partial\Omega) \geq c_8 2^{-j} \tag{7.81}$$

for all $j \in \mathbb{N}_0$. For short, as in 6.2, we call $\{\psi_{jl}\}$ a *family of approximate resolutions of unity* in Ω if one finds lattices $\{y^{jl}\}$ with all the properties (7.74)–(7.81).

We need the obvious counterparts of the notation introduced in 6.5. Let $\chi_{jl}(x)$ be the characteristic function of the ball K_{jl} given by (7.78) and let $\chi_{jl}^{(p)}(x)$ be given in analogy to (6.13). Let $0 < p \leq \infty$, $0 < q \leq \infty$, and

$$\lambda = \{\lambda_{jl} \in \mathbb{C} : j \in \mathbb{N}_0, \ l = 1, \ldots, L_j\}. \tag{7.82}$$

Then, as in (6.15),

$$f_{pq}^\Omega = \left\{ \lambda : \quad \|\lambda\,|f_{pq}^\Omega\| = \left\| \left(\sum_{j=0}^\infty \sum_{l=1}^{L_j} \left| \lambda_{jl}\, \chi_{jl}^{(p)}(\cdot) \right|^q \right)^{\frac{1}{q}} \,|L_p(\Omega) \right\| < \infty \right\} \quad (7.83)$$

are the counterparts of f_{pq} in (2.8). As in (6.11) we put

$$c_5 = 2^r, \qquad r \geq 0, \quad (7.84)$$

where c_5 is the constant in (7.78). Let $\varrho > r$. We introduce in analogy to (6.16)–(6.18),

$$\lambda = \{\lambda^\beta : \quad \beta \in \mathbb{N}_0^n\} \quad (7.85)$$

with

$$\lambda^\beta = \left\{ \lambda_{jl}^\beta \in \mathbb{C} : \quad j \in \mathbb{N}_0,\, l = 1, \ldots, L_j \right\} \quad (7.86)$$

and

$$\|\lambda\,|f_{pq}^\Omega\|_\varrho = \sup_{\beta \in \mathbb{N}_0^n} 2^{\varrho|\beta|}\, \|\lambda^\beta\,|f_{pq}^\Omega\|. \quad (7.87)$$

Some more information about this notation may be found in 6.2, 6.4, and 6.6. Definition 6.3 now must be modified as follows.

7.20 Definition

Let Ω be a bounded domain in $\mathbb{R}^n$. Let

$$s \in \mathbb{R}, \quad 0 < p \leq \infty, \quad \varkappa \in \mathbb{R}, \quad \text{and} \quad \beta \in \mathbb{N}_0^n.$$

Let $\{\psi_{jl}\}$ be a family of approximate resolutions of unity in Ω according to 7.19. Then

$$(\beta qu)_{jl}(y) = 2^{-j(s-\frac{n}{p})+j|\beta|}\, 2^{-k(\varkappa - s + \frac{n}{p})}\, (y - y^{jl})^\beta\, \psi_{jl}(y) \quad (7.88)$$

with $y \in \Omega$, $j \in \mathbb{N}_0$, and $l = 1, \ldots, L_j$, is called an $(s, p, \varkappa) - \beta$-quark where

$$k = k(j, l) \quad \text{such that} \quad y^{jl} \in \Omega_k \quad (7.89)$$

according to (7.21).

7.21 Discussion

By (7.21) and (7.77), (7.78) we have

$$k = k(j,l) \le j + c \tag{7.90}$$

for some $c > 0$. We rewrite (7.88) as

$$(\beta q u)_{jl}(y) = 2^{-\varkappa k}\, 2^{-(j-k)(s-\frac{n}{p})}\, (2^j y - 2^j y^{jl})^{\beta}\, \psi_{jl}(y) \tag{7.91}$$

and put temporarily

$$\psi^{\beta}_{jl}(y) = \left(2^j y - 2^j y^{jl}\right)^{\beta} \psi_{jl}(y), \quad \beta \in \mathbb{N}_0^n. \tag{7.92}$$

These $(s, p, \varkappa)$ – β-quarks are normalized building blocks for the spaces $F^s_{pq}(M, g^\varkappa)$ from Theorem 7.10: Under the restriction (7.37) we obtain by (7.38)

$$\left\|(\beta q u)_{jl}\,|F^s_{pq}(M, g^\varkappa)\right\| \sim 2^{-j(s-\frac{n}{p})} \left\|\varphi_{km}\, \psi^{\beta}_{jl}\,|F^s_{pq}(\mathbb{R}^n)\right\| + + \tag{7.93}$$

for some $m = m(j,l)$, where $++$ indicates a few other terms neighbouring k and m. By (7.39) and (7.90), (7.31), (7.80) it follows that

$$\left\|(\beta q u)_{jl}\,|F^s_{pq}(M, g^\varkappa)\right\| \le c(\beta), \tag{7.94}$$

independently of j and l. As for the dependence of $c(\beta)$ on β, we are by (7.78) and (7.84) in the same situation as in 6.4. We obtain

$$\left\|(\beta q u)_{jl}\,|F^s_{pq}(M, g^\varkappa)\right\| \le c\, 2^{r|\beta|}, \quad \beta \in \mathbb{N}_0^n, \tag{7.95}$$

for some $c > 0$ which is independent of j, l and β.

Recall that

$$L_1(\Omega, g^\tau) = \{f \in D'(\Omega) : \|g^\tau f\,|L_1(\Omega)\| < \infty\}, \quad \tau \in \mathbb{R}, \tag{7.96}$$

normed in the obvious way. Furthermore, σ_{pq} is given by (7.36).

7.22 Theorem

Let Ω be a bounded domain in $\mathbb{R}^n$. There is a family of approximate resolutions of unity $\{\psi_{jl}\}$ in Ω according to 7.19 with the following property. Let

$$0 < p \le \infty, \quad 0 < q \le \infty \quad \text{and} \quad s > \sigma_{pq} \tag{7.97}$$

(with $q = \infty$ if $p = \infty$). Let $\varkappa \in \mathbb{R}$ and let $F_{pq}^s(M, g^\varkappa)$ be the spaces introduced in Definition 7.8. Let $(\beta qu)_{jl}$ be the $(s, p, \varkappa)$-β-quarks according to Definition 7.20. Let $\varrho > r$ in (7.87), where r is given by (7.84). Then $f \in D'(\Omega)$ belongs to $F_{pq}^s(M, g^\varkappa)$ if, and only if, it can be represented as

$$f = \sum_{\beta \in \mathbb{N}_0^n} \sum_{j=0}^{\infty} \sum_{l=1}^{L_j} \lambda_{jl}^\beta (\beta qu)_{jl}(x) \tag{7.98}$$

with

$$\|\lambda \,|\, f_{pq}^\Omega\|_\varrho < \infty \tag{7.99}$$

(absolute and unconditional convergence in $L_1(\Omega, g^\tau)$ with $\tau = \varkappa - s + \frac{n}{p}$). Furthermore, the infimum in (7.99) over all admissible representations (7.98) is an equivalent quasi-norm in $F_{pq}^s(M, g^\varkappa)$.

Proof By Theorem 7.10 we have the equivalent quasi-norm (7.38). This can be compared with the equivalent quasi-norm (6.31) for the same parameters s, p, q. The corresponding β-quarks in (6.10) and (7.88) differ only by the extra factor $2^{j(\varkappa - s + \frac{n}{p})}$ in (7.38) (this is now the factor $2^{-k(\varkappa - s + \frac{n}{p})}$ in (7.88)). Otherwise one can take over the proof of Theorem 6.6. This applies also to the L_1-convergence, now with respect to $g^\tau f$ in place of f.

7.23 Optimal coefficients, other cases

Since the proof of the above theorem is the same as the proof of Theorem 6.6 one can carry over also the considerations from 6.7. In particular, in the understanding explained there, one can construct optimal coefficients $\lambda_{jl}^\beta = \lambda_{jl}^\beta(f)$ which depend linearly on f, in analogy to (6.35), (6.36).

Secondly, we recall the Riemannian lifting (7.57). Let

$$0 < p \leq \infty, \quad 0 < q \leq \infty, \quad s \in \mathbb{R}, \tag{7.100}$$

(with $q = \infty$ if $p = \infty$). Let $L \in \mathbb{N}$ such that $s + 2L > \sigma_{pq}$. Then one can apply Theorem 7.22 to

$$(-\Delta_g + \mu \,id)^{-L} f \in F_{pq}^{s+2L}(M, g^\varkappa),$$

where $\mu > 0$ is sufficiently large, which results finally in a quarkonial decomposition of f by the β-quarks

$$(-\Delta_g + \mu \,id)^L (\beta qu)_{jl},$$

where $(\beta qu)_{jl}$ are $(s + 2L, p, \varkappa)$-β-quarks.

7.24 Final comment, outlook

The three theorems in 7.10, 7.15, 7.22, might be considered as the main results of Section 7. They are formulated for arbitrary bounded domains Ω in $\mathbb{R}^n$, converted into a Riemannian manifold M or (M,g). However this point of view is too general for our later intentions to develop a spectral theory for some elliptic operators on these manifolds. One needs some more information about the quality of Ω. For that later purpose we introduced in 7.5 and 7.6 the so-called d-domains. It turns out that with this specification one can estimate the entropy numbers of compact embeddings discussed in Proposition 7.17 which, in turn, paves the way to a spectral theory of some elliptic operators in d-domains. This will be done in Section 21.

8 Taylor expansions of distributions

8.1 Introduction

In (1.18) we uglified the classical Taylor expansion of holomorphic functions in a domain Ω in the complex plane $\mathbb{C}$, which in our context will be better denoted as $\mathbb{R}^2$. But it provided an understanding of what can be expected when asking for Taylor expansions of (complex-valued) functions and distributions in $\mathbb{R}^n$ and in domains. This resulted in the question (1.23) with (1.22) and in the considerations given afterwards in 1.4 and 1.5. We discussed this problem in Section 2 for functions f belonging to spaces $B_{pq}^s(\mathbb{R}^n)$ or $F_{pq}^s(\mathbb{R}^n)$, where the smoothness s is restricted by

$$s > \sigma_p = n\left(\frac{1}{p} - 1\right)_+ \quad \text{or} \quad s > \sigma_{pq} = n\left(\frac{1}{\min(p,q)} - 1\right)_+, \tag{8.1}$$

respectively. The outcome is perfect. With the β-quarks $(\beta qu)_{\nu m}$ from Definition 2.4 we have

$$f(x) = \sum_{\beta,\nu,m} \lambda_{\nu m}^\beta (\beta qu)_{\nu m}(x), \quad x \in \mathbb{R}^n, \tag{8.2}$$

where we used the notation introduced in (2.74), based on the fact that we have absolute and unconditional convergence in $L_{\bar{p}}(\mathbb{R}^n)$ with $\bar{p} = \max(1,p)$. We refer to Definition 2.6, Theorem 2.9, and the discussions in 2.7 and 2.13. In case of arbitrary smoothness $s \in \mathbb{R}$ the situation seems to be less favourable. Instead of (8.2) we have by (3.12),

$$f = \sum_{\beta,\nu,m} \left(\eta_{\nu m}^\beta (\beta qu)_{\nu m} + \lambda_{\nu m}^\beta (\beta qu)_{\nu m}^L\right), \tag{8.3}$$

where the β-quarks $(\beta qu)_{\nu m}$ are of the same type as before, covered by Definition 2.4, and the β-quarks $(\beta qu)_{\nu m}^L$, introduced in Definition 3.2, satisfy some moment conditions. We refer to 3.3. By the discussion in 3.5 the series (8.3) converges unconditionally in $S'(\mathbb{R}^n)$, which justifies the formulation. On the one hand it is clear that for function spaces with, say, $s < 0$, moment conditions cannot be avoided. But on the other hand, one can ask whether one really needs all the η-terms in (8.3). This is not the case. The η-terms with $\nu = 0$ are sufficient and one arrives at a formulation comparable with (8.2). We discussed this situation in 3.9 with the outcome that the less elegant version (8.3) should be given preference in connection with applications and generalizations. However our aim in this section is different. We ask for Taylor expansions for distributions as explained above, comparable with (8.2). We concentrate on two cases, $\mathbb{R}^n$ and bounded C^∞ domains in $\mathbb{R}^n$. Function spaces serve now only as vehicles and will be simplified as much as possible. Otherwise we rely on Sections 2, 3 and 5, 6 as far as $\mathbb{R}^n$ and bounded C^∞ domains are concerned, respectively. We follow at least partly [Tri00a]. As far as $\mathbb{R}^n$ is concerned a few first remarks about Taylor expansions of tempered distributions may also be found in [Triδ], 14.13, pp. 100–101.

8.2 Weighted spaces and tempered distributions in $\mathbb{R}^n$

First we deal with distributions in $\mathbb{R}^n$. We use again the notation introduced in 2.1. In particular, $S(\mathbb{R}^n)$ is the Schwartz space, and $S'(\mathbb{R}^n)$ is the space of tempered distributions on $\mathbb{R}^n$, respectively. Let

$$1 \leq p \leq \infty, \quad s \in \mathbb{R}, \quad \alpha \in \mathbb{R}, \quad \langle x \rangle = (1+|x|^2)^{\frac{1}{2}} \quad \text{if} \quad x \in \mathbb{R}^n. \tag{8.4}$$

We put for brevity

$$B_p^s(\mathbb{R}^n) = B_{pp}^s(\mathbb{R}^n) \tag{8.5}$$

and introduce the weighted spaces

$$B_p^s(\mathbb{R}^n, \langle x \rangle^\alpha) = \{f \in S'(\mathbb{R}^n) : \langle x \rangle^\alpha f \in B_p^s(\mathbb{R}^n)\}, \tag{8.6}$$

obviously normed by

$$\|f \,|\, B_p^s(\mathbb{R}^n, \langle x \rangle^\alpha)\| = \|\langle x \rangle^\alpha f \,|\, B_p^s(\mathbb{R}^n)\|. \tag{8.7}$$

As for the Besov spaces in (8.5) we refer to Theorem 3.6. In particular by (3.23) and the B-counterpart of (7.19), (7.20),

$$\left(\sum_{j=0}^{\infty} 2^{jsp} \left\| (\varphi_j \widehat{f})^\vee(x) \langle x \rangle^\alpha \,|\, L_p(\mathbb{R}^n) \right\|^p \right)^{\frac{1}{p}} \tag{8.8}$$

is an equivalent norm in $B_p^s(\mathbb{R}^n, \langle x \rangle^\alpha)$. Details and further references may be found in [HaT94a], [HaT94b] and [ET96], Chapter 4. Hence we have the remarkable possibility to use likewise (8.6), (8.7) or (8.8). In particular, for fixed p, these spaces are monotonically ordered,

$$B_p^{s_1}(\mathbb{R}^n, \langle x \rangle^{\alpha_1}) \subset B_p^{s_2}(\mathbb{R}^n, \langle x \rangle^{\alpha_2}) \tag{8.9}$$

if

$$\text{either} \quad s_1 = s_2\,,\ \alpha_1 \geq \alpha_2 \quad \text{or} \quad s_1 \geq s_2\,,\ \alpha_1 = \alpha_2\,. \tag{8.10}$$

In addition we have the well-known embeddings between spaces of this type with different values of p. Recall that the topology of the locally convex spaces $S(\mathbb{R}^n)$ may be given by

$$\|\varphi\,|S(\mathbb{R}^n)\|_k = \sum_{|\gamma| \leq k} \sup_{x \in \mathbb{R}^n} \langle x \rangle^k |D^\gamma \varphi(x)|\,, \quad \varphi \in S(\mathbb{R}^n)\,, \tag{8.11}$$

where $k \in \mathbb{N}_0$. We have

$$\|\varphi\,|S(\mathbb{R}^n)\|_k \sim \sup_{x \in \mathbb{R}^n} \sum_{|\gamma| \leq k} \left| D^\gamma \langle x \rangle^k \varphi(x) \right|\,, \quad \varphi \in S(\mathbb{R}^n)\,. \tag{8.12}$$

By the above remarks and well-known embedding theorems for the spaces $B_p^s(\mathbb{R}^n)$, [Triβ], 2.7.1, p. 129, we obtain for any $k \in \mathbb{N}_0$,

$$\|\varphi\,|S(\mathbb{R}^n)\|_k \leq c_1 \,\|\varphi\,|B_p^s(\mathbb{R}^n, \langle x \rangle^k)\| \leq c_2 \,\|\varphi\,|S(\mathbb{R}^n)\|_l \tag{8.13}$$

with $k + \frac{n}{p} < s < l$, where by (8.8) the right-hand side is a consequence of

$$\begin{aligned}
&\|\varphi\,|B_p^s(\mathbb{R}^n, \langle x \rangle^k)\| \\
&\leq c \sup_{j \in \mathbb{N}_0} \sup_{x \in \mathbb{R}^n} 2^{j(s+\varepsilon)} |(\varphi_j \widehat{\varphi})^\vee(x)|\, \langle x \rangle^l \left(\sum_{m=0}^\infty 2^{-\varepsilon m p} \left\| \langle y \rangle^{-(l-k)}\,|L_p(\mathbb{R}^n) \right\|^p \right)^{\frac{1}{p}} \\
&\leq c' \,\|\langle x \rangle^l \varphi\,|B_\infty^{s+\varepsilon}(\mathbb{R}^n)\| \\
&\leq c'' \,\|\varphi\,|S(\mathbb{R}^n)\|_l\,, \tag{8.14}
\end{aligned}$$

(choosing $s < s + \varepsilon < l$) Now it follows easily by (8.9), (8.10) that for any fixed p with $1 \leq p \leq \infty$,

$$S(\mathbb{R}^n) = \bigcap_{s > 0} B_p^s(\mathbb{R}^n, \langle x \rangle^s)\,. \tag{8.15}$$

We denote the completion of $S(\mathbb{R}^n)$ in $B_p^s(\mathbb{R}^n, \langle x \rangle^s)$ by $\overset{\circ}{B}{}_p^s(\mathbb{R}^n, \langle x \rangle^s)$. Recall that

$$\overset{\circ}{B}{}_p^s(\mathbb{R}^n, \langle x \rangle^s) = B_p^s(\mathbb{R}^n, \langle x \rangle^s) \quad \text{if} \quad 1 \le p < \infty. \tag{8.16}$$

Hence, only $p = \infty$ is an exceptional case. By the usual dual pairing in $(S(\mathbb{R}^n), S'(\mathbb{R}^n))$ and by [Triβ], 2.11.2, formula (12) on p. 180, we get

$$\left(\overset{\circ}{B}{}_p^s(\mathbb{R}^n, \langle x \rangle^s) \right)' = B_{p'}^{-s}(\mathbb{R}^n, \langle x \rangle^{-s}), \tag{8.17}$$

where

$$\frac{1}{p} + \frac{1}{p'} = 1, \quad 1 \le p \le \infty, \quad s > 0. \tag{8.18}$$

Hence, for fixed p with $1 \le p \le \infty$,

$$S'(\mathbb{R}^n) = \bigcup_{s<0} B_p^s(\mathbb{R}^n, \langle x \rangle^s). \tag{8.19}$$

This reduces the question about Taylor expansions for tempered distributions to corresponding Taylor expansions for distributions belonging to the spaces on the right-hand side of (8.19). First we adapt the β-quarks introduced in Definition 3.2 not only to the weighted spaces considered now but also to the indicated possibility to reduce the η-sum in (3.12) to the terms with $\nu = 0$. Recall that $\langle m \rangle = (1+|m|^2)^{\frac{1}{2}}$, where $m \in \mathbb{Z}^n$. Again let $Q_{\nu m}$ be the cubes introduced at the beginning of 2.2.

8.3 Definition

Let ψ be a non-negative C^∞ function in $\mathbb{R}^n$ with

$$\operatorname{supp} \psi \subset \{ y \in \mathbb{R}^n : \ |y| < 2^r \} \tag{8.20}$$

for some $r \ge 0$ and

$$\sum_{m \in \mathbb{Z}^n} \psi(x-m) = 1 \quad \text{if} \quad x \in \mathbb{R}^n. \tag{8.21}$$

Let $\psi^\beta(x) = x^\beta \psi(x)$ where the x^β are the monomials (2.3). Let

$$s \in \mathbb{R}, \quad 0 < p \le \infty, \quad \beta \in \mathbb{N}_0^n, \quad \frac{L+1}{2} \in \mathbb{N}_0, \quad \text{and} \quad \alpha \in \mathbb{R}. \tag{8.22}$$

8. Taylor expansions of distributions

Then, for $m \in \mathbb{Z}^n$,

$$(\beta qu)_{\nu m}^{*L}(x) = 2^{-\nu(s-\frac{n}{p})} \langle 2^{-\nu}m \rangle^{-\alpha} \left((-\Delta)^{\frac{L+1}{2}} \psi^\beta \right)(2^\nu x - m), \quad x \in \mathbb{R}^n, \tag{8.23}$$

if $\nu \in \mathbb{N}$ and

$$(\beta qu)_{\nu m}^{*L}(x) = \langle m \rangle^{-\alpha} \psi^\beta(x-m), \quad x \in \mathbb{R}^n, \tag{8.24}$$

if $\nu = 0$, are called $(s,p,\alpha)^{*L} - \beta$-quarks related to $Q_{\nu m}$.

8.4 Discussion

We extended the Definitions 2.4 and 3.2 to the weighted case. We have again

$$\operatorname{supp}(\beta qu)_{\nu m}^{*L} \subset dQ_{\nu m} \quad \text{where} \quad \nu \in \mathbb{N}_0 \quad \text{and} \quad m \in \mathbb{Z}^n \tag{8.25}$$

for some $d > 0$ and

$$\left| (\beta qu)_{\nu m}^{*L}(x) \right| \le c_L \langle 2^{-\nu}m \rangle^{-\alpha} 2^{-\nu(s-\frac{n}{p})+r|\beta|} \quad \text{with} \quad \beta \in \mathbb{N}_0^n, \tag{8.26}$$

where c_L depends on L, but not on the other parameters involved, ν, m, β, α. This is the counterpart of (2.19) and (3.9). Of course, $\langle 2^{-\nu}m \rangle^{-\alpha}$ compensates the weight $\langle x \rangle^\alpha$ at $2^{-\nu}m$. Again we have the moment conditions (3.7) which justify how L appears in (8.23). If $L = -1$ in (8.23) and (8.24), then we have no moment conditions and we put, in analogy to 3.3,

$$(\beta qu)_{\nu m}^* = (\beta qu)_{\nu m}^{*-1} \quad \text{where} \quad \nu \in \mathbb{N}_0 \quad \text{and} \quad m \in \mathbb{Z}^n, \tag{8.27}$$

and call them $(s,p,\alpha)^*$-β-quarks. We have $c_{-1} = 1$ in (8.26). As for the weights, the construction of the β-quarks in (8.27) is very similar to that in Definition 7.20. But now we have (8.21) and the sequence $2^{-\nu}\mathbb{Z}^n$ with $\nu \in \mathbb{N}_0$ of regular lattices instead of the families of approximate resolutions of unity needed there. The * in the above definition distinguishes notationally the weighted β-quarks above from the weighted β-quarks in Definition 7.20. But there is a second reason. We have now the splitting $\nu \in \mathbb{N}$ in (8.23) and $\nu = 0$ in (8.24). This is new compared with other β-quarks introduced previously and the star * indicates this special construction. Finally we mention that the above β-quarks are normalized building blocks in $B_p^s(\mathbb{R}^n, \langle x \rangle^\alpha)$. Let

$$1 \le p \le \infty, \quad s > 0, \quad \alpha \in \mathbb{R}. \tag{8.28}$$

For any $\beta \in \mathbb{N}_0^n$ there are two positive constants c_1^β and c_2^β such that

$$c_1^\beta \le \left\| (\beta qu)_{\nu m}^* \,|\, B_p^s(\mathbb{R}^n, \langle x \rangle^\alpha) \right\| \le c_2^\beta \tag{8.29}$$

for all $\nu \in \mathbb{N}_0$ and $m \in \mathbb{Z}^n$. Let $\alpha = 0$. We apply (5.147) with $B_p^s(\mathbb{R}^n)$ in place of $\widetilde{F}_{pq}^s(B_1)$ and $\widetilde{F}_{pq}^s(B_\lambda)$ and $\lambda = 2^{-\nu}$ to $f(x) = \psi^\beta(2^\nu x - m)$ and obtain

$$\|(\beta qu)^*_{\nu m} | B_p^s(\mathbb{R}^n)\| \sim \|\psi^\beta | B_p^s(\mathbb{R}^n)\|. \qquad (8.30)$$

The case $\alpha \neq 0$ can be reduced to $\alpha = 0$ by pointwise multiplier assertions (a much stronger version than needed here may be found in 5.17). Hence for all p, s, α with (8.28) we have

$$\|(\beta qu)^*_{\nu m} | B_p^s(\mathbb{R}^n, \langle x \rangle^\alpha)\| \sim \|\psi^\beta | B_p^s(\mathbb{R}^n)\|, \qquad (8.31)$$

and consequently (8.29). By (8.20) the constant c_2^β in (8.29) can be estimated from above by

$$c_2^\beta \leq c_2 \, 2^{r|\beta|}, \quad \beta \in \mathbb{N}_0^n. \qquad (8.32)$$

In particular,

$$2^{-r|\beta|} (\beta qu)^*_{\nu m}(x), \quad \nu \in \mathbb{N}_0, m \in \mathbb{Z}^n, \beta \in \mathbb{N}_0^n, \qquad (8.33)$$

are normalized building blocks (uniform estimates from above). As for the sequence spaces used in Definitions 2.6 and 3.4 we have now the simple situation described in (2.9). Instead of b_{pp} or f_{pp} we write now b_p. We repeat briefly what we need, following the notation used in Definition 2.6. Let

$$0 < p \leq \infty \quad \text{and} \quad \varrho > r, \qquad (8.34)$$

where r is given by (8.20). Let

$$\lambda = \{\lambda^\beta : \beta \in \mathbb{N}_0^n\} \quad \text{with} \quad \lambda^\beta = \{\lambda^\beta_{\nu m} \in \mathbb{C} : \nu \in \mathbb{N}_0, \, m \in \mathbb{Z}^n\}. \qquad (8.35)$$

As in (2.9) and (2.23) we put

$$\|\lambda | b_p\|_\varrho = \sup_{\beta \in \mathbb{N}_0^n} 2^{\varrho|\beta|} \left(\sum_{\nu=0}^\infty \sum_{m \in \mathbb{Z}^n} |\lambda^\beta_{\nu m}|^p \right)^{\frac{1}{p}} \qquad (8.36)$$

(with the obvious modification if $p = \infty$). Furthermore, as in (8.2) we use the abbreviation (2.74) indicating always unconditional (sometimes absolute) convergence. Obviously, $L_p(\mathbb{R}^n, \langle x \rangle^\alpha)$ with $1 \leq p \leq \infty$ is given by (8.6), (8.7) with L_p in place of B_p^s.

8.5 Proposition

Let

$$1 \le p \le \infty, \quad s > 0, \quad \alpha \in \mathbb{R}. \tag{8.37}$$

Let $(\beta qu)^*_{\nu m}$ be $(s,p,\alpha)^* - \beta$-quarks according to Definition 8.3 and (8.27) with respect to a fixed function ψ. Let $\varrho > r$ where r has the same meaning as in (8.20). Then $B^s_p(\mathbb{R}^n, \langle x \rangle^\alpha)$ is the collection of all $f \in S'(\mathbb{R}^n)$ which can be represented as

$$f = \sum_{\beta,\nu,m} \lambda^\beta_{\nu m} (\beta qu)^*_{\nu m}(x) \tag{8.38}$$

with

$$\|\lambda \,|b_p\|_\varrho < \infty \tag{8.39}$$

(absolute and unconditional convergence in $L_p(\mathbb{R}^n, \langle x \rangle^\alpha)$). Furthermore, the infimum in (8.39) over all admissible representations (8.38) is an equivalent norm in $B^s_p(\mathbb{R}^n, \langle x \rangle^\alpha)$.

Proof The unweighted case $\alpha = 0$ is covered by Definition 2.6 and Theorem 2.9. The proof given there concentrates on the F-spaces. The situation for the spaces $B^s_p(\mathbb{R}^n)$ is much simpler and may be found in detail in [Triδ], 14.15, pp. 101–104. It is based on the theory of the (unweighted) spaces

$$\{f \in L_p(\mathbb{R}^n) : \operatorname{supp} \widehat{f} \subset B_{2^j}\}, \quad j \in \mathbb{N}_0, \tag{8.40}$$

where B_{2^j} is the ball centred at the origin and of radius 2^j. This theory can be extended to a large class of weighted L_p-spaces, in particular to spaces (8.40) with $L_p(\mathbb{R}^n, \langle x \rangle^\alpha)$ in place of $L_p(\mathbb{R}^n)$. We refer to [ST87], in particular 1.4 and 1.5. Otherwise one can follow the arguments in [Triδ], 14.15, which results in a proof of the proposition, including the absolute convergence of (8.38) in $L_p(\mathbb{R}^n, \langle x \rangle^\alpha)$.

8.6 Optimal coefficients, alternative proof

The assertions from 2.11 and Corollary 2.12 have appropriate counterparts. In particular there are optimal coefficients $\lambda^\beta_{\nu m} = \lambda^\beta_{\nu m}(f)$ in (8.38) in the understanding of (2.75) which depend linearly on f, in analogy to (2.82).

The above proof of Proposition 8.5 is based on the weighted counterpart of corresponding arguments in the unweighted case [Triδ], 14.15. Another possibility is to expand $\langle x \rangle^\alpha f \in B_p^s(\mathbb{R}^n)$ according to (2.24). Then one gets

$$f = \sum_{\beta,\nu,m} \zeta_{\nu m}^\beta \langle 2^{-\nu} m \rangle^{-\alpha} \left(\frac{\langle 2^{-\nu} m \rangle}{\langle x \rangle} \right)^\alpha (\beta qu)_{\nu m}(x), \qquad (8.41)$$

where $(\beta qu)_{\nu m}$ are $(s,p) - \beta$-quarks as in Definition 2.6. Now one can expand the analytic function $\langle 2^{-\nu} m \rangle \langle x \rangle^{-1}$ at $2^{-\nu} m$ and insert the resulting powers $(x - 2^{-\nu} m)^\gamma$ in the β-quarks. We do not go into detail. The direct way via the weighted counterparts of (8.40) seems to be more promising. In particular it can be applied also to other weights than $\langle x \rangle^\alpha$. Some references may be found in [Triδ], 14.14, p. 101.

8.7 Theorem

Let $1 \le p \le \infty$, the function ψ with (8.20), (8.21) and $\varrho > r$ be given.

(i)   For any $f \in S'(\mathbb{R}^n)$ there are numbers

$$s \in \mathbb{R}, \quad \alpha \in \mathbb{R}, \quad \text{and} \quad L \ge \max(-1, [-s]) \quad \text{with} \quad \frac{L+1}{2} \in \mathbb{N}_0, \qquad (8.42)$$

such that f can be expanded by

$$f = \sum_{\beta,\nu,m} \lambda_{\nu m}^\beta (\beta qu)_{\nu m}^{*L} \qquad (8.43)$$

(unconditional convergence in $S'(\mathbb{R}^n)$), where $(\beta qu)_{\nu m}^{*L}$ are $(s,p,\alpha)^{*L} - \beta$-quarks according to Definition 8.3 and

$$\|\lambda \,|b_p\|_\varrho < \infty \qquad (8.44)$$

according to (8.36).

(ii)   Let the numbers s, α, L and the sequence λ be given by (8.42) and (8.44), respectively. Then the right-hand side of (8.43) converges unconditionally in $S'(\mathbb{R}^n)$ and represents an element $f \in S'(\mathbb{R}^n)$.

Proof   *Step 1*   Let $f \in S'(\mathbb{R}^n)$. By (8.19) we may assume

$$f \in B_p^s(\mathbb{R}^n, \langle x \rangle^s) \qquad (8.45)$$

for some $s < 0$. Let $\eta \in S(\mathbb{R}^n)$ with

$$\eta(x) = 1 \quad \text{if} \quad |x| \le 2^{-K} \quad \text{and} \quad \eta(x) = 0 \quad \text{if} \quad |x| \ge \frac{3}{2} 2^{-K}, \qquad (8.46)$$

where $K \subset \mathbb{N}_0$ will be chosen later on. (If $K = 0$ then we have the function φ from (2.33)). Let

$$f = (\eta \widehat{f})^\vee + ((1-\eta)\widehat{f})^\vee = f_0 + f_1 \,. \tag{8.47}$$

We choose L with $\frac{L+1}{2} \in \mathbb{N}$ such that

$$\sigma = s + L + 1 > 0 \,. \tag{8.48}$$

In particular $L \geq \max(-1, [-s])$. Let

$$f_1 = \Delta^{\frac{L+1}{2}} g \quad \text{with} \quad g = (\Delta)^{-\frac{L+1}{2}} f_1 = c \left((1-\eta)(\xi) |\xi|^{-L-1} \widehat{f} \right)^\vee, \tag{8.49}$$

where $c = (-1)^{\frac{L+1}{2}}$. By the lifting properties in weighted B_p^s-spaces described in [ET96], Proposition 4.2.2, p. 158, it follows that

$$g \in B_p^\sigma(\mathbb{R}^n, \langle x \rangle^s) \tag{8.50}$$

and

$$\begin{aligned} \|g \,|\, B_p^\sigma(\mathbb{R}^n, \langle x \rangle^s)\| &\sim \|f_1 \,|\, B_p^s(\mathbb{R}^n, \langle x \rangle^s)\| \\ &\leq c \|f \,|\, B_p^s(\mathbb{R}^n, \langle x \rangle^s)\|, \end{aligned} \tag{8.51}$$

where c is independent of f. We apply Proposition 8.5 to g with σ and s in place of s and α, respectively. Then we have (8.38) with g in place of f, where we may assume that the summation over ν begins with $\nu = 1$ (instead of $\nu = 0$). This is also explicitly covered by (2.66). Hence,

$$g = \sum_{\beta \in \mathbb{N}_0^n} \sum_{\nu=1}^\infty \sum_{m \in \mathbb{Z}^n} \lambda_{\nu m}^\beta (\beta qu)_{\nu m}^*(x) \tag{8.52}$$

with (8.39) or (8.44), where the sum over ν in (8.36) is restricted now to $\nu \geq 1$, and the $(\sigma, p, s)^* - \beta$-quarks

$$(\beta qu)_{\nu m}^*(x) = 2^{-\nu(\sigma - \frac{n}{p})} \langle 2^{-\nu} m \rangle^{-s} \psi^\beta(2^\nu x - m), \quad x \in \mathbb{R}^n, \tag{8.53}$$

according to (8.23), (8.27). Recall that (8.52) converges absolutely and unconditionally in $L_p(\mathbb{R}^n, \langle x \rangle^s)$, and hence unconditionally in $S'(\mathbb{R}^n)$. We apply (8.49) and obtain

$$f_1 = \sum_{\beta \in \mathbb{N}_0^n} \sum_{\nu=1}^\infty \sum_{m \in \mathbb{Z}^n} \lambda_{\nu m}^\beta (\beta qu)_{\nu m}^{*L} \tag{8.54}$$

(unconditional convergence in $S'(\mathbb{R}^n)$), where $(\beta qu)_{\nu m}^{*L}$ are $(s, p, s)^{*L}$-β-quarks.

Step 2 We deal with f_0 in (8.47). Recall that f_0 is an entire analytic function. Hence it is quite clear how to proceed. By (8.21) we have

$$f_0(x) = \sum_{m \in \mathbb{Z}^n} \psi(x - m) f_0(x) \tag{8.55}$$

and we expand $f_0(x)$ at m. Then we get the desired term $\psi^\beta(x-m)$ in (8.24) with $\nu = 0$. But we must check whether the resulting coefficients fit in our scheme, in particular whether we have

$$\sup_{\beta \in \mathbb{N}_0^n} 2^{\varrho|\beta|} \left(\sum_{m \in \mathbb{Z}^n} |\lambda_{0m}^\beta|^p \right)^{\frac{1}{p}} < \infty, \tag{8.56}$$

as the relevant contribution to (8.44). We may assume $\varrho = K \in \mathbb{N}$ and now use (8.46). We follow the arguments from Step 2 of the proof of Theorem 2.9. We have obvious counterparts of (2.53)–(2.56) with f_0 on the left-hand of (2.53) and with $k = -K$. Using temporarily the same notation as there we have the simplified counterpart of (2.57),

$$\left(\sum_{m \in \mathbb{Z}^n} |\Lambda_{-K,m}|^p \langle m \rangle^s \right)^{\frac{1}{p}} \leq c \| f_0 | L_p(\mathbb{R}^n, \langle x \rangle^s) \|$$
$$\leq c \| f | B_p^s(\mathbb{R}^n, \langle x \rangle^s) \|. \tag{8.57}$$

We refer again to [ST87], 1.4, 1.5, especially Proposition 1.4.4 and Remark 1.5.2/1 on pp. 29 and 37, respectively. We choose also $k = -K$ in (2.58) and have the counterpart of (2.59) with f_0 on the left-hand side and $k = -K$ on the right-hand side. We use (2.63) with $k = -K$ and $\varrho = K$. Then we get the counterpart of (2.66) with f_0 on the left-hand side and the (unweighted) β-quarks

$$(\beta qu)_{0,m}(x) = \psi^\beta(x - m) \tag{8.58}$$

we wish to have. We incorporate the weights $\langle m \rangle^{-s}$ and obtain (8.24). Then the arguments after (2.66), now based on (8.57), give the desired result.

Step 3 Part (ii) follows from Proposition 8.5, Definition 3.4 and the discussion in 3.5. In particular, (3.20) coincides now with (8.48) and $L \geq [-s]$.

8.8 Remark

We relied on (8.45), hence $\alpha = s$. This simplifies (8.23). But this does not matter very much. Otherwise the above proof is a little bit curious. The main part in the splitting (8.47) is f_1. The surprisingly complicated arguments in Step 2 with respect to f_0 arose from the attempt to have (8.44) for arbitrary $\varrho > r$. In any case the singularity behaviour of f is described by (8.54), modulo entire analytic functions. Maybe the cases $p = 1$, $p = 2$ and $p = \infty$ are of special interest. We describe $p = \infty$.

8.9 Corollary

(Taylor expansions of tempered distributions in $\mathbb{R}^n$)

Let ψ be a non-negative C^∞ function in $\mathbb{R}^n$ with (8.20), (8.21), and again let $\psi^\beta(x) = x^\beta \psi(x)$ where $\beta \in \mathbb{N}_0^n$. Let $\varrho > r$. For any tempered distribution $f \in S'(\mathbb{R}^n)$ there are a natural number K and a positive number c (depending on ϱ) such that

$$f = \sum_{\beta \in \mathbb{N}_0^n} \sum_{m \in \mathbb{Z}^n} \lambda_{0m}^\beta \psi^\beta(x - m) \tag{8.59}$$

$$+ \sum_{\beta \in \mathbb{N}_0^n} \sum_{\nu=1}^{\infty} \sum_{m \in \mathbb{Z}^n} \lambda_{\nu m}^\beta \left(\Delta^K \psi^\beta\right)(2^\nu x - m), \quad x \in \mathbb{R}^n,$$

(unconditional convergence in $S'(\mathbb{R}^n)$) with $\lambda_{\nu m}^\beta \in \mathbb{C}$ and

$$2^{\varrho|\beta|} \left(2^\nu + |m|\right)^{-K} |\lambda_{\nu m}^\beta| \le c \quad \text{for all} \quad \beta \in \mathbb{N}_0^n, \, \nu \in \mathbb{N}_0, \, m \in \mathbb{Z}^n. \tag{8.60}$$

Proof We may choose in Theorem 8.7 and in (8.23)

$$s = \alpha = -K, \quad p = \infty, \quad \text{and} \quad L + 1 = 2K, \tag{8.61}$$

for some $K \in \mathbb{N}$. Then we have (8.48). We used (8.44), (8.36) with $p = \infty$ and incorporated the factors from (8.23) in the coefficients.

8.10 Discussion

The above corollary comes near to our heuristical discussion in 1.3. The exponential decay of the coefficients $\lambda_{\nu m}^\beta$ expressed by the factor $2^{\varrho|\beta|}$ has by Cauchy's formula a more or less obvious counterpart for holomorphic functions represented by (1.18). The constant c in (8.60) depends on ϱ. A related discussion may be found in 2.13. Let $p = \infty$ and $s = \alpha > 0$. Then, by Proposition 8.5, any

$$f \in B_\infty^s(\mathbb{R}^n, \langle x \rangle^s) = C^s(\mathbb{R}^n, \langle x \rangle^s) \tag{8.62}$$

can be represented by

$$f(x) = \sum_{\beta \in \mathbb{N}_0^n} \sum_{\nu=0}^{\infty} \sum_{m \in \mathbb{Z}^n} \lambda_{\nu m}^{\beta} \psi^{\beta}(2^{\nu} x - m) \qquad (8.63)$$

with, for some $c > 0$,

$$2^{\varrho|\beta|} (2^{\nu} + |m|)^s |\lambda_{\nu m}^{\beta}| \le c \quad \text{for all} \quad \beta \in \mathbb{N}_0^n, \; \nu \in \mathbb{N}_0, \; m \in \mathbb{Z}^n. \qquad (8.64)$$

Again $\varrho > r$ and c depends on ϱ. This coincides with (1.23) and is the best that can be expected. If the weight $\langle x \rangle^s$ is replaced by $\langle x \rangle^\alpha$ with $\alpha \in \mathbb{R}$, then one has to correct only the coefficients in (8.63), (8.64). If $f \in S'(\mathbb{R}^n)$ then moment conditions cannot be avoided and again (8.59), (8.60) seems to be the simplest possible case.

8.11 A remark on optimal coefficients

In 8.6 we added a remark on optimal coefficients in connection with the spaces covered by Proposition 8.5. The proof of Theorem 8.7 and, hence, of Corollary 8.9, is reduced by lifting to Proposition 8.5. Hence, there are optimal coefficients in (8.43) and (8.59) which depend linearly on f. However this applies only to those f for which we have the a priori information (8.45) for some s, which influences the calculations, and not uniformly to all tempered distributions.

8.12 Tempered distributions in domains

Let Ω be a bounded C^∞ domain in $\mathbb{R}^n$. As above, $D(\Omega)$ is the collection of all complex-valued C^∞ functions in Ω with compact support in Ω. Its dual $D'(\Omega)$ is the collection of all complex-valued distributions in Ω. Let $S(\Omega)$ be the collection of all $\varphi \in C^\infty(\overline{\Omega})$ such that for all multi-indices $\gamma \in \mathbb{N}_0^n$,

$$D^\gamma \varphi(y) = 0 \quad \text{if} \quad y \in \partial\Omega. \qquad (8.65)$$

With obvious interpretation $S(\Omega)$ can be identified with the subspace

$$\{ \varphi \in S(\mathbb{R}^n) : \; supp\, \varphi \subset \overline{\Omega} \} \qquad (8.66)$$

of $S(\mathbb{R}^n)$, furnished with the norms

$$\|\varphi | S(\Omega)\|_k = \sum_{|\gamma| \le k} \sup_{x \in \Omega} |D^\gamma \varphi(x)|, \quad k \in \mathbb{N}_0. \qquad (8.67)$$

8. Taylor expansions of distributions

One gets a complete locally convex space. Its strong topological dual is denoted by $S'(\Omega)$ and called the space of *tempered distributions* in Ω. With the usual interpretation we have the dense embeddings

$$D(\Omega) \subset S(\Omega) \subset S'(\Omega) \subset D'(\Omega). \tag{8.68}$$

The situation is similar to that in 8.2. In particular we are interested in the counterparts of (8.13), (8.15) and (8.19). Let $1 \leq p \leq \infty$ and $s \in \mathbb{R}$. As in (8.5) we put for brevity

$$B_p^s(\mathbb{R}^n) = B_{pp}^s(\mathbb{R}^n). \tag{8.69}$$

We denote the completion of $S(\mathbb{R}^n)$ in $B_p^s(\mathbb{R}^n)$ by $b_p^s(\mathbb{R}^n)$. We have

$$b_p^s(\mathbb{R}^n) = B_p^s(\mathbb{R}^n) \quad \text{if} \quad 1 \leq p < \infty, \quad s \in \mathbb{R}, \tag{8.70}$$

whereas $b_\infty^s(\mathbb{R}^n)$ does not coincide with $B_\infty^s(\mathbb{R}^n) = \mathcal{C}^s(\mathbb{R}^n)$. This modifies our notation after (8.15) and in (8.16) slightly, since we now wish to reserve $^\circ$ for corresponding spaces on the domain Ω. As for spaces on Ω we use the notation introduced in Definition 5.3, adapted to our rather special situation. Let again $1 \leq p \leq \infty$ and $s \in \mathbb{R}$. Then $B_p^s(\Omega)$ and $b_p^s(\Omega)$ is the restriction of $B_p^s(\mathbb{R}^n)$ and $b_p^s(\mathbb{R}^n)$ to Ω, respectively, normed according to Definition 5.3(i). As in Definition 5.3(ii) the completion of $D(\Omega)$ in $B_p^s(\Omega)$ is denoted by $\overset{\circ}{B}{}_p^s(\Omega)$. Let

$$1 \leq p \leq \infty, \quad s > 0. \tag{8.71}$$

Then

$$\widetilde{B}_p^s(\Omega) = \{ f \in B_p^s(\mathbb{R}^n) : \, supp\, f \subset \overline{\Omega} \, \} \tag{8.72}$$

and

$$\widetilde{b}_p^s(\Omega) = \{ f \in b_p^s(\mathbb{R}^n) : \, supp\, f \subset \overline{\Omega} \, \}, \tag{8.73}$$

according to Definition 5.3(iii) and Proposition 5.5. If $p < \infty$ then we have (8.70), and (8.73) with B in place of b gives the same spaces. Hence, also the spaces $\widetilde{B}_p^s(\Omega)$ and $\widetilde{b}_p^s(\Omega)$ coincide. We discussed in 5.21, 5.22, and 5.23 the relation between $\overset{\circ}{B}{}_p^s(\Omega)$ and $b_p^s(\Omega)$. Complementing the assertions described there by $p = 1$ and $p = \infty$, we get under the restriction (8.71),

$$\overset{\circ}{B}{}_p^s(\Omega) = \widetilde{b}_p^s(\Omega) \quad \text{if, and only if,} \quad s - \frac{1}{p} \notin \mathbb{N}_0. \tag{8.74}$$

As for the limiting cases $p = 1$ and $p = \infty$ we refer to [Tri99a], 2.4.2–2.4.4, pp. 703–705. Again by [Triβ], 2.7.1, p. 129, we have the counterpart of (8.13),

$$\|\varphi \,|S(\Omega)\|_k \leq c_1 \,\|\varphi \,|B_p^s(\Omega)\| \leq c_2 \,\|\varphi \,|S(\Omega)\|_l, \quad \varphi \in S(\Omega), \qquad (8.75)$$

with $1 \leq p \leq \infty$ and $k + \frac{n}{p} < s < l$. Of course, the positive constants c_1 and c_2 are independent of $\varphi \in S(\Omega)$. In particular, $S(\Omega)$ is dense in $\overset{\circ}{B}{}_p^s(\Omega)$ and $\widetilde{b}_p^s(\Omega)$. Furthermore, for any fixed p with $1 \leq p \leq \infty$ we have set-theoretically and topologically

$$S(\Omega) = \bigcap_{s>0} \overset{\circ}{B}{}_p^s(\Omega) = \bigcap_{s>0} \widetilde{b}_p^s(\Omega). \qquad (8.76)$$

Since both $D(\Omega)$ and $S(\Omega)$ are dense in $\overset{\circ}{B}{}_p^s(\Omega)$ and $\widetilde{b}_p^s(\Omega)$ one can interpret the dual spaces $\left(\overset{\circ}{B}{}_p^s(\Omega)\right)'$ and $\left(\widetilde{b}_p^s(\Omega)\right)'$ in the context of the dual pairings

$$(D(\Omega), D'(\Omega)) \quad \text{and} \quad (S(\Omega), S'(\Omega)). \qquad (8.77)$$

If

$$1 \leq p \leq \infty, \quad s > 0, \quad \text{and} \quad \frac{1}{p} + \frac{1}{p'} = 1, \qquad (8.78)$$

then

$$\left(\widetilde{b}_p^s(\Omega)\right)' = B_{p'}^{-s}(\Omega) \qquad (8.79)$$

in the above interpretation. We refer to [Triα], 2.10.5, 4.8.1, as far as $1 < p < \infty$ is concerned, and, including the limiting cases $p = 1$ and $p = \infty$, to [Triβ], 2.11.2, [Mar87b], [FrJ90], [RuS96], 2.1.5, p. 20. By (8.76), and (8.79), (8.74) it follows that for given p with $1 \leq p \leq \infty$,

$$S'(\Omega) = \bigcup_{s<0} B_p^s(\Omega), \qquad (8.80)$$

set-theoretically and topologically. Of course, one can replace $s < 0$ in (8.80) by $s \in \mathbb{R}$. Now we are in a similar situation as in 8.2. Next we need the counterpart of Definition 8.3. For this purpose we have to incorporate $(-\Delta)^{\frac{L+1}{2}}$ in the β-quarks from 6.3.

8.13 Definition

Let Ω be a bounded C^∞ domain in $\mathbb{R}^n$. Let $\{\varphi_{\nu m}\}$ be a family of approximate resolutions of unity in Ω according to 6.2. Let

$$s \in \mathbb{R}, \quad 0 < p \leq \infty, \quad \beta \in \mathbb{N}_0^n, \quad \text{and} \quad \frac{L+1}{2} \in \mathbb{N}_0. \tag{8.81}$$

Then

$$(\beta qu)^L_{\nu m}(x) = 2^{-\nu(s-\frac{n}{p})} \, 2^{-\nu(L+1)+\nu|\beta|} \, (-\Delta)^{\frac{L+1}{2}} \left[(x - x^{\nu,m})^\beta \, \varphi_{\nu m}(x) \right] \tag{8.82}$$

with $\nu \in \mathbb{N}_0$ and $m = 1, \ldots, M_\nu$, is called an $(s,p)^L - \beta$-quark.

8.14 Discussion, notation

This definition combines Definitions 8.3 and 6.3. We have again the moment conditions (3.7) which explains how L comes in. If $L = -1$ then there are no moment conditions and (8.82) coincides with (6.10).

Furthermore, $2^{-\nu(L+1)+\nu|\beta|}$ are the correct scaling factors compared with the somewhat simpler construction in (8.23). In particular, we have a counterpart of (8.30) and also of (8.33) as normalized building blocks with respect to all admitted parameters ν, m, β, where r has the same meaning as in (6.11), (6.12). The counterpart of (8.36) and (6.18) is given by

$$\|\lambda \, |b_p^\Omega\|_\varrho = \sup_{\beta \in \mathbb{N}_0^n} 2^{\varrho|\beta|} \left(\sum_{\nu=0}^\infty \sum_{m=1}^{M_j} |\lambda^\beta_{\nu m}|^p \right)^{\frac{1}{\nu}} \tag{8.83}$$

(with the obvious modification if $p = \infty$), where λ and λ^β have the same meaning as in (6.16), (6.17).

8.15 Theorem

Let Ω be a bounded C^∞ domain in $\mathbb{R}^n$. There is a family of approximate resolutions of unity according to 6.2 with the following property. Let

$$1 \leq p \leq \infty \quad \text{and} \quad \varrho > r, \tag{8.84}$$

where r is given by (6.11).

(i)   For any $f \in S'(\Omega)$ there are numbers $s \in \mathbb{R}$ and L with $\frac{L+1}{2} \in \mathbb{N}_0$ and $L \geq [-s]$ such that f can be represented as

$$f = \sum_{\beta \in \mathbb{N}_0^n} \sum_{\nu=0}^{\infty} \sum_{m=1}^{M_\nu} \lambda_{\nu m}^\beta (\beta q u)_{\nu m}^L \qquad (8.85)$$

(unconditional convergence in $S'(\Omega)$) with

$$\|\lambda \,|b_p^\Omega\|_\varrho < \infty, \qquad (8.86)$$

where the $(s,p)^L - \beta$-quarks $(\beta q u)_{\nu m}^L$ and (8.86) are given by (8.82) and (8.83), respectively.

(ii)   Let the numbers s, L, and the sequence λ be as in part (i) with (8.86). Then the right-hand side of (8.85) converges unconditionally in $S'(\Omega)$ and represents an element $f \in S'(\Omega)$.

Proof   *Step 1*   We need some preparation. Let

$$1 \leq p \leq \infty, \quad k \in \mathbb{N}, \quad \text{and} \quad \sigma > \frac{1}{p} + k - 1. \qquad (8.87)$$

Then the iterated Laplacian Δ^k maps

$$\{f \in B_p^\sigma(\Omega) : \quad D^\gamma f | \partial\Omega = 0 \text{ if } |\gamma| \leq k-1\} \qquad (8.88)$$

isomorphically onto $B_p^{\sigma-2k}(\Omega)$. If $1 < p < \infty$ then this well-known assertion is essentially covered by [Triα], 5.7.1, especially Remark 1 on p. 402 (95 edition). The full assertion, including $p = 1$ and $p = \infty$ (and even values with $p < 1$) follows from [FrR95] and may also be found in [RuS96], Chapter 3. We refer also to 6.9, where some remarks in this direction have been given. If, in addition, $p < \infty$ and $\sigma < \frac{1}{p} + k$, then the space in (8.88) coincides with $\overset{\circ}{B}{}_p^\sigma(\Omega)$. We refer to [Triβ], Corollary 3.4.3, p. 210. We discussed problems of this type also in 5.21–5.24. In particular, under these circumstances $\overset{\circ}{B}{}_p^s(\Omega)$ coincides with $\widetilde{B}_p^s(\Omega)$. In this form the resulting mapping property can be extended to $p = \infty$, given by (8.72) with $p = \infty$. In other words: Let

$$1 \leq p \leq \infty, \quad k \in \mathbb{N}, \quad \text{and} \quad \frac{1}{p} + k - 1 < \sigma < \frac{1}{p} + k. \qquad (8.89)$$

Then we have the isomorphic map

$$\Delta^k : \quad \text{from} \quad \widetilde{B}_p^\sigma(\Omega) \quad \text{onto} \quad B_p^{\sigma-2k}(\Omega). \qquad (8.90)$$

Step 2 We prove part (i). Let p and ϱ with (8.84) be fixed. Let $f \in S'(\Omega)$. By (8.80) we may assume

$$f \in B_p^s(\Omega) \quad \text{with} \quad s = \sigma - 2k \tag{8.91}$$

for some $k \in \mathbb{N}$, where σ is given by (8.89). Let $(\Delta^k)^{-1}$ be the inverse of the mapping (8.90). Then

$$(\Delta^k)^{-1} f \in \widetilde{B}_p^\sigma(\Omega) = \widetilde{F}_{pp}^\sigma(\Omega)$$

can be expanded according to Theorem 6.6 by

$$(\Delta^k)^{-1} f = \sum_{\beta \in \mathbb{N}_0^n} \sum_{\nu=0}^{\infty} \sum_{m=1}^{M_\nu} \lambda_{\nu m}^\beta (\beta qu)_{\nu m}(x) \tag{8.92}$$

with the $(s+2k, p) - \beta$-quarks

$$(\beta qu)_{\nu m}(x) = 2^{-\nu(s+2k-\frac{n}{p}) + \nu|\beta|} (x - x^{\nu, m})^\beta \varphi_{\nu m}(x) \tag{8.93}$$

and (8.83), where the latter coincides with (6.21). The series in (8.92) converges absolutely and unconditionally in $L_1(\Omega)$. Application of Δ^k gives (8.85) with the $(s, p)^L - \beta$-quarks (8.82) where

$$L + 1 = 2k = \sigma - s > -s, \quad \text{in particular} \quad L \geq [-s].$$

The series converges unconditionally in $S'(\Omega)$.

Step 3 Part (ii) follows from the discussion in 6.8, in particular in connection with (6.48)–(6.51). Here one needs $L \geq [-s]$, which is equivalent to $L+s+1 > 0$.

8.16 Corollary

(Taylor expansions of tempered distributions in domains)

Let Ω be a bounded C^∞ domain in $\mathbb{R}^n$. Let $\{\varphi_{\nu m}\}$ be the same family of approximate resolutions of unity according to 6.2 as in Theorem 8.15. For any $f \in S'(\Omega)$ there are a natural number K and a positive number c such that

$$f = \sum_{\beta \in \mathbb{N}_0^n} \sum_{\nu=0}^{\infty} \sum_{m=1}^{M_\nu} \lambda_{\nu m}^\beta \Delta^K \left[(x - x^{\nu,m})^\beta \varphi_{\nu m}(x) \right] \tag{8.94}$$

(unconditional convergence in $S'(\Omega)$) with $\lambda_{\nu m}^\beta \in \mathbb{C}$ and

$$2^{(\varrho\ \nu)|\beta|\ \nu(K-\frac{1}{2})} |\lambda_{\nu m}^\beta| \leq c \tag{8.95}$$

for all $\beta \in \mathbb{N}_0^n$, $\nu \in \mathbb{N}_0$, and $m = 1, \ldots, M_\nu$.

Proof We may choose in Theorem 8.15 and (8.82),

$$s = -K - \frac{1}{2}, \quad p = \infty \quad \text{and} \quad L + 1 = 2K, \tag{8.96}$$

for some $K \in \mathbb{N}$. Then we have $\sigma = K - \frac{1}{2}$ in (8.91), assuming that f belongs to $B_\infty^s(\Omega)$. Then we have (8.85), (8.86) with $p = \infty$. This coincides with (8.94), where we incorporated some factors from (8.82) in the newly-defined coefficients $\lambda_{\nu m}^\beta$.

8.17 Comment on Taylor expansions

The above corollary is the counterpart of Corollary 8.9. With the necessary modifications, the discussions from 8.10 and 8.11 can be extended to the situation considered in Corollary 8.16. In particular, in 1.3 we described in (1.18) the classical Taylor expansion for holomorphic functions adapted to our purpose with the coefficients (1.19), (1.21).

The representation (8.94) with the coefficients (8.95) might be considered as the appropriate extension of Taylor expansions from holomorphic functions to tempered distributions.

There is a new phenomenon, expressed by Δ^K in (8.94), compared with (1.18). It reflects the nature of singular distributions in relation to (holomorphic) functions or regular distributions. Then one needs elementary building blocks satisfying some moment conditions, and Δ^K might be considered as a simple way to incorporate them.

8.18 Generalizations

Theorems 8.7 and 8.15, and also Corollaries 8.9 and 8.16 deal with global Taylor expansions of tempered distributions in $\mathbb{R}^n$ and in bounded C^∞ domains. The considerations are based on the representations (8.19) and (8.80), respectively. It is well known that the topological structure of $D'(\mathbb{R}^n)$ and $D'(\Omega)$ is more complicated. In particular there are no representations of type (8.19) and (8.80) and nothing having the same elegance as Corollaries 8.9 and 8.16 can be expected. But there is a more or less obvious weak substitute. Let, for example, Ω be an arbitrary bounded domain in $\mathbb{R}^n$. Let Ω_j be given by (7.21), and let $\{\varphi_j\}_{j=0}^\infty$ be a resolution of unity,

$$\sum_{j=0}^\infty \varphi_j(x) = 1 \quad \text{if} \quad x \in \Omega. \tag{8.97}$$

One might think that φ_j is given by

$$\varphi_j(x) = \sum_{m=1}^{M_j} \varphi_{jm}(x), \quad x \in \Omega, \tag{8.98}$$

according to (7.29)–(7.31). If $f \in D'(\Omega)$ then

$$f_j = f\varphi_j \in S'(\mathbb{R}^n).$$

One can apply, for example, Corollary 8.9 or Corollary 8.16, to $f\varphi_j$. Then one gets via

$$f = \sum_{j=0}^{\infty} f\varphi_j, \quad \text{convergence in} \quad D'(\Omega), \tag{8.99}$$

something like a Taylor expansion for f. The outcome might be of some use in applications, but it is not satisfactory from an aesthetical point of view.

8.19 Final remarks

The first steps in the direction of Taylor expansions of tempered distributions in $\mathbb{R}^n$ were taken in [Triδ], 14.13, p. 101. The above Corollary 8.9 might be considered as a more detailed version of the remarks made there. Our considerations are based on some special weighted spaces introduced in 8.2. But this can be done on a much larger scale. As discussed briefly in 7.3 one can replace $\langle x \rangle^\alpha$ in 8.2 by weights w with (7.17), (7.18). Then one gets spaces of type

$$B_{pq}^s(\mathbb{R}^n, w(\cdot)) \quad \text{and} \quad F_{pq}^s(\mathbb{R}^n, w(\cdot)). \tag{8.100}$$

One has always the property that

$$f \mapsto w(\cdot)f \tag{8.101}$$

is an isomorphic map from these weighted spaces onto the corresponding unweighted spaces. This can be done in the framework of $S'(\mathbb{R}^n)$. Details may be found in [HaT94a], [HaT94b], and [ET96], Chapter 4. In particular any quarkonial assertion for unweighted spaces can be transferred via (8.101) to weighted spaces of the above type. This applies to the theory developed in Sections 2 and 3, but also to Theorem 8.7 and Corollary 8.9. On the other hand, if $w(x)$ is, for example, of growth $e^{\pm|x|^\varkappa}$ with $0 < \varkappa < 1$, then one needs ultra distributions. The related theory of the spaces (8.100) may be found in [ST87], 5.1. Again (8.101) gives an isomorphic map onto the related unweighted spaces. Hence, also in this case, the unweighted theory can be transferred to these spaces of ultra-distributions. It is somewhat surprising that this observation can even be extended to spaces with exponential weights of type $e^{\pm|x|}$. The theory of the related spaces of type (8.100) has been developed in [Scho98a] and [Scho98b], including the remarkable fact that (8.101) again provides an isomorphic map onto the related unweighted spaces. Hence even in the case of exponential weights one has counterparts of the corresponding assertions from Sections 2 and 3, and of Theorem 8.7 and Corollary 8.9.

As for spaces on domains, including Theorem 8.15 and Corollary 8.16, we relied on Sections 5 and 6, and hence on [Tri99a]. We followed [Tri00a]. There one finds also further results. In particular, in this paper we developed first the theory of Taylor expansions in bounded C^∞ domains. Then we used these results combined with localization assertions of type (5.13) to prove corresponding results for tempered distributions on $\mathbb{R}^n$. We refer to [Tri00a], Theorem 2.3.4. The outcome is somewhat different compared with Theorem 8.7 and Corollary 8.9.

9 Traces on sets, related function spaces and their decompositions

9.1 Introduction

In connection with fractal elliptic operators which will be considered in Chapter III we are particularly interested in d-sets and their perturbations, called (d, Ψ)-sets. A compact set Γ in $\mathbb{R}^n$ is called a d-set (in our notation) or a (d, Ψ)-set if there is a Radon measure μ in $\mathbb{R}^n$ with $\Gamma = \operatorname{supp} \mu$ such that

$$\mu(B(\gamma, r)) \sim r^d, \quad \text{or} \quad \mu(B(\gamma, r)) \sim r^d \Psi(r), \quad 0 < r < \frac{1}{2}, \tag{9.1}$$

respectively, where $B(\gamma, r)$ is a ball centred at $\gamma \in \Gamma$ and of radius r. Naturally $0 \le d \le n$ and $\Psi(r)$ is a perturbation where one might think typically of $\Psi(r) = |\log r|^b$ for some $b \in \mathbb{R}$. However, in connection with fractal drums in the plane, it is reasonable and desirable to admit more general sets Γ and related measures μ with $\operatorname{supp} \mu = \Gamma$. This leads naturally to the question under what circumstances functions f belonging to, say, some Sobolev spaces $H_p^s(\mathbb{R}^n)$ with $s > 0$ and $1 < p < \infty$, have traces $tr_\Gamma f$ on Γ such that

$$\|tr_\Gamma f \,|L_r(\Gamma)\| \le c \|f \,|H_p^s(\mathbb{R}^n)\|, \tag{9.2}$$

for some r with, say, $1 \le r < \infty$, and $c > 0$. Problems of this type have been studied with great intensity for more than thirty years, pioneered by Maz'ya, and then by D. R. Adams, Hedberg, Netrusov, Verbitsky and many other mathematicians, usually in the context of (linear and non-linear) potential theory. This means, instead of (9.2), one asks for inequalities of type

$$\|I_s f \,|L_r(\Gamma)\| \le c \|f \,|L_p(\mathbb{R}^n)\|, \quad f \in L_p(\mathbb{R}^n), \tag{9.3}$$

where again Γ is the support of a given measure μ in $\mathbb{R}^n$, and I_s is the Riesz potential,

$$(I_s f)(x) = (-\Delta)^{-\frac{s}{2}} f(x) = c \int_{\mathbb{R}^n} \frac{f(y)}{|x-y|^{n-s}}\, dy, \quad 0 < s < n. \tag{9.4}$$

Replacing the Riesz potential by the Bessel potential, which means $(-\Delta)^{-\frac{s}{2}}$ by $(id - \Delta)^{-\frac{s}{2}}$, in (9.4), one gets a reformulation of (9.2). The outstanding references in this field of research are [Maz85] and [AdH96]. In the survey [Ver99] one finds more recent results and additional references. Replacing the Riesz potentials by more general potentials, one arrives at

$$\|tr_\Gamma f \,|L_r(\Gamma)\| \le c \|f \,|F^s_{pq}(\mathbb{R}^n)\| \tag{9.5}$$

in generalization of (9.2). For a far-reaching treatment we refer in particular to [AdH96] and the literature mentioned there. It is not our aim to study problems of this type systematically. In connection with fractal drums in the plane we need later on some information about inequalities of type (9.2) in a Hilbert space setting, this means $r = p = 2$. But for the approach presented here there is no difference whether one deals with (9.2) where $r = p = 2$, or with (9.5) where

$$1 \le r < \infty, \quad 1 < p < \infty, \quad 1 < q < \infty. \tag{9.6}$$

This, in turn, might explain our restriction of the parameters to (9.6), whereas problem (9.5) makes sense also for

$$0 < p \le \infty, \quad 0 < q \le \infty, \quad s > \sigma_p = n\left(\frac{1}{p} - 1\right)_+, \tag{9.7}$$

($q = \infty$ if $p = \infty$), or even more general parameters p, q, s, and also $r < 1$ (but then one is leaving the realm of distributions) as far as the target space is concerned. Our first aim in this section is to demonstrate, under the restriction (9.6), the intimate relationship between

traces, duality, local means, and quarkonial decompositions.

We start with some preparations in 9.2 and give a necessary and sufficient criterion concerning inequalities of type (9.5) with (9.6) in 9.3. It is the main aim of the discussions afterwards to provide an understanding of what is going on. Then, of course, some formulations given are near or even coincide with what is known in the above-mentioned literature. It is a further aim of this section to study quarkonial decompositions of function spaces on fractals. This might be considered as the continuation of our studies in [Triδ], Sections 18 and 20. Needless to say, all parts of this section are closely related to each other, and we provide also some discussions in this direction.

9.2 The set-up: traces, duality, local means, measures

For given p, q, r with (9.6) and $s > 0$ we ask which measures μ in $\mathbb{R}^n$ have the trace property (9.5). Our arguments are based on duality. Then (9.6) is convenient. But it is quite clear that, based on extra arguments, the method can be extended to limiting situations, where p, q might be 1 and/or ∞. This will not be done here. Similarly we simplify the a priori assumptions for the underlying measure μ. We suppose that

μ *is a finite Radon measure in* $\mathbb{R}^n$ *with a compact support* $\Gamma = supp\,\mu$.

Then one has no problems with the density of smooth functions in the considered spaces, and with $L_1(\Gamma)$ there is a largest space in which everything happens (especially quarkonial decompositions). But it is quite clear that most of the assertions can be easily extended to locally finite measures in $\mathbb{R}^n$: all considerations are local, and all spaces involved can be decomposed in local parts. We refer to (5.13) for $F_{pq}^s(\mathbb{R}^n)$, and there is an obvious counterpart for $L_r(\Gamma)$. The assumption that μ is a finite Radon measure in $\mathbb{R}^n$ is rather obvious in our context, since we always wish to interpret μ as a tempered distribution

$$\mu(\varphi) = \int_\Gamma \varphi(\gamma)\,\mu(d\gamma), \quad \varphi \in S(\mathbb{R}^n). \tag{9.8}$$

By completion, μ can be interpreted as a (non-negative) linear continuous functional on the completion of $S(\mathbb{R}^n)$, restricted to Γ, in $L_\infty(\Gamma)$. But by the Riesz representation theorem there is a one-to-one relation between these finite Radon measures and these (non-negative) linear continuous functionals. We refer to [Mat95], p. 15, and for greater details, to [Mall95], Theorem 6.6 on p. 97, and [Lan93], Theorem 4.2, p. 268. Hence we can identify finite Radon measures, without ambiguity, with their tempered distributions (9.8), using at this moment the same letter μ. We have

$$\mu \in B_{1,\infty}^0(\mathbb{R}^n). \tag{9.9}$$

This is well known. For sake of completeness we give a short proof. We use (2.37). By (2.34) and the interpretation $\mu \in S'(\mathbb{R}^n)$ we have for $k \in \mathbb{N}$,

$$(\varphi_k \widehat{\mu})^{\vee}(x) = c\, 2^{kn} \int_{\Gamma} \varphi_1^{\vee}(2^{k-1}x - 2^{k-1}\gamma)\, \mu(d\gamma) \tag{9.10}$$

and hence, adding $j = 0$,

$$\|(\varphi_j \widehat{\mu})^{\vee}\,|L_1(\mathbb{R}^n)\| \le c\,\mu(\Gamma) < \infty, \quad j \in \mathbb{N}_0. \tag{9.11}$$

This proves (9.9). By embedding, [Triβ], 2.7.1, p. 129, we have

$$\mu \in B_{p\infty}^{\sigma}(\mathbb{R}^n), \quad 1 \le p \le \infty, \quad \sigma = -n\left(1 - \frac{1}{p}\right). \tag{9.12}$$

Hence we adopt the following point of view. If a finite measure μ in $\mathbb{R}^n$ is given, interpreted as a tempered distribution, then the quality of μ is judged by asking whether (9.12) can be improved in terms of function spaces $B_{pq}^s(\mathbb{R}^n)$ and $F_{pq}^s(\mathbb{R}^n)$.

Let μ be again a finite compactly supported Radon measure in $\mathbb{R}^n$. Put $\Gamma = \operatorname{supp} \mu$. Let p, q, r be given by (9.6), and let $s > 0$. If $\varphi \in S(\mathbb{R}^n)$ then we denote the pointwise trace of φ on Γ likewise as $tr_{\Gamma}\,\varphi$ (and call tr_{Γ} the trace operator), $\varphi|\Gamma$, or $\varphi(\gamma)$ with $\gamma \in \Gamma$. Let

$$\|g\,|L_r(\Gamma)\| = \left(\int_{\Gamma} |g(\gamma)|^r\, \mu(d\gamma)\right)^{\frac{1}{r}}, \quad g \in L_r(\Gamma), \tag{9.13}$$

where $L_r(\Gamma)$ has the usual meaning. We always write $L_r(\Gamma)$ instead of, maybe, $L_r(\mu)$ or $L_r(\Gamma, \mu)$. (It will never happen that we equip a given set Γ with two non-equivalent measures.) We ask whether there is a constant $c > 0$ such that

$$\|tr_{\Gamma}\,\varphi\,|L_r(\Gamma)\| \le c\,\|\varphi\,|F_{pq}^s(\mathbb{R}^n)\| \tag{9.14}$$

for all $\varphi \in S(\mathbb{R}^n)$. If this is the case then one can extend (9.14) from $S(\mathbb{R}^n)$ to $F_{pq}^s(\mathbb{R}^n)$ by completion, where we use that $S(\mathbb{R}^n)$ is dense in $F_{pq}^s(\mathbb{R}^n)$, [Triβ], Theorem 2.3.3, p. 48. In this way, any $f \in F_{pq}^s(\mathbb{R}^n)$ has a (uniquely determined) trace $tr_{\Gamma}\,f \in L_r(\Gamma)$, and

$$tr_{\Gamma}\,:\quad F_{pq}^s(\mathbb{R}^n) \mapsto L_r(\Gamma) \tag{9.15}$$

is denoted as trace operator. On the other hand, if $f \in L_r(\Gamma)$ is given, then f, or better the complex measure $f\mu$, can be interpreted in the usual way as a tempered distribution $id_{\Gamma}\,f$, given by

$$\begin{aligned}(id_{\Gamma}\,f)(\varphi) &= \int_{\Gamma} f(\gamma)\,\varphi(\gamma)\,\mu(d\gamma) \\ &= \int_{\Gamma} f(\gamma)\,(tr_{\Gamma}\,\varphi)(\gamma)\,\mu(d\gamma), \quad \varphi \in S(\mathbb{R}^n).\end{aligned} \tag{9.16}$$

We call id_Γ the *identification operator*. One can look at (9.16) also as dual pairings: the left-hand side in $S'(\mathbb{R}^n) - S(\mathbb{R}^n)$, and the right-hand side in $L_1(\Gamma) - L_\infty(\Gamma)$. This has the following consequence. Again let r, p, q be given by (9.6) and let, as usual,

$$\frac{1}{p} + \frac{1}{p'} = \frac{1}{q} + \frac{1}{q'} = \frac{1}{r} + \frac{1}{r'} = 1. \qquad (9.17)$$

Then, in the interpretation just mentioned,

$$(L_r(\Gamma))' = L_{r'}(\Gamma) \quad \text{and} \quad \left(F^\sigma_{pq}(\mathbb{R}^n)\right)' = F^{-\sigma}_{p'q'}(\mathbb{R}^n) \qquad (9.18)$$

for any $\sigma \in \mathbb{R}$. The first assertion is well known, the second one may be found in [Triβ], Theorem 2.11.2, p. 178. In particular, all spaces $F^\sigma_{pq}(\mathbb{R}^n)$ and also $L_r(\Gamma)$ with $1 < r < \infty$ are reflexive. By (9.16), the operators tr_Γ and id_Γ are dual to each other. Hence, (9.15) is equivalent to

$$id_\Gamma : \quad L_{r'}(\Gamma) \mapsto F^{-s}_{p'q'}(\mathbb{R}^n), \qquad (9.19)$$

and

$$tr'_\Gamma = id_\Gamma. \qquad (9.20)$$

If $r > 1$ then we have also

$$id'_\Gamma = tr_\Gamma. \qquad (9.21)$$

One could even define the trace operator tr_Γ as the dual of id_Γ under the above circumstances and $r > 1$. But we prefer the above version, completion of (9.14), since it makes sense also for a wider range of the parameters involved, for example as in (9.7).

We transformed the problem (9.15) into the equivalent problem (9.19). We deal with the latter question in terms of local means. Let

$$0 < u \leq \infty, \quad 0 < v \leq \infty, \quad \sigma < 0, \qquad (9.22)$$

(with $v = \infty$ if $u = \infty$). Let k be a C^∞ function in $\mathbb{R}^n$ with a compact support and

$$k(x) \geq 0, \quad \int_{\mathbb{R}^n} k(x)\, dx > 0. \qquad (9.23)$$

Let

$$k(2^{-\nu}, f)(x) = 2^{\nu n} \int_{\mathbb{R}^n} k(2^\nu (x-y))\, f(y)\, dy, \quad \nu \in \mathbb{N}_0, \qquad (9.24)$$

9. Traces on sets, related function spaces and their decompositions

be the local means of $f \in S'(\mathbb{R}^n)$ in the usual interpretation as dual pairings in $(S'(\mathbb{R}^n), S(\mathbb{R}^n))$. Then $f \in S'(\mathbb{R}^n)$ belongs to $F^\sigma_{uv}(\mathbb{R}^n)$ if, and only if,

$$\left\| \left(\sum_{\nu=0}^\infty 2^{\nu\sigma v} \left| k(2^{-\nu}, f)(\cdot) \right|^v \right)^{\frac{1}{v}} |L_u(\mathbb{R}^n) \right\| < \infty \qquad (9.25)$$

(equivalent quasi-norms) with the usual modification if $v = \infty$. This follows from [Triγ], 2.4.6, p. 122, (and 2.5.3, p. 138 if $u = v = \infty$) and the additional observation that, since $\sigma < 0$, one non-negative kernel k with (9.23) is sufficient. As for the latter point we refer to [Triγ], 2.4.1, pp. 100–101, where one may choose $s_1 = 0$ if $\sigma < 0$, and hence $N = 0$ in [Triγ], 2.4.6, p. 122. (At the time when [Triγ] was written, this observation was not of interest for us, and hence, neither checked nor formulated.) The first explicit formulation was given in the PhD-thesis [Win95], Theorem 3.1/3. But it may also be found in [AdH96], Corollary 4.3.8, p. 102 (restricted to $u > 1$, $v > 1$), and implicitly in [FrJ90] and [JPW90].

Finally we introduce some notation. Let $\varrho > 0$. As in 2.2 we denote by $\varrho Q_{\nu m}$ a cube in $\mathbb{R}^n$ with sides parallel to the axes of coordinates, centred at $2^{-\nu} m$ and with side length $\varrho 2^{-\nu}$ where $m \in \mathbb{Z}^n$ and $\nu \in \mathbb{N}_0$. Of course, $Q_{\nu m} = 1 Q_{\nu m}$. Again let μ be a finite Radon measure in $\mathbb{R}^n$ with compact $supp\, \mu = \Gamma$. If $f \in L_1(\Gamma)$ then we put

$$f_{\nu m} = \int_{2Q_{\nu m}} f(\gamma)\, \mu(d\gamma), \quad \nu \in \mathbb{N}_0, \quad m \in \mathbb{Z}^n. \qquad (9.26)$$

Of course, if $2Q_{\nu m} \cap \Gamma = \emptyset$ then $f_{\nu m} = 0$. Let $\chi_{\nu m}$ be the characteristic function of $Q_{\nu m}$.

9.3 Theorem

Let

$$1 < p < \infty, \quad 1 < q < \infty, \quad 1 \leq r < \infty, \quad \text{and} \quad s > 0. \qquad (9.27)$$

Let p' and r' be given by (9.17). Let μ be a finite Radon measure in $\mathbb{R}^n$ with compact support $\Gamma = supp\, \mu$. Then the trace operator tr_Γ,

$$tr_\Gamma : F^s_{pq}(\mathbb{R}^n) \mapsto L_r(\Gamma), \qquad (9.28)$$

exists according to the above explanations (9.14), (9.15), if, and only if,

$$\sup \sum_{\nu \in \mathbb{N}_0} \sum_{m \in \mathbb{Z}^n} 2^{-\nu p'(s-\frac{n}{p})} f^{p'}_{\nu m} < \infty, \qquad (9.29)$$

where the supremum is taken over all

$$f \in L_{r'}(\Gamma) \quad \text{with} \quad f \geq 0 \quad \text{and} \quad \|f\, |L_{r'}(\Gamma)\| \leq 1.$$

Proof *Step 1* We need some preparation. Let $f \in L_1(\Gamma)$ (the largest admitted space on Γ according to 9.2) and let $f \geq 0$. Choosing appropriate kernels k_1 and k_2 with the counterparts of (9.23), one has for the related means (9.24),

$$k_1(2^{-\nu}, f)(x) \geq 2^{\nu n} f_{\nu m} \geq k_2(2^{-\nu}, f)(x), \quad x \in Q_{\nu m}, \tag{9.30}$$

where $\nu \in \mathbb{N}_0$ and $m \in \mathbb{Z}^n$. It follows that

$$k_1(2^{-\nu}, f)(x) \geq 2^{\nu n} \sum_{m \in \mathbb{Z}^n} f_{\nu m} \chi_{\nu m}(x) \geq k_2(2^{-\nu}, f)(x), \quad x \in \mathbb{R}^n, \tag{9.31}$$

where $\nu \in \mathbb{N}_0$. Hence by (9.31), (9.25), and (9.16) the distribution $id_\Gamma f$ belongs to the positive cone $\overset{+}{F}{}^{\sigma}_{uv}(\mathbb{R}^n)$ of $F^{\sigma}_{uv}(\mathbb{R}^n)$ with (9.22) if, and only if,

$$\left\| \left(\sum_{\nu=0}^{\infty} \sum_{m \in \mathbb{Z}^n} 2^{\nu(n+\sigma)v} f^v_{\nu m} \chi_{\nu m}(\cdot) \right)^{\frac{1}{v}} \Big| L_u(\mathbb{R}^n) \right\| < \infty \tag{9.32}$$

(equivalent quasi-norm) with the usual modification if $v = \infty$. However for fixed $\sigma < 0$ and $0 < u \leq \infty$ all these positive cones coincide,

$$\overset{+}{F}{}^{\sigma}_{uv_1}(\mathbb{R}^n) = \overset{+}{F}{}^{\sigma}_{uv_2}(\mathbb{R}^n), \quad 0 < v_1 \leq v_2 \leq \infty. \tag{9.33}$$

The history of this remarkable property may be found in [AdH96], p. 126. A first proof, restricted to $1 < u < \infty$, $1 < v_1 \leq v_2 < \infty$, was given in [Ada89], which may also be found in [AdH96], Corollary 4.3.9, p. 103. As for the full assertion we refer to [JPW90] and [Net89a].

Step 2 After the preparations in 9.2 and in the above Step 1 one can prove the theorem rather quickly. Assuming first that tr_Γ is a bounded map according to (9.28). Then we have (9.19), and hence also (9.32) with $\sigma = -s$, $u = p'$, and any $0 < v \leq \infty$, uniformly with respect to

$$f \in L_{r'}(\Gamma), \quad f \geq 0, \quad \text{and} \quad \|f \,|L_{r'}(\Gamma)\| \leq 1.$$

Choosing $v = p'$ we get (9.29).

Step 3 Conversely, we assume that (9.29) holds, where $f \geq 0$ belongs to $L_{r'}(\Gamma)$. By (9.30) and (9.33) it follows that $id_\Gamma f \in \overset{+}{F}{}^{-s}_{p'q'}(\mathbb{R}^n)$ and

$$\left\| id_\Gamma f \,|F^{-s}_{p'q'}(\mathbb{R}^n) \right\| \leq c \, \|f \,|L_{r'}(\Gamma)\|. \tag{9.34}$$

Any $f \in L_{r'}(\Gamma)$ can be naturally decomposed by

$$f = f_1 - f_2 + if_3 - if_4 \quad \text{with} \quad f_l \geq 0. \tag{9.35}$$

Applying (9.34) to f_l, it follows that (9.34) holds for all $f \in L_{r'}(\Gamma)$. We obtain (9.19). But this is equivalent to (9.28).

9.4 Corollary

Let p, q, r, s be as in (9.27). Let μ be a finite Radon measure in $\mathbb{R}^n$ with compact support $\Gamma = \operatorname{supp} \mu$.

(i)   The trace operator tr_Γ with (9.28) exists if, and only if,

$$\sup \left\| \sup_{\nu \in \mathbb{N}_0, m \in \mathbb{Z}^n} 2^{\nu(n-s)} f_{\nu m} \chi_{\nu m}(\cdot) \,|L_{p'}(\mathbb{R}^n) \right\| < \infty, \qquad (9.36)$$

where the outer supremum is taken over all $f \in L_{r'}(\Gamma)$ with $f \geq 0$ and $\|f\,|L_{r'}(\Gamma)\| \leq 1$.

(ii)   If (9.36) (or equivalently (9.29)) is satisfied, then the existence of the (linear and bounded) trace operator tr_Γ according to (9.28) is independent of q.

Proof *Step 1* Let $0 < v \leq \infty$. By the proof of Theorem 9.3 condition (9.29) is equivalent to

$$\sup \left\| \left(\sum_{\nu=0}^{\infty} \sum_{m \in \mathbb{Z}^n} 2^{\nu(n-s)v} f_{\nu m}^v \chi_{\nu m}(\cdot) \right)^{\frac{1}{v}} |L_{p'}(\mathbb{R}^n) \right\| < \infty, \qquad (9.37)$$

with the usual modification if $v = \infty$, where the supremum is taken over all

$$f \in L_{r'}(\Gamma) \quad \text{with} \quad f \geq 0 \quad \text{and} \quad \|f\,|L_{r'}(\Gamma)\| \leq 1.$$

There are two distinguished cases, $v = p'$ leads to (9.29), and $v = \infty$ leads to (9.36). This proves part (i).

Step 2 In part (ii) we simply fix as an immediate consequence that the necessary and sufficient conditions (9.29) and (9.36) do not depend on q.

9.5 Discussion

We add a discussion from the point of view of the function spaces $F_{pq}^s(\mathbb{R}^n)$ and $B_{pq}^s(\mathbb{R}^n)$.

(i)   We exclude $r = \infty$ in (9.27) since it does not fit in the above scheme. But there is also no need to deal with target spaces of L_∞-type, since any space $F_{pq}^s(\mathbb{R}^n)$ or $B_{pq}^s(\mathbb{R}^n)$ which is continuously embedded in $L_\infty(\mathbb{R}^n)$ is automatically continuously embedded in $C(\mathbb{R}^n)$, the space of bounded continuous functions in $\mathbb{R}^n$, and hence traces can be taken pointwise. As for the claimed embedding we refer to [SiT95] and [RuS96], pp. 32/33.

(ii)   The case $p = 1$ can be incorporated in (9.27). The needed dual spaces according to (9.18) are of BMO-type and may be found in [Triβ], Theorem

2.11.2, p. 178, or [RuS96], p. 20. However an extension of Theorem 9.3 and Corollary 9.4 to values $p < 1$ by the above method is not possible (but desirable).

(iii) The fact that the existence of the trace operator tr_Γ in (9.28) is independent of q does not mean automatically that the trace spaces $tr_\Gamma F_{pq}^s(\mathbb{R}^n)$ are also independent of q (where s and p are assumed to be given). But this is the case under a mild extra condition on Γ with $|\Gamma| = 0$ and will be discussed in Theorem 9.21.

(iv) From the point of view of the above function spaces and their applications to partial differential operators there are two distinguished cases in connection with (9.28). First $r = p$ and secondly the question whether there is a trace at all within the given setting, which means in our situation that $r = 1$. If the latter is the case, then one can introduce a trace space

$$tr_\Gamma F_{pq}^s(\mathbb{R}^n) = \{g \in L_1(\Gamma) : \text{ there is an } f \in F_{pq}^s(\mathbb{R}^n) \text{ with } tr_\Gamma f = g\} \quad (9.38)$$

normed by

$$\|g \,|tr_\Gamma F_{pq}^s(\mathbb{R}^n)\| = \inf \|f \,|F_{pq}^s(\mathbb{R}^n)\| \quad (9.39)$$

where the infimum is taken over all $f \in F_{pq}^s(\mathbb{R}^n)$ with $tr_\Gamma f = g$. Now one can ask again whether for given s and p the Banach spaces $tr_\Gamma F_{pq}^s(\mathbb{R}^n)$ are independent of q and what can be said about their elements. We return to this problem several times. As an easy consequence of Theorem 9.3 we obtain in Theorem 9.9 a criterion for the existence of the trace spaces in (9.38). Later on we describe these spaces in terms of quarkonial decompositions in 9.29 and 9.33.

(v) Both (9.29) and (9.36) are special cases of (9.37), and for any given v with $0 < v \le \infty$, the trace property (9.28) and (9.37) are equivalent. Maybe $v = p'$, which results in (9.29), is the most handsome case, whereas (9.36) can be reformulated in terms of maximal functions. Let

$$(M_{\mu,s}f)(x) = \sup_{x \in Q_{vm}} 2^{v(n-s)} \int_{2Q_{vm}} |f(\gamma)| \, \mu(d\gamma), \quad f \in L_1(\Gamma), \quad (9.40)$$

be the maximal function related to (9.36). Then Corollary 9.4(i) can be reformulated as follows:

There is a constant $c_1 > 0$ with

$$\|tr_\Gamma f \,|L_r(\Gamma)\| \le c_1 \,\|f \,|F_{pq}^s(\mathbb{R}^n)\|, \quad f \in F_{pq}^s(\mathbb{R}^n), \quad (9.41)$$

if, and only if, there is a constant $c_2 > 0$ with

$$\|M_{\mu,s}g \,|L_{p'}(\mathbb{R}^n)\| \le c_2 \,\|g \,|L_{r'}(\Gamma)\|, \quad g \in L_{r'}(\Gamma). \quad (9.42)$$

But it is not clear whether one can take any advantage of this observation. On the other hand, the intimate relations between mapping properties of Riesz potentials (9.4) and also Bessel potentials and maximal inequalities including measures are well known. We refer to [AdH96], Section 3.6, and [Ver99], Section 4, and the literature mentioned there.

9.6 Corollary

(Necessary conditions)

Let p, q, r, s and the measure μ be as in Theorem 9.3. Let

$$s - \frac{n}{p} = -\frac{d}{r} \quad \text{for some} \quad d \in \mathbb{R}. \tag{9.43}$$

If tr_Γ exists according to (9.28), then there is a number $c > 0$ such that

$$\mu(Q_{\nu m}) \leq c\, 2^{-\nu d}, \quad \nu \in \mathbb{N}_0, \quad m \in \mathbb{Z}^n. \tag{9.44}$$

Proof Let $\nu \in \mathbb{N}_0$ and $m \in \mathbb{Z}^n$ with $\mu(Q_{\nu m}) > 0$. Let again $\chi_{\nu m}$ be the characteristic function of $Q_{\nu m}$ and

$$f(\gamma) = \mu^{-\frac{1}{r'}}(Q_{\nu m})\, \chi_{\nu m}(\gamma), \quad \gamma \in \Gamma. \tag{9.45}$$

Then, by (9.26),

$$\|f\,|L_{r'}(\Gamma)\| = 1 \quad \text{and} \quad f_{\nu m} = \mu(Q_{\nu m})^{\frac{1}{r}}. \tag{9.46}$$

Inserting (9.46) in (9.29) one gets (9.44).

9.7 Remark

If $s > \frac{n}{p}$ then the space $F^s_{pq}(\mathbb{R}^n)$ consists of continuous functions. Then one has pointwise traces and (9.44) is obvious. Hence one may assume $s \leq \frac{n}{p}$ in (9.43). If $d > n$ then it follows from (9.44) and measure properties that $\mu = 0$, which is always tacitly excluded. Finally, $d \leq 0$ makes sense, but as said, (9.44) is trivial. Hence, $s \leq \frac{n}{p}$ and $0 < d \leq n$ are the natural restrictions in (9.43).

9.8 Corollary

(Sufficient conditions)

Let p, q, s and the measure μ be as in Theorem 9.3.

(i)   Let $p \leq r < \infty$. If

$$\sum_{\nu=0}^{\infty} 2^{-\nu(s-\frac{n}{p})p'} \sup_{m \in \mathbb{Z}^n} \mu(2Q_{\nu m})^{\frac{p'}{r}} < \infty, \qquad (9.47)$$

then the trace operator tr_Γ with (9.28) exists.

(ii)   Let $p \leq r < \infty$, and let, in addition,

$$s - \frac{n}{p} > -\frac{d}{r} \quad \text{with} \quad 0 < d \leq n, \qquad (9.48)$$

and for some $c > 0$,

$$\mu(2Q_{\nu m}) \leq c\, 2^{-\nu d}, \quad \nu \in \mathbb{N}_0, \quad m \in \mathbb{Z}^n. \qquad (9.49)$$

Then the trace operator tr_Γ with (9.28) exists.

(iii)   Let $1 \leq r < p$. Let $\frac{1}{\varkappa} + \frac{r}{p} = 1$. If

$$\sum_{\nu=0}^{\infty} 2^{-\nu(s-\frac{n}{p})p'} \left(\sum_{m \in \mathbb{Z}^n} \mu(2Q_{\nu m})^{\varkappa} \right)^{\frac{p'}{\varkappa r}} < \infty, \qquad (9.50)$$

then the trace operator tr_Γ with (9.28) exists.

Proof   *Step 1*   We prove (i). Let

$$f \in L_{r'}(\Gamma) \quad \text{with} \quad \|f\,|L_{r'}(\Gamma)\| \leq 1 \quad \text{and} \quad f \geq 0.$$

Then, by (9.26) and Hölder's inequality,

$$f_{\nu m} \leq \mu(2Q_{\nu m})^{\frac{1}{r}} \left(\int_{2Q_{\nu m}} f^{r'}(\gamma)\, \mu(d\gamma) \right)^{\frac{1}{r'}}. \qquad (9.51)$$

Since $p' \geq r'$ we obtain

$$\sum_{m \in \mathbb{Z}^n} f_{\nu m}^{p'} \leq \sup_{l \in \mathbb{Z}^n} \mu(2Q_{\nu l})^{\frac{p'}{r}} \sum_{m \in \mathbb{Z}^n} \left(\int_{2Q_{\nu m}} f^{r'}(\gamma)\, \mu(d\gamma) \right)^{\frac{p'}{r'}}$$

$$\leq \sup_{l \in \mathbb{Z}^n} \mu(2Q_{\nu l})^{\frac{p'}{r}} \left(\sum_{m \in \mathbb{Z}^n} \int_{2Q_{\nu m}} f^{r'}(\gamma)\, \mu(d\gamma) \right)^{\frac{p'}{r'}}$$

$$\leq \sup_{l \in \mathbb{Z}^n} \mu(2Q_{\nu l})^{\frac{p'}{r}}. \qquad (9.52)$$

Inserting (9.52) in (9.29) we get (9.47).

Step 2 Part (ii) follows immediately from (9.48), (9.49) inserted in (9.47).

Step 3 We prove (iii). Let

$$f \in L_{r'}(\Gamma) \quad \text{and} \quad f \geq 0 \quad \text{with} \quad \|f\,|L_{r'}(\Gamma)\| \leq 1\,.$$

Since

$$\frac{1}{r\varkappa} + \frac{1}{r'} = \frac{1}{p'},\qquad (9.53)$$

it follows by Hölder's inequality that

$$\sum_{m \in \mathbb{Z}^n} f_{\nu m}^{p'} \leq \sum_{m \in \mathbb{Z}^n} \mu(2Q_{\nu m})^{\frac{p'}{r}} \left(\int_{2Q_{\nu m}} f^{r'}(\gamma)\,\mu(d\gamma) \right)^{\frac{p'}{r'}}$$

$$\leq \left(\sum_{m \in \mathbb{Z}^n} \mu(2Q_{\nu m})^{\varkappa} \right)^{\frac{p'}{\varkappa r}} \left(\sum_{m \in \mathbb{Z}^n} \int_{2Q_{\nu m}} f^{r'}(\gamma)\,\mu(d\gamma) \right)^{\frac{p'}{r'}}$$

$$\leq c \left(\sum_{m \in \mathbb{Z}^n} \mu(2Q_{\nu m})^{\varkappa} \right)^{\frac{p'}{\varkappa r}} \qquad (9.54)$$

for some $c > 0$ which is independent of ν. Now part (iii) is a consequence of (9.29) and (9.54).

9.9 Theorem

(Necessary and sufficient conditions)

(i) (D. R. Adams) *Let p, q, r, s, and the measure μ be as in Theorem 9.3. Let, in addition, $r > p$, and*

$$s - \frac{n}{p} = -\frac{d}{r} \quad \text{with} \quad 0 < d \leq n\,. \qquad (9.55)$$

Then the trace operator tr_Γ with (9.28) exists if, and only if, there is a number $c > 0$ with

$$\mu(2Q_{\nu m}) \leq c\,2^{-\nu d}, \quad \nu \in \mathbb{N}_0\,,\quad m \in \mathbb{Z}^n\,. \qquad (9.56)$$

(ii) *Let p, q, s, and the measure μ be as in Theorem 9.3. Let $\frac{1}{p} + \frac{1}{p'} = 1$. Then the trace space $\mathrm{tr}_\Gamma F_{pq}^s(\mathbb{R}^n)$ as the range of the continuous mapping*

$$\mathrm{tr}_\Gamma\,:\quad F_{pq}^{s}(\mathbb{R}^n) \mapsto L_1(\Gamma)\,, \qquad (9.57)$$

according to (9.38) exists if, and only if,

$$\sum_{\nu \in \mathbb{N}_0} 2^{-\nu p'(s-\frac{n}{p})} \sum_{m \in \mathbb{Z}^n} \mu(2Q_{\nu m})^{p'} < \infty\,. \qquad (9.58)$$

Proof *Step 1* We prove (i). By Corollary 9.6, condition (9.56) is necessary for the existence of tr_Γ with (9.28). By Corollary 9.8(ii) condition (9.56) is sufficient in case of the strict inequality (9.48) in place of (9.55). The limiting case (9.55) is much deeper and essentially covered by the following famous beautiful observation of D. R. Adams in the context of mapping properties of Riesz potentials: Under the above circumstances, $\infty > r > p > 1$, (9.55), and (9.56), there is a number $c > 0$ with (9.3) for all $f \in L_p(\mathbb{R}^n)$, where the Riesz potential is given by (9.4). For proofs we refer to [Maz85], Theorem 2 on p. 52, and [AdH96], Theorem 7.2.2 in combination with Proposition 5.1.2, pp. 193, 131. The original proof goes back to [Ada71] and [Ada73]. One can see easily that the Riesz potentials $(-\Delta)^{-\frac{s}{2}} f$ can be replaced in (9.3) by the Bessel potentials $(id - \Delta)^{-\frac{s}{2}} f$. This follows also from a Fourier multiplier assertion for $|\xi|^s (1+|\xi|^2)^{-\frac{s}{2}}$ in $L_p(\mathbb{R}^n)$ with $1 < p < \infty$. But (9.3) with the Bessel potential in place of the Riesz potential is equivalent to

$$\|tr_\Gamma f \,|L_r(\Gamma)\| \le c\, \|f\,|H^s_p(\mathbb{R}^n)\|, \quad f \in H^s_p(\mathbb{R}^n), \tag{9.59}$$

where $H^s_p(\mathbb{R}^n) = F^s_{p,2}(\mathbb{R}^n)$ are the (fractional) Sobolev spaces. Hence the trace operator tr_Γ exists for this special case and (9.29) is satisfied. Finally, by Corollary 9.4(ii) the existence of tr_Γ with (9.28) is independent of q.

Step 2 Part (ii) is a direct consequence of the necessary and sufficient criterion (9.29) with $r' = \infty$. One has to insert $f(\gamma) = c > 0$ with $\gamma \in \Gamma$ as the worst case.

9.10 Comments: the case $r > p$

The arguments concerning part (i) of the above theorem rely decisively on the quoted results by D. R. Adams and were not derived directly from the necessary and sufficient criteria (9.29) or (9.36). It is not clear whether this can be done on the same (rather simple) level of arguments as in Corollaries 9.6 and 9.8. We return to this point in 9.12–9.14 and discuss in some detail the criterion (9.42) in terms of the maximal function (9.40) under the above circumstances. This will shed some light on the parameters involved. At the moment we only mention that the condition $r > p$ in part (i) is crucial. The assertion of part (i) is wrong in general if $p = r$. We give a counter-example. Let $\Gamma = \partial\Omega$ be the boundary of a bounded C^∞ domain in $\mathbb{R}^n$ and $\mu = \mathcal{H}^{n-1}|\Gamma$. Then we have (9.56) with $d = n-1$, and hence (9.55) with $r = p$ and $s = \frac{1}{p}$. But there is no trace of $F^{\frac{1}{p}}_{pq}(\mathbb{R}^n)$ with $1 < p < \infty$ and $0 < q < \infty$ into $L_p(\partial\Omega)$. We refer to Proposition 5.6. Hence part (i) of the above theorem cannot be extended to $p = r$. Another counterargument may be found in [AdH96], in the remark after Theorem 7.2.2 on p. 193. Finally we mention the close connection to fractal geometry: The existence of a measure μ with (9.56) is equivalent to $\mathcal{H}^d(\Gamma) > 0$, where $\mathcal{H}^d$ is the Hausdorff measure in $\mathbb{R}^n$. This is Frostman's lemma and may be found in [Mat95], Theorem 8.8, p. 112.

9.11 Comments, references, further results: the case $r \leq p$

(i)   The criterion given in Theorem 9.3 applies to all values of the parameters p, q, r, s with (9.27). However by the literature and also to some extent by the discussions and assertions in 9.5–9.10, it is quite clear that it is reasonable to distinguish in connection with the problem (9.28) between the three cases

$$r > p, \quad r = p, \quad \text{and} \quad r < p. \tag{9.60}$$

If $r > p$ then one has D. R. Adams' beautiful necessary and sufficient condition (9.56). By 9.10 this cannot be extended to the cases $r \leq p$.

(ii)   The case $r = p$ is of special interest for us, where one has (restricted to $p = r$) the necessary and sufficient conditions (9.29), (9.36), (9.42), the necessary condition (9.44), and the sufficient condition (9.47). Obviously this case $p = r$ attracted a lot of attention in literature for more than three decades in the context of potential theory and preferably related to embeddings of type (9.3) for Riesz potentials and Bessel potentials. Necessary and sufficient conditions for the case $r \geq p$, and especially for $r = p$, expressed in terms of capacity, may be found in [Maz85], in particular Section 2.3, Corollary 2.3.3, p. 113 (even in the larger context of Orlicz spaces), and Chapter 8, and in [AdH96], Theorem 7.2.1, pp. 191–192. More recent necessary and sufficient conditions in the case $r = p$ for embeddings of type (9.3), but with the Bessel potential in place of the Riesz potential, have been given in [MaV95] and [MaN95]. A short description, further results and references can be found in [AdH96], 7.6, pp. 208–213.

(iii)   As for the case $r < p$ we refer again to [Maz85], especially Theorem 2.3.5, p. 120, and to [MaN95]. Furthermore in the note sections in [Maz85] and [AdH96] (called comments in [Maz85]) and in the survey [Ver99] one finds the relevant references to the original papers. Nearest to us are the assertions obtained in the survey [Ver99] and in the underlying paper [COV99]. We give a brief description of those results in [Ver99] and [COV99] which are directly related to our own approach presented above. In particular we wish to emphasize that the criterion (9.58) is also covered by [Ver99], Theorems 15 and 16, pp. 261, 263, and [COV99]. Furthermore we compare the sufficient condition (9.50) with a corresponding (necessary and sufficient) criterion in [Ver99], Theorems 15 and 16, and [COV99]. The starting point is the Hedberg-Wolff potential, which goes back to [HeW83], and which plays a crucial role in this field of research. We refer to [AdH96], pp. 109, 167, and the comments and references on p. 126. In particular, Wolff's inequality in [AdH96], Theorem 4.5.2, p. 109, turns out to be equivalent with (9.58), as was pointed out in [Ver99], p. 262. Let $\varrho Q_{\nu m}$ be the same cubes as at the end of 9.2 and let $\widetilde{\chi}_{\nu m}(x)$ be the characteristic function of $2Q_{\nu m}$. Let again μ be a compactly supported Radon measure in $\mathbb{R}^n$ and

$$1 < p < \infty, \quad \frac{1}{p} + \frac{1}{p'} = 1, \quad s > 0. \tag{9.61}$$

Then the dyadic version of the Hedberg-Wolff potential adapted to our (inhomogeneous) situation is given by

$$\widetilde{W}_{s,p}\mu(x) = \sum_{\nu=0}^{\infty} \sum_{m \in \mathbb{Z}^n} \mu(2Q_{\nu m})^{p'-1} 2^{\nu(n-sp)(p'-1)} \widetilde{\chi}_{\nu m}(x) \tag{9.62}$$

$$= \sum_{\nu=0}^{\infty} \sum_{m \in \mathbb{Z}^n} 2^{-\nu(s-\frac{n}{p})p'} \mu(2Q_{\nu m})^{\frac{p'}{p}} \widetilde{\chi}_{\nu m}(x), \quad x \in \mathbb{R}^n.$$

Let, as above,

$$1 \le r < p \quad \text{and} \quad \sigma = \frac{r(p-1)}{p-r}. \tag{9.63}$$

Then one has the embedding (9.3) with the Bessel potential in place of the Riesz potential, or, equivalently, (9.2), if, and only if,

$$\widetilde{W}_{s,p}\mu \in L_\sigma(\Gamma). \tag{9.64}$$

This is the inhomogeneous (dyadic) version of [Ver99], Theorems 15 and 16, pp. 261, 263. A detailed proof may be found in [COV99]. If $r = 1$, then $\sigma = 1$, and the criterion (9.64) coincides with the criterion (9.58) (as it should be). We discuss briefly the case $p > r > 1$. Then we have the sufficient condition (9.50) and the criterion (9.64). The numbers $\varkappa$ and σ are related by

$$\sigma \frac{p'}{p} + 1 = \frac{r}{p-r} + 1 = \frac{p}{p-r} = \varkappa \quad \text{and} \quad \frac{1}{\sigma} = \frac{p'(p-r)}{pr} = \frac{p'}{r\varkappa}. \tag{9.65}$$

We obtain

$$\left\| \sum_{m \in \mathbb{Z}^n} \mu(2Q_{\nu m})^{\frac{p'}{p}} \widetilde{\chi}_{\nu m} \,|\, L_\sigma(\Gamma) \right\| \le c \left(\sum_{m \in \mathbb{Z}^n} \mu(2Q_{\nu m})^{\varkappa} \right)^{\frac{p'}{\varkappa r}}. \tag{9.66}$$

Now by (9.63), $\sigma \ge 1$, and by the triangle inequality, (9.64) is a consequence of (9.50), as it should be.

9. Traces on sets, related function spaces and their decompositions

(iv)   As said in 9.1 it is not our aim to deal systematically with the flourishing field of research which has been treated so far in this section. Our goal is rather modest: Since we need in the later applications a closer look at more general sets and measures than d-sets and (d, Ψ)-sets we wanted to provide an understanding of what is going on. Our arguments are characterized by duality, independence of the positive cone

$$\overset{+}{F}{}^s_{pq}(\mathbb{R}^n) \quad \text{of} \quad F^s_{pq}(\mathbb{R}^n) \quad \text{with} \quad s < 0,$$

on q, and related maximal inequalities. We could not find the assertion of Theorem 9.3 or the equivalence of (9.41) and (9.42) explicitly stated in the literature. On the other hand, the discussions in 9.10 and in this subsection make clear that some of the above necessary or/and sufficient conditions for (9.3) or (9.2) are known, in particular in connection with the two parts of Theorem 9.9. Also the interplay of the symbiotic relationship between duality, positive cones, and maximal inequalities is not new. It may be found in several places in [AdH96] and rather explicitly in [JPW90]. In the following subsections 9.12–9.14 we comment briefly on the equivalence of (9.41) and (9.42) for d-sets. But otherwise we return to the main subject of this book. This means in the above context, we are not only interested in whether the embedding (9.28) exists, but also in a description of the trace spaces and their properties. This is the point where the quarkonial decompositions, considered in the previous sections, enter.

9.12 Maximal functions related to d-sets

A set Γ in $\mathbb{R}^n$ is called a d-set if there are a Borel measure μ in $\mathbb{R}^n$ and positive numbers c_1 and c_2 such that $\operatorname{supp} \mu = \Gamma$ and

$$c_1 t^d \leq \mu(B(\gamma, t)) \leq c_2 t^d \quad \text{for all} \quad 0 < t < 1, \tag{9.67}$$

and $\gamma \in \Gamma$, where $B(\gamma, t)$ is the ball centred at $\gamma \in \Gamma$ and of radius t. Then μ is a Radon measure and, up to equivalences, uniquely determined. In particular it can be identified with the restriction $\mathcal{H}^d|\Gamma$ of the Hausdorff measure $\mathcal{H}^d$ on Γ. The Hausdorff dimension of Γ is d. Obviously, $0 \leq d \leq n$. More details and a short proof of the mentioned equivalence may be found in [Triδ], pp. 5–6. This notation coincides with [JoW84] and differs from that in fractal geometry as used in [Fal85], p. 8 and [Mat95], where one finds on p. 92 further comments and references.

Let Γ be a compact d-set. We apply Theorem 9.9. As far as part (i) is concerned we are interested in the limiting situation

$$1 < p < r < \infty, \quad 0 < d \leq n, \quad s - \frac{n}{p} = -\frac{d}{r}, \tag{9.68}$$

and, so far, $1 < q < \infty$. Then the trace operator tr_Γ according to (9.28) exists. But this is equivalent to the maximal inequality (9.42) with the fractional maximal function $M_{\mu,s}f$ given by (9.40), which can be identified under the above circumstances with

$$(M_\mu^\alpha f)(x) = \sup_{x \in Q_{\nu m}} 2^{\nu\alpha} \frac{1}{\mu(4Q_{\nu m})} \int_{2Q_{\nu m}} |f(\gamma)|\, \mu(d\gamma) \qquad (9.69)$$

where

$$\alpha = n - s - d = \frac{n}{p'} - \frac{d}{r'}. \qquad (9.70)$$

For given $x \in \mathbb{R}^n$ the supremum on the right-hand side of (9.69) can be restricted to those $\nu \in \mathbb{N}_0$ and $m \in \mathbb{Z}^n$ such that $x \in Q_{\nu m}$ and $2Q_{\nu m} \cap \Gamma \neq \emptyset$. Then $\mu(4Q_{\nu m}) \sim 2^{-\nu d}$. In particular, if $x \notin \Gamma$ then only values $\nu \in \mathbb{N}$ with $2^\nu \leq c\, dist(x, \Gamma)^{-1}$ for some $c > 0$ are of interest.

9.13 Proposition

Let Γ be a compact d-set in $\mathbb{R}^n$ with $0 < d \leq n$ and let μ be a related Radon measure with $supp\, \mu = \Gamma$ and (9.67) (which is uniquely determined up to equivalences).

(i) *Let p, q, s be as in Theorem 9.3. Then the trace space $tr_\Gamma F_{pq}^s(\mathbb{R}^n)$ according to (9.38) and Theorem 9.9 (ii) exists if, and only if, $s > \frac{n-d}{p}$.*

(ii) *Let*

$$1 < v < u < \infty, \quad \alpha = \frac{n}{u} - \frac{d}{v}, \qquad (9.71)$$

and let $M_\mu^\alpha f$ be given by (9.69). There is a number $c > 0$ such that

$$\|M_\mu^\alpha f \mid L_u(\mathbb{R}^n)\| \leq c \|f \mid L_v(\Gamma)\| \qquad (9.72)$$

for all $f \in L_v(\Gamma)$.

Proof *Step 1* We prove part (i). The number of cubes $Q_{\nu m}$ with $2Q_{\nu m} \cap \Gamma \neq \emptyset$ can be estimated from above and from below by positive constants multiplied with $2^{\nu d}$. There are at least $c_1 2^{\nu d}$ such cubes with $\mu(2Q_{\nu m}) \geq c_2 2^{-\nu d}$ for some $c_1 > 0$ and $c_2 > 0$. Then (9.58) is equivalent to

$$\sum_{\nu \in \mathbb{N}_0} 2^{-\nu p'(s-\frac{n}{p})} 2^{-\nu dp'} 2^{\nu d} = \sum_{\nu \in \mathbb{N}_0} 2^{-\nu p'(s-\frac{n-d}{p})} < \infty. \qquad (9.73)$$

This proves part (i).

Step 2 Part (ii) follows from the considerations in 9.12 with $u = p'$ and $v = r'$.

9.14 Remark

Maybe the proof of the curious maximal inequality in (9.72) is more interesting than the assertion itself: It reduces (9.72) with $u = p'$ and $v = r'$, via the independence of $\overset{+}{F}{}^s_{pq}(\mathbb{R}^n)$ with $s<0$ on q, to (9.29), and hence to (9.28), and to Adams' observation in Theorem 9.9(i). Since these reductions are even equivalences it follows that, at least in general, there are no maximal inequalities (9.72) if $u=v$. We refer to the counterexample in 9.10. On the other hand, (9.72) with (9.71) is sharp. To make clear what is meant we look at some special functions. Let $0 \in \Gamma$, $\varkappa \in \mathbb{R}$,

$$f(\gamma) = |\gamma|^{-\frac{d}{v}} |\log|\gamma||^{\varkappa}, \quad \gamma \in \Gamma, \quad |\gamma| \leq \frac{1}{2}, \qquad (9.74)$$

and $f(\gamma) = 0$ otherwise. Then it follows that

$$f \in L_v(\Gamma) \quad \text{if, and only if,} \quad v\varkappa < -1. \qquad (9.75)$$

Let $x \in \mathbb{R}^n$ and $c_1 2^{-\nu} \leq |x| \leq c_2 2^{-\nu}$ for two small positive numbers c_1, c_2 and $\nu \in \mathbb{N}$, $\nu \geq \nu_0$. Let $Q_\nu = Q_{\nu m}$ with $m = 0$. Then for some $c > 0$,

$$\begin{aligned}
\left(M_\mu^\alpha f\right)(x) &\geq c\, 2^{\nu\alpha}\, 2^{\nu d} \int_{2Q_\nu} |\gamma|^{-\frac{d}{v}} |\log|\gamma||^{\varkappa} \mu(d\gamma) \\
&\geq c'\, 2^{\nu(\alpha + d + \frac{d}{v} - d)}\, \nu^{\varkappa} \\
&= c'\, 2^{\nu \frac{n}{u}}\, \nu^{\varkappa} \\
&\geq c''\, |x|^{-\frac{n}{u}} |\log|x||^{\varkappa}, \qquad (9.76)
\end{aligned}$$

where c, c', c'' are positive numbers. Comparing this estimate with (9.72) it is clear that α in (9.71) cannot be improved and also $u \geq v$ is indispensable. The different arguments at the beginning of this subsection excluded also $u=v$ (which is not covered by the more elementary calculations in (9.74)–(9.76)).

9.15 Trace spaces

By Theorem 9.3 and Corollary 9.4(ii) we know that the existence of the trace operator tr_Γ in (9.28) is independent of q. We described our point of view in 9.5(iv). In particular we are interested in the trace space $tr_\Gamma F^s_{pq}(\mathbb{R}^n)$ according to (9.38) and the problem whether it is independent of q. We do not know under what (necessary and sufficient) circumstances this is true in general, but there is an affirmative answer for a large class of measures μ in dependence on the geometry of $\Gamma = supp\,\mu$. First we recall the nowadays classical assertion about the traces of $F^s_{pq}(\mathbb{R}^n)$ on $\Gamma = \mathbb{R}^{n-1}$, interpreted as the hyperplane $x_n = 0$

in $\mathbb{R}^n$, where $x = (x', x_n)$ with $x' \in \mathbb{R}^{n-1}$ (and equipped with the Lebesgue measure). Let $n \geq 2$,

$$0 < p < \infty, \quad 0 < q \leq \infty, \quad s - \frac{1}{p} > (n-1)\left(\frac{1}{p} - 1\right)_+. \tag{9.77}$$

Then

$$tr_{\mathbb{R}^{n-1}} F^s_{pq}(\mathbb{R}^n) = B^{s-\frac{1}{p}}_{pp}(\mathbb{R}^{n-1}). \tag{9.78}$$

In particular, the trace space is independent of q. The first full proof of this assertion is due to B. Jawerth, [Jaw77]. We dealt several times with this problem, [Triβ], Theorem 2.7.2, p. 132, and [Triγ], 4.4.2, p. 213. There one finds also further references. In our context, where Γ is more general, but p, q, s are restricted by

$$1 < p < \infty, \quad 1 < q < \infty, \quad s > 0, \tag{9.79}$$

at least the existence of the trace $tr_{\mathbb{R}^{n-1}}$ follows from Corollary 9.8(ii) with $p = r$, $d = n - 1$ and hence $s > \frac{1}{p}$ and also from Proposition 9.13(i). One can give a new and comparatively simple proof of (9.78) with (9.77) based on the quarkonial technique developed in Sections 2 and 3. We will not stress this point since we are interested here in general measures μ and sets $\Gamma = supp\,\mu$ according to 9.2 and Theorem 9.3. Basically one has to adapt the arguments given in the proof of Theorem 9.21 below to (9.78), (9.77). We give an outline in 9.23 of how to do this.

9.16 Definition

(Ball condition)
A non-empty Borel set Γ in $\mathbb{R}^n$ is said to satisfy the ball condition if there is a number $0 < \eta < 1$ with the following property: For any ball $B(x, t)$ centred at $x \in \mathbb{R}^n$ and of radius $0 < t < 1$ there is a ball $B(y, \eta t)$ (centred at $y \in \mathbb{R}^n$ and of radius ηt) with

$$B(y, \eta t) \subset B(x, t) \quad \text{and} \quad B(y, \eta t) \cap \overline{\Gamma} = \emptyset. \tag{9.80}$$

9.17 Remarks and some notation

This formulation coincides essentially with [Triδ], 18.10, p. 142. Of course, one can replace balls by cubes (with sides parallel to the axes), and one can restrict the above definition to balls (or cubes) centred at Γ. Furthermore, $|\Gamma| = 0$ for the Lebesgue measure $|\Gamma|$ of Γ. We give a short proof. Let Q be a cube (with sides parallel to the axes) having side-length 1 and let, without restriction of generality, $\Gamma \subset Q$. Let $k \in \mathbb{N}$. We divide Q naturally in 2^{kn} sub-cubes having side-lengths 2^{-k}. If k is large, then one of these sub-cubes has an empty intersection with Γ. We apply this procedure iteratively to the remaining $2^{kn} - 1$ sub-cubes and get

$$|\Gamma| \leq \lim_{l \to \infty} \left(1 - 2^{-kn}\right)^l = 0. \tag{9.81}$$

Furthermore, replacing, if necessary, η in (9.80) by $\frac{\eta}{2}$ we may complement (9.80) by

$$dist\,(B(y, \eta t), \overline{\Gamma}) \geq \eta t, \quad 0 < t < 1. \tag{9.82}$$

Conditions of this or modified type are well known and have been used on many occasions in mathematics. In connection with irregular boundaries of bounded domains in $\mathbb{R}^n$ we refer to [TrW96], 3.2, and [ET96], 2.5, pp. 58–59. There one finds also a few references to the literature where notation of this type occurs.

We are interested here in the interplay between the quality of a Radon measure μ and the geometry of its support $\Gamma = supp\,\mu$. Let again μ be a finite Radon measure in $\mathbb{R}^n$ with compact support $\Gamma = supp\,\mu$ and let

$$c_1 \Phi(r) \leq \mu(B(\gamma, r)) \leq c_2 \Phi(r), \quad 0 < r < 1, \quad \gamma \in \Gamma, \tag{9.83}$$

where $0 < c_1 \leq c_2$ are suitable numbers. Typical examples are given by (9.1), related to d-sets and (d, Ψ)-sets. Let $\Phi(0) = 0$, $\Phi(r)$ continuous and, obviously, monotone increasing in $[0, 1)$, and $\Phi(r) > 0$ if $0 < r < 1$. Then μ is non-atomic (or *diffuse* according to [Bou56], §5.10, p. 61 or [Die75], 13.18, p. 215), and satisfies the doubling condition

$$\mu(B(\gamma, 2r)) \leq c\,\mu(B(\gamma, r)), \quad \gamma \in \Gamma, \quad 0 < r < 1, \tag{9.84}$$

where c is a suitable constant. The latter follows from (9.83) and the fact that there is a number $N \in \mathbb{N}$ such that Γ intersected with a given ball of radius $2r$ in $\mathbb{R}^n$ centred at Γ, can be covered by at most N balls of radius r centred at Γ. Hence at the expense of some constants one can replace the balls $B(\gamma, r)$ in (9.83) by balls $B(\gamma, ar)$ with $a > 0$ and also by cubes with side-length ar.

9.18 Proposition

Let μ be a finite Radon measure in $\mathbb{R}^n$ with compact support $\Gamma = supp\,\mu$ and with (9.83), where $\Phi(r)$ is a continuous monotonically increasing function on $[0, 1)$ with $\Phi(r) > 0$ if $0 < r < 1$ and $\Phi(0) = 0$. Then Γ satisfies the ball condition according to 9.16 if, and only if, there are two positive numbers c and λ such that

$$\Phi(2^{-\nu}) \le c\, 2^{(n-\lambda)\varkappa}\, \Phi(2^{-\nu-\varkappa}) \quad \text{for all} \quad \nu \in \mathbb{N}_0 \quad \text{and} \quad \varkappa \in \mathbb{N}_0. \qquad (9.85)$$

Proof *Step 1* We assume that Γ satisfies the ball condition. Let Q^ν with $\nu \in \mathbb{N}_0$ be a cube with side-length $2^{-\nu}$ centred at some point $\gamma \in \Gamma$. We subdivide Q^ν naturally in $2^{\varkappa n}$ cubes with side-length $2^{-\nu-\varkappa}$, where $\varkappa \in \mathbb{N}$. If $\varkappa$ is large then at least one of these sub-cubes has an empty intersection with Γ. We apply this argument to each of the remaining

$$2^{\varkappa n} - 1 = 2^{(n-\lambda)\varkappa} \qquad (9.86)$$

cubes with side-length $2^{-\nu-\varkappa}$. By iteration, $\Gamma \cap Q^\nu$ can be covered by $2^{(n-\lambda)\varkappa l}$ cubes with side-length $2^{-\nu-\varkappa l}$. Switching to balls and using (9.84) we have

$$\begin{aligned}
\Phi(2^{-\nu}) &\le c_1\, \mu(B(\gamma, 2^{-\nu})) & (9.87)\\
&\le c_2\, 2^{(n-\lambda)\varkappa l}\, \mu(B(\gamma, 2^{-\nu-\varkappa l})) \\
&\le c_3 2^{(n-\lambda)\varkappa l}\, \Phi(2^{-\nu-\varkappa l}), \quad l \in \mathbb{N},
\end{aligned}$$

where c_1, c_2, c_3 are independent of $\nu \in \mathbb{N}_0$ and $l \in \mathbb{N}$. By (9.84) and its Φ-counterpart we obtain (9.85).

Step 2 Conversely, we assume that (9.85) holds. Let again Q^ν be a cube with $\nu \in \mathbb{N}_0$ and side-length $2^{-\nu}$. Again we subdivide Q^ν naturally in $2^{\varkappa n}$ cubes with side-length $2^{-\nu-\varkappa}$, where $\varkappa \in \mathbb{N}$. Let $N_\varkappa$ be the maximal number of these cubes having a non-empty intersection with Γ. There are two numbers $N \in \mathbb{N}$ and $c \ge 1$ such that $N_\varkappa$ cubes centred at Γ and with side length $c\, 2^{-\nu-\varkappa}$ cover $\Gamma \cap Q^\nu$ at most N times. Hence, by (9.85),

$$N_\varkappa \Phi(2^{-\nu-\varkappa}) \le c_1\, \mu(c_2 Q^\nu) \le c_3\, \Phi(2^{-\nu}) \le c_4\, 2^{(n-\lambda)\varkappa}\, \Phi(2^{-\nu-\varkappa}), \qquad (9.88)$$

where the positive numbers c_1, c_2, c_3, c_4 are independent of ν and $\varkappa$. If $\varkappa$ is large then $N_\varkappa < 2^{\varkappa n}$. This means that at least one of the $2^{\varkappa n}$ sub-cubes of Q^ν with side-length $2^{-\nu-\varkappa}$ has an empty intersection with Γ. But this is equivalent to the ball condition.

9.19 Remark

Although the arguments are not so complicated, the outcome is a little bit surprising. In particular, there is no need to assume that a given d-set with $d < n$ has in addition this property as was done, for example, in [Triδ], 18.12, p. 142. If $d < n$ in (9.1) then one may choose $\lambda = n - d$ in (9.85), and hence Γ

9. Traces on sets, related function spaces and their decompositions

satisfies the ball condition. In case of d-sets the above proposition was brought to the attention of the author by A. Caetano in 1998 and may be found in his paper [Cae99], but as mentioned in this paper it is also covered by [Jon93b], Proposition 2, p. 288. We also refer in this context to [Cae00], where criteria for anisotropic fractals satisfying the ball condition are given.

9.20 Some preparation

The definition of the trace operator tr_Γ in (9.15), based on (9.13) and the inequality (9.14) for smooth functions makes sense for all

$$1 \le r < \infty, \quad 0 < p < \infty, \quad 0 < q < \infty, \tag{9.89}$$

since $S(\mathbb{R}^n)$ is dense in all spaces $F^s_{pq}(\mathbb{R}^n)$. If Γ satisfies the ball condition according to 9.16, then $q = \infty$ can be incorporated in the definition of tr_Γ as follows: Let $f \in F^s_{p\infty}(\mathbb{R}^n)$ be given by its quarkonial decomposition, say, (3.12), (3.16). Then f can be written as $f = f_1 + f_2$, where f_1 collects all those β-quarks with

$$\Gamma \cap supp\,(\beta qu)_{\nu m} \ne \emptyset \quad \text{and} \quad \Gamma \cap supp\,(\beta qu)^L_{\nu m} \ne \emptyset.$$

Hence, f_1 is the trace-relevant part of f and, as will be shown in the proof of Theorem 9.21 below in detail, $f_1 \in F^s_{pv}(\mathbb{R}^n)$ for any v with $0 < v \le \infty$. Assuming that $tr_\Gamma f_1$ makes sense according to (9.13)–(9.15) with (9.89), then we put $tr_\Gamma f = tr_\Gamma f_1$ also in case of $q = \infty$. We return later on in 9.29 and 9.33 to the possibility of defining traces and trace spaces in terms of β-quarks. Recall that $H^s_p(\mathbb{R}^n) = F^s_{p,2}(\mathbb{R}^n)$ are the (fractional) Sobolev spaces.

9.21 Theorem

Let

$$1 < p < \infty, \quad 0 < q \le \infty, \quad 1 \le r < \infty, \quad \text{and} \quad s > 0. \tag{9.90}$$

Let

$$\frac{1}{p} + \frac{1}{p'} = \frac{1}{r} + \frac{1}{r'} = 1.$$

Let μ be a finite Radon measure in $\mathbb{R}^n$ such that the support $\Gamma = supp\,\mu$ is compact and satisfies the ball condition according to Definition 9.16. Then the trace operator tr_Γ,

$$tr_\Gamma : F^s_{pq}(\mathbb{R}^n) \mapsto L_r(\Gamma), \tag{9.91}$$

exists according to 9.20 if, and only if,

$$\sup \sum_{\nu \in \mathbb{N}_0} \sum_{m \in \mathbb{Z}^n} 2^{-\nu p'(s-\frac{n}{p})} f_{\nu m}^{p'} < \infty, \qquad (9.92)$$

where the supremum is taken over all

$$f \in L_{r'}(\Gamma) \quad \text{with} \quad f \geq 0 \quad \text{and} \quad \|f\,|L_{r'}(\Gamma)\| \leq 1$$

and where $f_{\nu m}$ is given by (9.26). Furthermore, the range of tr_Γ according to (9.91) is independent of all q with $0 < q \leq \infty$ and of all r satisfying (9.92):

$$tr_\Gamma F_{pq}^s(\mathbb{R}^n) = \{g \in L_1(\Gamma) : \text{there is an } f \in H_p^s(\mathbb{R}^n) \text{ with } tr_\Gamma f = g\}. \qquad (9.93)$$

Proof *Step 1* Let $q < \infty$. Then one does not need for the proof of the independence of the trace space $tr_\Gamma F_{pq}^s(\mathbb{R}^n)$ of r that Γ satisfies the ball condition. It follows simply from the definition of $tr_\Gamma f$ according to 9.20 as the limit of some $tr_\Gamma \varphi$ with $\varphi \in S(\mathbb{R}^n)$ in $L_r(\Gamma)$, provided that one has the inequality (9.14) for all $\varphi \in S(\mathbb{R}^n)$, and usual measure-theoretical arguments.

Step 2 We have to extend the assertion of Theorem 9.3 to all q with $0 < q \leq \infty$ and to prove that the resulting trace spaces $tr_\Gamma F_{pq}^s(\mathbb{R}^n)$ are independent of q (and of r, but this has been done in Step 1). Now we rely on the assumption that Γ satisfies the ball condition. Let $f \in F_{pq}^s(\mathbb{R}^n)$. We use the quarkonial decomposition (3.12),

$$f = \sum_{\beta,\nu,m} \left(\eta_{\nu m}^\beta (\beta qu)_{\nu m} + \lambda_{\nu m}^\beta (\beta qu)_{\nu m}^L \right), \qquad (9.94)$$

with (3.16), (3.17), which converges unconditionally in $L_p(\mathbb{R}^n)$ by the discussion in 3.5. (If $q \geq 1$, then we could even use the simpler version (2.31)). We decompose (9.94) in

$$f = f_1 + f_2 = \sum_{\beta,\nu,m}{}' (\cdots) + \sum_{\beta,\nu,m}{}'' (\cdots), \qquad (9.95)$$

where $\sum'_{\beta,\nu,m} (\cdots)$ collects all those (ν, m) with

$$dQ_{\nu m} \cap \Gamma \neq \emptyset, \qquad (9.96)$$

and where $dQ_{\nu m}$ has the same meaning as in (3.8). Of course, $\sum''_{\beta,\nu,m} (\cdots)$ collects the remaining couples (ν, m). We may assume $q < \infty$ (how to incorporate $q = \infty$ has been explained in 9.20). Then $tr_\Gamma f$ makes sense according to (9.14) where finite sums in (9.94) can be taken as approximating functions $\varphi \in S(\mathbb{R}^n)$. In particular we have

$$tr_\Gamma f = tr_\Gamma f_1 = g \in L_r(\Gamma) \subset L_1(\Gamma). \tag{9.97}$$

Since Γ satisfies the ball condition we find by 9.16 and (9.82) for any cube $Q_{\nu m}$ with (9.96) a ball $B_{\nu m}$ with

$$B_{\nu m} \subset Q_{\nu m}, \quad dist(B_{\nu m}, \Gamma) \geq c\, 2^{-\nu}, \tag{9.98}$$

for some $c > 0$ (independently of the couples (ν, m) involved). We may assume that all these balls have pairwise disjoint supports (or that there is a number $N \in \mathbb{N}$ such that at most N of these balls have non-empty intersection). By the discussions in 3.8 and 2.15 one can use in the sequence spaces f_{pq} in (3.17) the characteristic functions of the balls $B_{\nu m}$ instead of, say, the characteristic functions of $Q_{\nu m}$. We refer in particular to (2.102). Restricted to the above couples (ν, m) with (9.98) the related quasi-norms of f_{pq} are independent of q (equivalent quasi-norms). This applies to f_1 with the following outcome. Let $0 < q_0 \leq q_1 \leq \infty$. By (3.17) and suitably chosen $f \in F^s_{pq_1}(\mathbb{R}^n)$ we obtain

$$\|f_1 | F^s_{pq_0}(\mathbb{R}^n)\| \leq c \|f | F^s_{pq_1}(\mathbb{R}^n)\| \leq c' \|g | tr_\Gamma F^s_{pq_1}(\mathbb{R}^n)\|. \tag{9.99}$$

Hence, by (9.97), $g \in tr_\Gamma F^s_{pq_0}(\mathbb{R}^n)$. The converse is obvious (by the monotonicity of the spaces F^s_{pq} with respect to q) and we obtain

$$tr_\Gamma F^s_{pq_0}(\mathbb{R}^n) = tr_\Gamma F^s_{pq_1}(\mathbb{R}^n) \tag{9.100}$$

(equivalent quasi-norms). The independence of r follows from Step 1

9.22 Remark

Let μ be a finite Radon measure such that $\Gamma = supp\,\mu$ is compact. If the embedding (9.91) exists for some fixed r_0, then it exists also for all r with $1 \leq r \leq r_0$, and by Step 1 of the above proof the corresponding trace spaces $tr_\Gamma F^s_{pq}(\mathbb{R}^n)$ are independent of r. The assumption that Γ satisfies the ball condition is not needed for the independence of r. If, in addition, Γ satisfies the ball condition, then $tr_\Gamma F^s_{pq}(\mathbb{R}^n)$ is also independent of q. But it is not clear what happens if Γ does not satisfy the ball condition. If Γ has a non-empty interior, then one has the same situation as for spaces $F^s_{pq}(\Omega)$ in bounded domains Ω, introduced in Definition 5.3. They depend on q (for given s, p). The situation for compact sets with empty interior or with $|\Gamma| = 0$ not satisfying the ball condition is unclear. But one may consult in this context [FrJ90], Theorem 13.7.

9.23 Proof of (9.78) with (9.77): Outline

One expands $f \in F^s_{pq}(\mathbb{R}^n)$ by (9.94). Of course, $\mathbb{R}^{n-1}$, interpreted as the hyperplane $x_n = 0$, satisfies the ball condition. Furthermore, according to Definition 3.2 the $(s,p)^L - \beta$-quarks $(\beta qu)^L_{\nu m}(x)$ in $\mathbb{R}^n$ become $(s - \frac{1}{p}, p) - \beta$-quarks $(\beta qu)_{\nu m}(x', 0)$ on $\mathbb{R}^{n-1}$ (or atomic derivations where moment conditions are no longer needed). Then it follows easily that

$$f(x', 0) \in B^{s-\frac{1}{p}}_{pp}(\mathbb{R}^{n-1}).$$

Conversely, let $g(x') \in B^{s-\frac{1}{p}}_{pp}(\mathbb{R}^{n-1})$. Then one has quarkonial decompositions of g with optimal coefficients which depend linearly on g. These expansions can be extended to

$$f = ext\, g, \quad f \in F^s_{pq}(\mathbb{R}^n), \quad f(x', 0) = g(x') \in B^{s-\frac{1}{p}}_{pp}(\mathbb{R}^{n-1}), \qquad (9.101)$$

where ext is a linear and bounded extension operator.

9.24 The quarkonial approach: basic notation

With the help of the quarkonial decompositions described in Sections 2 and 3 one has a rather direct access to trace spaces. We introduce the necessary notation where we follow first the scheme developed in 2.14. Let Γ be a compact set in $\mathbb{R}^n$ and let

$$\Gamma_\varepsilon = \{x \in \mathbb{R}^n : \, dist\,(x, \Gamma) < \varepsilon\}, \quad \varepsilon > 0, \qquad (9.102)$$

be the ε-neighbourhood of Γ. Let $k \in \mathbb{N}_0$ and let

$$\{\gamma^{k,m} : m = 1, \ldots, M_k\} \subset \Gamma \quad \text{and} \quad \{\psi^{k,m} : m = 1, \ldots, M_k\} \qquad (9.103)$$

be *approximate lattices* and *subordinated resolutions of unity* with the following properties: There are positive numbers c_1, c_2, c_3 with

$$\left|\gamma^{k,m_1} - \gamma^{k,m_2}\right| \geq c_1\, 2^{-k}, \quad k \in \mathbb{N}_0, \quad m_1 \neq m_2, \qquad (9.104)$$

and

$$\Gamma_{\varepsilon_k} \subset \bigcup_{m=1}^{M_k} B\left(\gamma^{k,m}, c_2\, 2^{-k}\right), \quad k \in \mathbb{N}_0, \qquad (9.105)$$

where $\varepsilon_k = c_3\, 2^{-k}$. Here $B(x,c)$ has the same meaning as in (2.96): a ball centred at $x \in \mathbb{R}^n$ and of radius $c > 0$. Furthermore, $\psi^{k,m}(x)$ are non-negative C^∞ functions in $\mathbb{R}^n$ with

$$supp\, \psi^{k,m} \subset B\left(\gamma^{k,m}, c_2\, 2^{-k+1}\right), \quad k \in \mathbb{N}_0, \, m = 1, \ldots, M_k, \qquad (9.106)$$

$$\left|D^\alpha \psi^{k,m}(x)\right| \leq c_\alpha\, 2^{k|\alpha|}, \quad x \in \mathbb{R}^n, \, k \in \mathbb{N}_0, \, m = 1, \ldots, M_k, \qquad (9.107)$$

9. Traces on sets, related function spaces and their decompositions   145

for all $\alpha \in \mathbb{N}_0^n$ and suitable constants c_α, and

$$\sum_{m=1}^{M_k} \psi^{k,m}(x) = 1, \quad k \in \mathbb{N}_0, \quad x \in \Gamma_{\varepsilon_k}. \tag{9.108}$$

We always assume that the approximate lattices $\{\gamma^{k,m}\}$ and the subordinated resolutions of unity $\{\psi^{k,m}\}$ can be extended to $\mathbb{R}^n$ such that one gets approximate lattices $\{x^{k,m}\}$ and related resolutions of unity $\{\psi^{k,m}\}$ according to 2.14 with (2.93)–(2.98), (2.100) (with sufficiently large K in (2.100)). In other words, of interest are those approximate lattices $\{x^{k,m}\}$ in $\mathbb{R}^n$ and subordinated resolutions of unity $\{\psi^{k,m}\}$ according to 2.14 which are adapted near Γ in the way described above.

Let again μ be a finite Radon measure in $\mathbb{R}^n$ with compact support $\Gamma = supp\,\mu$. By Corollary 9.8 and Theorem 9.9 it is quite clear what type of conditions for μ might be helpful in connection with traces of, say, $B_{pq}^s(\mathbb{R}^n)$ on Γ. By Theorem 9.21 and Remark 9.22 it is also quite clear that it is reasonable to switch now from $F_{pq}^s(\mathbb{R}^n)$ to $B_{pq}^s(\mathbb{R}^n)$. The latter spaces are not only technically simpler but they produce also a richer scale of trace spaces on Γ.

9.25 Definition

Let μ be a finite Radon measure in $\mathbb{R}^n$ with compact support $\Gamma = supp\,\mu$. Let

$$B_{km} = B\left(\gamma^{k,m}, c_2\, 2^{-k+1}\right), \quad k \in \mathbb{N}_0, \quad m = 1, \ldots, M_k, \tag{9.109}$$

be the above balls with (9.105). Let

$$1 < u \leq \infty, \quad 0 < v \leq \infty, \quad \text{and} \quad t \geq 0. \tag{9.110}$$

Then

$$\mu_{uv}^t = \left(\sum_{k=0}^{\infty} 2^{tkv} \left(\sum_{m=1}^{M_k} \mu(B_{km})^u\right)^{\frac{v}{u}}\right)^{\frac{1}{v}} \tag{9.111}$$

(with the obvious modification if u and/or v are infinity) and, with $\frac{1}{u} + \frac{1}{u'} = 1$,

$$t_u = \sup\left\{t : \mu_{u'\infty}^t < \infty\right\}. \tag{9.112}$$

9.26 Discussion, properties, examples

The motivation for introducing the numbers μ_{uv}^t comes from Corollary 9.8 and Theorem 9.9. Of interest are those parameters u, v, t with $\mu_{uv}^t < \infty$. These references explain also why duality in (9.112) arises. We discuss the conditions (9.110) and add a few monotonicity assertions. Let $1 < u_0 < u_1 \le \infty$. Then

$$\left(\sum_{m=1}^{M_k} \mu(B_{km})^{u_1} \right)^{\frac{1}{u_1}} \le \left(\sum_{m=1}^{M_k} \mu(B_{km})^{u_0} \right)^{\frac{1}{u_0}} \le \sum_{m=1}^{M_k} \mu(B_{km}) \sim 1 \qquad (9.113)$$

(independently of k). This makes clear that the restriction $u > 1$ in (9.111) is natural. Since μ is a measure and since for $a_{lj} \ge 0$,

$$\left(\sum_{l=1}^{L} \sum_{j=1}^{J} a_{lj}^u \right)^{\frac{1}{u}} \le \left(\sum_{l=1}^{L} \left(\sum_{j=1}^{J} a_{lj} \right)^u \right)^{\frac{1}{u}}, \qquad (9.114)$$

it follows that

$$\left(\sum_{m=1}^{M_k} \mu(B_{km})^u \right)^{\frac{1}{u}} \text{ is bounded,} \quad k \in \mathbb{N}, \qquad (9.115)$$

and it is reasonable to ask what happens if $k \to \infty$. This justifies $t \ge 0$ in (9.111). Obviously,

$$\mu_{u_1 v_1}^t \le \mu_{u_0 v_0}^t \quad \text{for} \quad u_0 \le u_1 \text{ and } v_0 \le v_1, \qquad (9.116)$$

and, if $0 \le t_1 < t_0$,

$$\mu_{uv_1}^{t_1} \le \mu_{uv_0}^{t_0} \quad \text{for} \quad 1 < u \le \infty, \, 0 < v_0 \le \infty, \, 0 < v_1 \le \infty. \qquad (9.117)$$

In particular, t_u in (9.112) is reasonable: if one replaces $\mu_{u'\infty}^t$ by $\mu_{u'v}^t$ then one gets again t_u. Finally, since $M_k \le c\, 2^{kn}$, it follows that

$$0 < c \le \sum_{m=1}^{M_k} \mu(B_{km}) \le \left(\sum_{m=1}^{M_k} \mu(B_{km})^u \right)^{\frac{1}{u}} 2^{\frac{kn}{u'}}. \qquad (9.118)$$

Hence, if $t > \frac{n}{u'}$ then $\mu_{uv}^t = \infty$. In other words, only values $0 \le t \le \frac{n}{u'}$ are of interest. As an example we assume that Γ is a d-set according to (9.1) with $0 \le d \le n$. Then $M_k \sim 2^{kd}$ and

$$\mu_{uv}^t \sim \left(\sum_{k=0}^{\infty} 2^{tkv}\, 2^{-k\frac{v}{u'}d} \right)^{\frac{1}{v}}. \qquad (9.119)$$

Hence by (9.111), (9.112),

$$\mu_{u\infty}^t < \infty \quad \text{if, and only if,} \quad t \le \frac{d}{u'}, \quad \text{and, hence,} \quad t_u = \frac{d}{u}. \qquad (9.120)$$

Next we wish to introduce β-quarks on Γ. We modify (2.99). The normalizing factor $s - \frac{n}{p}$ on $\mathbb{R}^n$ must be replaced by $s - \frac{d}{p}$ on d-sets. This follows at least in a somewhat indirect way from our considerations in [Triδ], Sections 18 and 20. Taking d-sets as a guide, (9.120) suggests that $s - t_p$ might be the right substitute. Recall that

$$\gamma^\beta = \gamma_1^{\beta_1} \cdots \gamma_n^{\beta_n} \quad \text{where} \quad \gamma \in \Gamma \subset \mathbb{R}^n \quad \text{and} \quad \beta \in \mathbb{N}_0^n.$$

9.27 Definition

Let μ be a finite Radon measure in $\mathbb{R}^n$ with compact support $\Gamma = \operatorname{supp} \mu$. Let $\{\psi^{k,m}\}$ be the resolution of unity introduced in 9.24 with (9.106)–(9.109). Let $1 < p \leq \infty$ and $s \geq 0$. Then

$$(\beta qu)_{km}(\gamma) = 2^{-k(s-t_p)} 2^{k|\beta|} (\gamma - \gamma^{k,m})^\beta \psi^{k,m}(\gamma), \quad \gamma \in \Gamma, \tag{9.121}$$

with

$$\beta \in \mathbb{N}_0^n, \quad k \in \mathbb{N}_0 \quad \text{and} \quad m = 1, \ldots, M_k,$$

is called an $(s,p) - \beta$-quark related to the ball B_{km} in (9.109).

9.28 Discussion, preparation, motivation

Obviously, we use $(\beta qu)_{km}(x)$ from (9.121), naturally extended to $x \in \mathbb{R}^n$, also as a β-quark in $\mathbb{R}^n$. Comparison of the exponents of the normalizing factors in (2.99) and (9.121) shows that there is a shifting of the smoothness index

$$s \quad \text{on} \quad \Gamma \quad \text{to} \quad s + \frac{n}{p} - t_p \quad \text{on} \quad \mathbb{R}^n. \tag{9.122}$$

To get a feeling for what is going on and to provide some motivation for how to proceed, we assume again that Γ is a compact d-set in $\mathbb{R}^n$ according to (9.1) with $0 < d < n$. By (9.120) the index-shifting (9.122) specifies

$$s \quad \text{on} \quad d\text{-set} \quad \Gamma \quad \text{and} \quad s + \frac{n-d}{p} \quad \text{on} \quad \mathbb{R}^n, \quad 0 < d < n. \tag{9.123}$$

Furthermore, by [Triδ], Theorem 18.6, p. 139, we have for these compact d-sets Γ,

$$tr_\Gamma B_{pq}^{\frac{n-d}{p}}(\mathbb{R}^n) = L_p(\Gamma), \quad 0 < d < n, \quad 1 < p < \infty, \quad 0 < q \leq 1, \tag{9.124}$$

(equivalent quasi-norms). Taking (9.124) as a starting point we introduced in [Triδ], Definition 20.2, p. 159, on these compact d-sets Γ with $0 < d < n$, the spaces

$$B_{pq}^s(\Gamma) = tr_\Gamma B_{pq}^{s+\frac{n-d}{p}}(\mathbb{R}^n), \quad s > 0, \quad 1 < p < \infty, \quad 0 < q \leq \infty. \tag{9.125}$$

In contrast to [Triδ], 20.2, we now restrict p by $p > 1$. Then we do not have the notational difficulty discussed in [Triδ], 20.3, pp. 160–161, and we can use the notation $B^s_{pq}(\Gamma)$ without any ambiguity. We take (9.125) and the comparison between (9.123) and (9.122) as a motivation in our more general situation. But in contrast to [Triδ] we first introduce corresponding spaces $B^s_{pq}(\Gamma)$ intrinsically and prove afterwards the counterpart of (9.125). As mentioned and justified at the end of 9.24 we now give preference to B^s_{pq}-spaces, compared with F^s_{pq}-spaces. Then one needs the counterparts of the sequence spaces b_{pq} from (2.7), of the (quasi-)norms (2.23), and the numbers ϱ and r with $\varrho > r$ and r given by (2.14). As for $r \geq 0$ we must estimate the influence of $|\beta|$ on (9.121), hence, by (9.106) with $2c_2 \geq 1$,

$$2^{k|\beta|}\,|\gamma - \gamma^{k,m}|^{|\beta|}\,\psi^{k,m}(\gamma) \leq (2c_2)^{|\beta|} = 2^{r|\beta|}. \tag{9.126}$$

In other words,

$$2^r = 2c_2 \quad \text{with} \quad c_2 \quad \text{from} \quad (9.106) \tag{9.127}$$

is a good choice, where we assume in addition $2c_2 \geq 1$. The counterpart of (2.6) and (2.7) is given by

$$\lambda = \{\lambda_{km} \in \mathbb{C} : \quad k \in \mathbb{N}_0,\; m = 1, \ldots, M_k\}, \tag{9.128}$$

where M_k has the above meaning and

$$b^\Gamma_{pq} = \left\{\lambda : \|\lambda \,|b^\Gamma_{pq}\| = \left(\sum_{k=0}^{\infty}\left(\sum_{m=1}^{M_k}|\lambda_{km}|^p\right)^{\frac{q}{p}}\right)^{\frac{1}{q}} < \infty\right\}, \tag{9.129}$$

where we may now assume $1 < p < \infty$ and $0 < q \leq \infty$ (with the obvious modification of (9.129) if $q = \infty$). As usual $\frac{1}{q} + \frac{1}{q'} = 1$ with $q' = \infty$ if $q \leq 1$.

9.29 Definition

Let μ be a finite Radon measure in $\mathbb{R}^n$ with compact support $\Gamma = \operatorname{supp}\mu$. Let $1 < p < \infty$ and let t_p be given by (9.112).

(i) Let $s > 0$ and $0 < q \leq \infty$. Let $(\beta qu)_{km}$ be the $(s,p) - \beta$-quarks on Γ according to Definition 9.27. We put

$$\lambda = \{\lambda^\beta : \beta \in \mathbb{N}_0^n\} \quad \text{with} \quad \lambda^\beta = \{\lambda^\beta_{km} \in \mathbb{C} : k \in \mathbb{N}_0,\; m = 1, \ldots, M_k\}. \tag{9.130}$$

9. Traces on sets, related function spaces and their decompositions

Let $\varrho > r$, where r is given by (9.127), $\lambda^\beta \in b_{pq}^\Gamma$ and

$$\|\lambda\,|b_{pq}^\Gamma\|_\varrho = \sup_{\beta \in \mathbb{N}_0^n} 2^{\varrho|\beta|}\,\|\lambda^\beta\,|b_{pq}^\Gamma\| < \infty. \tag{9.131}$$

Then $B_{pq}^s(\Gamma)_\varrho$ is the collection of all $g \in L_1(\Gamma)$ which can be represented as

$$g(\gamma) = \sum_{\beta \in \mathbb{N}_0^n} \sum_{k=0}^{\infty} \sum_{m=1}^{M_k} \lambda_{km}^\beta\,(\beta qu)_{km}(\gamma), \quad \gamma \in \Gamma, \tag{9.132}$$

with (9.131). Furthermore,

$$\|g\,|B_{pq}^s(\Gamma)_\varrho\| = \inf \|\lambda\,|b_{pq}^\Gamma\|_\varrho, \tag{9.133}$$

where the infimum is taken over all admissible representations.

(ii)   Let $0 < q \le \infty$ and let $\mu_{p'q'}^{t_p} < \infty$ according to (9.111). Let $(\beta qu)_{km}$ be $(0,p)-\beta$-quarks. Then $B_{pq}^0(\Gamma)_\varrho$ is the collection of all $g \subset L_1(\Gamma)$ which can be represented as (9.132) with (9.131) and (9.133).

9.30   Remark

This is the counterpart of Definition 2.6. We shall see that the series in (9.132) converges absolutely in $L_1(\Gamma)$ and we write in analogy to (2.31)

$$g = \sum_{\beta,k,m} \lambda_{km}^\beta\,(\beta qu)_{km}(\gamma). \tag{9.134}$$

We discuss briefly the role of t_p. By 9.26 we have always $t_p \le \frac{n}{p}$. Hence $s+\frac{n}{p}-t_p$ in (9.122) is positive if

$$\text{either} \quad t_p < \frac{n}{p},\ s \ge 0 \quad \text{or} \quad t_p = \frac{n}{p},\ s > 0.$$

If $t_p = \frac{n}{p}$ then, by (9.118), only in case of $q' = \infty$ it might happen that $\mu_{p'q'}^{t_p} < \infty$. The resulting spaces $B_{pq}^0(\Gamma)_\varrho$ with $q \le 1$ make sense, but as we shall see later on, they do not fit in our scheme since $s + \frac{n}{p} - t_p = 0$. In our model case when Γ is a compact d-set with $0 < d < n$, then we have $t_p = \frac{d}{p}$ by (9.120). In this case the above spaces $B_{pq}^s(\Gamma)_\varrho$ coincide with the spaces in (9.125) ; furthermore $\mu_{p'\infty}^{t_p} < \infty$ and by (9.124) we shall get for compact d-sets with $0 < d < n$,

$$B_{pq}^0(\Gamma)_\varrho = L_p(\Gamma), \quad 1 < p < \infty, \quad 0 < q \le 1. \tag{9.135}$$

Whether something like this is valid for the more general situation according to the above definition is not clear. We return to this point in 9.34(i). To ensure

(9.135) one has to find substitutes for the arguments in [Triδ], formula (18.11) and Theorem 18.6 on pp. 137, 139–141. If one replaces Γ in (9.135) by $\mathbb{R}^n$ and if $B^0_{pq}(\mathbb{R}^n)$ has the usual meaning, then (9.135) is wrong. Hence in case of $\mathbb{R}^n$ (or domains in $\mathbb{R}^n$) there are no representations of type (9.134): One needs moment conditions.

So far all the spaces $B^s_{pq}(\Gamma)_\varrho$ seem to depend on the chosen β-quarks and on ϱ. But as we shall see this is not the case, at least if $s + \frac{n}{p} - t_p > 0$. But first we clarify the technical side.

9.31 Proposition

Under the hypotheses of Definition 9.29 the spaces $B^s_{pq}(\Gamma)_\varrho$ are quasi-Banach spaces for all admitted parameters s, p, q (Banach spaces if $q \geq 1$). The series (9.132) converges absolutely in $L_1(\Gamma)$ and for all spaces one has

$$B^s_{pq}(\Gamma)_\varrho \subset L_1(\Gamma) \tag{9.136}$$

(continuous embedding).

Proof Let $1 < p < \infty$, $0 < q \leq \infty$, $s = 0$ and $\mu^{t_p}_{p'q'} < \infty$. Let g be given by (9.132), where $(\beta qu)_{km}$ are $(0, p) - \beta$-quarks according to (9.121). With r given by (9.126) one gets by Hölder's inequality

$$\int_\Gamma |g(\gamma)| \mu(d\gamma)$$
$$\leq \sum_{\beta,k,m} 2^{kt_p} 2^{r|\beta|} \mu(B_{km}) |\lambda^\beta_{km}|$$
$$\leq \sum_\beta 2^{r|\beta|} \|\lambda^\beta \, |b^\Gamma_{pq}\| \left(\sum_{k=0}^\infty 2^{kt_p q'} \left(\sum_{m=1}^{M_k} \mu(B_{km})^{p'} \right)^{\frac{q'}{p'}} \right)^{\frac{1}{q'}}$$
$$\leq \mu^{t_p}_{p'q'} \sum_\beta 2^{r|\beta|} \|\lambda^\beta \, |b^\Gamma_{pq}\|. \tag{9.137}$$

In case of $q \leq 1$, one uses first Hölder's inequality with $q = 1$ and $q' = \infty$ and afterwards the monotonicity of b^Γ_{pq} with respect to q. Since $\varrho > r$ one can estimate the right-hand side of (9.137) by the quasi-norm in (9.131). This proves (9.136) in this case. If $s > 0$ then one has to use $(s, p) - \beta$-quarks according to (9.121) and by (9.117) the situation is even better and one gets the counterpart of (9.137). By the completeness of the sequence spaces involved and standard arguments of Banach space theory, one proves that all the spaces $B^s_{pq}(\Gamma)_\varrho$ are complete.

9.32 Trace spaces, revisited

Let again μ be a finite Radon measure in $\mathbb{R}^n$ with compact support $\Gamma = \operatorname{supp} \mu$. We described our point of view concerning trace spaces of $F_{pq}^s(\mathbb{R}^n)$ in (9.38), (quasi-)normed by (9.39) with the underlying inequality (9.14) for $\varphi \in S(\mathbb{R}^n)$, where $r = 1$. By Theorem 9.21 these trace spaces $tr_\Gamma F_{pq}^s(\mathbb{R}^n)$ are independent of q if Γ satisfies the ball condition. Hence it is reasonable to switch from F_{pq}^s to B_{pq}^s. Although there is hardly anything new, we describe briefly this set-up in terms of B_{pq}^s-spaces. Let

$$1 < p < \infty, \quad 0 < q \leq \infty, \quad \text{and} \quad s > 0. \tag{9.138}$$

First one asks whether there is a constant $c > 0$ such that

$$\|tr_\Gamma \varphi \,|L_1(\Gamma)\| \leq c \|\varphi \,|B_{pq}^s(\mathbb{R}^n)\| \tag{9.139}$$

for all $\varphi \in S(\mathbb{R}^n)$. If $q < \infty$ then $S(\mathbb{R}^n)$ is dense in $B_{pq}^s(\mathbb{R}^n)$ and we define $tr_\Gamma f$ with $f \in B_{pq}^s(\mathbb{R}^n)$ by completion. If $q = \infty$ then we shall always have the situation that tr_Γ exists even for the larger space $B_{pp}^{s-\varepsilon}(\mathbb{R}^n)$ for some $\varepsilon > 0$. In particular, $tr_\Gamma f$ makes sense also for $f \in B_{p\infty}^s(\mathbb{R}^n)$. Hence we can introduce in analogy to (9.38), (9.39),

$$tr_\Gamma B_{pq}^s(\mathbb{R}^n) = \{g \in L_1(\Gamma) : \text{there is an } f \in B_{pq}^s(\mathbb{R}^n) \text{ with } tr_\Gamma f = g\} \tag{9.140}$$

quasi-normed by

$$\|g \,|tr_\Gamma B_{pq}^s(\mathbb{R}^n)\| = \inf \|f \,|B_{pq}^s(\mathbb{R}^n)\|, \tag{9.141}$$

where the infimum is taken over all $f \in B_{pq}^s(\mathbb{R}^n)$ with $tr_\Gamma f = g$.

9.33 Theorem

Let μ be a finite Radon measure in $\mathbb{R}^n$ with compact support $\Gamma = \operatorname{supp} \mu$.

(i) Let

$$1 < p < \infty, \quad 0 < q \leq \infty, \quad s > 0, \tag{9.142}$$

and let t_p be given by (9.112). Then the spaces $B_{pq}^s(\Gamma)_\varrho$, introduced in Definition 9.29, are independent of all admitted resolutions of unity (9.106)–(9.108) and of all admitted numbers ϱ (equivalent quasi-norms), and will be denoted now simply by $B_{pq}^s(\Gamma)$. Furthermore,

$$B_{pq}^s(\Gamma) = tr_\Gamma B_{pq}^{s+\frac{n}{p}-t_p}(\mathbb{R}^n) \tag{9.143}$$

(equivalent quasi-norms).

(ii) *Let*

$$1 < p < \infty, \quad 0 < q < \infty, \quad t_p < \frac{n}{p}, \quad \text{and} \quad \mu_{p'q'}^{t_p} < \infty, \tag{9.144}$$

according to (9.111). Then assertions of part (i) hold also for the spaces $B_{pq}^0(\Gamma)_\varrho$, now denoted by $B_{pq}^0(\Gamma)$. In particular,

$$B_{pq}^0(\Gamma) = tr_\Gamma B_{pq}^{\frac{n}{p}-t_p}(\mathbb{R}^n), \tag{9.145}$$

(equivalent quasi-norms).

Proof By the discussion in 9.30 we always have $s + \frac{n}{p} - t_p > 0$. Hence we can apply Definition 2.6 to the above spaces $B_{pq}^{s+\frac{n}{p}-t_p}(\mathbb{R}^n)$, where we now assume that the original $(s + \frac{n}{p} - t_p, p) - \beta$-quarks in the quarkonial decomposition (2.24) for $f \in B_{pq}^{s+\frac{n}{p}-t_p}(\mathbb{R}^n)$ are replaced by the generalized versions in (2.99). In addition we assume that the resolutions of unity (2.97) used there are adapted to Γ in the way described in 9.24. Then $g = tr_\Gamma f$ is just the quarkonial decomposition (9.132) now with (s,p)-quarks on Γ. Hence, $tr_\Gamma f \in B_{pq}^s(\Gamma)_\varrho$. By (9.136) we have in particular (9.139) with $s + \frac{n}{p} - t_p$ in place of s. In particular $tr_\Gamma B_{pq}^{s+\frac{n}{p}-t_p}(\mathbb{R}^n)$ makes sense and

$$\|tr_\Gamma f \,|B_{pq}^s(\Gamma)_\varrho\| \le c \, \|f\,|B_{pq}^{s+\frac{n}{p}-t_p}(\mathbb{R}^n)\|_{\psi,\varrho} \tag{9.146}$$

in the notation (2.25). Conversely, let g be given by (9.132), (9.133) with respect to $(s,p) - \beta$-quarks on Γ. By the discussion in 9.28 we interpret this decomposition in $\mathbb{R}^n$ with the index shifting (9.122). Denoting the function obtained in this way by $f = ext\, g$ then

$$\|f\,|B_{pq}^{s+\frac{n}{p}-t_p}(\mathbb{R}^n)\| \le c \,\|g\,|B_{pq}^s(\Gamma)_\varrho\|, \quad tr_\Gamma f = g. \tag{9.147}$$

But then we obtain (9.143) and (9.145) in all cases. Furthermore, by Theorem 2.9 (in the version of 2.14), all (quasi-)norms for all admitted resolutions of unity and all admitted values of ϱ are equivalent to each other.

9.34 Miscellany: critical comments, references, further results

We collect in this rather long subsection some further topics related to this section. But otherwise the diverse ten points (or subsubsections) are largely independent of each other. They will be denoted by 9.34(i)–9.34(x).

9.34(i) Recall that a compact set Γ in $\mathbb{R}^n$ is called a d-set with $0 < d < n$ if there is a Radon measure μ with

$$\operatorname{supp}\mu = \Gamma, \quad \mu(B(\gamma,r)) \sim r^d, \quad 0 < r < 1, \tag{9.148}$$

where again $B(\gamma, r)$ is a ball in $\mathbb{R}^n$ centred at $\gamma \in \Gamma$ and of radius r. We have (9.120), in particular $t_p = \frac{d}{p}$ with $1 < p < \infty$, (9.124), and (9.125), where the latter spaces now coincide with (9.143). In other words, taking d-sets as a model case, our definition of the spaces $B_{pq}^s(\Gamma)$ in 9.29 seems to be justified by Theorem 9.33. Comparing (9.119), (9.144), (9.145) with (9.124) we obtain, as announced in (9.135), for compact d-sets with $0 < d < n$,

$$B_{pq}^0(\Gamma) = L_p(\Gamma), \quad 1 < p < \infty, \quad 0 < q \le 1. \tag{9.149}$$

However for more general measures in the above Theorem 9.33 there is little hope of finding assertions of type (9.149). We discuss an example which we will treat later on in Sections 22 and 23 in connection with the spectral theory of fractal drums. There we are interested in (d, Ψ)-sets according to (9.1). A typical example is a compact set Γ in $\mathbb{R}^n$ for which there is a Radon measure μ with

$$\operatorname{supp}\mu = \Gamma, \quad \mu(B(\gamma,r)) \sim r^d |\log r|^b, \quad 0 < r < \frac{1}{2}, \tag{9.150}$$

where $0 < d < n$ and $\Psi(r) = |\log r|^b$ with $b \in \mathbb{R}$. Then Γ satisfies the ball condition introduced in 9.16. This follows from Proposition 9.18 and will be proved in Proposition 22.6. Hence one can apply Theorem 9.21. Similarly as for d-sets we have also here $t_p = \frac{d}{p}$, in modification of (9.119), the spaces $B_{pq}^s(\Gamma)$ from Definition 9.29 make sense and we get (9.143), but not necessarily (9.149). If one wishes to incorporate $L_p(\Gamma)$ as a trace space, then one must modify the smoothness $\sigma = s + \frac{n-d}{p}$ in $B_{pq}^\sigma(\mathbb{R}^n)$ by a perturbed smoothness (σ, Ψ^a) with $a \in \mathbb{R}$, resulting in spaces $B_{pq}^{(\sigma, \Psi^{a+\frac{1}{p}})}(\mathbb{R}^n)$ and corresponding trace spaces $B_{pq}^{(s, \Psi^a)}(\Gamma)$, incorporating $L_p(\Gamma)$. We return to this subject in Section 22, where one finds also precise definitions and results. We refer in this context to [EdT98], [EdT99a], and the detailed studies of these spaces in [Mou99], [Mou01b] and [Bri99], where the latter paper deals with atomic and quarkonial decompositions of the spaces $B_{pq}^{(s, \Psi^a)}(\Gamma)$ on Γ. At the moment we wish only to make clear that

for more general sets Γ and related measures μ than d-sets, there might be several different interesting scales of trace spaces which must be distinguished carefully.

9.34(ii) Both d-sets and, more generally, (d, Ψ)-sets are isotropic: there are no distinguished directions in $\mathbb{R}^n$. On the other hand, in fractal geometry one creates wonderful trees, ferns and other fractal beauties with the help of so-called iterated function systems. The resulting compact fractal sets Γ are usually not isotropic. A discussion including the necessary references may be found in [Triδ], Sections 4 and 5. There we introduced also anisotropic and nonisotropic compact d-sets Γ in the plane and used the outcome later on for a spectral theory. Obviously, one can apply Definition 9.29 to introduce the spaces $B_{pq}^s(\Gamma)$ on general fractals and one again gets (9.143) and (9.145). But it is even more doubtful than in the above step from (isotropic) d-sets to (d, Ψ)-sets whether the resulting spaces $B_{pq}^s(\Gamma)$ are well adapted to these general fractals. Again there is little hope of finding something like (9.149) in this context. On the other hand in the anisotropic case, respecting the axes of coordinates, a corresponding theory for anisotropic spaces has been developed in [Far00], [Far99] and [Far98]. In the first paper one finds the anisotropic counterpart of the theory of the spaces in $\mathbb{R}^n$ developed in Section 2, in the two other papers the relations to anisotropic fractals, related spaces, including an anisotropic counterpart of (9.149) and applications to spectral theory. Again there are several interesting scales of spaces of B_{pq}^s-type on anisotropic fractals Γ.

9.34(iii) The first systematic study of B_{pq}^s-spaces on d-sets is due to H. Wallin and A. Jonsson. They summarized their results in [JoW84]. The corresponding spaces $B_{pq}^s(\Gamma)$ are defined with the help of first and higher differences $\Delta_h^M f(\gamma)$ and approximation procedures by polynomials. Sometimes the so-called Markov property, see [JoW84], Chapter 2, of the underlying d-set is required (but not always). Spaces of B_{pq}^s-type on more general sets, based on differences, wavelets, and atoms, were considered later on in [Jon93b], [Jon93a], [Jon94], [JoW95], [Byl94], [Jon98a], [Jon98b]. Of course, beyond (isotropic) d-sets one has always the ambiguities described above in (i) and (ii). Nevertheless it would be of interest to compare the approach in the quoted papers with our own method. As for the quarkonial approach to the spaces $B_{pq}^s(\Gamma)$ according to Definition 9.29 and Theorem 9.33 we refer also to [Bri01]. This thesis covers also spaces $B_{pq}^s(\Gamma)$ with $p \leq 1$, in particular in connection with d-sets and (d, Ψ)-sets according to (9.1).

9.34(iv) The question of how to define traces always attracted a lot of attention. Our own method is rather crude: we ask first for inequalities of type (9.14) or (9.139), say, for $\varphi \in S(\mathbb{R}^n)$, and declare the rest to be a matter of completion (with a minor struggle if $S(\mathbb{R}^n)$ is not dense in the space considered). But there are more subtle possibilities, consisting, roughly speaking, in diverse refinements of Lebesgue points of measurable functions. Detailed studies may be found in [Maz85], [Zie89], [AdH96], and also in [JoW84], Chapter 8.

9.34(v) Let Ω be a bounded domain in $\mathbb{R}^n$. We introduced the spaces $B^s_{pq}(\Omega)$ and $F^s_{pq}(\Omega)$ in Definition 5.3 as restrictions of the corresponding spaces in $\mathbb{R}^n$ on Ω for the full range of the parameters $0 < p \le \infty$ (with $p < \infty$ for the F-spaces), $0 < q \le \infty$, $s \in \mathbb{R}$. On the other hand with $\Gamma = \overline{\Omega}$,

$$F^s_{pq}(\overline{\Omega}) = tr_{\overline{\Omega}} F^s_{pq}(\mathbb{R}^n) \quad \text{and} \quad B^s_{pq}(\overline{\Omega}) = tr_{\overline{\Omega}} B^s_{pq}(\mathbb{R}^n) \qquad (9.151)$$

fit in the scheme of this section, where the parameters p, q, s are restricted by (9.27) (with, say, $r = 1$) and (9.142), respectively. One may ask how the spaces on Ω and on $\overline{\Omega}$ are related to each other. This results in the question of whether there are intrinsic quarkonial decompositions for the spaces on Ω in dependence on geometric properties of Ω. We discussed this problem in detail in [TrW96] on an atomic level. A brief description may also be found in [ET96], 2.5, pp. 57–65. However there is no problem in extending these considerations to quarkonial decompositions, at least in those cases, where no moment conditions are needed. We do not go into detail. But we describe the outcome briefly. As in [ET96], p.58, we *denote by $MR(n)$ (minimally regular) the collection of all bounded domains Ω in $\mathbb{R}^n$ with $\Omega = int(\overline{\Omega})$* (here $int(\overline{\Omega})$ means the interior of $\overline{\Omega}$). Then, in obvious modification of [ET96], 2.5, one has intrinsic quarkonial decompositions for all spaces $B^s_{pq}(\Omega)$ with

$$0 < p \le \infty, \quad 0 < q \le \infty, \quad s > \sigma_p = n\left(\frac{1}{p} - 1\right)_+. \qquad (9.152)$$

Restricted to our scheme, given by (9.142), we obtain,

$$B^s_{pq}(\overline{\Omega}) = B^s_{pq}(\Omega), \quad 1 < p < \infty, \, 0 < q \le \infty, \, s > 0, \qquad (9.153)$$

where the quarkonial decomposition in question is just that one from Definition 9.29 with $\Gamma = \overline{\Omega}$. The above assertion means that this can be extended to the parameter values in (9.152). As for the spaces $F^s_{pq}(\Omega)$ the assumption $\Omega \in MR(n)$ is not sufficient. We follow again [TrW96] and [ET96], p.59, and denote by $IR(n)$ *(interior regular) the collection of all domains $\Omega \in MR(n)$ for which there is a positive number c such that for any ball B centred on $\partial\Omega$ with side length less than 1,*

$$|\Omega \cap B| \ge c|B|. \qquad (9.154)$$

There is a counterpart $f^{\overline{\Omega}}_{pq}$ of $b^{\overline{\Omega}}_{pq}$ in (9.129) adapting (2.13) to $\Gamma = \overline{\Omega}$, which may also be found in detail in [TrW96] or [ET96], p.62. Then one has for all spaces $F^s_{pq}(\Omega)$ with

$$0 < p < \infty, \quad 0 < q \le \infty, \quad s > \sigma_{pq} = n\left(\frac{1}{\min(p,q)} - 1\right)_+, \qquad (9.155)$$

intrinsic quarkonial decompositions in analogy to the atomic decompositions in [TrW96] and [ET96], p. 64. Restricted to our scheme, given by (9.27), we obtain

$$F^s_{pq}(\overline{\Omega}) = F^s_{pq}(\Omega), \quad 1 < p < \infty, \, 1 < q < \infty, \, s > 0. \tag{9.156}$$

9.34(vi) Closely related to the discussion in (v) is the following question. Again let $\Omega \in MR(n)$ and let $\Gamma = \partial\Omega$. If we have (9.139) for all $\varphi \in S(\mathbb{R}^n)$, then $tr_\Gamma B^s_{pq}(\mathbb{R}^n)$ with (9.140), (9.141) makes sense. One can ask the same questions with Ω in place of $\mathbb{R}^n$ in (9.139)–(9.141). Let $C^\infty(\overline{\Omega})$ be the restriction of $S(\mathbb{R}^n)$ to $\overline{\Omega}$. Then the counterpart of (9.139) is given by

$$\|tr_\Gamma \varphi \,|L_1(\Gamma)\| \le c \,\|\varphi \,|B^s_{pq}(\Omega)\|, \quad \varphi \in C^\infty(\overline{\Omega}). \tag{9.157}$$

In case of affirmative answers of the question (9.157) one has obvious counterparts of (9.140), (9.141) with Ω in place of $\mathbb{R}^n$. Denoting by $\mathbb{R}^n tr_\Gamma g$ and by $\Omega tr_\Gamma f$ the traces of $g \in B^s_{pq}(\mathbb{R}^n)$ and $f \in B^s_{pq}(\Omega)$ (if they exist) then one would like to have

$$tr_\Gamma B^s_{pq}(\mathbb{R}^n) = tr_\Gamma B^s_{pq}(\Omega) \quad \text{and} \quad \Omega tr_\Gamma f = \mathbb{R}^n tr_\Gamma g \tag{9.158}$$

with $f = g|\Omega$ (restriction of g to Ω). Since $\Omega \in MR(n)$ one gets (9.158) as an easy consequence of (9.153). More precisely: Under the hypotheses of Theorem 9.33(i) one gets with $\Gamma = \partial\Omega$,

$$B^s_{pq}(\Gamma) = tr_\Gamma B^{s+\frac{n}{p}-t_p}_{pq}(\mathbb{R}^n) = tr_\Gamma B^{s+\frac{n}{p}-t_p}_{pq}(\Omega) \tag{9.159}$$

and individually

$$\Omega tr_\Gamma f = \mathbb{R}^n tr_\Gamma g \quad \text{where} \quad f = g|\Omega. \tag{9.160}$$

As for the F^s_{pq}-spaces it seems to be reasonable to combine (9.154) with the ball condition introduced in 9.16. We say that $\Gamma = \partial\Omega$ satisfies the *interior ball condition* if it satisfies the ball condition according to Definition 9.16 with (9.82) and, additionally,

$$B(y, \eta t) \subset \Omega, \tag{9.161}$$

in the notation used there. Then we have in particular (9.154). Furthermore by Theorem 9.21 the trace spaces in (9.93) are independent of q. This applies also to $tr_\Gamma F^s_{pq}(\Omega)$, where now the extension of q in (9.156) to $q \le 1$ and $q = \infty$ does not cause any problem. More precisely: Let Ω be a bounded domain in $\mathbb{R}^n$

9. Traces on sets, related function spaces and their decompositions   157

satisfying the interior ball condition. Then we can complement (9.159) under the hypotheses of Theorem 9.33(i) by

$$B^s_{pp}(\Gamma) = tr_\Gamma F^{s+\frac{n}{p}-t_p}_{pq}(\mathbb{R}^n) = tr_\Gamma F^{s+\frac{n}{p}-t_p}_{pq}(\Omega), \qquad (9.162)$$

again with (9.160).

9.34(vii) In connection with quarkonial decompositions of the spaces $B^s_{pq}(\mathbb{R}^n)$ and $F^s_{pq}(\mathbb{R}^n)$ in Definition 2.6 and Theorem 2.9, we obtained in Corollary 2.12 universal optimal coefficients (2.82), (2.83) which depend linearly on $f \in S'(\mathbb{R}^n)$. In other words, if $f \in S'(\mathbb{R}^n)$ then one can first calculate the coefficients $\lambda^\beta_{\nu m}$ in (2.82) and then one can ask in the indicated way to which spaces $B^s_{pq}(\mathbb{R}^n)$ or $F^s_{pq}(\mathbb{R}^n)$ this distribution belongs. This applies not only to the spaces $B^s_{pq}(\mathbb{R}^n)$ and $F^s_{pq}(\mathbb{R}^n)$ according to Definition 2.6 with the restrictions for s in (2.21), (2.26), but also to the general situation treated in Section 3. We refer to 3.7. The situation is similar for spaces on domains and on manifolds. The respective remarks may be found in 6.7 and 7.23. In case of Taylor expansions of arbitrary tempered distributions $f \in S'(\mathbb{R}^n)$ the situation is slightly different. As indicated in 8.11, the optimal coefficients depend linearly on f but they are no longer universal. From this point of view (optimal coefficients depending linearly on the considered function and being, at the best, universal), the situation treated in this section, and reflected by Definition 9.29 and Theorem 9.33, is unsatisfactory. However it is questionable whether one can expect something like a counterpart of Corollary 2.12. But a few assertions can be made. In case of Hilbert spaces we put $H^s = B^s_{2,2}$ both on Γ according to Definition 9.29 and on $\mathbb{R}^n$. Then we have by (9.143),

$$H^s(\Gamma) = tr_\Gamma H^{s+\frac{n}{2}-t_2}(\mathbb{R}^n), \quad s > 0. \qquad (9.163)$$

By Hilbert space arguments (which have nothing to do with the specific situation considered here) it follows that there is an isomorphic map, denoted by ext, from $H^s(\Gamma)$ onto the orthogonal complement of

$$\{f \in H^{s+\frac{n}{2}-t_2}(\mathbb{R}^n) : tr_\Gamma f = 0\} = H^{s+\frac{n}{2}-t_2}_\Gamma(\mathbb{R}^n), \qquad (9.164)$$

which results in the Weyl decomposition

$$H^{s+\frac{n}{2}-t_2}(\mathbb{R}^n) = H^{s+\frac{n}{2}-t_2}_\Gamma(\mathbb{R}^n) \oplus ext\, H^s(\Gamma). \qquad (9.165)$$

Expanding $f \in H^{s+\frac{n}{2}-t_2}(\mathbb{R}^n)$ according to Corollary 2.12 optimally and linearly and reducing these quarkonial expansions via $id = tr_\Gamma \circ ext$ to $H^s(\Gamma)$ as in the proof of Theorem 9.33, one gets optimal quarkonial decompositions of $g \in H^s(\Gamma)$ where the coefficients depend linearly on g (but they are not universal in the above sense). If $B^s_{pq}(\Gamma)$ is not a Hilbert space then the situation is less favourable. By the above arguments one finds optimal coefficients λ_{km}^β in the quarkonial decomposition (9.132) which depend linearly on g if there is a linear and bounded extensions operator from $B_{pq}^s(\Gamma)$ into $B_{pq}^{s+\frac{n}{p}-t_p}(\mathbb{R}^n)$. If Γ is a compact d-set in $\mathbb{R}^n$ with $0 < d < n$ and if $1 < p < \infty$, $1 \le q \le \infty$ and $0 < s \notin \mathbb{N}$ then, by [JoW84], Theorem 3 on p. 155, and [Jon96] there is such a linear and bounded extension operator,

$$ext : \quad B_{pq}^s(\Gamma) \mapsto B_{pq}^{s+\frac{n-d}{p}}(\mathbb{R}^n). \tag{9.166}$$

Hence there are optimal coefficients λ_{km}^β in (9.132) which depend linearly on g (but they are not universal). Other cases may be found in [Bri01].

9.34(viii) Again let $\Gamma = \operatorname{supp}\mu$ be a compact d-set according to (9.1) with $0 < d < n$. Let $1 < p < \infty$. Dualising (9.124) one gets (with the identification according to (9.16))

$$id_\Gamma L_p(\Gamma) = B_{p,\infty}^{-\frac{n-d}{p'},\Gamma}(\mathbb{R}^n), \quad \frac{1}{p} + \frac{1}{p'} = 1, \tag{9.167}$$

where

$$B_{pq}^{s,\Gamma}(\mathbb{R}^n) = \{f \in B_{pq}^s(\mathbb{R}^n) : f(\varphi) = 0 \text{ if } \varphi \in S(\mathbb{R}^n) \text{ and } \varphi|\Gamma = 0\}, \tag{9.168}$$

$s \in \mathbb{R}$, $0 < q \le \infty$. We refer for the definition (9.168) and the result (9.167) to [Triδ], 17.2, p. 125, and 18.2, p. 136, respectively. By the above discussion in point (i) nothing like (9.167) can be expected in the general case. But (9.168) makes sense for any, say, compact set Γ in $\mathbb{R}^n$ with $|\Gamma| = 0$. Then $s \le 0$ (otherwise the space is trivial) and

$$B_{pq}^{s,\Gamma}(\mathbb{R}^n) \subset \widetilde{B}_{pq}^{s,\Gamma}(\mathbb{R}^n) = \{f \in B_{pq}^s(\mathbb{R}^n) : \operatorname{supp} f \subset \Gamma\}. \tag{9.169}$$

In general these two spaces do not coincide. But we have

$$B_{pq}^{s,\Gamma}(\mathbb{R}^n) = \widetilde{B}_{pq}^{s,\Gamma}(\mathbb{R}^n), \quad s < 0, \ 1 < p < \infty, \ 0 < q \le \infty, \tag{9.170}$$

if the compact set Γ with $|\Gamma| = 0$ preserves the Markov inequality. For such a set Γ one has $D^\alpha \varphi|\Gamma = 0$ for all $\alpha \in \mathbb{N}_0^n$ if $\varphi \in S(\mathbb{R}^n)$ and $\varphi|\Gamma = 0$. We refer to [JoW84], pp. 34–35, 41. This is sufficient to prove (9.170). Any d-set with $d > n - 1$ has this property, [JoW84], p. 39. This observation came out in discussions with M. Bricchi (Jena, 2000). Details may be found in [Bri00] and in his thesis [Bri01]. In this book we do not rely on Markov inequalities. But the following characterization due to P. Wingren, [Wig88], Proposition 2, p. 430, or [JoW97], Theorem 2, p. 193, might be of some interest also for our later discussions on Weyl measures in 19.18(i). *A compact set Γ in $\mathbb{R}^n$*

preserves the Markov inequality if, and only if, there is a number $c > 0$ such that for every ball $B(\gamma, r)$ centred at $\gamma \in \Gamma$ and of radius r, $0 < r < 1$, there are $n+1$ points $\gamma_j \in \Gamma \cap B(\gamma, r)$ such that the ball inscribed in the convex hull $\text{conv}(\gamma_1, \ldots, \gamma_{n+1})$ has a radius not less than cr.

9.34(ix) Again let μ be a finite Radon measure in $\mathbb{R}^n$ with compact support $\Gamma = \text{supp}\,\mu$. By Theorems 9.9(ii) and 9.33 we have satisfactory answers to the questions under which conditions trace spaces exist. Excluding $q = \infty$ in (9.142), it makes sense to ask under what conditions $D(\mathbb{R}^n\backslash\Gamma)$ is dense in

$$B_{pq,\Gamma}^{s+\frac{n}{p}-t_p}(\mathbb{R}^n) = \left\{ f \in B_{pq}^{s+\frac{n}{p}-t_p}(\mathbb{R}^n) : tr_\Gamma f = 0 \right\}, \tag{9.171}$$

where all the notation has the same meaning as in Theorem 9.33. Problems of this type have a long history and in the classical case when Γ is smooth, for example $\Gamma = \partial\Omega$ is the boundary of a C^∞ domain in $\mathbb{R}^n$, one has final answers which may be found e.g., in [Triα], 4.7.1, p. 330. If Γ is non-smooth or even an arbitrary set in $\mathbb{R}^n$ the situation is much more complicated. In the context of *spectral synthesis*, where conditions are expressed in terms of capacities, one has deep and definitive answers. We refer to [AdH96], Theorems 9.1.3 and 10.1.1, pp. 234, 281. There one finds also historical comments. This theory goes back to Hedberg, Netrusov and their co-workers. Our aim here is different. We return to the above problem about the density of $D(\mathbb{R}^n\backslash\Gamma)$ in some spaces of type (9.171), later on in connection with the Dirichlet problem in fractal domains, see 19.5.

9.34(x) A quasi-metric ϱ on a set X is a function $X \times X \mapsto [0, \infty)$ with

$$\varrho(x, y) = \varrho(y, x) \quad \text{for all} \quad x \in X, \, y \in X, \tag{9.172}$$

$$\varrho(x, y) = 0 \quad \text{if, and only if,} \quad x = y, \tag{9.173}$$

$$\varrho(x, y) \leq c \left(\varrho(x, z) + \varrho(z, y)\right) \quad \text{for} \quad x \in X, \, y \in X, \, z \in X, \tag{9.174}$$

for some $c > 0$. By [CoW71] a space of homogeneous type (X, ϱ, μ) is a set X equipped with a quasi-metric ϱ (generating a topology) and a positive locally finite diffuse measure μ satisfying the doubling condition in analogy to (9.84). On spaces of such a type one can develop a substantial analysis, based on Calderón's reproducing formula and Littlewood-Paley techniques. Homogeneous spaces of type $\dot{B}_{pq}^s$ and $\dot{F}_{pq}^s$ with $1 \leq p, q \leq \infty$, $|s| \leq \varepsilon$, on (X, ϱ, μ) were introduced on these bases in [HaS94]. This theory has been gradually extended to homogeneous spaces with $p < 1$ in [Han94], [Han98], [HaL99], and to inhomogeneous spaces in [HLY99a] and [HLY99b]. This includes atomic characterizations, $T1$ theorems, and Calderón-Zygmund integral operators. We do not go into detail and refer to the quoted papers. If, as always in this section, $X = \Gamma = supp\,\mu$ is the compact support of a finite Radon measure μ in $\mathbb{R}^n$, naturally equipped with a metric (and maybe satisfying the doubling condition (9.84)), then we have at least three possibilities to define spaces of type B^s_{pq} and F^s_{pq}: the quarkonial approach according to Definition 9.29, Theorem 9.33 (maybe complemented by an F^s_{pq}-counterpart), the Jonsson-Wallin approach briefly mentioned in 9.34(iii), and the just indicated possibility. It would be of interest to study the interplay of these diverse possibilities more closely than has been done so far.

Chapter II
Sharp Inequalities

10 Introduction: Outline of methods and results

Let again $\mathbb{R}^n$ be euclidean n-space and let

$$0 < p < \infty, \quad 0 < q < \infty, \quad s > \sigma_p = n\left(\frac{1}{p} - 1\right)_+. \tag{10.1}$$

Then both $B^s_{pq}(\mathbb{R}^n)$ and $F^s_{pq}(\mathbb{R}^n)$ are not only subspaces of $S'(\mathbb{R}^n)$ (the collection of all tempered distributions in $\mathbb{R}^n$) but also subspaces of $L^{loc}_1(\mathbb{R}^n)$ (the collection of all complex-valued locally Lebesgue-integrable functions in $\mathbb{R}^n$, interpreted in the usual way as distributions). Let A^s_{pq} be either B^s_{pq} or F^s_{pq} and let $f \in A^s_{pq}(\mathbb{R}^n)$ with (10.1). Of interest is the singularity behaviour of f, usually expressed in terms of the distribution function μ_f, the non-increasing rearrangement f^* of f and its maximal function f^{**}, which are given by

$$\mu_f(\lambda) = |\{x \in \mathbb{R}^n \,:\, |f(x)| > \lambda\}|, \quad \lambda \geq 0, \tag{10.2}$$

$$f^*(t) = \inf\{\lambda \,:\, \mu_f(\lambda) \leq t\}, \quad t \geq 0, \tag{10.3}$$

and

$$f^{**}(t) = \frac{1}{t} \int_0^t f^*(\tau)\,d\tau, \quad t > 0, \tag{10.4}$$

respectively. Of interest are only those spaces $A^s_{pq}(\mathbb{R}^n)$ for which there exist functions $f \in A^s_{pq}(\mathbb{R}^n)$ such that $f^*(t)$ tends to infinity if $t > 0$ tends to zero. As indicated in Fig. 10.1 one has to distinguish between three cases. If $s > \frac{n}{p}$ then all spaces $B^s_{pq}(\mathbb{R}^n)$ and $F^s_{pq}(\mathbb{R}^n)$ are continuously embedded in $L_\infty(\mathbb{R}^n)$

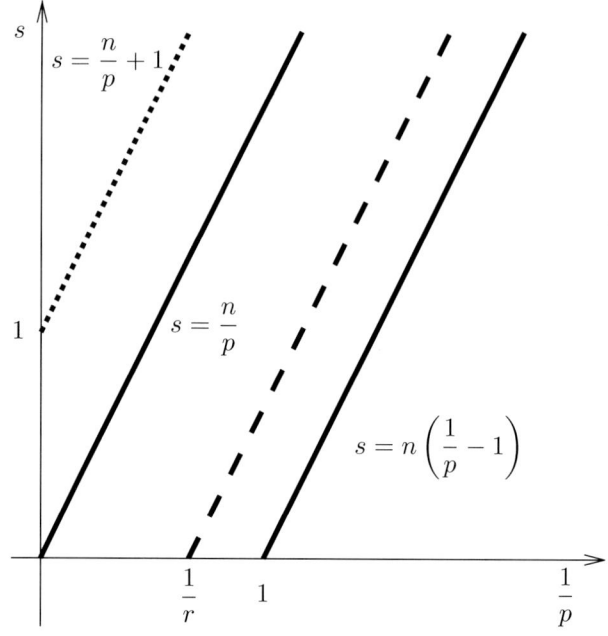

Fig. 10.1

and, hence, all functions $f^*(t)$ with $f \in A_{pq}^s(\mathbb{R}^n)$ are bounded. This is of no interest for us (at least as far as only the growth of functions is considered). The two remaining cases are called:

$$\text{sub-critical} \quad \text{if} \quad 0 < p < \infty, \quad 0 < q \leq \infty, \quad \sigma_p < s < \frac{n}{p}, \tag{10.5}$$

and

$$\text{critical} \quad \text{if} \quad 0 < p < \infty, \quad 0 < q \leq \infty, \quad s = \frac{n}{p}. \tag{10.6}$$

In all spaces belonging to the sub-critical case there are essentially unbounded functions f with $f^*(t) \to \infty$ if $t \downarrow 0$. In the critical case the situation is more delicate. It depends on the parameters p, q and whether A_{pq}^s is B_{pq}^s or F_{pq}^s. We give in Section 11 a detailed and definitive description of the relevant scenery surrounding the embedding of the spaces $B_{pq}^s(\mathbb{R}^n)$ and $F_{pq}^s(\mathbb{R}^n)$ in $L_1^{loc}(\mathbb{R}^n)$, $L_\infty(\mathbb{R}^n)$ and other classical target spaces. Let $A_{pq}^s(\mathbb{R}^n)$ be a space which is embedded in $L_1^{loc}(\mathbb{R}^n)$ but not in $L_\infty(\mathbb{R}^n)$. Then (in temporarily somewhat vague terms) the *growth envelope function* $\mathcal{E}_G A_{pq}^s$ is a function $t \mapsto \mathcal{E}_G A_{pq}^s(t)$ which is equivalent to

$$\mathcal{E}_G |A_{pq}^s(t) = \sup \left\{ f^*(t) \, : \, \|f\,|A_{pq}^s(\mathbb{R}^n)\| \leq 1 \right\}, \quad 0 < t < \varepsilon, \tag{10.7}$$

10. Introduction: Outline of methods and results

where $0 < \varepsilon < 1$ is a given number. The notation $\mathcal{E}_G | A_{pq}^s$ indicates that this function is taken with respect to a given quasi-norm $\| \cdot | A_{pq}^s(\mathbb{R}^n) \|$. If

$$\| \cdot | A_{pq}^s(\mathbb{R}^n) \|_1 \sim \| \cdot | A_{pq}^s(\mathbb{R}^n) \|_2 \tag{10.8}$$

are two equivalent quasi-norms in a given space $A_{pq}^s(\mathbb{R}^n)$, then (in obvious notation)

$$\mathcal{E}_G |_1 A_{pq}^{\varepsilon}(t) \sim \mathcal{E}_G |_2 A_{pq}^{\varepsilon}(t), \quad 0 < t \le \varepsilon, \tag{10.9}$$

are also equivalent. Since we never distinguish between equivalent quasi-norms of a given space $A_{pq}^s(\mathbb{R}^n)$ it is reasonable to extend this point of view to what we wish to call later on the *growth envelope function* $\mathcal{E}_G A_{pq}^s$. By definition one has the sharp inequality

$$\sup_{0 < t < \varepsilon} \frac{f^*(t)}{\mathcal{E}_G | A_{pq}^s(t)} \le c \, \| f | A_{pq}^s(\mathbb{R}^n) \|, \quad f \in A_{pq}^s(\mathbb{R}^n). \tag{10.10}$$

One may look at the left-hand side of (10.10) as the quasi-norm $L_\infty(I_\varepsilon, \mu)$ for a suitable Borel measure μ on the interval $I_\varepsilon = (0, \varepsilon]$. We wish to strengthen (10.10), replacing $L_\infty(I_\varepsilon, \mu)$ by $L_u(I_\varepsilon, \mu)$. Let $\mathcal{E}_G(t)$ be an unbounded positive, continuous, monotonically decreasing function on I_ε (this will apply in particular to all growth envelope functions with which we deal later on). Then the associated Borel measure $\mu = \mu_\Psi$ with respect to the distribution function $\Psi(t) = -\log \mathcal{E}_G(t)$ is the natural and distinguished choice for the above purpose. If $g(t)$ is a non-negative monotonically decreasing function on I_ε (with $g = f^*$ where $f \in A_{pq}^s(\mathbb{R}^n)$ as a typical example) then the corresponding quasi-norms are monotone,

$$\sup_{0 < t < \varepsilon} \frac{g(t)}{\mathcal{E}_G(t)} \le c_1 \left(\int_0^\varepsilon \left(\frac{g(t)}{\mathcal{E}_G(t)} \right)^{u_1} \mu_\Psi(dt) \right)^{\frac{1}{u_1}}$$

$$\le c_0 \left(\int_0^\varepsilon \left(\frac{g(t)}{\mathcal{E}_G(t)} \right)^{u_0} \mu_\Psi(dt) \right)^{\frac{1}{u_0}}, \tag{10.11}$$

where $0 < u_0 < u_1 < \infty$. We refer to Proposition 12.2. Hence by (10.10) it makes sense to ask for the smallest number $u = u(A_{pq}^s)$ with $0 < u \le \infty$ such that

$$\left(\int_0^\varepsilon \left(\frac{f^*(t)}{\mathcal{E}_G(t)} \right)^u \mu_\Psi(dt) \right)^{\frac{1}{u}} \le c \, \| f | A_{pq}^s(\mathbb{R}^n) \|, \quad f \in A_{pq}^s(\mathbb{R}^n), \tag{10.12}$$

where

$$\mathcal{E}_G(t) = \mathcal{E}_G A_{pq}^s(t) \quad \text{and} \quad \Psi(t) = -\log \mathcal{E}_G A_{pq}^s(t).$$

We denote provisionally the couple

$$\mathfrak{E}_G(A^s_{pq}) = \bigl(\mathcal{E}_G A^s_{pq}(t), u\bigr) \tag{10.13}$$

as the *growth envelope* of $A^s_{pq}(\mathbb{R}^n)$. We calculate the growth envelopes for all relevant spaces with the typical outcome

$$\mathfrak{E}_G(A^s_{pq}) = (t^{-\frac{1}{r}}, u) \quad \text{and} \quad \mathfrak{E}_G(A^s_{pq}) = (|\log t|^v, u) \tag{10.14}$$

in the sub-critical and critical case, respectively. Here r has the same meaning as in Fig. 10.1, whereas $0 < u \le \infty$ and $0 < v \le 1$ are suitable numbers depending on the parameters in B^s_{pq} and F^s_{pq}. In particular, the growth envelope exists in all cases. It is a natural and very precise description of the growth of functions belonging to $A^s_{pq}(\mathbb{R}^n)$. Inserting (10.14) in (10.12) one gets quasi-norms of the same type as in the Lorentz spaces $L_{ru}(I_\varepsilon)$ or in the special Lorentz-Zygmund spaces $L_{\infty,u}(\log L)_a(I_\varepsilon)$. However in the above outlined context neither spaces of this nor any other type are prescribed as target spaces in the course of setting up the required inequalities, and asking only afterwards for best parameters. (Of course all the spaces $L_{ru}(\log L)_a(I_\varepsilon)$ are reasonable refinements of classical target spaces like L_p, C, etc.) Here they emerge naturally and, hence, they are at the heart of the matter of the described singularity theory.

If $s > \frac{n}{p}$ then all spaces $A^s_{pq}(\mathbb{R}^n)$ are embedded in $L_\infty(\mathbb{R}^n)$ and a singularity theory in the above argument does not make much sense. However in the context of continuity there is one case which has attracted special attention in recent times and which corresponds to the line $s = 1 + \frac{n}{p}$ in Fig. 10.1. Hence we complement (10.5) and (10.6) by the case called:

$$\text{super-critical} \quad \text{if} \quad 0 < p < \infty, \quad 0 < q \le \infty, \quad s = 1 + \frac{n}{p}. \tag{10.15}$$

First we recall that

$$\omega(f,t) = \sup_{x \in \mathbb{R}^n, |h| \le t} |f(x+h) - f(x)| \quad \text{and} \quad \widetilde{\omega}(f,t) = \frac{\omega(f,t)}{t} \tag{10.16}$$

with $t > 0$, are the usual modulus of continuity and the divided modulus of continuity, respectively. Then $Lip(\mathbb{R}^n)$ is the collection of all complex-valued functions in $\mathbb{R}^n$ such that

$$\|f\,|Lip(\mathbb{R}^n)\| = \sup_{x \in \mathbb{R}^n} |f(x)| + \sup_{0 < t < 1} \widetilde{\omega}(f,t) < \infty. \tag{10.17}$$

We have the remarkable fact (explained in greater detail in Section 11) that

$$A^{\frac{n}{p}}_{pq}(\mathbb{R}^n) \subset L_\infty(\mathbb{R}^n) \quad \text{if, and only if,} \quad A^{1+\frac{n}{p}}_{pq}(\mathbb{R}^n) \subset Lip(\mathbb{R}^n). \tag{10.18}$$

10. Introduction: Outline of methods and results

This makes (almost) clear that in the critical case (10.6) the growth envelope function $\mathcal{E}_G A_{pq}^{\frac{n}{p}}$ is unbounded if, and only if, the *continuity envelope function* $\mathcal{E}_C A_{pq}^{1+\frac{n}{p}}$, defined as an equivalent function to

$$\mathcal{E}_C | A_{pq}^{1+\frac{n}{p}}(t) = \sup\left\{\widetilde{\omega}(f,t) \; : \; \|f\,|A_{pq}^{1+\frac{n}{p}}(\mathbb{R}^n)\| \leq 1\right\}, \quad 0 < t < \varepsilon, \quad (10.19)$$

for some $\varepsilon > 0$, is unbounded. Obviously, we rely here on the same notational agreement as in connection with (10.7), (10.8), (10.9). Furthermore, up to equivalences, the continuity envelope function is positive, continuous and monotonically decreasing in the interval $(0, \varepsilon]$. Then one has an immediate counterpart of (10.11), (10.12), which justifies the introduction of the *continuity envelope*

$$\mathfrak{E}_C(A_{pq}^{1+\frac{n}{p}}) = \left(\mathcal{E}_C A_{pq}^{1+\frac{n}{p}}(t), u\right) \quad (10.20)$$

in analogy to (10.13). We feel that the outcome is beautiful and perfect: One has in all cases of interest, this means all cases not covered by (10.18),

$$\mathfrak{E}_G(A_{pq}^{\frac{n}{p}}) = \mathfrak{E}_C(A_{pq}^{1+\frac{n}{p}}). \quad (10.21)$$

We shall deal first with the critical case and afterwards lift not only the inequalities but also some extremal functions, responsible for the sharpness, by 1 to the super-critical case. In the one-dimensional case this is based on the simple but rather effective observation,

$$\widetilde{\omega}(f,t) \leq c|f'|^{**}(t), \quad 0 < t < 1, \quad (10.22)$$

which provides at least an understanding of the method. (We use the notation introduced in (10.4)). There is a counterpart in $\mathbb{R}^n$, but it is more complicated. It may be found in 12.16.

The close connection between inequalities of Hardy type and rearrangement inequalities hidden in (10.12), (10.14), is based on the well-known observation,

$$\int_{\mathbb{R}^n} b^p(x)\,|f(x)|^p\,dx \leq \int_0^\infty b^{*p}(t)\,f^{*p}(t)\,dt, \quad 0 < p < \infty, \quad (10.23)$$

where $b(x)$ is a non-negative compactly supported weight function. This approach to Hardy inequalities has the advantage that both the singularity behaviour of the fixed weight function $b(x)$ and also of f, belonging to a given function space, are considered on a global scale. In particular, $b(x)$ may degenerate not only in points, hyper-planes, or smooth surfaces, but also on rather irregular sets. Let, as an example, Γ be a compact d-set in $\mathbb{R}^n$ according to (9.67) with $0 < d < n$, and let

$$D(x) = dist(x, \Gamma), \quad x \in \mathbb{R}^n, \tag{10.24}$$

be the distance of $x \in \mathbb{R}^n$ to Γ. Then

$$b(x) = D(x)^a |\log D(x)|^b, \quad a < 0, \quad b \in \mathbb{R}, \quad 0 < D(x) < 1, \tag{10.25}$$

is a typical weight function in our context. One obtains, for example, in the critical case (10.6) for spaces $F_{pq}^{\frac{n}{p}}(\mathbb{R}^n)$ not covered by (10.18) (this means $1 < p < \infty$) the Hardy inequality

$$\int_{D(x)<\varepsilon} \left| \frac{f(x)}{\log D(x)} \right|^p \frac{dx}{D^{n-d}(x)} \leq c \, \|f \, | F_{pq}^{\frac{n}{p}}(\mathbb{R}^n)\|^p \tag{10.26}$$

where $0 < \varepsilon < 1$ and $0 < q \leq \infty$. Recall that this applies in particular to the Sobolev spaces

$$H_p^{\frac{n}{p}}(\mathbb{R}^n) = F_{p,2}^{\frac{n}{p}}(\mathbb{R}^n), \quad 1 < p < \infty. \tag{10.27}$$

If $d = 0$, then one may choose $\Gamma = \{0\}$ and gets

$$\int_{|x|<\varepsilon} \left| \frac{f(x)}{\log |x|} \right|^p \frac{dx}{|x|^n} \leq c \, \|f \, | F_{pq}^{\frac{n}{p}}(\mathbb{R}^n)\|^p, \quad 1 < p < \infty, \tag{10.28}$$

$0 < q \leq \infty$, $0 < \varepsilon < 1$, where again (10.27) is an outstanding example. The indicated approach, which means the reduction of Hardy inequalities via (10.23) to rearrangement inequalities, is universal in our context. In particular it applies to all sub-critical cases and those critical cases which are not covered by (10.18). However the outcome is not always satisfactory. There seems to be a tricky interplay between weights, the geometry of Γ, and possible measures on Γ. We will discuss this point later on, although there are no final answers. For example, (10.26) looks better than it really is. On the other hand in case of $\Gamma = \{0\}$ one gets sharp assertions: The functions responsible for the sharpness of the rearrangement inequalities are also extremal functions for the related Hardy inequalities. Roughly speaking, these extremal functions convert the inequality (10.23) into an equivalence.

This chapter is organized as follows. In Section 11 we set the scene and collect (mostly without proofs) a few well-known classical embedding assertions in all three cases. Section 12 deals both with growth and continuity envelopes. Here we rely, at least partly, on recent work of D. D. Haroske, [Har01], where she introduced the notation of envelopes used above and studied growth and continuity envelopes systematically and, in particular, in the context of more general spaces. We restrict ourselves here to those properties which are more or less of direct use for later applications to the spaces $B^s_{pq}(\mathbb{R}^n)$ and $F^s_{pq}(\mathbb{R}^n)$. On this basis we study in Sections 13, 14 and 15 the critical, super-critical and sub-critical case, respectively. Hardy inequalities in the setting outlined above, will be considered in Section 16. Finally we collect in Section 17 some additional material and references.

11 Classical inequalities

11.1 Some notation

We use the notation introduced in the previous sections. In particular, let $\mathbb{R}^n$ be again euclidean n-space where $n \in \mathbb{N}$. The Schwartz space $S(\mathbb{R}^n)$, its dual $S'(\mathbb{R}^n)$, and the spaces $L_p(\mathbb{R}^n)$ with $0 < p \leq \infty$ have the same meaning as in 2.1, the latter quasi-normed by (2.1). Let $L_1^{loc}(\mathbb{R}^n)$ be the collection of all complex-valued locally Lebesgue-integrable functions in $\mathbb{R}^n$. Any $f \in L_1^{loc}(\mathbb{R}^n)$ is interpreted in the usual way as a regular distribution. Conversely, as usual, a distribution on $\mathbb{R}^n$ is called *regular* if, and only if, it can be identified (as a distribution) with a locally integrable function on $\mathbb{R}^n$. If $A(\mathbb{R}^n)$ is a collection of distributions on $\mathbb{R}^n$, then

$$A(\mathbb{R}^n) \subset L_1^{loc}(\mathbb{R}^n) \tag{11.1}$$

simply means that any element f of $A(\mathbb{R}^n)$ is a regular distribution $f \in L_1^{loc}(\mathbb{R}^n)$. Then, in particular, the distribution function $\mu_f(\lambda)$, the rearrangement $f^*(t)$ and its maximal function $f^{**}(t)$ in (10.2)–(10.4) make sense accepting that they might be infinite. If $A_1(\mathbb{R}^n)$ and $A_2(\mathbb{R}^n)$ are two quasi-normed spaces, continuously embedded in $S'(\mathbb{R}^n)$, then

$$A_1(\mathbb{R}^n) \subset A_2(\mathbb{R}^n) \tag{11.2}$$

always means that there is a constant $c > 0$ such that

$$\|f \,|\, A_2(\mathbb{R}^n)\| \leq c \,\|f \,|\, A_1(\mathbb{R}^n)\| \quad \text{for all} \quad f \in A_1(\mathbb{R}^n), \tag{11.3}$$

(*continuous embedding*). On the other hand we do not use the word embedding in connection with inequalities of type (10.12).

The spaces $B^s_{pq}(\mathbb{R}^n)$ and $F^s_{pq}(\mathbb{R}^n)$ for the full scale of parameters

$$0 < p \leq \infty, \quad 0 < q \leq \infty, \quad s \in \mathbb{R}, \tag{11.4}$$

(with $p < \infty$ in the F-case) were introduced in Definitions 2.6 and 3.4 or in a more traditional (this means Fourier-analytical) way in connection with Theorems 2.9 and 3.6. A list of special cases, including Sobolev spaces, classical Sobolev spaces, Hölder-Zygmund spaces, and classical Besov spaces, may be found in 1.2. Although they are not special cases of the two scales $B^s_{pq}(\mathbb{R}^n)$ and $F^s_{pq}(\mathbb{R}^n)$ we need also the spaces $C(\mathbb{R}^n)$, $C^1(\mathbb{R}^n)$, and $Lip(\mathbb{R}^n)$. The latter space, including the modulus of continuity and the divided modulus of continuity were introduced in (10.17) and (10.16), respectively. Recall that $C(\mathbb{R}^n)$ is the space of all complex-valued, bounded, uniformly continuous functions in $\mathbb{R}^n$, normed by

$$\|f\,|C(\mathbb{R}^n)\| = \sup_{x \in \mathbb{R}^n} |f(x)|, \qquad (11.5)$$

whereas

$$C^1(\mathbb{R}^n) = \left\{ f \in C(\mathbb{R}^n) \,:\, \frac{\partial f}{\partial x_j} \in C(\mathbb{R}^n) \text{ with } j = 1,\ldots,n \right\} \qquad (11.6)$$

is the obviously normed related space of differentiable functions. Then $C^1(\mathbb{R}^n)$ is a closed subspace of $Lip(\mathbb{R}^n)$ and by the mean value theorem,

$$\|f\,|C^1(\mathbb{R}^n)\| \sim \|f\,|Lip(\mathbb{R}^n)\|, \quad f \in C^1(\mathbb{R}^n), \qquad (11.7)$$

(equivalent norms). First we clarify under what conditions $B^s_{pq}(\mathbb{R}^n)$ and $F^s_{pq}(\mathbb{R}^n)$ consist of regular distributions according to (11.1).

The word *classical* in the heading of this Section 11 has a double meaning. In Theorems 11.2 and 11.4 we collect (mainly without proofs) sharp embeddings in *classical* target spaces such as

$$L_1^{loc}(\mathbb{R}^n), \quad L_r(\mathbb{R}^n), \quad C(\mathbb{R}^n), \quad C^1(\mathbb{R}^n), \quad Lip(\mathbb{R}^n),$$

whereas Theorem 11.7 describes those *classical* refined inequalities in limiting situations (from the middle of the 1960s up to around 1980) which are the roots of recent research and, in particular, of our further intentions in this chapter.

11.2 Theorem

(i) Let

$$0 < p < \infty, \quad 0 < q \le \infty, \quad s \in \mathbb{R}. \qquad (11.8)$$

Then

$$F^s_{pq}(\mathbb{R}^n) \subset L_1^{loc}(\mathbb{R}^n) \qquad (11.9)$$

if, and only if,

$$\text{either} \quad 0 < p < 1, \quad s \geq n\left(\frac{1}{p} - 1\right), \quad 0 < q \leq \infty, \quad (11.10)$$

$$\text{or} \quad 1 \leq p < \infty, \quad s > 0, \quad 0 < q \leq \infty, \quad (11.11)$$

$$\text{or} \quad 1 \leq p < \infty, \quad s = 0, \quad 0 < q \leq 2. \quad (11.12)$$

(ii) *Let*

$$0 < p \leq \infty, \quad 0 < q \leq \infty, \quad s \in \mathbb{R}. \quad (11.13)$$

Then

$$B^s_{pq}(\mathbb{R}^n) \subset L^{loc}_1(\mathbb{R}^n) \quad (11.14)$$

if, and only if,

$$\text{either} \quad 0 < p \leq \infty, \quad s > n\left(\frac{1}{p} - 1\right)_+, \quad 0 < q \leq \infty, \quad (11.15)$$

$$\text{or} \quad 0 < p \leq 1, \quad s = n\left(\frac{1}{p} - 1\right), \quad 0 < q \leq 1, \quad (11.16)$$

$$\text{or} \quad 1 < p \leq \infty, \quad s = 0, \quad 0 < q \leq \min(p, 2). \quad (11.17)$$

11.3 Remark

This theorem coincides with Theorem 3.3.2 in [SiT95], where one finds also a proof. We refer also to [RuS96], pp. 32–35, where some of the key ideas of the proof are outlined. This theorem clarifies in a final way for which spaces $F^s_{pq}(\mathbb{R}^n)$ and $B^s_{pq}(\mathbb{R}^n)$ it makes sense to look at the distribution function $\mu_f(\lambda)$ in (10.2) and at the rearrangement $f^*(t)$ in (10.3) and to ask the questions sketched in the introduction in Section 10. We restrict ourselves in the sequel to spaces $B^s_{pq}(\mathbb{R}^n)$ and $F^s_{pq}(\mathbb{R}^n)$ with (10.1). In other words we exclude all borderline cases covered by Theorem 11.2 where either $p = \infty$ or $s = n(\frac{1}{p}-1)_+$. It would be of interest to have a closer look at these excluded spaces and also at a few other spaces not treated here, for example $F^s_{\infty q}(\mathbb{R}^n)$, including in particular $bmo(\mathbb{R}^n)$. Later on we return to these excluded spaces and add in 13.7 some comments and give a few references.

Next we wish to clarify the embedding of the spaces $B^s_{pq}(\mathbb{R}^n)$ and $F^s_{pq}(\mathbb{R}^n)$ in the sub-critical, critical, super-critical case, according to (10.5), (10.6), (10.15), respectively, in distinguished target spaces; this means $L_r(\mathbb{R}^n)$ in the sub-critical case (where r has the same meaning as in Fig. 10.1), $L_\infty(\mathbb{R}^n)$ and $C(\mathbb{R}^n)$ in the critical case, $Lip(\mathbb{R}^n)$ and $C^1(\mathbb{R}^n)$ in the super-critical case. These are the only cases of interest for us in the context outlined in Section 10. We do not repeat other assertions of the substantial embedding theory of the spaces $B^s_{pq}(\mathbb{R}^n)$ and $F^s_{pq}(\mathbb{R}^n)$, including their special cases as Sobolev and Hölder-Zygmund spaces. The classical part of this theory may be found in [Triα], 2.8, and the almost classical part in [Triβ], 2.7.1. The final clarification of this type of embeddings goes back to [SiT95]. Descriptions of these results may be found in [RuS96], 2.2 and in [ET96], 2.3.3. As said, we restrict ourselves here to those special assertions which are directly related to the problems outlined in Section 10. Recall that the spaces $C(\mathbb{R}^n)$, $C^1(\mathbb{R}^n)$, and $Lip(\mathbb{R}^n)$ were introduced in 11.1 and (10.17).

11.4 Theorem

(i) (Sub-critical case) *Let*

$$1 < r < \infty, \quad s > 0, \quad s - \frac{n}{p} = -\frac{n}{r}, \quad \text{and} \quad 0 < q \leq \infty, \quad (11.18)$$

(the dashed line in Fig. 10.1). Then

$$B^s_{pq}(\mathbb{R}^n) \subset L_r(\mathbb{R}^n) \quad \text{if, and only if,} \quad 0 < q \leq r, \quad (11.19)$$

and

$$F^s_{pq}(\mathbb{R}^n) \subset L_r(\mathbb{R}^n) \quad \text{for all} \quad 0 < q \leq \infty. \quad (11.20)$$

(ii) (Critical case) *Let*

$$0 < p < \infty, \quad 0 < q \leq \infty, \quad s = \frac{n}{p}. \quad (11.21)$$

Then

$$B^{\frac{n}{p}}_{pq}(\mathbb{R}^n) \subset C(\mathbb{R}^n) \quad \text{if, and only if,} \quad 0 < p < \infty, \quad 0 < q \leq 1, \quad (11.22)$$

and

$$F^{\frac{n}{p}}_{pq}(\mathbb{R}^n) \subset C(\mathbb{R}^n) \quad \text{if, and only if,} \quad 0 < p \leq 1, \quad 0 < q \leq \infty. \quad (11.23)$$

In (11.22) and (11.23) one can replace $C(\mathbb{R}^n)$ by $L_\infty(\mathbb{R}^n)$.

(iii) (Super-critical case) *Let*

$$0 < p < \infty, \quad 0 < q \leq \infty, \quad s = 1 + \frac{n}{p}, \quad (11.24)$$

(*the dotted line in Fig. 10.1*). Then

$$B_{pq}^{1+\frac{n}{p}}(\mathbb{R}^n) \subset C^1(\mathbb{R}^n) \quad \text{if, and only if,} \quad 0 < p < \infty, \quad 0 < q \leq 1, \quad (11.25)$$

and

$$F_{pq}^{1+\frac{n}{p}}(\mathbb{R}^n) \subset C^1(\mathbb{R}^n) \quad \text{if, and only if,} \quad 0 < p \leq 1, \quad 0 < q \leq \infty. \quad (11.26)$$

In (11.25) and (11.26) one can replace $C^1(\mathbb{R}^n)$ by $Lip(\mathbb{R}^n)$.

Proof (of part (iii)) Detailed proofs of parts (i) and (ii) may be found in [SiT95], Theorems 3.2.1 and 3.3.1, Remark 3.3.5; short descriptions are given in [RuS96], 2.2, and in [ET96], 2.3.3. Part (iii) is essentially the lifting of part (ii) by 1. But this is by no means obvious and must be justified. First we remark that

$$f \in F_{pq}^{1+\frac{n}{p}}(\mathbb{R}^n) \quad \text{if, and only if,} \quad f \in F_{pq}^{\frac{n}{p}}(\mathbb{R}^n) \quad \text{and} \quad \frac{\partial f}{\partial x_j} \in F_{pq}^{\frac{n}{p}}(\mathbb{R}^n)$$
(11.27)

where $j = 1, \ldots, n$ and (equivalent quasi-norms)

$$\|f \mid F_{pq}^{1+\frac{n}{p}}(\mathbb{R}^n)\| \sim \|f \mid F_{pq}^{\frac{n}{p}}(\mathbb{R}^n)\| + \sum_{j=1}^{n} \left\|\frac{\partial f}{\partial x_j} \mid F_{pq}^{\frac{n}{p}}(\mathbb{R}^n)\right\|. \quad (11.28)$$

We refer to [Triβ], Theorem 2.3.8, pp. 58/59. Hence, the if-part of (11.26) follows from (11.23). Conversely, assume

$$F_{pq}^{1+\frac{n}{p}}(\mathbb{R}^n) \subset C^1(\mathbb{R}^n) \quad \text{for some} \quad 0 < p < \infty, \quad 0 < q \leq \infty. \quad (11.29)$$

We construct special Fourier multipliers and introduce some cones in $\mathbb{R}^n$,

$$K_t = \{\xi = (\xi', \xi_n) \in \mathbb{R}^n : \xi_n > 0, \ |\xi'| < t\xi_n\}, \quad t > 0, \quad (11.30)$$

where obviously $\xi' = (\xi_1, \ldots, \xi_{n-1}) \in \mathbb{R}^{n-1}$. Let φ be a C^∞ function in $\mathbb{R}^n \setminus \{0\}$ with

$$\varphi(\xi) = \varphi\left(\frac{\xi}{|\xi|}\right), \quad \varphi(\xi) = 1 \quad \text{if} \quad \xi \in K_t \quad \text{and} \quad \text{supp}\, \varphi \subset \overline{K_{2t}}. \quad (11.31)$$

Let $\psi(\xi)$ be a C^∞ function in $\mathbb{R}^n$ which is identically 1 if $|\xi| \geq 1$ and $0 \notin supp\,\psi$. Then, by [Triβ], Theorem 2.3.7 on p. 57,

$$\varphi^j(\xi) = \varphi(\xi)\,\psi(\xi)\,\xi_j \xi_n^{-1}, \quad \text{where} \quad j = 1, \ldots, n, \quad (11.32)$$

are Fourier multipliers in all spaces $F_{pq}^s(\mathbb{R}^n)$ and $B_{pq}^s(\mathbb{R}^n)$. Let

$$f \in F_{pq}^{\frac{n}{p}}(\mathbb{R}^n) \quad \text{with} \quad supp\, \widehat{f} \subset K_t \cap \{\xi \in \mathbb{R}^n \,:\, |\xi| > 1\}, \tag{11.33}$$

and

$$g(x) = \left(\xi_n^{-1} \widehat{f}(\xi)\right)^{\vee}(x) = \left(\xi_n^{-1} \varphi^n(\xi)\, \widehat{f}(\xi)\right)^{\vee}(x), \quad x \in \mathbb{R}^n, \tag{11.34}$$

where we used the notation introduced in 2.8. Since $\varphi^j(\xi)$ are Fourier multipliers it follows by (11.28) that

$$g \in F_{pq}^{1+\frac{n}{p}}(\mathbb{R}^n) \quad \text{and} \quad \frac{\partial g}{\partial x_n}(x) = i\, f(x). \tag{11.35}$$

Hence if we assume (11.29) for some p and q, then it follows for functions f with (11.33) that $f \in C(\mathbb{R}^n)$ and

$$\|f\,|C(\mathbb{R}^n)\| \le c\, \|f\,|F_{pq}^{\frac{n}{p}}(\mathbb{R}^n)\|. \tag{11.36}$$

An arbitrary function $f \in F_{pq}^{\frac{n}{p}}(\mathbb{R}^n)$ can be decomposed in finitely many functions of type (11.33) and a harmless function $\left((1-\psi)\widehat{f}\right)^{\vee}$, where ψ has the above meaning. Then we have (11.36) for those p, q with (11.29). This is the converse we are looking for, and it proves (11.26). Finally we must show that one can replace $C^1(\mathbb{R}^n)$ in (11.26) by $Lip(\mathbb{R}^n)$. Since $Lip(\mathbb{R}^n)$ is the larger space we must disprove

$$F_{pq}^{1+\frac{n}{p}}(\mathbb{R}^n) \subset Lip(\mathbb{R}^n) \tag{11.37}$$

if $p > 1$ and $0 < q \le \infty$. By the monotonicity of the spaces $F_{pq}^{1+\frac{n}{p}}(\mathbb{R}^n)$ with respect to q we may assume $q < \infty$. Then $S(\mathbb{R}^n)$ is dense in $F_{pq}^{1+\frac{n}{p}}(\mathbb{R}^n)$ and we have (11.7) for $f \in S(\mathbb{R}^n)$. By completion it follows (11.37) with $C^1(\mathbb{R}^n)$ in place of $Lip(\mathbb{R}^n)$. But this contradicts (11.26) since $p > 1$. The proof for the B-spaces is the same.

11.5 Remark

Usually, $s - \frac{n}{p}$ is called the *differential dimension* of the spaces $B_{pq}^s(\mathbb{R}^n)$ and $F_{pq}^s(\mathbb{R}^n)$. In particular, $-\frac{n}{r}$ is the differential dimension of $L_r(\mathbb{R}^n)$. This notion can obviously be extended to the above target spaces $C(\mathbb{R}^n)$, $L_\infty(\mathbb{R}^n)$ (differential dimension 0) and $C^1(\mathbb{R}^n)$, $Lip(\mathbb{R}^n)$ (differential dimension 1). Continuous embeddings between function spaces with the same differential dimension are often called *limiting embeddings*. The above theorem deals exclusively with embeddings of this type in the indicated specific situations:

sub-critical :   differential dimension $-\frac{n}{r}$,

critical :   differential dimension 0,

super-critical :   differential dimension 1.

Furthermore, (10.18) is now an immediate consequence of the above theorem. As explained in Section 10, in connection with the growth envelope $\mathfrak{E}_G(A_{pq}^{\frac{n}{p}})$ in (10.13) for the critical case and the continuity envelope $\mathfrak{E}_C(A_{pq}^{1+\frac{n}{p}})$ in (10.20) for the super-critical case, we are interested only in those spaces which are not covered by (10.18), this means by the parts (ii) and (iii) of the above theorem in the spaces

$$B_{pq}^{\frac{n}{p}}(\mathbb{R}^n), \quad B_{pq}^{1+\frac{n}{p}}(\mathbb{R}^n) \quad \text{with} \quad 0 < p < \infty, \quad 1 < q \leq \infty, \tag{11.38}$$

and

$$F_{pq}^{\frac{n}{p}}(\mathbb{R}^n), \quad F_{pq}^{1+\frac{n}{p}}(\mathbb{R}^n) \quad \text{with} \quad 1 < p < \infty, \quad 0 < q \leq \infty. \tag{11.39}$$

If $q = 2$ in (11.39) then we get by (1.9) the Sobolev spaces

$$H_p^{\frac{n}{p}}(\mathbb{R}^n), \quad H_p^{1+\frac{n}{p}}(\mathbb{R}^n) \quad \text{with} \quad 1 < p < \infty. \tag{11.40}$$

On the other hand, we have the famous Sobolev embedding

$$H_p^s(\mathbb{R}^n) \subset L_r(\mathbb{R}^n), \quad 1 < p < \infty, \quad -\frac{n}{r} = s - \frac{n}{p} < 0, \tag{11.41}$$

as a special case of (11.20). But it is just the failure to extend (11.41) from the sub-critical case $1 < r < \infty$, to the critical case $r = \infty$, which triggered the search for adequate substitutes. Then we are back to the 1960s. At the same time refinements of sub-critical embeddings according to part (i) of Theorem 11.4 for the Sobolev spaces $H_p^s(\mathbb{R}^n)$ and the classical Besov spaces $B_{pq}^s(\mathbb{R}^n)$ in terms of Lorentz spaces L_{ru} have been studied (in the West inspired by interpolation theory). Around 1980 further refinements in the critical case and the first steps in the super-critical case for the Sobolev spaces in (11.40) were taken. We shall describe this nowadays historical part of these refined embeddings in Theorem 11.7 below, including in 11.8 the respective references. Mostly for this reason we discuss in 11.6 the relevant target spaces, whereas later on we prefer to formulate our results in terms of inequalities.

11.6 Lorentz-Zygmund spaces

For the reasons just explained we restrict ourselves to a brief description. The standard reference for Lorentz spaces and Zygmund spaces is [BeS88]. Their combination, the Lorentz-Zygmund spaces, were introduced in [BeR80]. Let $0 < \varepsilon < 1$ and let $I_\varepsilon = (0, \varepsilon]$. We use the rearrangement $f^*(t)$ of a complex-valued measurable function $f(t)$ on I_ε as introduced in (10.2), (10.3), temporarily with I_ε in place of $\mathbb{R}^n$.

(i) *Lorentz spaces* Let $0 < r < \infty$ and $0 < u \le \infty$. Then $L_{ru}(I_\varepsilon)$ is the set of all measurable complex-valued functions f on I_ε such that

$$\int_0^\varepsilon \left(t^{\frac{1}{r}} f^*(t) \right)^u \frac{dt}{t} < \infty \quad \text{if} \quad 0 < u < \infty, \tag{11.42}$$

and

$$\sup_{t \in I_\varepsilon} t^{\frac{1}{r}} f^*(t) < \infty \quad \text{if} \quad u = \infty. \tag{11.43}$$

Of course, $L_{rr}(I_\varepsilon) = L_r(I_\varepsilon)$ with $0 < r < \infty$, are the usual Lebesgue spaces.

(ii) *Zygmund spaces* Let $0 < r < \infty$ and $a \in \mathbb{R}$. Then $L_r(\log L)_a(I_\varepsilon)$ is the set of all measurable complex-valued functions f on I_ε such that

$$\int_0^\varepsilon |f(t)|^r \log^{ar}(2 + |f(t)|) \, dt < \infty. \tag{11.44}$$

Let $a < 0$. Then $L_\infty(\log L)_a(I_\varepsilon)$ is the set of all measurable complex-valued functions f on I_ε such that there is a number $\lambda > 0$ with

$$\int_0^\varepsilon \exp\left\{ (\lambda |f(t)|)^{-\frac{1}{a}} \right\} dt < \infty. \tag{11.45}$$

When $r < \infty$, then this notation resembles that in [BeS88], pp. 252–253. The alternative notation $L_{\exp, -a}(I_\varepsilon)$ for $L_\infty(\log L)_a(I_\varepsilon)$ is closer to that employed in [BeS88]. The somewhat curious looking expression (11.45) may be justified by the following unified alternative representation, where $f \in L_r(\log L)_a(I_\varepsilon)$ if, and only if,

$$\left(\int_0^\varepsilon |\log t|^{ar} f^{*r}(t) \, dt \right)^{\frac{1}{r}} < \infty, \quad \text{when} \quad 0 < r < \infty, \quad a \in \mathbb{R}, \tag{11.46}$$

and

$$\sup_{t \in I_\varepsilon} |\log t|^a f^*(t) < \infty, \quad \text{when} \quad r = \infty, \quad a < 0. \tag{11.47}$$

We refer to [BeS88], p. 252, or [ET96], p. 66. In [ET96], 2.6.1, 2.6.2, one finds also another unifying representation, further properties and references. This way of looking at these spaces fits also in the scheme developed in the following Section 12. Both (11.46) and (11.47) are quasi-norms.

(iii) *Lorentz-Zygmund spaces*   The combination of the above Lorentz spaces and Zygmund spaces results in the Lorentz-Zygmund spaces studied in detail in [BeR80]. Let

$$0 < r < \infty, \quad 0 < u \leq \infty \quad \text{and} \quad a \in \mathbb{R}.$$

Then $L_{ru}(\log L)_a(I_\varepsilon)$ is the set of all measurable complex-valued functions f on I_ε such that

$$\left(\int_0^\varepsilon \left(t^{\frac{1}{r}} |\log t|^a f^*(t) \right)^u \frac{dt}{t} \right)^{\frac{1}{u}} < \infty \quad \text{if} \quad 0 < u < \infty, \tag{11.48}$$

and

$$\sup_{t \in I_\varepsilon} t^{\frac{1}{r}} |\log t|^a f^*(t) < \infty \quad \text{if} \quad u = \infty. \tag{11.49}$$

Again (11.48) and (11.49) are quasi-norms. Note that

$$L_{rr}(\log L)_a(I_\varepsilon) = L_r(\log L)_a(I_\varepsilon) \quad \text{where} \quad 0 < r < \infty, \quad a \in \mathbb{R},$$

and

$$L_{ru}(\log L)_0(I_\varepsilon) = L_{ru}(I_\varepsilon) \quad \text{where} \quad 0 < r < \infty, \quad 0 < u \leq \infty.$$

For details we refer to [BeS88], p. 253, and, in particular to [BeR80]. This reference covers also the interesting case $r = \infty$, hence $L_{\infty,u}(\log L)_a(I_\varepsilon)$. If $r = u = \infty$ then (11.49) coincides with (11.47), and one needs $a < 0$. If $r = \infty$ and $0 < u < \infty$, then one needs in (11.48) that $au < -1$, hence $a + \frac{1}{u} < 0$. Otherwise, if $r = \infty$ and $au \geq -1$, then there are no non-trivial functions f with (11.48). Some well-known embeddings between these spaces will come out later on in 12.4 in a natural way.

We are not so much interested in the Lorentz-Zygmund spaces for their own sake. We formulate later on our results in terms of inequalities using rearrangements. This will also be done in the next theorem where we collect the historical roots of the theory outlined in the introductory Section 10, although the original formulations looked sometimes quite different. This applies in particular when the spaces $L_r(\log L)_a(I_\varepsilon)$ in the original versions (11.44), (11.45) are involved. Not only the spaces themselves but also their equivalent characterizations via (11.46), (11.47) can be traced back to Hardy-Littlewood,

Zygmund, Lorentz and Bennett. The corresponding references may be found in the Note Sections in [BeS88], pp. 288, 180–181, and also in [BeR80], in connection with Corollary 10.2 and Theorem 10.3. This resulted finally in the Lorentz-Zygmund spaces quasi-normed by (11.48), (11.49) in [BeR80]. In particular, all the equivalent (quasi-)norms mentioned above were known around 1980 (and often long before).

Recall that $H_p^s(\mathbb{R}^n)$ are the Sobolev spaces, $1 < p < \infty$, $s > 0$, according to (1.9) with the classical Sobolev spaces $W_p^s(\mathbb{R}^n)$ as special cases. The classical Besov spaces are normed by (1.14). The divided differences $\widetilde{\omega}(f,t)$ are given by (10.16). Let $0 < \varepsilon < 1$.

11.7 Theorem

(Classical refined inequalities in limiting situations)

(i) (Sub-critical case, Lorentz spaces, dashed line in Fig. 10.1)
Let $s > 0$,

$$1 < p < \infty, \quad -\frac{n}{r} = s - \frac{n}{p} < 0, \quad 1 \le q \le \infty. \qquad (11.50)$$

Then there is a constant $c > 0$ such that

$$\left(\int_0^\varepsilon \left(t^{\frac{1}{r}} f^*(t) \right)^p \frac{dt}{t} \right)^{\frac{1}{p}} \le c \, \|f \,|H_p^s(\mathbb{R}^n)\| \quad \text{for all} \quad f \in H_p^s(\mathbb{R}^n), \qquad (11.51)$$

and

$$\left(\int_0^\varepsilon \left(t^{\frac{1}{r}} f^*(t) \right)^q \frac{dt}{t} \right)^{\frac{1}{q}} \le c \, \|f \,|B_{pq}^s(\mathbb{R}^n)\| \quad \text{for all} \quad f \in B_{pq}^s(\mathbb{R}^n) \qquad (11.52)$$

(with (11.43) if $q = \infty$).

(ii) (Sub-critical case, Zygmund spaces, dashed line in Fig. 10.1)
Let $s > 0$,

$$1 < p < \infty, \quad -\frac{n}{r} = s - \frac{n}{p} < 0, \quad r < q \le \infty, \quad -\infty < a < \frac{1}{q} - \frac{1}{r}. \qquad (11.53)$$

Then there is a constant $c > 0$ such that

$$\left(\int_0^\varepsilon |\log t|^{ar} f^{*r}(t) \, dt \right)^{\frac{1}{r}} \le c \, \|f \,|B_{pq}^s(\mathbb{R}^n)\| \quad \text{for all} \quad f \in B_{pq}^s(\mathbb{R}^n). \qquad (11.54)$$

(iii) **(Critical case)** *Let*

$$1 < p < \infty, \quad \frac{1}{p} + \frac{1}{p'} = 1. \tag{11.55}$$

Then there is a constant $c > 0$ such that

$$\sup_{0 < t < \varepsilon} \frac{f^*(t)}{|\log t|^{\frac{1}{p'}}} \leq c \, \|f \, | H_p^{\frac{n}{p}}(\mathbb{R}^n)\| \quad \text{for all} \quad f \in H_p^{\frac{n}{p}}(\mathbb{R}^n), \tag{11.56}$$

and

$$\left(\int_0^\varepsilon \left(\frac{f^*(t)}{|\log t|} \right)^p \frac{dt}{t} \right)^{\frac{1}{p}} \leq c \, \|f \, | H_p^{\frac{n}{p}}(\mathbb{R}^n)\| \quad \text{for all} \quad f \in H_p^{\frac{n}{p}}(\mathbb{R}^n). \tag{11.57}$$

(iv) **(Super-critical case)** *Let*

$$1 < p < \infty, \quad \frac{1}{p} + \frac{1}{p'} = 1, \quad 1 + \frac{n}{p} = k \in \mathbb{N}. \tag{11.58}$$

Then there is a constant $c > 0$ such that

$$\sup_{0 < t < \varepsilon} \frac{\widetilde{\omega}(f, t)}{|\log t|^{\frac{1}{p'}}} \leq c \, \|f \, | W_p^k(\mathbb{R}^n)\| \quad \text{for all} \quad f \in W_p^k(\mathbb{R}^n). \tag{11.59}$$

11.8 Historical references and comments

We tried to collect in the above theorem those refined inequalities which we believe are the roots of the programme outlined in the introductory Section 10. "Refined" must be understood in comparison with Theorem 11.4 asking for a tuning of the admissible target spaces finer than used there. The spaces described in 11.6 may be considered as a reasonable choice for such an undertaking. A balanced or even comprehensive history of inequalities of this type seems to be rather complicated. Many mathematicians contributed to this flourishing field of research, and parallel developments in the East (in the Russian literature) have often passed unnoticed in the West, but also vice versa. In a sequence of points, denoted as 11.8(i) etc., we collect related papers, comment on a few topics, and try to clarify to what extent the above inequalities fit in our scheme. We shift more recent references to a later occasion (with exception of a few surveys which in turn describe the history) and restrict ourselves to those papers which stand for the early development of this theory (although this covers a period of some 20 years).

11.8(i) (Sub-critical case, Lorentz spaces)   The inequalities (11.51), (11.52) came into being in the middle of the 1960s with the advent of the interpolation theory: there was no escape, as we outline in the next point. But there are more direct approaches, especially in connection with a wider scale of Besov spaces (generalized moduli of continuity), vector-valued Besov spaces, and the question about the sharpness of these inequalities. We refer to [Pee66], [Str67], [Her68], [Bru72], [Bru76], [Gol85], [Gol86], and the surveys [Kol89], [Kol98], [Liz86], which describe especially what has been done in the Russian literature.

11.8(ii) (Sub-critical case, interpolation) We use without further explanations the real interpolation $(A_0, A_1)_{\theta,q}$ of two (quasi-)Banach spaces A_0 and A_1, and $0 < \theta < 1$, $0 < q \leq \infty$. We refer to [Triα], [BeL76] or [BeS88], where one finds all that one needs. Recall that

$$(L_{r_0}, L_{r_1})_{\theta,p} = L_{rp}, \quad 1 < r_0 < r_1 < \infty, \quad \frac{1}{r} = \frac{1-\theta}{r_0} + \frac{\theta}{r_1}, \quad 0 < p \leq \infty, \tag{11.60}$$

on I_ε or on $\mathbb{R}^n$, where L_{rq} are the Lorentz spaces from 11.6(i). Lifting of (11.60) on $\mathbb{R}^n$ gives

$$\left(H_{p_0}^s, H_{p_1}^s\right)_{\theta,p} = H_p^s, \quad 1 < p_0 < p_1 < \infty, \quad \frac{1}{p} = \frac{1-\theta}{p_0} + \frac{\theta}{p_1}, \quad s \in \mathbb{R}. \tag{11.61}$$

Furthermore, again on $\mathbb{R}^n$,

$$\left(B_{p,1}^{s_0}, B_{p,1}^{s_1}\right)_{\theta,q} = B_{pq}^s, \quad 0 < s_0 < s_1 < \infty, \quad s = (1-\theta)s_0 + \theta s_1, \tag{11.62}$$

$0 < q \leq \infty$. Now (11.20) with $H_p^s = F_{p,2}^s$, and the interpolations (11.61), (11.60) give (11.51), whereas (11.19), and the interpolations (11.62), (11.60) result in (11.52).

11.8(iii) (Sub-critical case, Zygmund spaces) The inequality (11.54) is less satisfactory than the inequalities (11.51) and (11.52). We explain the reason in the next point. In addition, the above Zygmund spaces are not naturally linked with the spaces $B_{pq}^s(\mathbb{R}^n)$ and $F_{pq}^s(\mathbb{R}^n)$ in the reasoning of Section 10 in sub-critical situations, in contrast to the Lorentz spaces. Although we could not find a direct formulation of (11.54) in the literature, assertions of this type are apparently included (in a somewhat hidden way) in a larger theory of embeddings of the form

$$B_{pq}^{\omega(\cdot)} \subset L_\Phi \quad 1 \leq p < \infty, \quad 0 < q \leq \infty, \tag{11.63}$$

where ω is a generalized modulus of continuity and L_Φ stands for an Orlicz space. This was studied in the 1980s in great detail in the Russian literature. In [Gol86], Theorem 5.4, one finds necessary and sufficient conditions for embeddings of type (11.63). A corresponding formulation may also be found in

[KaL87], Theorem 8.5, p. 27. This paper surveys what has been done by the Russian school of the theory of function spaces. With some modifications it is the English version of [Liz86] where the same result (11.63) may be found in D.1.8. At least in some cases $L_r(\log L)_a$ can be identified with an Orlicz space. We refer to [BeS88], p. 266, Example 8.3(e), with $\Phi(t) = t^r |\log t|^a$. Taking together all these remarks then some assertions of type (11.54) for classical Besov spaces are hidden in [Gol86], [Liz86], [KaL87].

11.8(iv) (Sub-critical case, Hölder inequalities) As said in the previous point, (11.54) with (11.53) does not fit optimally in our context. Furthermore, this estimate follows from (11.52) and Hölder's inequality: Let again $0 < \varepsilon < 1$,

$$1 < r < q \leq \infty, \quad \frac{1}{r} = \frac{1}{q} + \frac{1}{u}, \tag{11.64}$$

and $h(t) > 0$ if $0 < t \leq \varepsilon$. Then by Hölder's inequality,

$$\left(\int_0^\varepsilon h(t)^r \, f^{*r}(t) \, dt \right)^{\frac{1}{r}} = \left(\int_0^\varepsilon \left(t^{\frac{1}{r}} h(t) f^*(t) \right)^r \frac{dt}{t} \right)^{\frac{1}{r}}$$

$$\leq \left(\int_0^\varepsilon \left(t^{\frac{1}{r}} f^*(t) \right)^q \frac{dt}{t} \right)^{\frac{1}{q}} \left(\int_0^\varepsilon h(t)^u \frac{dt}{t} \right)^{\frac{1}{u}}. \tag{11.65}$$

The last factor with $h(t) = |\log t|^a$ converges if, and only if, $au < -1$. This proves the if-part of:

$$\left(\int_0^\varepsilon |\log t|^{ar} f^{*r}(t) \, dt \right)^{\frac{1}{r}} \leq c \left(\int_0^\varepsilon \left(t^{\frac{1}{r}} f^*(t) \right)^q \frac{dt}{t} \right)^{\frac{1}{q}} \tag{11.66}$$

if, and only if, $a < \frac{1}{q} - \frac{1}{r}$. To disprove (11.66) if $a \geq \frac{1}{q} - \frac{1}{r}$ one may choose $a = \frac{1}{q} - \frac{1}{r}$ and, as a counter-example,

$$f(t) = t^{-\frac{1}{r}} |\log t|^{-\frac{1}{q}} (\log |\log t|)^{-\frac{1}{r}}, \quad 0 < t < \varepsilon, \tag{11.67}$$

assuming that $\varepsilon > 0$ is sufficiently small. But this makes clear that $|\log t|^u$ in (11.54) is a distinguished but not natural choice. There are better functions $h(t)$ such that the last factor in (11.65) converges. Maybe a systematic treatment in this direction would result in problems of type (11.63). Nevertheless we return to inequalities of type (11.54) later on in Corollary 15.4.

11.8(v) (Critical case) The inequality (11.56) has a rich history reflecting especially well parallel developments in East and West. First we recall that by the equivalence of (11.47) and (11.45) for some $\lambda > 0$, the left-hand side of (11.56) is finite if, and only if,

$$\int_0^\varepsilon \exp\left\{(\lambda |f(t)|)^{p'}\right\} dt < \infty \quad \text{for some} \quad \lambda > 0. \tag{11.68}$$

In this version, (11.56) is due to R. S. Strichartz, [Str72], including a sharpness assertion. Corresponding results for the classical Sobolev spaces

$$W_p^k(\mathbb{R}^n) = H_p^{\frac{n}{p}}(\mathbb{R}^n) \quad \text{with} \quad 1 < p < \infty, \quad \frac{n}{p} = k \in \mathbb{N}, \tag{11.69}$$

(especially if $k = 1$) in the version of (11.68) had been obtained before in [Tru67] and may also be found in [GiT77], Theorem 7.15 and (7.40) on p. 155. This paper by N. S. Trudinger made problems of this type widely known and influenced further development, in terms of spaces of type (11.68) and more general Orlicz spaces. For classical Sobolev spaces with the fixed norm (1.4), especially if $k = 1$, hence $W_n^1(\mathbb{R}^n)$, it makes sense to ask for the best constant λ in (11.68). This may be found in [Mos71]. As noted above there was a parallel and independent development in the East. We refer in particular to [Poh65] and the even earlier forerunner [Yud61]. Best constants for the embeddings of Sobolev spaces according to (11.69) in spaces of type (11.68), extending [Mos71] to all $k \in \mathbb{N}$, may be found in [Ada88]. This paper contains also a balanced history of this subject, including the Russian literature. The natural counterpart of (11.56) for classical Besov spaces $B_{pq}^{\frac{n}{p}}(\mathbb{R}^n)$, normed by (1.14), is given by

$$\sup_{0 < t < \varepsilon} \frac{f^*(t)}{|\log t|^{\frac{1}{q'}}} \leq c \, \|f \,|B_{pq}^{\frac{n}{p}}(\mathbb{R}^n)\|, \quad 1 < p < \infty, \quad 1 < q \leq \infty, \tag{11.70}$$

and $\frac{1}{q} + \frac{1}{q'} = 1$. Limiting embeddings of spaces $B_{pq}^{\frac{n}{p}}(\mathbb{R}^n)$ in Orlicz spaces were considered by J. Peetre in [Pee66]. If one takes his assertion in [Pee66], Theorem 9.1 on p. 303, and reformulates it in terms of later developments in the 1980s and 1990s, and which may be found in [ET96], 2.6.2, then one arrives at (11.70) or equivalently at (11.68) with q' in place of p'. As for (11.57) we first remark that

$$\sup_{0 < t < \varepsilon} \frac{f^*(t)}{|\log t|^{\frac{1}{p'}}} \leq c \left(\int_0^\varepsilon \left(\frac{f^*(t)}{|\log t|} \right)^p \frac{dt}{t} \right)^{\frac{1}{p}}. \tag{11.71}$$

This is an assertion of type (10.11) and will be considered in detail in 12.4, especially (12.28). Hence (11.57) is sharper than (11.56). This improvement goes back to [Has79] and [BrW80] (as a consequence of Theorem 2 on p. 781), including sharpness assertion in [Has79]. We refer in this context also to [Zie89], 2.10.5, 2.10.6, pp. 100–103, where one finds improved and more detailed versions of the arguments in [BrW80].

11.8(vi) (Super-critical case) The first direct proof of (11.59) may be found in [BrW80], Corollary 5, p. 786. On the other hand, accepting (10.22), its n-dimensional counterpart, and that in (11.51), (11.52), (11.56), (11.57), f^* can be replaced by f^{**}, then all the inequalities in the super-critical case can be obtained by lifting from the critical case, in particular (11.59) by lifting of a special case of (11.56), and a stronger and more general version of (11.59) by lifting of (11.57), an inequality which has also been proved in [BrW80]. Maybe this connection passed unnoticed not only at that time but also in recent research dealing separately with the critical and super-critical case. This close interdependence is also well reflected by the parts (ii) and (iii) in Theorem 11.4, and by its short version (10.18).

12 Envelopes

12.1 Rearrangement and the growth of functions

In this section we introduce the concept of growth envelopes and continuity envelopes as outlined in the introductory Section 10. We restrict ourselves to those spaces $B^s_{pq}(\mathbb{R}^n)$ and $F^s_{pq}(\mathbb{R}^n)$ which are of interest for us. On this basis we prove in the subsequent Sections 13–16 the main results of this chapter. The new notion of growth and continuity envelopes makes sense for a much wider range of function spaces and has been introduced and studied recently by D. D. Haroske in [Har01]. We take over a few results obtained there, including the relevant notation of envelopes and envelope functions.

First we recall again what is meant by rearrangement. For our purpose it is sufficient to assume that $f \in L_r(\mathbb{R}^n)$ with $1 \le r < \infty$. Then as in (10.2), (10.3), the distribution function μ_f and the non-increasing rearrangement f^* of f are given by

$$\mu_f(\lambda) = |\{x \in \mathbb{R}^n : |f(x)| > \lambda\}|, \quad \lambda \ge 0, \tag{12.1}$$

and

$$f^*(t) = \inf\{\lambda : \mu_f(\lambda) \le t\}, \quad t \ge 0. \tag{12.2}$$

We wish to measure the growth of functions f belonging to $B^s_{pq}(\mathbb{R}^n)$ and $F^s_{pq}(\mathbb{R}^n)$ either in the sub-critical situation according to Theorem 11.4(i) or in the critical situation for those parameters p, q not covered by Theorem 11.4(ii). Let $A_{pq}^s(\mathbb{R}^n)$ be such a space and let $f \in A_{pq}^s(\mathbb{R}^n)$. Then the behaviour of the rearrangement $f^*(t)$ if $t \downarrow 0$ indicates how singular the function f might be on the global scale, the whole of $\mathbb{R}^n$. Inspired by the spaces in 11.6 we try to measure the possible growth of f in terms of

$$\left(\int_0^\varepsilon (f^*(t)\, w(t))^u \, \frac{dt}{t} \right)^{\frac{1}{u}} \tag{12.3}$$

for some fixed $0 < \varepsilon < 1$, $0 < u \leq \infty$ (appropriately modified if $u = \infty$) and some positive continuous monotonically increasing weight functions $w(t)$ on $[0, \varepsilon)$ with $w(0) = 0$, $w(t) > 0$ if $0 < t < \varepsilon$, and

$$w(2^{-j-1}) \sim w(2^{-j}), \quad j \in \mathbb{N}, \quad j \geq J(\varepsilon). \tag{12.4}$$

Recall that we use the equivalence sign "$\sim$" as explained before, for example in (7.10). Then we have

$$\left(\int_0^\varepsilon (f^*(t)\, w(t))^u \, \frac{dt}{t} \right)^{\frac{1}{u}} \sim \left(\sum_j (f^*(2^{-j})\, w(2^{-j}))^u \right)^{\frac{1}{u}}, \tag{12.5}$$

where the equivalence constants are independent of f. Hence (12.3) behaves like a sequence space ℓ_u, including the well-known monotonicity with respect to u.

The following reformulation of this type of singularity measurement might be helpful for a better understanding. We assume that the reader is familiar with the basic facts concerning rearrangement. They may be found in [BeS88]. In particular, rearrangement is measure-preserving,

$$|\{t > 0 : \tau_0 \geq f^*(t) \geq \tau_1\}| = |\{x \in \mathbb{R}^n : \tau_0 \geq |f(x)| \geq \tau_1\}|, \tag{12.6}$$

where $0 \leq \tau_1 \leq \tau_0 < \infty$. We refer also to [Tri$\alpha$], p. 132. Here "$\tau_0 \geq$" and/or "$\geq \tau_1$" can be replaced by "$\tau_0 >$" and/or "$> \tau_1$", respectively. Furthermore, the Lebesgue measure on $\mathbb{R}^n$ is divisible: If M is a Lebesgue-measurable set in $\mathbb{R}^n$ with $0 < |M| < \infty$ and if λ is a positive number with $0 < \lambda < |M|$, then there is a Lebesgue-measurable set M_λ with

$$M_\lambda \subset M \quad \text{and} \quad |M_\lambda| = \lambda. \tag{12.7}$$

Let again $f \in A_{pq}^s(\mathbb{R}^n)$ as above. By (12.6), (12.7) there is a set M with $|M| = 1$ and

$$M \subset \{x \in \mathbb{R}^n : |f(x)| \geq f^*(1)\}, \quad \mathbb{R}^n \setminus M \subset \{x \in \mathbb{R}^n : |f(x)| \leq f^*(1)\}. \tag{12.8}$$

The set M can be decomposed by

$$M = \bigcup_{j=1}^{\infty} M_j, \quad M_j \cap M_k = \emptyset \text{ if } j \neq k, \quad |M_j| = 2^{-j} \text{ if } j \in \mathbb{N}, \qquad (12.9)$$

such that

$$M_j \subset \{x \in \mathbb{R}^n : f^*(2^{-j+1}) \leq |f(x)| \leq f^*(2^{-j})\}, \quad j \in \mathbb{N}. \qquad (12.10)$$

In other words, if (12.3), (12.4) or (12.5) is finite for all $f \in A^s_{pq}(\mathbb{R}^n)$, then we get some information on how rapidly $|f(x)|$ might grow on a sequence of sets M_j or M^j with $|M_j| = 2^{-j}$ or $|M^j| = 2^{-j+1}$ where $M^j = \cup_{l=j}^{\infty} M_l$.

For a closer look at (12.3) and (12.5) in connection with the above spaces $A^s_{pq}(\mathbb{R}^n)$ and inspired by the rearrangement formulations of Lorentz-Zygmund spaces according to 11.6, we adopt a slightly more general point of view. Let ψ be a real continuous monotonically increasing function on the interval $[0, \varepsilon]$, where $0 < \varepsilon < 1$, with

$$\psi(0) = 0 \quad \text{and} \quad \psi(t) > 0 \quad \text{if} \quad 0 < t \leq \varepsilon. \qquad (12.11)$$

Let $\Psi(t) = \log \psi(t)$ and let μ_Ψ be the associated Borel measure. We refer to [Lan93], especially p. 285, for details of this notation. In particular, all the integrals below with respect to the distribution function $\Psi(t)$, but also with respect to the other distribution functions needed below, can be interpreted as Riemann-Stieltjes integrals (defined in the usual way via Riemann-Stieltjes sums). If, in addition, $\psi(t)$ is differentiable in $(0, \varepsilon)$ then

$$\mu_\Psi(dt) = \frac{\psi'(t)}{\psi(t)} dt \quad \text{in} \quad (0, \varepsilon). \qquad (12.12)$$

Let $\psi_1(t)$ and $\psi_2(t)$ be two real continuous monotonically increasing functions on $[0, \varepsilon]$ with the counterpart of (12.11). Then, again, ψ_1 and ψ_2 are said to be equivalent, $\psi_1 \sim \psi_2$, if there are positive numbers c_1 and c_2 with

$$c_1 \psi_1(t) \leq \psi_2(t) \leq c_2 \psi_1(t), \quad 0 \leq t \leq \varepsilon. \qquad (12.13)$$

12.2 Proposition

(i) Let ψ be a real continuous monotonically increasing function on the interval $[0, \varepsilon]$ with (12.11). Let $0 < u_0 < u_1 < \infty$. There are two positive numbers

c_0 and c_1 such that

$$\sup_{0<t\le\varepsilon} \psi(t)\,g(t) \le c_1 \left(\int_0^\varepsilon (\psi(t)\,g(t))^{u_1}\,\mu_\Psi(dt)\right)^{\frac{1}{u_1}}$$

$$\le c_0 \left(\int_0^\varepsilon (\psi(t)\,g(t))^{u_0}\,\mu_\Psi(dt)\right)^{\frac{1}{u_0}} \quad (12.14)$$

for all non-negative monotonically decreasing functions g on $(0,\varepsilon]$.

(ii) *Let ψ_1 and ψ_2 be two such functions which are equivalent according to (12.13), and let $\Psi_j(t) = \log\psi_j(t)$ and μ_{Ψ_j} be the corresponding distribution functions and measures ($j=1,2$). Let $0 < u \le \infty$. Then*

$$\left(\int_0^\varepsilon (\psi_1(t)\,g(t))^u\,\mu_{\Psi_1}(dt)\right)^{\frac{1}{u}} \sim \left(\int_0^\varepsilon (\psi_2(t)\,g(t)))^u\,\mu_{\Psi_2}(dt)\right)^{\frac{1}{u}} \quad (12.15)$$

(with the sup-norm if $u=\infty$) for all non-negative monotonically decreasing functions g on $(0,\varepsilon]$, where the equivalence constants are independent of g.

Proof *Step 1* We begin with some preparation. Let $0 < u < \infty$. We claim

$$\psi^u(t) \sim \int_0^t \psi^u(\tau)\,\mu_\Psi(d\tau), \quad 0 < t \le \varepsilon, \quad (12.16)$$

where the equivalence constants are independent of t and of all admitted ψ according to (i). Let $\psi(t)$ be fixed and let

$$2^{-l} < \psi(t) \le 2^{-l+1} \quad \text{for some} \quad l \in \mathbb{Z} \quad (12.17)$$

(where, as in 2.1, $\mathbb{Z}$ is the collection of all integers). Let a_j be a decreasing sequence of positive numbers, tending to zero with $\psi(a_j) = 2^{-j}$. Then

$$a_l < t \le a_{l-1} \quad (12.18)$$

(with the replacement of a_{l-1} by ε if $\psi(\varepsilon) < 2^{-l+1}$). Since $\Psi(a_j) - \Psi(a_{j+1}) = 1$ it follows that

$$\int_0^{a_l} \psi^u(\tau)\,\mu_\Psi(d\tau) = \sum_{j=l}^\infty \int_{a_{j+1}}^{a_j} \psi^u(\tau)\,\mu_\Psi(d\tau)$$

$$\sim \sum_{j=l}^\infty 2^{-ju} \sim 2^{-lu} \sim \psi^u(t). \quad (12.19)$$

Similarly with a_{l-1} (respectively ε) in place of a_l on the left-hand side. This proves (12.16).

Step 2 We prove (i). Let $0 < u < \infty$. By (12.16) and the monotonicity of the non-negative function g it follows that

$$\psi(t)\, g(t) \leq c \left(\int_0^t g^u(\tau)\, \psi^u(\tau)\, \mu_\Psi(d\tau) \right)^{\frac{1}{u}}$$
$$\leq c \left(\int_0^\varepsilon g^u(\tau)\, \psi^u(\tau)\, \mu_\Psi(d\tau) \right)^{\frac{1}{u}}, \quad 0 < t \leq \varepsilon. \quad (12.20)$$

This is even sharper than the first estimate in (12.14). Let $0 < u_0 < u_1 < \infty$. Then

$$\int_0^\varepsilon (\psi(t)\, g(t))^{u_1}\, \mu_\Psi(dt) \leq \int_0^\varepsilon (\psi(t)\, g(t))^{u_0}\, \mu_\Psi(dt) \cdot \left(\sup_{0 < \tau \leq \varepsilon} \psi(\tau)\, g(\tau) \right)^{u_1 - u_0}. \quad (12.21)$$

Using (12.20) with u_0 in place of u we get the second inequality in (12.14).

Step 3 We prove (ii). This is obvious if $u = \infty$ (it is the left-hand side of (12.14)). Let $0 < u < \infty$. We begin with some preparation. Let $H(t)$ be a continuous monotonically increasing distribution function on $[0, \varepsilon]$ with $H(0) = 0$. Let μ_H be the associated measure. Let $G(t)$ be a bounded non-negative monotonically decreasing function on $[0, \varepsilon]$. Then the Riemann-Stieltjes sums

$$\sum_{j=0}^{N-1} G(b_j)(H(b_{j+1}) - H(b_j)) = \sum_{j=1}^{N-1} H(b_j)(G(b_{j-1}) - G(b_j)) + H(\varepsilon) G(b_{N-1}) \quad (12.22)$$

tend for suitable subdivisions

$$0 = b_0 < b_1 < \cdots < b_{N-1} < b_N = \varepsilon \quad (12.23)$$

of $[0, \varepsilon]$ to the Riemann-Stieltjes integral

$$\int_0^\varepsilon G(t)\, \mu_H(dt). \quad (12.24)$$

If one has two such distribution functions $H_1(t)$ and $H_2(t)$ and related measures μ_{H_1} and μ_{H_2} which are equivalent, then by (12.22) and $G(t) \geq 0$ the corresponding Riemann-Stieltjes sums are also equivalent (with the same equivalence constants, independently of G) and this extends to the integrals (12.24).

With these facts established, one can prove the equivalence (12.15) as follows. We put $G(t) = g^u(t)$ (so far assumed to be bounded) and

$$\mu_H(dt) = \psi_1^u(t)\,\mu_{\Psi_1}(dt)\,.$$

The corresponding distribution function $H(t)$ is the right-hand side of (12.16) with ψ_1, Ψ_1 in place of ψ, Ψ. By (12.16) this distribution function is equivalent to the distribution function $\psi_1^u(t)$. Hence the corresponding integrals of type (12.24) are also equivalent. Since $\psi_1 \sim \psi_2$ this assertion extends to ψ_2 and we obtain (12.15). Unbounded functions $g(t)$ can be approximated by bounded ones.

12.3 Discussion

The monotonicity (12.14) is the refined and more systematic version of what follows from (12.5). In the applications in the following sections we identify $\psi^{-1}(t)$ with

the growth envelope function $\mathcal{E}_G|A_{pq}^s(t)$ or

the continuity envelope function $\mathcal{E}_C|A_{pq}^{1+\frac{n}{p}}(t)$

from (10.7) or (10.19), respectively. They have essentially the required properties. However by our general point of view we do not distinguish between equivalent quasi-norms in a given space $A_{pq}^s(\mathbb{R}^n)$. With a few exceptions such as $L_p(\mathbb{R}^n)$ or, to a lesser extent, the classical Sobolev spaces $W_p^k(\mathbb{R}^n)$, there is no *primus inter pares* among the equivalent quasi-norms. The situation is much the same as in the final slogan in G. Orwell's novel, *Animal Farm*, [Orw51], p. 114, which reads, adapted to our situation, as follows,

All equivalent quasi-norms are equal
but some equivalent quasi-norms are more
equal than others.

In other words, any notation of relevance must be checked to see what happens when a quasi-norm is replaced by an equivalent one. This is the reason why we included part (ii) in the above proposition.

12.4 Examples

We discuss a few examples which, on the one hand, will be useful for our later considerations, and, on the other hand, shed new light on the Lorentz-Zygmund spaces mentioned in 11.6.

Example 1 Let $0 < r < \infty$ and $\psi(t) = t^{\frac{1}{r}}$. By (12.12) we have

$$\mu_\Psi(dt) = \frac{1}{r}\frac{dt}{t}, \quad 0 < t \leq \varepsilon. \tag{12.25}$$

Let $0 < u_0 < u_1 < \infty$. Then (12.14) results in

$$\sup_{0<t\leq\varepsilon} t^{\frac{1}{r}} g(t) \leq c_1 \left(\int_0^\varepsilon \left(t^{\frac{1}{r}} g(t)\right)^{u_1} \frac{dt}{t}\right)^{\frac{1}{u_1}} \leq c_0 \left(\int_0^\varepsilon \left(t^{\frac{1}{r}} g(t)\right)^{u_0} \frac{dt}{t}\right)^{\frac{1}{u_0}} \tag{12.26}$$

for all non-negative monotonically decreasing functions g on $(0, \varepsilon]$. With $g(t) = f^*(t)$ we have the Lorentz spaces $L_{ru}(I_\varepsilon)$ introduced in 11.6(i), and the well-known monotonicity with respect to u (for fixed r), [BeS88], p. 217.

Example 2 Let $b > 0$ and $\psi(t) = |\log t|^{-b}$ where $0 < t \leq \varepsilon < 1$. By (12.12) we have

$$\mu_\Psi(dt) = b\frac{dt}{t|\log t|}, \quad 0 < t \leq \varepsilon < 1. \tag{12.27}$$

Let $0 < u_0 < u_1 < \infty$. Then (12.14) results in

$$\sup_{0<t<\varepsilon} \frac{g(t)}{|\log t|^b} \leq c_1 \left(\int_0^\varepsilon \left(\frac{g(t)}{|\log t|^b}\right)^{u_1} \frac{dt}{t|\log t|}\right)^{\frac{1}{u_1}}$$
$$\leq c_0 \left(\int_0^\varepsilon \left(\frac{g(t)}{|\log t|^b}\right)^{u_0} \frac{dt}{t|\log t|}\right)^{\frac{1}{u_0}} \tag{12.28}$$

for all non-negative monotonically decreasing functions g on $(0, \varepsilon]$. Let $a = -b$ and $g(t) = f^*(t)$. Then the left-hand side of (12.28) coincides with (11.47). As mentioned there the corresponding spaces $L_\infty(\log L)_a(I_\varepsilon)$ can be equivalently described by (11.45). The right-hand side of (12.28), say, with $u = u_1$, fits in the scheme of the special Lorentz-Zygmund spaces $L_{\infty u}(\log L)_a(I_\varepsilon)$ in 11.6(iii) with $a = -b - \frac{1}{u}$. Then the requirement mentioned there, $au < -1$, coincides with $b > 0$, and looks more natural. Inequalities of type (12.28) in terms of $L_{\infty u}(\log L)_a(I_\varepsilon)$ may be found in [BeR80], Theorem 9.5, where the notation *diagonal* comes from the natural combination $a + \frac{1}{u} = -b$. As stated above, we are not so much interested in the spaces $L_{ru}(\log L)_a(I_\varepsilon)$ for their own sake. We formulate our assertions in terms of inequalities of the same type as in Theorem 11.7. In particular, (11.71) follows from (12.28) with $u_1 = p$ and $b = \frac{1}{p'}$.

Example 3 Let $0 < r < \infty$, $a \in \mathbb{R}$ and $\psi(t) = t^{\frac{1}{r}} |\log t|^a$. By (12.12),

$$\mu_\Psi(dt) \sim \frac{dt}{t}, \quad 0 < t \leq \varepsilon, \tag{12.29}$$

where $\varepsilon > 0$ is chosen so small that $\psi(t)$ is monotone in the interval $[0, \varepsilon]$. Let $0 < u_0 < u_1 < \infty$. Then (12.14) results in

$$\sup_{0<t<\varepsilon} t^{\frac{1}{r}} |\log t|^a g(t) \leq c_1 \left(\int_0^\varepsilon \left(t^{\frac{1}{r}} |\log t|^a g(t) \right)^{u_1} \frac{dt}{t} \right)^{\frac{1}{u_1}}$$

$$\leq c_0 \left(\int_0^\varepsilon \left(t^{\frac{1}{r}} |\log t|^a g(t) \right)^{u_0} \frac{dt}{t} \right)^{\frac{1}{u_0}} \tag{12.30}$$

for all non-negative monotonically decreasing functions g on $(0, \varepsilon]$.
With $g = f^*$ we have (11.48), (11.49), and hence the Lorentz-Zygmund spaces $L_{ru}(\log L)_a(I_\varepsilon)$ introduced there. The inequality (12.30) is known and may be found in [BeR80], Theorem 9.3.

12.5 Growth envelope functions

The concept of growth envelope functions $\mathcal{E}_G | A_{pq}^s$ (outlined so far in (10.7), (10.9) modulo equivalences) makes sense for all spaces $A_{pq}^s(\mathbb{R}^n)$ (where $A_{pq}^s(\mathbb{R}^n)$ always means either $B_{pq}^s(\mathbb{R}^n)$ or $F_{pq}^s(\mathbb{R}^n)$) which are covered by Theorem 11.2. But we exclude borderline cases where $p = \infty$ or $s = n(\frac{1}{p} - 1)_+$. Hence we always assume

$$0 < p < \infty, \quad 0 < q \leq \infty, \quad s > \sigma_p = n\left(\frac{1}{p} - 1\right)_+. \tag{12.31}$$

Furthermore the growth envelope function $\mathcal{E}_G | A_{pq}^s$ is designed to be a sharp instrument to measure on a global scale how singular (with respect to its growth) a function belonging to $A_{pq}^s(\mathbb{R}^n)$ can be. Hence it is reasonable to restrict the considerations to those spaces $A_{pq}^s(\mathbb{R}^n)$ with (12.31) which are, in addition, not embedded in $L_\infty(\mathbb{R}^n)$. To make clear which spaces are meant one must complement Theorem 11.4 by non-limiting embeddings. Since by Theorem 11.4(ii) one has in the critical case both embeddings and non-embeddings in $C(\mathbb{R}^n)$, one can combine this assertion with elementary embeddings for $A_{pq}^s(\mathbb{R}^n)$ with fixed p and variable s, q of type as in [Triβ], Proposition 2 on p. 47, to get a final answer. This results in all spaces in sub-critical situations (11.18) and in those spaces in critical situations $s = \frac{n}{p}$ which are not covered by (11.22) and (11.23). To avoid any misunderstanding we give a precise formulation which spaces we wish to exclude:

Under the assumption (12.31) the following three assertions (i), (ii), (iii) are equivalent to each other:

12.5(i) $\quad A^s_{pq}(\mathbb{R}^n) \subset L_\infty(\mathbb{R}^n)$,

12.5(ii) $\quad A^s_{pq}(\mathbb{R}^n) \subset C(\mathbb{R}^n)$,

12.5(iii)

$$either \quad s > \frac{n}{p},$$

$$or \quad A^s_{pq}(\mathbb{R}^n) = B^s_{pq}(\mathbb{R}^n) \quad with \quad 0 < p < \infty, \ s = \frac{n}{p}, \ 0 < q \leq 1,$$

$$or \quad A^s_{pq}(\mathbb{R}^n) = F^s_{pq}(\mathbb{R}^n) \quad with \quad 0 < p \leq 1, \ s = \frac{n}{p}, \ 0 < q \leq \infty.$$

A full proof of this assertion has been given in [SiT95], Theorem 3.3.1. A short description may be found in [RuS96], 2.2.4, p. 32 - 33.

Obviously, the concept of growth envelope functions makes sense for a much larger scale of function spaces than considered here. It has been studied recently in [Har01]. We do not go into detail, but we have a brief look at $L_r(\mathbb{R}^n)$ with $1 \leq r < \infty$, obviously normed by (2.1) (we recall the Orwellian confession at the end of 12.3) and put

$$\mathcal{E}_G | L_r(t) = \sup\{f^*(t) : \|f\,|L_r(\mathbb{R}^n)\| \leq 1\}, \quad 0 < t < \varepsilon. \tag{12.32}$$

Then

$$\mathcal{E}_G | L_r(t) \sim t^{-\frac{1}{r}}, \quad 0 < t < \varepsilon. \tag{12.33}$$

The estimate of $\mathcal{E}_G|L_r(t)$ from above by $t^{-\frac{1}{r}}$ follows from (12.26) with $u_1 = r$ and $g = f^*$. For the estimate from below one can choose the function $t^{-\frac{1}{r}}\chi_M(x)$, where $\chi_M(x)$ is the characteristic function of a set M with $|M| = t$. As far as the growth envelope function $\mathcal{E}_G|A^s_{pq}$ for one of the spaces $A^s_{pq}(\mathbb{R}^n)$ of interest is concerned, we have first a closer look at $\mathcal{E}_G|A^s_{pq}$ with respect to a given quasi-norm $\|\cdot|A^s_{pq}(\mathbb{R}^n)\|$.

12.6 Proposition

(B) *Let either*

$$1 < r < \infty, \quad s > 0, \quad s - \frac{n}{p} = -\frac{n}{r}, \quad 0 < q \leq \infty, \tag{12.34}$$

(sub-critical case, dashed line in Fig. 10.1) or

$$0 < p < \infty, \quad s = \frac{n}{p}, \quad 1 < q \leq \infty, \tag{12.35}$$

(critical case) for the spaces $B^s_{pq}(\mathbb{R}^n)$.

(F) Let either s, p, q be as in (12.34) or

$$1 < p < \infty, \quad s = \frac{n}{p}, \quad 0 < q \leq \infty, \tag{12.36}$$

(critical case) for the spaces $F^s_{pq}(\mathbb{R}^n)$.
Let $A^s_{pq}(\mathbb{R}^n)$ be either $B^s_{pq}(\mathbb{R}^n)$ with (B) or $F^s_{pq}(\mathbb{R}^n)$ with (F). Let, by definition, $\mathcal{E}_G|A^s_{pq}$,

$$\mathcal{E}_G|A^s_{pq}(t) = \sup\{f^*(t) : \|f\,|A^s_{pq}(\mathbb{R}^n)\| \leq 1\}, \quad 0 < t < \varepsilon, \tag{12.37}$$

where ε is a given number. Then $\mathcal{E}_G|A^s_{pq}$ is a positive, monotonically decreasing, unbounded function on the interval $(0, \varepsilon]$ with

$$\mathcal{E}_G|A^s_{pq}(2^{-j}) \sim \mathcal{E}_G|A^s_{pq}(2^{-j+1}), \quad j = J, J+1, \ldots, \tag{12.38}$$

(where the equivalence constants are independent of j). Furthermore, in the sub-critical case given by (12.34) we have

$$\mathcal{E}_G|A^s_{pq}(t) \leq c\,t^{-\frac{1}{r}}, \quad 0 < t \leq \varepsilon, \tag{12.39}$$

for some $c > 0$, and in the critical case given by (12.35) or (12.36),

$$\mathcal{E}_G|A^s_{pq}(t) \leq c_\eta\, t^{-\eta}, \quad 0 < t \leq \varepsilon, \tag{12.40}$$

for any $\eta > 0$ and a suitable constant $c_\eta > 0$.

Proof *Step 1* Obviously, $\mathcal{E}_G|A^s_{pq}(t)$ is monotonically decreasing (this means non-increasing) and positive for all $t > 0$. Assume that $\mathcal{E}_G|A^s_{pq}(t)$ is bounded. By (12.2) we have

$$\|f\,|L_\infty(\mathbb{R}^n)\| = f^*(0) \leq \sup_{0<t<\varepsilon} \mathcal{E}_G|A^s_{pq}(t) \tag{12.41}$$

for all $f \in A^s_{pq}(\mathbb{R}^n)$ with $\|f\,|A^s_{pq}(\mathbb{R}^n)\| = 1$, and hence

$$\|f\,|L_\infty(\mathbb{R}^n)\| \leq \left(\sup_{0<t<\varepsilon} \mathcal{E}_G|A^s_{pq}(t)\right) \|f\,|A^s_{pq}(\mathbb{R}^n)\|, \quad f \in A^s_{pq}(\mathbb{R}^n). \tag{12.42}$$

However (B) and (F) collect just those cases with (12.31) which are not covered by 12.5(i)–12.5(iii). Hence $\mathcal{E}_G|A^s_{pq}(t)$ is unbounded if $t \downarrow 0$.

Step 2 Let s, p, q be given by (12.34). We prove (12.39). As remarked in 11.8(ii), the inequality (11.52) with $q = \infty$, hence

$$\sup_{0<t<\varepsilon} t^{\frac{1}{r}} f^*(t) \leq c\,\|f\,|B^s_{p\infty}(\mathbb{R}^n)\| \leq c'\,\|f\,|A^s_{pq}(\mathbb{R}^n)\|, \tag{12.43}$$

is very classical, and taken for granted here. The second inequality is an elementary embedding, [Triβ], Proposition 2 on p. 47. This proves (12.39) in all sub-critical cases. As for the critical cases (12.35) and (12.36) we note the elementary non-limiting embedding

$$A_{pq}^{\frac{n}{p}}(\mathbb{R}^n) \subset L_r(\mathbb{R}^n) \quad \text{for any} \quad \max(p,1) < r < \infty. \tag{12.44}$$

Now (12.40) follows from (12.33) and, as a consequence of (12.44),

$$\mathcal{E}_G | A_{pq}^{\frac{n}{p}}(t) \leq c\, \mathcal{E}_G | L_r(t). \tag{12.45}$$

Step 3   We prove (12.38). Let

$$f \in A_{pq}^s(\mathbb{R}^n) \quad \text{with} \quad \|f\,|A_{pq}^s(\mathbb{R}^n)\| \leq 1$$

and let $g(x) = f(2^{-\frac{1}{n}}x)$ where $x \in \mathbb{R}^n$. By (12.1) we have

$$\begin{aligned}\mu_g(\lambda) &= |\{x \in \mathbb{R}^n : |f(2^{-\frac{1}{n}}x)| > \lambda\}| \\ &= 2|\{x \in \mathbb{R}^n : |f(x)| > \lambda\}| = 2\mu_f(\lambda), \quad \lambda > 0.\end{aligned} \tag{12.46}$$

Hence, by (12.2) (and by (12.39), (12.40)),

$$f^*(2^{-j}) = g^*(2^{-j+1}), \quad j = J+1, \ldots. \tag{12.47}$$

Furthermore with some $c > 0$ (independent of f)

$$c\,\|g\,|A_{pq}^s(\mathbb{R}^n)\| \leq \|f\,|A_{pq}^s(\mathbb{R}^n)\| \leq 1. \tag{12.48}$$

Now, by (12.37), and (12.47), (12.48), it follows that

$$\mathcal{E}_G | A_{pq}^s(2^{-j+1}) \geq c\, \mathcal{E}_G | A_{pq}^s(2^{-j}), \quad j = J+1, \ldots, \tag{12.49}$$

with the same c as in (12.48). Since the converse inequality is obvious we obtain (12.38).

12.7 Equivalence classes of growth envelope functions

If one puts

$$w(t) = \frac{1}{\mathcal{E}_G | A_{pq}^s(t)}, \quad 0 < t \leq \varepsilon, \tag{12.50}$$

then (12.38) coincides with (12.4) and we have (12.5). This was one of our motivations. The refinement of this point of view at the end of 12.1, which resulted in Proposition 12.2 and in the discussion in 12.3, requires for the underlying monotonically increasing distribution function $w(t) = \psi(t)$ with (12.11) that it is in addition continuous. However one can circumvent the possibly somewhat delicate question as to whether or not $\mathcal{E}_G|A^s_{pq}(t)$ is continuous. First we remark that for two equivalent quasi-norms,

$$\| \cdot |A^s_{pq}(\mathbb{R}^n)\|_1 \sim \| \cdot |A^s_{pq}(\mathbb{R}^n)\|_2, \tag{12.51}$$

of a given space $A^s_{pq}(\mathbb{R}^n)$ we have (in obvious notation)

$$\mathcal{E}_G|_1 A^s_{pq}(t) \sim \mathcal{E}_G|_2 A^s_{pq}(t), \quad 0 < t \leq \varepsilon, \tag{12.52}$$

as an immediate consequence of (12.37). Equivalence must always be understood according to (12.13) adapted to the above situation. This fits in our Orwellian point of view confessed at the end of 12.3.

The collection of all positive unbounded monotonically decreasing functions on the interval $(0, \varepsilon]$ can be subdivided into equivalence classes, where a class consists of all those admitted functions which are equivalent to one (and hence to all) functions in the given class.

By (12.52) all *growth envelope functions* for a space $A^s_{pq}(\mathbb{R}^n)$ covered by Proposition 12.6 belong to the same equivalence class, denoted by $[\mathcal{E}_G A^s_{pq}]$. This class contains also representatives which are continuous on $(0, \varepsilon]$ (in addition to the other required properties). For example, one can start with a fixed growth envelope function $\mathcal{E}_G|A^s_{pq}$ and define $\mathcal{E}_G A^s_{pq}$ (without the midline) as the polygonal line with

$$\mathcal{E}_G A^s_{pq}(2^{-j}) = \mathcal{E}_G|A^s_{pq}(2^{-j}), \quad j = J, J+1, \ldots, \tag{12.53}$$

and linear in the intervals $2^{-j-1} \leq t \leq 2^{-j}$ (modification at ε). Then one can apply Proposition 12.2 with

$$\psi(t) = \mathcal{E}_G A^s_{pq}(t)^{-1}, \quad 0 < t \leq \varepsilon.$$

One can even use (12.12).

12.8 Definition

Let $A^s_{pq}(\mathbb{R}^n)$ be either $B^s_{pq}(\mathbb{R}^n)$ with (B) or $F^s_{pq}(\mathbb{R}^n)$ with (F) according to Proposition 12.6. Let $[\mathcal{E}_G A^s_{pq}]$ be the equivalence class associated to $A^s_{pq}(\mathbb{R}^n)$ according to 12.7. Let

$$\mathcal{E}_G A^s_{pq} \in [\mathcal{E}_G A^s_{pq}] \quad \text{be a continuous representative.} \tag{12.54}$$

Let
$$\psi(t) = \mathcal{E}_G A_{pq}^s(t)^{-1} \quad \text{and} \quad \Psi(t) = \log \psi(t) = -\log \mathcal{E}_G A_{pq}^s(t), \quad (12.55)$$

$0 < t \le \varepsilon$, according to 12.1 and let μ_Ψ be the associated Borel measure on $[0, \varepsilon]$. Let $0 < u \le \infty$. Then the couple

$$\mathfrak{E}_G A_{pq}^s = ([\mathcal{E}_G A_{pq}^s], u) \quad (12.56)$$

is called the growth envelope for $A_{pq}^s(\mathbb{R}^n)$ when

$$\left(\int_0^\varepsilon (\psi(t) f^*(t))^v \, \mu_\Psi(dt) \right)^{\frac{1}{v}} \le c \, \|f \,|\, A_{pq}^s(\mathbb{R}^n)\| \quad (12.57)$$

(modified as on the left-hand side of (12.14) if $v = \infty$) holds for some $c = c_v > 0$ and all $f \in A_{pq}^s(\mathbb{R}^n)$ if, and only if, $u \le v \le \infty$.

12.9 Discussion and notational agreement

First we recall that under the restriction (12.31) (excluding borderline cases $p = \infty$ or $s = \sigma_p$) the conditions (B) and (F) cover all cases for which this concept is reasonable. Furthermore, the definition of the number u in (12.56) makes sense and is independent of the chosen representative $\mathcal{E}_G A_{pq}^s$. This follows from both parts of Proposition 12.2. However we must add a remark. By definition we have always

$$\sup_{0<t\le\varepsilon} \psi(t) f^*(t) = \sup_{0<t\le\varepsilon} \frac{f^*(t)}{\mathcal{E}_G A_{pq}^s(t)} \le c \, \|f \,|\, A_{pq}^o(\mathbb{R}^n)\| \quad (12.58)$$

for some $c > 0$ and all $f \in A_{pq}^s(\mathbb{R}^n)$. Hence by Proposition 12.2 it always makes sense to put

$$u = \inf\{v : (12.57) \text{ holds}\}. \quad (12.59)$$

But it is not clear from the very beginning whether (12.57) remains valid with u in place of v. However this will be always the case for all spaces considered here. This may justify the incorporation here of this additional information immediately in the definition. Furthermore we wish to simplify (12.56) by

$$\mathfrak{E}_G A_{pq}^s = (\mathcal{E}_G A_{pq}^s(t), u), \quad (12.60)$$

where $\mathcal{E}_G A_{pq}^s$ is a continuous representative according to (12.54). The situation is similar to the usual simplification of writing $f \in L_p(\mathbb{R}^n)$ instead of $[f] \in$

$L_p(\mathbb{R}^n)$, where $[f]$ stands for the equivalence class of all measurable functions g which coincide with f almost everywhere. This is also justified by the typical examples in (10.14). Hence we prefer, for example,

$$\mathcal{E}_G A_{pq}^s = (t^{-\frac{1}{r}}, u) \quad \text{compared with} \quad ([t^{-\frac{1}{r}}], u), \tag{12.61}$$

or even more cumbersome versions avoiding the explicit appearance of the variable t. (Of course the use of $[\cdot]$ is much the same as above in $f \in L_p(\mathbb{R}^n)$ compared with $[f] \in L_p(\mathbb{R}^n)$.) Next we collect a few rather simple properties which make clear what type of sharp inequalities can be expected.

12.10 Proposition

Let $A_{pq}^s(\mathbb{R}^n)$ be either $B_{pq}^s(\mathbb{R}^n)$ with (B) or $F_{pq}^s(\mathbb{R}^n)$ with (F) according to Proposition 12.6. Let $0 < \varepsilon < 1$. Let $\mathcal{E}_G A_{pq}^s$ be a continuous growth envelope function as in (12.54), let, in notational modification of (12.55),

$$E(t) = -\log \mathcal{E}_G A_{pq}^s(t), \quad 0 < t \le \varepsilon, \tag{12.62}$$

and let μ_E be the associated Borel measure on $(0, \varepsilon]$.

(i) Let $\varkappa(t)$ be a positive function on $(0, \varepsilon]$. Then there is a number $c > 0$ such that

$$\sup_{0 < t \le \varepsilon} \frac{\varkappa(t) f^*(t)}{\mathcal{E}_G A_{pq}^s(t)} \le c \, \|f \, |A_{pq}^s(\mathbb{R}^n)\| \quad \text{for all} \quad f \in A_{pq}^s(\mathbb{R}^n) \tag{12.63}$$

if, and only if, $\varkappa$ is bounded.

(ii) Let $\varkappa(t)$ be a positive monotonically decreasing function on $(0, \varepsilon]$ and let for some $0 < v < \infty$ and some $c > 0$

$$\left(\int_0^\varepsilon \left(\frac{f^*(t)}{\mathcal{E}_G A_{pq}^s(t)} \right)^v \mu_E(dt) \right)^{\frac{1}{v}} \le c \, \|f \, |A_{pq}^s(\mathbb{R}^n)\| \tag{12.64}$$

for all $f \in A_{pq}^s(\mathbb{R}^n)$. Then for some $c' > 0$,

$$\left(\int_0^\varepsilon \left(\frac{\varkappa(t) f^*(t)}{\mathcal{E}_G A_{pq}^s(t)} \right)^v \mu_E(dt) \right)^{\frac{1}{v}} \le c' \, \|f \, |A_{pq}^s(\mathbb{R}^n)\| \tag{12.65}$$

for all $f \in A_{pq}^s(\mathbb{R}^n)$ if, and only if, $\varkappa$ is bounded.

Proof *Step 1* The proof of (i) is simple. On the one hand we have (12.58). On the other hand, if (12.63) holds for some $\varkappa$, then for any fixed t with $0 < t \leq \varepsilon$,

$$\frac{\varkappa(t)\, f^*(t)}{\mathcal{E}_G A^s_{pq}(t)} \leq c \quad \text{for all } f \text{ with } \|f\, |A^s_{pq}(\mathbb{R}^n)\| \leq 1. \tag{12.66}$$

Now by (12.54) and (12.37) it follows $\varkappa(t) \leq c'$ uniformly with respect to t.

Step 2 We prove (ii). The function $g(t) = \varkappa(t)\, f^*(t)$ is non-negative and monotonically decreasing on $(0, \varepsilon]$. Hence, by (12.14),

$$\frac{\varkappa(t)\, f^*(t)}{\mathcal{E}_G A^s_{pq}(t)} \leq c\, \|f\, |A^s_{pq}(\mathbb{R}^n)\|, \quad 0 < t \leq \varepsilon. \tag{12.67}$$

Then (ii) follows from (i).

12.11 Discussion

In part (ii) we assumed that $\varkappa$ is monotonically decreasing. This is natural in our context, where we ask for (12.65) under the assumption (12.64), and also in connection with the definition of the growth envelope in (12.60). On the other hand, if $\varkappa$ is non-negative on $(0, \varepsilon]$ and, maybe, wildly oscillating (or monotonically increasing), then at least formally the question (12.65) makes sense without assuming that (12.64) holds. To look at the discretised version of this problem we assume that the numbers a_l have the same meaning as in Step 1 of the proof of Proposition 12.2 with $\psi^{-1}(t) = \mathcal{E}_G A^s_{pq}(t)$. Then the discrete twin of the left-hand side of (12.65) is given by

$$\left(\sum_{j=J}^{\infty} 2^{-jv} f^*(a_j)^v \int_{a_{j+1}}^{a_j} \varkappa(t)^v \, \mu_E(dt) \right)^{\frac{1}{v}}. \tag{12.68}$$

This suggests that not so much the pointwise behaviour of $\varkappa(t)$ but the behaviour of the indicated integral means is of interest. However we do not study problems of this type in the sequel.

12.12 Moduli of continuity

We outlined in Section 10 our methods and results. As explained there in connection with (10.22) we deal with the super-critical case by lifting the results obtained in the critical case. In rough terms, the role played by $f^*(t)$ in critical (and sub-critical) situations is taken over in super-critical cases by the divided modulus of continuity $\widetilde{\omega}(f,t)$. First we recall what we need in the sequel.

Let $f(x) \in C(\mathbb{R}^n)$, where $C(\mathbb{R}^n)$ has been introduced in 11.1 as the set of all complex-valued, bounded, uniformly continuous functions in $\mathbb{R}^n$. Then

$$\omega(f,t) = \sup_{x \in \mathbb{R}^n, |h| \le t} |f(x+h) - f(x)|, \quad 0 \le t < \infty, \tag{12.69}$$

is called the modulus of continuity. Let $f \in C(\mathbb{R}^n)$ be fixed. Then $\omega(f,t)$ is a non-negative and monotonically increasing (this means non-decreasing) continuous function on $[0, \infty)$; in particular,

$$\omega(f,t) \to \omega(0) = 0 \quad \text{if} \quad t \downarrow 0. \tag{12.70}$$

Furthermore, $\omega(f,t)$ is almost concave in the following sense: Let $\overline{\omega}(f,t)$ be the least concave majorant of $\omega(f,t)$. Then

$$\frac{1}{2}\overline{\omega}(f,t) \le \omega(f,t) \le \overline{\omega}(f,t). \tag{12.71}$$

We refer to [DeL93], Ch. 2, §6, pp. 40–44, where one finds proofs of all these properties. Let

$$\widetilde{\omega}(f,t) = \frac{\omega(f,t)}{t}, \quad t > 0, \tag{12.72}$$

be the divided modulus of continuity. By (12.71) we have

$$\widetilde{\omega}(f,t) \sim \frac{\overline{\omega}(f,t)}{t}. \tag{12.73}$$

Since $\overline{\omega}(f,t)$ is concave and continuous on $[0, \infty)$ and $\overline{\omega}(f,0) = 0$ it follows that the right-hand side of (12.73) is monotonically decreasing on $(0, \infty)$. Hence $\widetilde{\omega}(f,t)$ is at least equivalent to a monotonically decreasing function. This is sufficient for our purpose. The concept of moduli of continuity has been widely used in the theory of function spaces. Our goal here is rather limited. We are interested exclusively in the super-critical case according to (10.15), and, even more restrictive, only in those spaces $B_{pq}^{1+\frac{n}{p}}(\mathbb{R}^n)$ and $F_{pq}^{1+\frac{n}{p}}(\mathbb{R}^n)$ which are not continuously embedded in $Lip(\mathbb{R}^n)$. This means by Theorem 11.4(iii),

$$B_{pq}^{1+\frac{n}{p}}(\mathbb{R}^n) \quad \text{with} \quad 0 < p < \infty, \quad 1 < q \le \infty, \tag{12.74}$$

and

$$F_{pq}^{1+\frac{n}{p}}(\mathbb{R}^n) \quad \text{with} \quad 1 < p < \infty, \quad 0 < q \le \infty. \tag{12.75}$$

This is in good agreement with (10.18) on the one hand and (12.35), (12.36) on the other hand. We remark that

$$t \mapsto \sup\{\widetilde{\omega}(f,t) \,:\, \|f\,|A_{pq}^{1+\frac{n}{p}}(\mathbb{R}^n)\| \leq 1\} \tag{12.76}$$

is a bounded function on the interval $(0,1)$ if, and only if, $A_{pq}^s(\mathbb{R}^n)$ is continuously embedded in $Lip(\mathbb{R}^n)$. Hence, (12.74) and (12.75) cover just those cases, where (12.76) is unbounded. Now we are very much in the same situation as in Proposition 12.6 with the following outcome.

12.13 Proposition

Let $A_{pq}^{1+\frac{n}{p}}(\mathbb{R}^n)$ be either the space in (12.74) or the space in (12.75). Let, for some $\varepsilon > 0$, the continuity envelope function $\mathcal{E}_C|A_{pq}^{1+\frac{n}{p}}$, be defined by

$$\mathcal{E}_C|A_{pq}^{1+\frac{n}{p}}(t) = \sup\{\widetilde{\omega}(f,t) \,:\, \|f\,|A_{pq}^{1+\frac{n}{p}}(\mathbb{R}^n)\| \leq 1\}, \quad 0 < t < \varepsilon. \tag{12.77}$$

Then $\mathcal{E}_C|A_{pq}^{1+\frac{n}{p}}$ is a positive, continuous, unbounded function on the interval $(0,\varepsilon]$ with

$$\mathcal{E}_C|A_{pq}^{1+\frac{n}{p}}(2^{-j}) \sim \mathcal{E}_C|A_{pq}^{1+\frac{n}{p}}(2^{-j+1}), \quad j = J, J+1, \ldots, \tag{12.78}$$

(where the equivalence constants are independent of j). Furthermore, $\mathcal{E}_C|A_{pq}^{1+\frac{n}{p}}$ is equivalent to a monotonically decreasing function, and for any $\eta > 0$ there is a number $c_\eta > 0$ such that

$$\mathcal{E}_C|A_{pq}^{1+\frac{n}{p}}(t) \leq c_\eta\, t^{-\eta}, \quad 0 < t \leq \varepsilon. \tag{12.79}$$

Proof By the above remarks, $\mathcal{E}_C|A_{pq}^{1+\frac{n}{p}}$ is positive, unbounded, and equivalent to a monotonically decreasing function. By [DeL93], p. 41, we have

$$\omega(f,2t) \leq 2\,\omega(f,t) \quad \text{and} \quad |\omega(f,t_1+t_2) - \omega(f,t_1)| \leq \omega(f,t_2). \tag{12.80}$$

This proves (12.78) and the continuity of $\mathcal{E}_C|A_{pq}^{1+\frac{n}{p}}$. Finally, for given η, $1 > \eta > 0$, we have the non-limiting embedding

$$\sup_{0 < t < \varepsilon} \frac{\omega(f,t)}{t^{1-\eta}} \leq c\,\|f\,|A_{pq}^{1+\frac{n}{p}}(\mathbb{R}^n)\|, \tag{12.81}$$

[Triβ], 2.7.1, p. 131, formula (12). This proves (12.79).

12.14 Definition

Let $A_{pq}^{1+\frac{n}{p}}(\mathbb{R}^n)$ be either the space $B_{pq}^{1+\frac{n}{p}}(\mathbb{R}^n)$ from (12.74) or the space $F_{pq}^{1+\frac{n}{p}}(\mathbb{R}^n)$ from (12.75). Let $0 < \varepsilon < 1$. Then $[\mathcal{E}_C A_{pq}^{1+\frac{n}{p}}]$ is the equivalence class of all continuous monotonically decreasing functions on the interval $(0, \varepsilon]$ which are equivalent to one (and hence to all) continuity envelope function $\mathcal{E}_C | A_{pq}^{1+\frac{n}{p}}$ according to (12.77). Let

$$\mathcal{E}_C A_{pq}^{1+\frac{n}{p}} \in [\mathcal{E}_C A_{pq}^{1+\frac{n}{p}}],$$

$$\psi(t) = \mathcal{E}_C A_{pq}^{1+\frac{n}{p}}(t)^{-1} \quad \text{and} \quad \Psi(t) = \log \psi(t) = -\log \mathcal{E}_C A_{pq}^{1+\frac{n}{p}}(t), \quad (12.82)$$

$0 < t \le \varepsilon$, according to 12.1 and let μ_Ψ be the associated Borel measure on $[0, \varepsilon]$. Let $0 < u \le \infty$. Then the couple

$$\mathfrak{E}_C A_{pq}^{1+\frac{n}{p}} = ([\mathcal{E}_C A_{pq}^{1+\frac{n}{p}}], u) \tag{12.83}$$

is called the continuity envelope for $A_{pq}^{1+\frac{n}{p}}(\mathbb{R}^n)$ when

$$\left(\int_0^\varepsilon (\psi(t) \, \widetilde{\omega}(f,t))^v \, \mu_\Psi(dt) \right)^{\frac{1}{v}} \le c \, \|f \,|\, A_{pq}^{1+\frac{n}{p}}(\mathbb{R}^n)\| \tag{12.84}$$

(modified as on the left-hand side of (12.14) if $v = \infty$) holds for some $c = c_v > 0$ and all $f \in A_{pq}^{1+\frac{n}{p}}(\mathbb{R}^n)$ if, and only if, $u \le v \le \infty$.

12.15 Remark and notational agreement

This definition is the same as Definition 12.8, mutatis mutandis. In particular, all that had been said before 12.8 in 12.7, but also afterwards in 12.9, in Proposition 12.10, and in 12.11, has respective counterparts which will not be repeated here. But we mention that, much as in (12.60), we simplify (12.83) by

$$\mathfrak{E}_C A_{pq}^{1+\frac{n}{p}} = (\mathcal{E}_C A_{pq}^{1+\frac{n}{p}}(t), u), \tag{12.85}$$

where $\mathcal{E}_C A_{pq}^{1+\frac{n}{p}}$ is a representative of $[\mathcal{E}_C A_{pq}^{1+\frac{n}{p}}]$. As stated above, we reduce later on the super-critical case to the critical case by lifting. If $n = 1$, then one has (10.22). In higher dimensions the situation is more complicated. In the next proposition we prove what we need later. Recall that

$$\nabla f(x) = \left(\frac{\partial f}{\partial x_1}(x), \ldots, \frac{\partial f}{\partial x_n}(x) \right), \quad x \in \mathbb{R}^n, \tag{12.86}$$

and, hence,
$$|\nabla f(x)| = \left(\sum_{n=1}^{n}\left|\frac{\partial f}{\partial x_j}(x)\right|^2\right)^{\frac{1}{2}} \sim \sum_{j=1}^{n}\left|\frac{\partial f}{\partial x_j}(x)\right|. \tag{12.87}$$

Furthermore we need the rearrangement $|\nabla f|^*(t)$ and its maximal function $|\nabla f|^{**}(t)$ according to (10.3) and (10.4) with $|\nabla f|$ in place of f. Let $\omega(f,t)$ and $\widetilde{\omega}(f,t)$ be the modulus of continuity and the divided modulus of continuity introduced in (12.69) and (12.72). Finally, $C^1(\mathbb{R}^n)$ has the same meaning as in (11.6).

12.16 Proposition

(i) Let $0 < \varepsilon < 1$. There is a number $c > 0$ such that
$$\widetilde{\omega}(f,t) \le c\,|\nabla f|^{**}(t^{2n-1}) + 3 \sup_{0<\tau\le t^2} \tau^{-\frac{1}{2}}\omega(f,\tau) \tag{12.88}$$
for all $0 < t < \varepsilon$ and all $f \in C^1(\mathbb{R}^n)$.

(ii) Let $0 < p \le \infty$, $v > 0$, and $0 < \varepsilon < 1$. There is a number $c > 0$ such that
$$\int_0^\varepsilon \left(\frac{\widetilde{\omega}(f,t)}{|\log t|^v}\right)^p \frac{dt}{t|\log t|} \le c \int_0^\varepsilon \left(\frac{|\nabla f|^*(t)}{|\log t|^v}\right)^p \frac{dt}{t|\log t|} \quad \text{if } p < \infty, \tag{12.89}$$
and
$$\sup_{0<t<\varepsilon} \frac{\widetilde{\omega}(f,t)}{|\log t|^v} \le c \sup_{0<t<\varepsilon} \frac{|\nabla f|^*(t)}{|\log t|^v} \quad \text{if } p = \infty, \tag{12.90}$$
for all $f \in C^1(\mathbb{R}^n)$.

Proof *Step 1* We prove (i). Let t with $0 < t < \varepsilon$ be fixed. Replacing $f(x)$ by $\varrho f(x)$ for some $\varrho > 0$ we may assume that the supremum in (12.88) equals 1, hence
$$|f(x+y) - f(x)| \le \tau^{\frac{1}{2}} \quad \text{for all } x \in \mathbb{R}^n \text{ and } y \in \mathbb{R}^n \text{ with } |y| \le \tau, \tag{12.91}$$
where $\tau \le t^2$. Then (12.88) is equivalent to
$$t^{-1}|f(x+y) - f(x)| \le c\,|\nabla f|^{**}(t^{2n-1}) + 3 \tag{12.92}$$
for all $x \in \mathbb{R}^n$ and $y \in \mathbb{R}^n$ with $|y| \le t$. Of course it is sufficient to concentrate on those x and y for which the left-hand side of (12.92) is larger than 3. Without restriction of generality we may assume $x = 0$ and $y = y^1 = (y_1, 0, \ldots, 0)$. Hence,
$$A = |f(y^1) - f(0)| \ge 3t \quad \text{with} \quad y^1 = (y_1, 0, \ldots, 0), \quad 0 < y_1 \le t. \tag{12.93}$$

Let $y^2 = (0, y_2, \ldots, y_n) = (0, y')$ with $y' \in \mathbb{R}^{n-1}$ and $|y^2| = |y'| = \tau \leq t^2$. With $y = y^1 + y^2$ we obtain by (12.93) and (12.91),

$$|f(y) - f(y^2)| \geq |f(y^1) - f(0)| - |f(y) - f(y^1)| - |f(y^2) - f(0)|$$
$$\geq A - 2t \geq \frac{A}{3}.$$
(12.94)

Similarly one can estimate $|f(y) - f(y^2)|$ from above by $2A$. By construction y and y^2 differ only with respect to the first component. We fix $y' \in \mathbb{R}^{n-1}$ with $|y'| \leq t^2$ and obtain

$$|f(y) - f(y^2)| = \left| \int_0^{y_1} \frac{\partial f}{\partial x_1}(\sigma, y') \, d\sigma \right| \leq \int_0^t |\nabla f(\sigma, y')| \, d\sigma. \quad (12.95)$$

The left-hand side is equivalent to A. We integrate over $y' \in \mathbb{R}^{n-1}$ with $|y'| \leq t^2$. Then we have for some $c > 0$,

$$t^{2n-2} A \leq c \int_T |\nabla f(x)| \, dx, \quad (12.96)$$

where $T = [0, t] \times \{y' : |y'| \leq t^2\}$ is a tube in $\mathbb{R}^n$ with the volume $|T| = t^{2n-1}$. By standard arguments for rearrangements we obtain (switching to arbitrary $x \in \mathbb{R}^n$ and $y \in \mathbb{R}^n$ with $|y| \leq t$ and the counterpart of (12.93))

$$\frac{|f(x+y) - f(x)|}{t} \leq \frac{c}{t^{2n-1}} \int_0^{t^{2n-1}} |\nabla f|^*(\sigma) \, d\sigma \leq c |\nabla f|^{**}(t^{2n-1}). \quad (12.97)$$

This proves (12.88).

Step 2 We prove (ii). Let $p < \infty$. By (i) we have

$$\int_0^\varepsilon \left(\frac{\widetilde{\omega}(f, t)}{|\log t|^v} \right)^p \frac{dt}{t |\log t|}$$

$$\leq c \left(\sup_{0 < t \leq \varepsilon} t^{\frac{1}{2}} \widetilde{\omega}(f, t) \right)^p + c \int_0^\varepsilon \left(\frac{|\nabla f|^{**}(t^{2n-1})}{|\log t|^v} \right)^p \frac{dt}{t |\log t|}$$

$$\leq c' \int_0^\varepsilon \left(t^{\frac{1}{2}} \widetilde{\omega}(f, t) \right)^p \frac{dt}{t} + c' \int_0^\varepsilon \left(\frac{|\nabla f|^{**}(t)}{|\log t|^v} \right)^p \frac{dt}{t |\log t|}. \quad (12.98)$$

Here we used that $\widetilde{\omega}(f, t)$ is equivalent to a monotonically decreasing function. Then application of (12.26) justifies the first term on the right-hand side of

(12.98). In connection with the second term we used the transformation $\tau = t^{2n-1}$. Let $g(t) = |\nabla f|^*(t)$. Then $|\nabla f|^{**}(t) = Mg(t)$ is the maximal function of $g(t)$ according to (10.4). We wish to prove that

$$\int_0^\varepsilon \left(\frac{Mg(t)}{|\log t|^v}\right)^p \frac{dt}{t|\log t|} \leq c \int_0^\varepsilon \left(\frac{g(t)}{|\log t|^v}\right)^p \frac{dt}{t|\log t|}. \tag{12.99}$$

Let $\varepsilon \sim 2^{-J}$ and $t \sim 2^{-j}$ with $j \geq J$. Then

$$Mg(2^{-j}) \sim Mg(t) = \frac{1}{t}\int_0^t g(\tau)\,d\tau \sim \sum_{l=0}^\infty 2^{-l} g(2^{-l-j}). \tag{12.100}$$

Let $u = vp + 1$ and $q < p$. Then the left-hand side of (12.99) can be estimated from above by

$$c \sum_{j=J}^\infty \frac{Mg(2^{-j})^p}{j^u} \leq c' \sum_{j=J}^\infty \sum_{l=0}^\infty \frac{(j+l)^u}{j^u} 2^{-lq} \frac{g^p(2^{-j-l})}{(j+l)^u}$$

$$\leq c'' \sum_{k=J}^\infty \frac{g^p(2^{-k})}{k^u} \sum_{l=0}^{k-J} \frac{k^u}{(k-l)^u} 2^{-lq}. \tag{12.101}$$

Since $\frac{k}{k-l}$ can be estimated from above by $1 + \frac{l}{k-l} \leq 1 + l$, it follows that the last factor in (12.101) can be estimated from above by a constant, which is independent of J. Then the right-hand side of (12.101) is equivalent to the right-hand side of (12.99). This proves (12.99). We return to (12.98) and remark in addition that for any $\eta > 0$ there is an ε_0, $0 < \varepsilon_0 < 1$, such that

$$t^{\frac{1}{2}} \leq \eta |\log t|^{-u} \quad \text{if} \quad 0 < t \leq \varepsilon_0, \tag{12.102}$$

where $u = vp+1$ has the above meaning. Inserting (12.99) with $g(t) = |\nabla f|^*(t)$ and (12.102) with a small η in the right-hand side of (12.98), then we have on the right-hand side the desired term from the right-hand side of (12.89) and in addition the same term as on the left-hand side with a factor, say, $\frac{1}{2}$. This proves (12.89) under the additional assumption $0 < \varepsilon \leq \varepsilon_0$. We remove this restriction. Let $0 < \varepsilon < 1$ and let $0 < \varkappa < 1$. Since $\widetilde{\omega}(f,t)$ is equivalent to a monotonically decreasing function it follows that

$$\int_0^\varepsilon \left(\frac{\widetilde{\omega}(f,t)}{|\log t|^v}\right)^p \frac{dt}{t|\log t|} \leq c\int_0^\varepsilon \left(\frac{\widetilde{\omega}(f,\varkappa t)}{|\log t|^v}\right)^p \frac{dt}{t|\log t|}$$

$$\leq c' \int_0^{\varkappa \varepsilon} \left(\frac{\widetilde{\omega}(f,\tau)}{|\log \tau|^v}\right)^p \frac{d\tau}{\tau|\log \tau|}. \tag{12.103}$$

This reduces the case $0 < \varepsilon < 1$ to $0 < \varepsilon \leq \varepsilon_0$. Then we obtain (12.89). If $p = \infty$ then one can follow the above arguments with the necessary modifications and arrives at (12.90).

12.17 Remark

In the one-dimensional case, (12.88) with $n = 1$ reduces to

$$\widetilde{\omega}(f,t) \leq c |f'|^{**}(t), \quad 0 < t < \varepsilon, \quad f \in C^1(\mathbb{R}). \tag{12.104}$$

This follows from (12.95). It coincides with (10.22). The situation in $\mathbb{R}^n$ with $n \geq 2$ seems to be more complicated. Whether there is a direct counterpart of (12.104) with $|\nabla f|^{**}(t^n)$ is not so clear. On the other hand, the choice of $\tau^{-\frac{1}{2}} \omega(f, \tau)$ in the second term on the right-hand side of (12.88) is convenient and sufficient for us, but it can be modified. If one replaces $\tau^{-\frac{1}{2}} \omega(f, \tau)$ by $\varkappa(\tau) \tau^{-1} \omega(f, \tau)$ where $\varkappa(\tau)$ is a positive, say, monotonically increasing function with $\varkappa(\tau) \to 0$ if $\tau \to 0$, then one ends up with $\varrho(t) t^n$ in place of t^{2n-1} in the first term on the right-hand side of (12.88) with $\varrho(t) \to 0$ arbitrarily slowly if $t \to 0$. However if one wishes to apply modified versions of (12.88) to get (12.89) one needs a counterpart of (12.102) with $\varkappa(t)$ in place of $t^{\frac{1}{2}}$.

13 The critical case

13.1 Introduction

By the terminology of (10.6) the critical case covers the spaces

$$B_{pq}^{\frac{n}{p}}(\mathbb{R}^n) \quad \text{and} \quad F_{pq}^{\frac{n}{p}}(\mathbb{R}^n) \quad \text{with} \quad 0 < p < \infty \quad \text{and} \quad 0 < q \leq \infty. \tag{13.1}$$

This corresponds to the line of slope n in Fig. 10.1 starting from the origin. Generally in this Chapter II we are interested exclusively in spaces $B_{pq}^s(\mathbb{R}^n)$ and $F_{pq}^s(\mathbb{R}^n)$ which are not only subspaces of $S'(\mathbb{R}^n)$ but also of $L_1^{loc}(\mathbb{R}^n)$ (and, hence, consist entirely of regular distributions). We refer to Section 10 where we outlined our intentions. Theorem 11.2 clarifies under what conditions $B_{pq}^s(\mathbb{R}^n)$ and $F_{pq}^s(\mathbb{R}^n)$ are subspaces of $L_1^{loc}(\mathbb{R}^n)$. Recall that in all cases considered here (critical, super-critical, sub-critical) we always exclude borderline situations, which means in general

$$p = \infty \quad \text{and/or} \quad s = \sigma_p = n \left(\frac{1}{p} - 1 \right)_+ \quad \text{if} \quad 0 < p < \infty, \tag{13.2}$$

and especially according to Theorem 11.2,

$$B_{\infty,q}^0(\mathbb{R}^n) \quad \text{with} \quad 0 < q \leq 2, \tag{13.3}$$

in the critical situation $s = \frac{n}{p}$. A further distinguished borderline space in connection with the critical situation not treated in this section is $bmo(\mathbb{R}^n) = F^0_{\infty,2}(\mathbb{R}^n)$. Here we add at least a brief remark at the end of this section in 13.7. Otherwise as a further restriction of (13.1) we are interested only in those spaces which are not continuously embedded in $L_\infty(\mathbb{R}^n)$ (or, which is the same, in $C(\mathbb{R}^n)$); this means by Theorem 11.4, and as has been detailed in 11.5, especially (11.38), (11.39), we deal only with the spaces

$$B^{\frac{n}{p}}_{pq}(\mathbb{R}^n) \quad \text{with} \quad 0 < p < \infty, \quad 1 < q \leq \infty, \tag{13.4}$$

and

$$F^{\frac{n}{p}}_{pq}(\mathbb{R}^n) \quad \text{with} \quad 1 < p < \infty, \quad 0 < q \leq \infty. \tag{13.5}$$

This covers in particular the respective Sobolev spaces mentioned in (11.40). As outlined in the introductory Section 10 we wish to measure the singularity behaviour of functions belonging to the spaces in (13.4), (13.5) in terms of the growth envelope as introduced in Definition 12.8. Instead of $\mathfrak{E}_G A^s_{pq}$ in (12.56) we use the more handsome version (12.60). In the theorem below we calculate explicitly the growth envelopes for all spaces in (13.4) and (13.5). By Proposition 12.10 it is clear that one gets rather sharp assertions concerning the singularity behaviour of elements of these spaces in a very condensed form. Hence, it seems to be reasonable, after proving the theorem, to discuss what this means in detail. Finally we add references in 13.5 and, as said, a remark about $bmo(\mathbb{R}^n)$ in 13.7. Let $1 \leq v \leq \infty$. As usual v' is given by $\frac{1}{v} + \frac{1}{v'} = 1$.

13.2 Theorem

(i) Let

$$0 < p < \infty, \quad 1 < q \leq \infty. \tag{13.6}$$

Then

$$\mathfrak{E}_G B^{\frac{n}{p}}_{pq} = (|\log t|^{\frac{1}{q'}}, q). \tag{13.7}$$

(ii) Let

$$1 < p < \infty, \quad 0 < q \leq \infty. \tag{13.8}$$

Then

$$\mathfrak{E}_G F^{\frac{n}{p}}_{pq} = (|\log t|^{\frac{1}{p'}}, p). \tag{13.9}$$

Proof We break the rather long proof into 7 steps. Here is a guide. In Step 1 and Step 2 we prove those sharp inequalities which correspond to the right-hand sides of (13.7) and (13.9), respectively. In Step 3 we formulate what this means in terms of the growth envelope functions: They can be estimated from above by $|\log t|^{\frac{1}{q'}}$ and $|\log t|^{\frac{1}{p'}}$, respectively. To prove the sharpness we need extremal functions. They will be constructed in Steps 4 and 5. The outcome is of self-contained interest, also in connection with the super-critical case considered in Section 14, and will be formulated separately in Corollary 13.4. In Steps 6 and 7 we prove that $|\log t|^{\frac{1}{q'}}$ and $|\log t|^{\frac{1}{p'}}$ are envelope functions, and that q and p, respectively, are the correct exponents according to (13.7) and (13.9).

Step 1 Let p and q be given by (13.6), and let, as always, $0 < \varepsilon < 1$. We prove that there is a number $c > 0$ such that

$$\left(\int_0^\varepsilon \left(\frac{f^*(t)}{|\log t|} \right)^q \frac{dt}{t} \right)^{\frac{1}{q}} \leq c \, \|f \, | B_{pq}^{\frac{n}{p}}(\mathbb{R}^n)\| \tag{13.10}$$

for all $f \in B_{pq}^{\frac{n}{p}}(\mathbb{R}^n)$ with the interpretation

$$\sup_{0 < t \leq \varepsilon} \frac{f^*(t)}{|\log t|} \leq c \, \|f \, | B_{p\infty}^{\frac{n}{p}}(\mathbb{R}^n)\|$$

in case of $q = \infty$. Let $0 < p_1 < p_2 < \infty$. Then

$$B_{p_1 q}^{\frac{n}{p_1}}(\mathbb{R}^n) \subset B_{p_2 q}^{\frac{n}{p_2}}(\mathbb{R}^n), \tag{13.11}$$

[Triβ], Theorem 2.7.1, p. 129. Hence it is sufficient to prove (13.10) for large values of p, in particular, we may assume

$$1 < p < \infty, \quad 1 < q \leq \infty. \tag{13.12}$$

We rely on atomic decompositions for the spaces $B_{pq}^{\frac{n}{p}}(\mathbb{R}^n)$. Details (and also references to the original papers) may be found in [Triδ], Sections 13. (One could also use corresponding quarkonial decompositions according to Definition 2.6 and Theorem 2.9, but atoms are sufficient at the moment.) By [Triδ], Theorem 13.8, any $f \in B_{pq}^{\frac{n}{p}}(\mathbb{R}^n)$ can be optimally decomposed in atoms $a_{jm}(x)$ and complex numbers b_{jm} such that

$$f(x) = \sum_{j=0}^\infty f_j(x) \quad \text{with} \quad f_j(x) = \sum_{m \in \mathbb{Z}^n} b_{jm} \, a_{jm}(x) \tag{13.13}$$

and

$$A = \left(\sum_{j=0}^{\infty} \left(\sum_{m\in\mathbb{Z}^n} |b_{jm}|^p\right)^{\frac{q}{p}}\right)^{\frac{1}{q}} \sim \|f\,|B_{pq}^{\frac{n}{p}}(\mathbb{R}^n)\| \qquad (13.14)$$

(obviously modified by $\sup_j$ if $q = \infty$). The equivalence constants are independent of f. Recall that the atoms $a_{jm}(x)$ have the following properties:

$$\operatorname{supp} a_{jm} \subset \{y \in \mathbb{R}^n : |y - 2^{-j}m| < d\, 2^{-j}\}, \qquad (13.15)$$

$$|D^\gamma a_{jm}(x)| \le 2^{j|\gamma|} \quad \text{for all} \quad \gamma \in \mathbb{N}_0^n \quad \text{with} \quad |\gamma| \le \left[\frac{n}{p}\right] + 1, \qquad (13.16)$$

for some $d > 0$ and all $j \in \mathbb{N}_0$ and $m \in \mathbb{Z}^n$. Let $\chi_{jl}(t)$ be the characteristic function of the interval $[C2^{-jn}(l-1), C2^{-jn}l)$ on $\mathbb{R}_+ = [0,\infty)$, where $C > 0$, $j \in \mathbb{N}_0$, and $l \in \mathbb{N}$. For fixed $j \in \mathbb{N}_0$ let b_{jl}^* with $l \in \mathbb{N}$ be the (decreasing) rearrangement of b_{jm} with $m \in \mathbb{Z}^n$. If $C > 0$ and $c > 0$ are chosen appropriately then

$$f_j^*(t) \le c \sum_{l=1}^{\infty} b_{jl}^* \chi_{jl}(t), \quad \text{where} \quad t > 0 \quad \text{and} \quad j \in \mathbb{N}_0. \qquad (13.17)$$

Let $C2^{-jn}(l-1) < t \le C2^{-jn}l$. Then

$$f_j^{**}(t) = \frac{1}{t}\int_0^t f^*(\tau)\,d\tau \le \frac{c}{l}\sum_{k=1}^{l} b_{jk}^* = c\, b_{jl}^{**}. \qquad (13.18)$$

Since $\{b_{jk}^*\}_{k=1}^{\infty}$ is monotonically decreasing and $1 < p < \infty$ we have

$$\sum_{l=1}^{\infty} b_{jl}^{*p} \le \sum_{l=1}^{\infty} b_{jl}^{**p} \le c \sum_{l=1}^{\infty} b_{jl}^{*p} = c \sum_{m\in\mathbb{Z}^n} |b_{jm}|^p = C_j^p. \qquad (13.19)$$

The left-hand side is obvious since $b_{jl}^* \le b_{jl}^{**}$. The second estimate is the sequence version of the Hardy-Littlewood maximal inequality and can easily be reduced to the usual formulation of this maximal inequality. (A formulation and a proof of the latter may be found in [Ste70], p. 5.) Let

$$C2^{-(k+1)n} \le t < C2^{-kn} \quad \text{with} \quad k \in \mathbb{N}.$$

By the additivity property of f^{**} according to [BeS88], Theorem 3.4 on p. 55, and (13.13), (13.18) we obtain

$$f^{**}(t) \le \sum_{j=0}^{\infty} f_j^{**}(t) \le c\sum_{j=0}^{k} b_{j1}^* + c\sum_{j=k+1}^{\infty} b_{j,2^{(j-k)n}}^{**} \qquad (13.20)$$

where we used $b_{j1}^{**} = b_{j1}^*$. If $1 < q < \infty$, then

$$\int_0^\varepsilon \left(\frac{f^*(t)}{|\log t|}\right)^q \frac{dt}{t} \le c_1 \sum_{k=1}^\infty \left(\frac{f^{**}(C2^{-kn})}{k}\right)^q$$

$$\le c_2 \sum_{k=1}^\infty \left(\frac{1}{k}\sum_{j=0}^k b_{j1}^*\right)^q + c_2 \sum_{k=1}^\infty \left(\frac{1}{k}\sum_{j=k+1}^\infty b_{j,2(j-k)n}^{**}\right)^q = A_1^q + A_2^q. \quad (13.21)$$

Again we can apply the sequence version of the Hardy-Littlewood maximal inequality to the first sum A_1 and obtain

$$A_1 \le c_1 \left(\sum_{j=0}^\infty b_{j1}^{*q}\right)^{\frac{1}{q}} \le c_2 A \sim \|f\,|B_{pq}^{\frac{n}{p}}(\mathbb{R}^n)\|, \quad (13.22)$$

where we used (13.14). If $q = \infty$ then (13.21) must be replaced by

$$\sup_{0<t\le\varepsilon} \frac{f^*(t)}{|\log t|} \le c_2 \sup_k \frac{1}{k}\sum_{j=0}^k b_{j1}^* + c_2 \sup_k \frac{1}{k}\sum_{j=k+1}^\infty b_{j,2(j-k)n}^{**} = A_1 + A_2. \quad (13.23)$$

The term A_1 can be estimated from above by $\sup_j b_{j1}^*$, and hence by the right-hand side of (13.22). We estimate A_2. Since for fixed j the sequence b_{jl}^{**} is monotonically decreasing we have by (13.19),

$$\sum_{l=1}^\infty 2^{ln} b_{j,2^{ln}}^{**p} \le c C_j^p, \quad j \in \mathbb{N}_0, \quad (13.24)$$

and, hence,

$$b_{j,2^{ln}}^{**} \le c'\, 2^{-\frac{ln}{p}} C_j, \quad j \in \mathbb{N}_0, \quad l \in \mathbb{N}. \quad (13.25)$$

It follows that

$$\sum_{j=k+1}^\infty b_{j,2(j-k)n}^{**} \le c \sum_{j=k+1}^\infty 2^{-\frac{(j-k)n}{p}} C_j \le c' \sup_l C_l. \quad (13.26)$$

Now we get in both cases, $q = \infty$ by (13.23) and $1 < q < \infty$ by (13.21),

$$A_2 \le c \sup_l C_l \le c' A \sim \|f\,|B_{pq}^{\frac{n}{p}}(\mathbb{R}^n)\|. \quad (13.27)$$

Here we used again (13.14). Now (13.23) if $q = \infty$ and (13.21) if $1 < q < \infty$, and the estimates (13.22) and (13.27) prove (13.10).

Step 2 Let p and q be given by (13.8). Let again $0 < \varepsilon < 1$. We prove that there is a number $c > 0$ such that

$$\left(\int_0^\varepsilon \left(\frac{f^*(t)}{|\log t|} \right)^p \frac{dt}{t} \right)^{\frac{1}{p}} \le c \, \| f \, | F_{pq}^{\frac{n}{p}}(\mathbb{R}^n) \| \tag{13.28}$$

for all $f \in F_{pq}^{\frac{n}{p}}(\mathbb{R}^n)$. We reduce this case to (13.10) using the following consequence of an observation by Ju. V. Netrusov, [Net89a], Theorem 1.1 and Remark 4 on p. 191 (in the English translation): For any $f \in F_{p\infty}^{\frac{n}{p}}(\mathbb{R}^n)$ there is a function $g \in B_{pp}^{\frac{n}{p}}(\mathbb{R}^n)$ such that

$$|f(x)| \le g(x) \text{ a.e. in } \mathbb{R}^n, \text{ and } \| g \, | B_{pp}^{\frac{n}{p}}(\mathbb{R}^n) \| \le c \, \| f \, | F_{p\infty}^{\frac{n}{p}}(\mathbb{R}^n) \|, \tag{13.29}$$

where c is independent of f and g. Since $1 < p < \infty$ we can apply (13.10) to g and $B_{pp}^{\frac{n}{p}}(\mathbb{R}^n)$. Together with (13.29) we obtain for $f \in F_{pq}^{\frac{n}{p}}(\mathbb{R}^n)$,

$$\int_0^\varepsilon \left(\frac{f^*(t)}{|\log t|} \right)^p \frac{dt}{t} \le \int_0^\varepsilon \left(\frac{g^*(t)}{|\log t|} \right)^p \frac{dt}{t} \le c_1 \, \| g \, | B_{pp}^{\frac{n}{p}}(\mathbb{R}^n) \|^p$$

$$\le c_2 \| f \, | F_{p\infty}^{\frac{n}{p}}(\mathbb{R}^n) \|^p \le c_3 \, \| f \, | F_{pq}^{\frac{n}{p}}(\mathbb{R}^n) \|^p, \tag{13.30}$$

where we used in addition the monotonicity of the F-spaces with respect to the q-index. This proves (13.28).

Step 3 Let p, q be given by (13.6) and let $b = \frac{1}{q'}$ in Example 2 in 12.4. Then we obtain by (12.28) and (13.10),

$$\sup_{0 < t < \varepsilon} \frac{f^*(t)}{|\log t|^{\frac{1}{q'}}} \le c_1 \left(\int_0^\varepsilon \left(\frac{f^*(t)}{|\log t|^{\frac{1}{q'}}} \right)^q \frac{dt}{t |\log t|} \right)^{\frac{1}{q}}$$

$$= c_1 \left(\int_0^\varepsilon \left(\frac{f^*(t)}{|\log t|} \right)^q \frac{dt}{t} \right)^{\frac{1}{q}} \le c_2 \, \| f \, | B_{pq}^{\frac{n}{p}}(\mathbb{R}^n) \|. \tag{13.31}$$

If p, q are given by (13.8) then it follows in a similar way by (13.28) that

$$\sup_{0 < t < \varepsilon} \frac{f^*(t)}{|\log t|^{\frac{1}{p'}}} \le c \, \| f \, | F_{pq}^{\frac{n}{p}}(\mathbb{R}^n) \|. \tag{13.32}$$

Let $\mathcal{E}_G B_{pq}^{\frac{n}{p}}$ and $\mathcal{E}_G F_{pq}^{\frac{n}{p}}$ be the respective growth envelope functions according to Definition 12.8 and (12.37). Then it follows by (13.31) and (13.32) that

$$\mathcal{E}_G B_{pq}^{\frac{n}{p}}(t) \leq |\log t|^{\frac{1}{q'}}, \quad 0 < p < \infty, \quad 1 < q \leq \infty, \tag{13.33}$$

and

$$\mathcal{E}_G F_{pq}^{\frac{n}{p}}(t) \leq |\log t|^{\frac{1}{p'}}, \quad 1 < p < \infty, \quad 0 < q \leq \infty. \tag{13.34}$$

Step 4 To prove the converse of (13.33), (13.34) and to show that q, p are the correct numbers in (13.7), (13.9), respectively, we need some extremal functions. Let $\psi(x)$ be a non-trivial, non-negative, compactly supported C^∞ function in $\mathbb{R}^n$, for example,

$$\psi(x) = e^{-\frac{1}{1-|x|^2}} \text{ if } |x| < 1 \text{ and } \psi(x) = 0 \text{ if } |x| \geq 1. \tag{13.35}$$

Let

$$1 < p < \infty, \quad 1 \leq q \leq \infty. \tag{13.36}$$

Let $b = \{b_j\}_{j=1}^\infty$ be a sequence of non-negative numbers with

$$b_1 \geq b_2 \geq \cdots \geq b_j \geq b_{j+1} \geq \cdots \text{ and } \sum_{j=1}^\infty b_j^p < \infty, \tag{13.37}$$

and let

$$f(x) = \sum_{j=1}^\infty b_j \psi(2^{j-1}x). \tag{13.38}$$

We wish to prove

$$\sum_{j=1}^\infty b_j^p \sim \int_0^\varepsilon \left(\frac{f^*(t)}{|\log t|}\right)^p \frac{dt}{t} \sim \|f\,|F_{pq}^{\frac{n}{p}}\mathbb{R}^n)\|^p, \tag{13.39}$$

where the equivalence constants are independent of b. We remark that (13.38) is an atomic or quarkonial decomposition in $F_{pq}^{\frac{n}{p}}(\mathbb{R}^n)$ according to [Triδ], Theorem 13.8, p. 75, or the above Definition 2.6, respectively. With the sequence space f_{pq}, given by (2.8), we obtain (in obvious notation)

$$\|f\,|F_{pq}^{\frac{n}{p}}(\mathbb{R}^n)\| \leq c \, \|b\,|f_{pq}\| \sim \left(\sum_{j=1}^\infty b_j^p\right)^{\frac{1}{p}}. \tag{13.40}$$

The inequality in (13.40) is covered by the above references. The equivalence in (13.40) follows from the special structure of f in (13.38) and the modifications of f_{pq} indicated in 2.15 (which show that under the above circumstances q in f_{pq} is unimportant). Next we remark that $f(x)$ with, say, (13.35), is non-negative, rotationally invariant, and monotonically decreasing in radial directions. We have

$$f(x) \sim \sum_{j=1}^{k} b_j \quad \text{if} \quad |x| \sim 2^{-k} \quad \text{where} \quad k \in \mathbb{N}, \tag{13.41}$$

and, hence,

$$f^*(t) \sim \sum_{j=1}^{k} b_j \quad \text{if} \quad t \sim 2^{-kn} \quad \text{where} \quad k \in \mathbb{N}. \tag{13.42}$$

It follows that

$$\int_0^\varepsilon \left(\frac{f^*(t)}{|\log t|}\right)^p \frac{dt}{t} \sim \sum_{k=K}^{\infty} \left(\frac{1}{k}\sum_{j=1}^{k} b_j\right)^p \sim \sum_{k=1}^{\infty} b_k^p, \tag{13.43}$$

where we used in the second equivalence again the sequence version of the Hardy-Littlewood maximal inequality as in connection with (13.19) and the monotonicity of the numbers b_j according to (13.37); (the number K is related to ε, but otherwise unimportant). Now (13.39) follows from (13.43), (13.28), (13.40). Similarly, but technically more simply, one obtains for $0 < p < \infty$,

$$\sum_{j=1}^{\infty} b_j^q \sim \int_0^\varepsilon \left(\frac{f^*(t)}{|\log t|}\right)^q \frac{dt}{t} \sim \|f\,|B_{pq}^{\frac{n}{p}}(\mathbb{R}^n)\|^q \quad \text{if} \quad 1 < q < \infty, \tag{13.44}$$

and

$$b_1 = \sup_j b_j \sim \sup_{0 < t \leq \varepsilon} \frac{f^*(t)}{|\log t|} \sim \|f\,|B_{p\infty}^{\frac{n}{p}}(\mathbb{R}^n)\| \tag{13.45}$$

as follows: One has (13.40) with B in place of F and with q on the right-hand side, $1 < q \leq \infty$. The first equivalences in (13.44), (13.45) follow as in (13.43), including $q = \infty$. Together with (13.10) one gets (13.44) and (13.45).

Step 5 The extremal functions $f(x)$ in (13.38) apply to all cases for the B-spaces, but, so far only to the F-spaces with (13.36). If $q < 1$ is small then the $\psi(2^{j-1}x)$ are no longer atoms or quarks in $F_{pq}^{\frac{n}{p}}(\mathbb{R}^n)$. One needs moment conditions. We describe the respective repair. Let again ψ and $b = \{b_j\}_{j=1}^{\infty}$ be given by, say, (13.35) and (13.37) with $1 < p < \infty$. Let $x^0 \neq 0$. We modify (13.38) by

$$f(x) = \sum_{j=1}^{\infty} b_j \chi(2^{j-1} x) = \sum_{j=1}^{\infty} b_j \left(\psi(2^{j-1} x) - \psi(2^{j-1} x - x^0) \right). \tag{13.46}$$

Although not really necessary one may choose x^0 such that the supports of $\psi(2^{j-1} x - x^0)$ are pairwise disjoint. Furthermore the function $\chi(x)$ satisfies the first moment condition

$$\int_{\mathbb{R}^n} \chi(x)\, dx = 0. \tag{13.47}$$

Otherwise (13.40)–(13.43) remain unchanged and we get (13.39) for those q for which first moment conditions in the related atoms are sufficient. If higher moment conditions

$$\int_{\mathbb{R}^n} x^\beta \chi(x)\, dx = 0 \quad \text{for} \quad |\beta| \le L \tag{13.48}$$

are needed, then the construction in (13.46) must be modified by

$$\chi(x) = \psi(x) - \psi_L(x - x^0), \tag{13.49}$$

where ψ_L is a suitable C^∞ function with a compact support, say, in the unit ball in $\mathbb{R}^n$, and $x^0 \neq 0$ chosen in such a way that the supports of $\psi_L(2^{j-1} x - x^0)$ are pairwise disjoint. An explicit construction of such a function may be found in [TrW96], pp. 665–666. We refer also to Corollary 13.4 below and its proof where we have for later purposes a second and more detailed look at constructions of this type. After this modification we get (13.39) now for all p, q with (13.8).

Step 6 We prove the converse of (13.33), (13.34). Let p, q, and ψ be given by (13.36) and (13.35), respectively, and let

$$f_J(x) = J^{-\frac{1}{p}} \sum_{j=1}^{J} \psi(2^{j-1} x), \quad x \in \mathbb{R}^n, \quad J \in \mathbb{N}. \tag{13.50}$$

Then by (13.42) and (13.39),

$$f_J^*(2^{-Jn}) \sim J^{\frac{1}{p'}} \quad \text{and} \quad \|f_J \,|\, F_{pq}^{\frac{n}{p}}(\mathbb{R}^n)\| \sim 1 \tag{13.51}$$

uniformly in J. Hence by (12.54) and (12.37),

$$\mathcal{E}_G F_{pq}^{\frac{n}{p}}(2^{-Jn}) \ge f_J^*(2^{-Jn}) \sim J^{\frac{1}{p'}}, \quad J \in \mathbb{N}. \tag{13.52}$$

This proves the converse of (13.34). If $q < 1$ then one has to replace f in (13.38) as indicated in Step 5. Similarly for $B_{pq}^{\frac{n}{p}}(\mathbb{R}^n)$. Hence $|\log t|^{\frac{1}{q'}}$ and $|\log t|^{\frac{1}{p'}}$ are the growth envelope functions for $B_{pq}^{\frac{n}{p}}(\mathbb{R}^n)$ and $F_{pq}^{\frac{n}{p}}(\mathbb{R}^n)$, respectively.

Step 7 We must prove that q and p are the correct numbers in (13.7) and (13.9), respectively. Since we know already (13.31) and its F-counterpart (13.28) we must prove that q, respectively p, cannot be improved. Assume that there is a number v with $v < q$ and

$$\int_0^\varepsilon \left(\frac{f^*(t)}{|\log t|^{\frac{1}{q'}}}\right)^v \frac{dt}{t|\log t|} \leq c \,\|f\,|B_{pq}^{\frac{n}{p}}(\mathbb{R}^n)\|^v. \tag{13.53}$$

Let, according to (13.38),

$$f(x) = \sum_{j=2}^\infty b_j \,\psi(2^{j-1}x) \quad \text{with} \quad b_j = j^{-\frac{1}{q}} (\log j)^{-\frac{1}{v}}. \tag{13.54}$$

By (13.44) we have $f \in B_{pq}^{\frac{n}{p}}(\mathbb{R}^n)$. On the other hand, by (13.42), we can estimate the left-hand side of (13.53) from below for some $c > 0$ by

$$c \sum_{k=K}^\infty (k\, b_k)^v \, k^{-\frac{v}{q'}-1} = c \sum_{k=K}^\infty k^{-1} (\log k)^{-1} = \infty. \tag{13.55}$$

We get a contradiction. This proves (13.7). Similarly one obtains (13.9).

13.3 Inequalities

The above theorem covers all cases of interest (excluding borderline situations according to (13.2)). It describes in a rather condensed way very sharp inequalities. It seems be reasonable to make clear the outcome. We use Example 2 in 12.4, Definition 12.8 and Proposition 12.10. Let $0 < \varepsilon < 1$.

13.3(i) *The B-spaces* Let $\varkappa(t)$ be a positive monotonically decreasing function on $(0, \varepsilon]$. Let $0 < u \leq \infty$. Let p and q be given by (13.6). Then

$$\left(\int_0^\varepsilon \left(\frac{\varkappa(t)\, f^*(t)}{|\log t|^{\frac{1}{q'}}}\right)^u \frac{dt}{t|\log t|}\right)^{\frac{1}{u}} \leq c \,\|f\,|B_{pq}^{\frac{n}{p}}(\mathbb{R}^n)\| \tag{13.56}$$

for some $c > 0$ and all $f \in B_{pq}^{\frac{n}{p}}(\mathbb{R}^n)$ if, and only if, $\varkappa$ is bounded and $q \leq u \leq \infty$ (with the modification (13.59) below if $u = \infty$). In particular, if $1 < q < \infty$, then

$$\sup_{0<t<\varepsilon} \frac{f^*(t)}{|\log t|^{\frac{1}{q'}}} \le c_0 \left(\int_0^\varepsilon \left(\frac{f^*(t)}{|\log t|}\right)^q \frac{dt}{t}\right)^{\frac{1}{q}} \le c_1 \|f\,|B^{\frac{n}{p}}_{pq}(\mathbb{R}^n)\| \qquad (13.57)$$

are the two end-point cases according to (12.28). If $q = \infty$ then one has

$$\sup_{0<t<\varepsilon} \frac{f^*(t)}{|\log t|} \le c \|f\,|B^{\frac{n}{p}}_{p\infty}(\mathbb{R}^n)\|. \qquad (13.58)$$

Let $\varkappa(t)$ be an (arbitrary) positive function on $(0, \varepsilon]$ and again let $1 < q \le \infty$. Then

$$\sup_{0<t<\varepsilon} \frac{\varkappa(t)\,f^*(t)}{|\log t|^{\frac{1}{q'}}} \le c \|f\,|B^{\frac{n}{p}}_{pq}(\mathbb{R}^n)\| \qquad (13.59)$$

for some $c > 0$ and all $f \in B^{\frac{n}{p}}_{pq}(\mathbb{R}^n)$ if, and only if, $\varkappa$ is bounded. However the difference between the assumptions for $\varkappa$ in (13.56) and in (13.59) is rather immaterial. We discussed this point in 12.11.

13.3(ii) *The F-spaces* Let $\varkappa(t)$ be a positive monotonically decreasing function on $(0, \varepsilon]$. Let $0 < u \le \infty$. Let p and q be given by (13.8). Then

$$\left(\int_0^\varepsilon \left(\frac{\varkappa(t)\,f^*(t)}{|\log t|^{\frac{1}{p'}}}\right)^u \frac{dt}{t|\log t|}\right)^{\frac{1}{u}} \le c \|f\,|F^{\frac{n}{p}}_{pq}(\mathbb{R}^n)\| \qquad (13.60)$$

for some $c > 0$ and all $f \in F^{\frac{n}{p}}_{pq}(\mathbb{R}^n)$ if, and only if, $\varkappa$ is bounded and $p \le u \le \infty$, with the modification

$$\sup_{0<t<\varepsilon} \frac{\varkappa(t)\,f^*(t)}{|\log t|^{\frac{1}{p'}}} \le c \|f\,|F^{\frac{n}{p}}_{pq}(\mathbb{R}^n)\| \quad \text{if} \quad u = \infty. \qquad (13.61)$$

In particular,

$$\sup_{0<t<\varepsilon} \frac{f^*(t)}{|\log t|^{\frac{1}{p'}}} \le c_0 \left(\int_0^\varepsilon \left(\frac{f^*(t)}{|\log t|}\right)^p \frac{dt}{t}\right)^{\frac{1}{p}} \le c_1 \|f\,|F^{\frac{n}{p}}_{pq}(\mathbb{R}^n)\| \qquad (13.62)$$

are the two end-point cases according to (12.28). As above, if $\varkappa$ is an arbitrary positive function, then we have (13.61) if, and only if, $\varkappa$ is bounded.

Let $a \in \mathbb{R}$. Then $a_+ = \max(0, a)$ and $[a]$ stands for the largest integer smaller than or equal to a.

13.4 Corollary

(i) Let $0 < \delta \le \frac{1}{4}$ and let for $y \in \mathbb{R}$,

$$h(y) = e^{-\frac{1}{\delta^2 - y^2}} \quad \text{if } |y| < \delta \quad \text{and} \quad h(y) = 0 \text{ if } |y| \ge \delta. \tag{13.63}$$

Let $L \in \mathbb{N}_0$ and

$$h_L(y) = h(y) - \sum_{l=0}^{L} \varrho_l \, h^{(l)}(y-1). \tag{13.64}$$

There are numbers $\varrho_l \in \mathbb{R}$ such that (moment conditions)

$$\int_{\mathbb{R}} y^k \, h_L(y) \, dy = 0 \quad \text{if} \quad k = 0, \ldots, L. \tag{13.65}$$

(ii) Let $L+1 \in \mathbb{N}_0$ and let h_L with (13.65) be complemented by $h_{-1} = h$ (then (13.65) is empty). Let

$$f_b(x) = \sum_{j=1}^{\infty} b_j \, h_L(2^{j-1} x_1) \prod_{m=2}^{n} h(2^{j-1} x_m), \quad x = (x_1, \ldots, x_n) \in \mathbb{R}^n, \tag{13.66}$$

where $b = \{b_j\}_{j=1}^{\infty}$ is a sequence of non-negative numbers with

$$b_1 \ge b_2 \ge \cdots \ge b_j \ge b_{j+1} \ge \cdots. \tag{13.67}$$

Let p, q be given by (13.6) in the B-case, by (13.8) in the F-case, and

$$L_B = -1, \quad L_F = \max\left(-1, \left[n\left(\frac{1}{\min(p,q)} - 1\right)_+ - \frac{n}{p}\right]\right). \tag{13.68}$$

Let $L+1 \in \mathbb{N}_0$ with $L \ge L_B$ in the B-case and $L \ge L_F$ in the F-case. Let $0 < \varepsilon < 1$. If $b \in \ell_q$, then

$$\left(\sum_{j=1}^{\infty} b_j^q\right)^{\frac{1}{q}} \sim \left(\int_0^{\varepsilon} \left(\frac{f_b^*(t)}{|\log t|}\right)^q \frac{dt}{t}\right)^{\frac{1}{q}} \sim \|f \,|\, B_{pq}^{\frac{n}{p}}(\mathbb{R}^n)\| \tag{13.69}$$

(usual modification if $q = \infty$) and, if $b \in \ell_p$, then

$$\left(\sum_{j=1}^{\infty} b_j^p\right)^{\frac{1}{p}} \sim \left(\int_0^{\varepsilon} \left(\frac{f_b^*(t)}{|\log t|}\right)^p \frac{dt}{t}\right)^{\frac{1}{p}} \sim \|f \,|\, F_{pq}^{\frac{n}{p}}(\mathbb{R}^n)\|, \tag{13.70}$$

where the equivalence constants are independent of b.

Proof *Step 1* If one inserts (13.64) in (13.65) then one gets a triangular matrix for ϱ_l from which these coefficients can be uniquely calculated.

Step 2 By the product structure of the terms in (13.66) we have

$$\int_{\mathbb{R}^n} x^\beta f_b(x)\, dx = 0 \quad \text{for} \quad |\beta| \leq L \tag{13.71}$$

(where (13.71) is empty if $L = -1$): Since the sequence b is bounded, all respective sums for $x^\beta f_b(x)$ converge at least in $L_1(\mathbb{R}^n)$. Recall that one needs moment conditions (13.71) for atoms in $B^s_{pq}(\mathbb{R}^n)$ and $F^s_{pq}(\mathbb{R}^n)$ up to order L, where

$$L \geq \max\left(-1, [\sigma_p - s]\right) \quad \text{and} \quad L \geq \max\left(-1, [\sigma_{pq} - s]\right), \tag{13.72}$$

respectively, with σ_p and σ_{pq} given by (2.20). We refer to [Triδ], Theorem 13.8 on p. 75. Here we have $s = \frac{n}{p}$, hence $L \geq -1$ for the B-spaces and $L \geq L_F$ for the F-spaces. This formalizes what we said in Step 5 of the proof of Theorem 13.2. Otherwise the proof of the corollary is covered by Steps 4 and 5 of this proof.

13.5 Further references and comments

We described in Theorem 11.7 and in (11.70) the classical inequalities related to the critical case considered in Theorem 13.2. Recall that

$$H_p^{\frac{n}{p}}(\mathbb{R}^n) = F_{p,2}^{\frac{n}{p}}(\mathbb{R}^n), \quad 1 < p < \infty, \tag{13.73}$$

are the Sobolev spaces with the classical Sobolev spaces $W_p^k(\mathbb{R}^n)$ in (11.69) as special cases. In 11.8(v) we tried to collect the historical references of (11.56), (11.57), and (11.70). Obviously all these cases are covered by Theorem 13.2 and by 13.3. In more recent times, inequalities of type (11.57) have again attracted some attention, mostly restricted to the case of classical Sobolev spaces according to (11.69), but in the context of general rearrangement-invariant (quasi-)norms. We refer in particular to [CwP98], [EKP00], and [Pic99]. The last paper surveys some aspects of embeddings of classical Sobolev spaces $W_p^k(\mathbb{R}^n)$, especially of $W_p^1(\mathbb{R}^n)$, in rearrangement-invariant spaces. Furthermore, there is a connection between inequalities of type (11.56), (11.57) and capacity estimates in function spaces. Details may be found in [EKP00] and [Pic99] with references to Maz'ya's results in this direction, especially in [Maz85], pp. 105, 109. As mentioned above, parallel or earlier developments in the East have often passed unnoticed in the West. In particular, Ju. V. Netrusov proved in [Net87b], Theorem 3 on p. 108, assertions, which are related to [CwP98] and [EKP00], in the framework of spaces of type F_{pq}^s, including optimality of range spaces. He generalizes earlier results in the Russian literature by Brudnyi, Kaljabin, and especially by Gold'man. A good description of and detailed references to this earlier work may be found in [Liz86], D.1.8 and D.1.9, pp. 398–404. Our own contributions started in [Tri93] and were repeated in a slightly improved form in [ET96], Theorem 2.7.1, p. 82, and Theorem 2.7.3, p. 93. The main new point is the construction of extremal functions f belonging both to $H_p^{\frac{n}{p}}(\mathbb{R}^n)$ and $B_{pp}^{\frac{n}{p}}(\mathbb{R}^n)$ with $1 < p < \infty$ and having the singularity behaviour

$$f(x) = |\log|x||^{\frac{1}{p'}} (\log|\log|x||)^{-\sigma} \quad \text{where} \quad \sigma > \frac{1}{p}, \tag{13.74}$$

near the origin. This is now essentially covered by the function f given by (13.38) with, say, $b = \{b_j\}_{j=2}^\infty$,

$$b_j = j^{-\frac{1}{p}} |\log j|^{-\sigma}; \quad j = 2, 3, \ldots . \tag{13.75}$$

Then $b \in \ell_p$, and hence we have on the one hand (13.39) especially for $H_p^{\frac{n}{p}}(\mathbb{R}^n)$, and (13.44), especially for $B_{pp}^{\frac{n}{p}}(\mathbb{R}^n)$. On the other hand, if $|x| \sim 2^{-k}$, then it follows by (13.41),

$$f(x) \sim \sum_{j=2}^k b_j \sim \int_2^k y^{-\frac{1}{p}} (\log y)^{-\sigma} \, dy \sim k^{\frac{1}{p'}} (\log k)^{-\sigma}$$
$$\sim |\log|x||^{\frac{1}{p'}} (\log|\log|x||)^{-\sigma} . \tag{13.76}$$

At the same time it is now clear that the functions in (13.38) improve the earlier developments in [Tri93] and [ET96]. The equivalence (13.39) in Step 4 of the above proof coincides essentially with [EdT99b], Theorem 2.1. This paper might be considered as a forerunner of Theorem 13.2, restricted to $H_p^{\frac{n}{p}}(\mathbb{R}^n)$ and $B_{pp}^{\frac{n}{p}}(\mathbb{R}^n)$. Even worse, we used there (11.57), going back to [Has79] and [BrW80], as a starting point and derived the corresponding inequality for $B_{pp}^{\frac{n}{p}}(\mathbb{R}^n)$, this means (13.56) with $u = p = q$ and $\varkappa = 1$, via non-linear interpolation from (11.57). Otherwise the sharpness in [EdT99b] is on the $\varkappa$-level as described in 13.3. All other parts of Theorem 13.2 and its proof are new and published here for the first time. Especially the concept of growth envelopes in the above context came out very recently in collaboration with D. D. Haroske, [Har01]. Finally we mention the extension of the related results in [Tri93] and in [ET96], Theorem 2.7.1, to spaces with dominating mixed derivatives in [KrS96], including optimality results via extremal functions.

13.6 Spaces on domains

Let Ω be a domain in $\mathbb{R}^n$. The spaces $B^s_{pq}(\Omega)$ and $F^s_{pq}(\Omega)$ have been introduced in Definition 5.3 for all admitted s, p, q. The concept of the *growth envelope* and the *growth envelope function* according to Definition 12.8 and the notational agreement (12.60) can be carried over under the same natural restrictions as there to the respective spaces $A^s_{pq}(\Omega)$. We denote them by

$$\mathfrak{E}_{G,\Omega} A^s_{pq} = (\mathcal{E}_{G,\Omega} A^s_{pq}(t), u). \tag{13.77}$$

In the critical case, considered in this section, Theorem 13.2 can be extended to spaces on domains: If p, q are given by (13.6), then

$$\mathfrak{E}_{G,\Omega} B^{\frac{n}{p}}_{pq} = \mathfrak{E}_G B^{\frac{n}{p}}_{pq} = (|\log t|^{\frac{1}{q'}}, q) \tag{13.78}$$

and, if p, q are given by (13.8), then

$$\mathfrak{E}_{G,\Omega} F^{\frac{n}{p}}_{pq} = \mathfrak{E}_G F^{\frac{n}{p}}_{pq} = (|\log t|^{\frac{1}{p'}}, p). \tag{13.79}$$

To justify these assertions we remark first

$$\mathcal{E}_{G,\Omega} A^s_{pq}(t) \leq \mathcal{E}_G A^s_{pq}(t), \quad 0 < t \leq \varepsilon, \tag{13.80}$$

as a more or less immediate consequence of the definition of spaces on domains by restriction of corresponding spaces on $\mathbb{R}^n$. On the other hand, the construction of extremal functions in Steps 4 and 5 of the proof of Theorem 13.2 is strictly local. Hence the arguments in Steps 6 and 7 of this proof can be carried over from $\mathbb{R}^n$ to Ω. Then one obtains (13.78) and (13.79).

13.7 The space bmo

We always exclude borderline situations. In our context, described by Theorem 11.2, this means in general (13.2), and with respect to the critical case, (13.3). Furthermore, we excluded in all our considerations so far the spaces $F^s_{\infty q}(\mathbb{R}^n)$. If $1 < q < \infty$, these spaces were introduced in [Tri78], 2.5.1, p. 118, and may also be found in [Triβ], 2.3.4, p. 50. This has been modified and, in particular, extended to all q, $0 < q < \infty$, in [FrJ90], Section 5. In the critical situation we have $s = 0$. At least some of these spaces fit in the scheme (11.9),

$$F^0_{\infty q}(\mathbb{R}^n) \subset F^0_{\infty 2}(\mathbb{R}^n) = bmo(\mathbb{R}^n) \subset L^{loc}_1(\mathbb{R}^n) \quad \text{if} \quad 0 < q \leq 2, \tag{13.81}$$

where $bmo(\mathbb{R}^n)$ is the inhomogeneous space consisting of those locally integrable functions with bounded mean oscillation for which

$$\|f \,|bmo(\mathbb{R}^n)\| = \sup_{|Q| \leq 1} \frac{1}{|Q|} \int_Q |f(x) - f_Q|\, dx + \sup_{|Q| > 1} \frac{1}{|Q|} \int_Q |f(x)|\, dx < \infty. \tag{13.82}$$

Here Q stands for cubes in $\mathbb{R}^n$ and f_Q is the mean value of f with respect to Q. We refer for details and further information to [Triβ], 2.2.2, p. 37, and 2.5.8, p. 93. Let $\psi(x)$ be a C^∞ function with a compact support near the origin, for example ψ from (13.35). It is well known and can be checked easily that $\psi(x)|\log|x||$ belongs to $bmo(\mathbb{R}^n)$. But this is a local matter and can be extended by (13.82) to

$$f(x) = \sum_{m \in \mathbb{Z}^n} \psi(x-m)|\log|x-m|| \in bmo(\mathbb{R}^n). \tag{13.83}$$

This makes clear that there is no growth envelope function $\mathcal{E}_G bmo$ according to Definition 12.8 and (12.37), or in other words,

$$\mathcal{E}_G bmo(t) = \infty \quad \text{for all} \quad 0 < t < \infty. \tag{13.84}$$

However in sharp contrast to the situation described in 13.6 if $p < \infty$, the growth envelope and the growth envelope function are reasonable for the spaces $bmo(\Omega)$, where Ω is a bounded domain in $\mathbb{R}^n$ and where $bmo(\Omega)$ is again defined by restriction of $bmo(\mathbb{R}^n)$ on Ω. Let, for example, $\Omega = Q$ be a cube with $|Q| = \varepsilon < 1$. A detailed study of the spaces $bmo(Q)$ may be found in [BeS88], Chapter 5, Section 7. In particular by [BeS88], Corollary 7.11, on p. 383, we have

$$bmo(Q) \subset L_\infty(\log L)_{-1}(Q) \tag{13.85}$$

where we used that L_{exp} according to 11.6(ii) (again with reference to [BeS88]) coincides with the space on the right-hand side of (13.85). In particular,

$$\sup_{0 < t \le \varepsilon} \frac{f^*(t)}{|\log t|} \le c \, \|f\,|bmo(Q)\|. \tag{13.86}$$

On the other hand, J. Marschall proved in [Mar95], Lemma 16 on p. 253 (with a forerunner in [Mar87b])

$$B^{\frac{n}{p}}_{p\infty}(\mathbb{R}^n) \subset F^0_{\infty q}(\mathbb{R}^n), \quad 0 < p < \infty, \quad 0 < q \le \infty, \tag{13.87}$$

in particular,

$$B^{\frac{n}{p}}_{p\infty}(\mathbb{R}^n) \subset bmo(\mathbb{R}^n), \quad 0 < p < \infty. \tag{13.88}$$

However by (13.7), 13.6 (and the notation introduced there) and (13.86) one gets

$$\mathcal{E}_{G,Q} bmo = (|\log t|, \infty) \tag{13.89}$$

In any case in borderline situations one has to distinguish carefully between global and local singularity behaviour.

14 The super-critical case

14.1 Introduction

By the terminology of (10.15) the super-critical case covers the spaces

$$B_{pq}^{1+\frac{n}{p}}(\mathbb{R}^n) \quad \text{and} \quad F_{pq}^{1+\frac{n}{p}}(\mathbb{R}^n) \quad \text{with } 0 < p < \infty \text{ and } 0 < q \leq \infty. \tag{14.1}$$

This corresponds to the dotted line in Fig. 10.1. Recall that we always exclude in this chapter borderline situations as described in (13.2). This means in the super-critical case that we do not deal with the spaces $B_{\infty q}^1(\mathbb{R}^n)$ and also not with the spaces $F_{\infty q}^1(\mathbb{R}^n)$ briefly mentioned in 13.7. As a further restriction of (14.1) we are interested only in those spaces which are not continuously embedded in $C^1(\mathbb{R}^n)$ (or, which is the same, in $Lip(\mathbb{R}^n)$); this means by Theorem 11.4, and has been detailed in 11.5, especially in (11.38), (11.39), we deal with the spaces

$$B_{pq}^{1+\frac{n}{p}}(\mathbb{R}^n) \quad \text{with} \quad 0 < p < \infty, \quad 1 < q \leq \infty, \tag{14.2}$$

and

$$F_{pq}^{1+\frac{n}{p}}(\mathbb{R}^n) \quad \text{with} \quad 1 < p < \infty, \quad 0 < q \leq \infty. \tag{14.3}$$

This covers in particular the Sobolev spaces mentioned in (11.40). As outlined in the introductory Section 10 we wish to measure the continuity of functions belonging to the spaces (14.2), (14.3) in terms of the continuity envelope as introduced in Definition 12.14. Instead of $\mathfrak{E}_C A_{pq}^{1+\frac{n}{p}}$ in (12.83) we use the more handsome version (12.85). In the theorem below we calculate explicitly the continuity envelope for all spaces in (14.2) and (14.3). Afterwards we describe what this means in detail. Finally we add a few references. Otherwise we try to keep the presentation of the super-critical case as close as possible in its formulations to the critical case considered in the previous section (this applies also to this introduction compared with 13.1). In rough terms, using Proposition 12.16 as a vehicle, we lift Theorem 13.2 from the critical to the super-critical situation. Let $1 \leq v \leq \infty$. As usual, v' is given by $\frac{1}{v} + \frac{1}{v'} = 1$.

14.2 Theorem

(i)   Let

$$0 < p < \infty, \quad 1 < q \leq \infty. \tag{14.4}$$

Then

$$\mathfrak{E}_C B_{pq}^{1+\frac{n}{p}} = (|\log t|^{\frac{1}{q'}}, q). \tag{14.5}$$

(ii) Let
$$1 < p < \infty, \quad 0 < q \le \infty. \tag{14.6}$$

Then
$$\mathfrak{E}_C F_{pq}^{1+\frac{n}{p}} = (|\log t|^{\frac{1}{p'}}, p). \tag{14.7}$$

Proof *Step 1* Recall that

$$\|f\,|B_{pq}^{1+\frac{n}{p}}(\mathbb{R}^n)\| \sim \|f\,|B_{pq}^{\frac{n}{p}}(\mathbb{R}^n)\| + \sum_{j=1}^n \|\frac{\partial f}{\partial x_j}\,|B_{pq}^{\frac{n}{p}}(\mathbb{R}^n)\| \tag{14.8}$$

and similarly for $F_{pq}^{1+\frac{n}{p}}(\mathbb{R}^n)$, [Triβ], Theorem 2.3.8, pp. 58–59. Let p, q be given by (14.4). Using (12.87) we obtained by Theorem 13.2 and (13.57) with $0 < \varepsilon < 1$,

$$\left(\int_0^\varepsilon \left(\frac{|\nabla f|^*(t)}{|\log t|} \right)^q \frac{dt}{t} \right)^{\frac{1}{q}} \le c\,\|f\,|B_{pq}^{1+\frac{n}{p}}(\mathbb{R}^n)\| \tag{14.9}$$

(obviously modified according to (13.58) if $q = \infty$). Similarly for $F_{pq}^{1+\frac{n}{p}}(\mathbb{R}^n)$ if p, q are given by (14.6), based on (13.62). We apply Proposition 12.16 and obtain, if $q < \infty$, by completion

$$\left(\int_0^\varepsilon \left(\frac{\widetilde{\omega}(f,t)}{|\log t|} \right)^q \frac{dt}{t} \right)^{\frac{1}{q}} \le c\,\|f\,|B_{pq}^{1+\frac{n}{p}}(\mathbb{R}^n)\|, \quad f \in B_{pq}^{1+\frac{n}{p}}(\mathbb{R}^n), \tag{14.10}$$

(and again similarly in the F-case). If $q = \infty$ then we wish to have

$$\sup_{0 < t < \varepsilon} \frac{\widetilde{\omega}(f,t)}{|\log t|} \le c\,\|f\,|B_{p\infty}^{1+\frac{n}{p}}(\mathbb{R}^n)\|, \quad f \in B_{p\infty}^{1+\frac{n}{p}}(\mathbb{R}^n). \tag{14.11}$$

Let $f \in B_{p\infty}^{1+\frac{n}{p}}(\mathbb{R}^n)$ and let φ be as in (2.33). Then we can apply (12.90) to

$$f_j = (\varphi(2^{-j}\cdot)\widehat{f})^\vee.$$

We obtain (14.11) with f_j in place of f, where the corresponding right-hand sides can be estimated uniformly with respect to j; hence

$$\sup_{0 < t < \varepsilon} \frac{\widetilde{\omega}(f_j,t)}{|\log t|} \le c\,\|f\,|B_{p\infty}^{1+\frac{n}{p}}(\mathbb{R}^n)\|, \quad f \in B_{p\infty}^{1+\frac{n}{p}}(\mathbb{R}^n). \tag{14.12}$$

By elementary embedding, $f_j(x)$ converges pointwise to $f(x)$. Then (14.11) follows from (14.12) and $j \to \infty$. Similarly for $F_{p\infty}^{1+\frac{n}{p}}(\mathbb{R}^n)$. Since $\widetilde{\omega}(f,t)$ is equivalent to a monotonically decreasing function we are now in the same situation as in Step 3 of the proof of Theorem 13.2. We get

$$\mathcal{E}_C B_{pq}^{1+\frac{n}{p}}(t) \leq |\log t|^{\frac{1}{q'}}, \quad 0 < p < \infty, \quad 1 < q \leq \infty, \tag{14.13}$$

and

$$\mathcal{E}_C F_{pq}^{1+\frac{n}{p}}(t) \leq |\log t|^{\frac{1}{p'}}, \quad 1 < p < \infty, \quad 0 < q \leq \infty. \tag{14.14}$$

Step 2 To construct extremal functions we rely on Corollary 13.4 and put

$$h^L(y) = \int_{-\infty}^{y} h_L(z)\,dz, \quad y \in \mathbb{R}, \tag{14.15}$$

where h_L has the same meaning as in part (i) of this corollary with $L \in \mathbb{N}_0$. Then h^L is a compactly supported C^∞ function. Integration by parts and (13.65) prove that

$$\int_{\mathbb{R}} y^k h^L(y)\,dy = \frac{1}{k+1}\int_{\mathbb{R}}(y^{k+1})'\int_{-\infty}^{y} h_L(z)\,dz\,dy = 0 \quad \text{if} \quad k = 0,\ldots,L-1, \tag{14.16}$$

(if $L = 0$ then (14.16) is empty). We replace $f_b(x)$ in (13.66) by

$$f^b(x) = \sum_{j=1}^{\infty} b_j 2^{-j+1} h^L(2^{j-1}x_1) \prod_{m=2}^{n} h(2^{j-1}x_m), \quad x = (x_1,\ldots,x_n) \in \mathbb{R}^n, \tag{14.17}$$

where $b = \{b_j\}_{j=1}^{\infty}$ is a sequence with $b_j \geq 0$ and (13.67). This can be interpreted as an atomic decomposition in $B_{pq}^{1+\frac{n}{p}}(\mathbb{R}^n)$ and $F_{pq}^{1+\frac{n}{p}}(\mathbb{R}^n)$, where the necessary moment conditions according to (13.72), now with $s = 1 + \frac{n}{p}$, may be assumed to be satisfied by the above construction. Let $0 < \varepsilon < 1$. We claim that we have in analogy to (13.69) and (13.70),

$$\left(\sum_{j=1}^{\infty} b_j^q\right)^{\frac{1}{q}} \sim \left(\int_0^\varepsilon \left(\frac{\widetilde{\omega}(f^b,t)}{|\log t|}\right)^q \frac{dt}{t}\right)^{\frac{1}{q}} \sim \|f^b\,|B_{pq}^{1+\frac{n}{p}}(\mathbb{R}^n)\| \tag{14.18}$$

if $b \in \ell_q$ (usual modification when $q = \infty$) and

$$\left(\sum_{j=1}^{\infty} b_j^p\right)^{\frac{1}{p}} \sim \left(\int_0^{\varepsilon} \left(\frac{\widetilde{\omega}(f^b, t)}{|\log t|}\right)^p \frac{dt}{t}\right)^{\frac{1}{p}} \sim \|f^b \,|\, F_{pq}^{1+\frac{n}{p}}(\mathbb{R}^n)\|. \qquad (14.19)$$

Of course we always assume that (14.4) and (14.6) are satisfied. First we remark that

$$\|f^b \,|\, F_{pq}^{1+\frac{n}{p}}(\mathbb{R}^n)\| \le c \left(\sum_{j=1}^{\infty} b_j^p\right)^{\frac{1}{p}} \qquad (14.20)$$

in analogy to (13.40). Secondly we claim

$$\widetilde{\omega}(f^b, t) \sim \sum_{j=1}^{k} b_j \quad \text{if} \quad t \sim 2^{-k} \quad \text{where} \quad k \in \mathbb{N}, \qquad (14.21)$$

in analogy (but also in slight modification) of (13.42). Let $\eta > 0$ be small and $k \in \mathbb{N}$. Then one obtains by (14.17), and (14.15), (13.64),

$$f^b(0) - f^b(-\eta 2^{-k}, 0, \ldots, 0) \ge \sum_{j=1}^{k} b_j \, (h^L)'(z_{j,k}) \eta 2^{-k}, \qquad (14.22)$$

with $-\frac{\delta}{2} < z_{j,k} < 0$, where δ has the same meaning as in (13.63). Since $(h^L)' = h_L$, all factors $h_L(z_{j,k}) \ge c > 0$ for some c which is independent of j and k. Hence the left-hand side of (14.21) can be estimated from below by its right-hand side. To prove the converse we note that the terms with $j \ge k$ in (14.17) are harmless. Together with the monotonicity (13.67) of the coefficients b_j, and the converse of (14.22) we get (14.21). But now we are very much in the same situation as in Step 4 of the proof of Theorem 13.2. The counterparts of (13.43) and (13.39) prove (14.19). Similarly one obtains (14.18). We are now in the same situation as in Steps 6 and 7 of the proof of Theorem 13.2. First we get equality in (14.13), (14.14) and that q and p are the correct numbers in (14.5) and (14.7), respectively.

14.3 Inequalities

The above Theorem 14.2 is the counterpart of Theorem 13.2. Even more, with Proposition 12.16 as a vehicle, 14.2 is a consequence of 13.2. It covers all cases of interest (excluding borderline situations as described in (13.2), which means here $p = \infty$). It describes in a rather condensed way very sharp inequalities. In analogy to 13.3 we discuss the outcome, where now the harvest is even richer, since we have not only inequalities in terms of moduli of continuity hidden in Theorem 14.2, but even sharper inequalities of type (14.9). As in 13.3 we formulate the corresponding assertions for the B-spaces in (i) and for the F-spaces in (ii), but in contrast to 13.3 a few explanations and justifications are needed. This will be done afterwards in (iii). We always assume that $0 < \varepsilon < 1$.

14.3(i) *The B-spaces* Let $\varkappa(t)$ be a positive monotonically decreasing function on $(0, \varepsilon]$. Let $0 < u \leq \infty$. Let p and q be given by (14.4). Then: (i_1)

$$\left(\int_0^\varepsilon \left(\frac{\varkappa(t)\,\widetilde{\omega}(f,t)}{|\log t|^{\frac{1}{q'}}} \right)^u \frac{dt}{t|\log t|} \right)^{\frac{1}{u}} \leq c \, \|f\,|B_{pq}^{1+\frac{n}{p}}(\mathbb{R}^n)\| \qquad (14.23)$$

for some $c > 0$ and all $f \in B_{pq}^{1+\frac{n}{p}}(\mathbb{R}^n)$ if, and only if, $\varkappa$ is bounded and $q \leq u \leq \infty$ (with the modification (14.28) below if $u = \infty$) and (i_2)

$$\left(\int_0^\varepsilon \left(\frac{\varkappa(t)\,|\nabla f|^*(t)}{|\log t|^{\frac{1}{q'}}} \right)^u \frac{dt}{t|\log t|} \right)^{\frac{1}{u}} \leq c \, \|f\,|B_{pq}^{1+\frac{n}{p}}(\mathbb{R}^n)\| \qquad (14.24)$$

for some $c > 0$ and all $f \in B_{pq}^{1+\frac{n}{p}}(\mathbb{R}^n)$ if, and only if, $\varkappa$ is bounded and $q \leq u \leq \infty$ (again with the indicated modification if $u = \infty$). In particular, if $1 < q < \infty$, then

$$\sup_{0<t<\varepsilon} \frac{\widetilde{\omega}(f,t)}{|\log t|^{\frac{1}{q'}}} \leq c_0 \left(\int_0^\varepsilon \left(\frac{\widetilde{\omega}(f,t)}{|\log t|} \right)^q \frac{dt}{t} \right)^{\frac{1}{q}} \leq c_1 \, \|f\,|B_{pq}^{1+\frac{n}{p}}(\mathbb{R}^n)\| \qquad (14.25)$$

(and similarly with $|\nabla f|^*(t)$ in place of $\widetilde{\omega}(f,t)$) are the two end-point cases according to (12.28). The two types of inequalities (14.23) and (14.24) are connected by

$$\left(\int_0^\varepsilon \left(\frac{\widetilde{\omega}(f,t)}{|\log t|^{\frac{1}{q'}}} \right)^u \frac{dt}{t|\log t|} \right)^{\frac{1}{u}} \leq c_0 \left(\int_0^\varepsilon \left(\frac{|\nabla f|^*(t)}{|\log t|^{\frac{1}{q'}}} \right)^u \frac{dt}{t|\log t|} \right)^{\frac{1}{u}}$$

$$\leq c_1 \, \|f\,|B_{pq}^{1+\frac{n}{p}}(\mathbb{R}^n)\| \qquad (14.26)$$

for some $c_0 > 0$, $c_1 > 0$, and all $f \in B_{pq}^{1+\frac{n}{p}}(\mathbb{R}^n)$, where again $q \leq u \leq \infty$ (with the modification (14.28) below if $u = \infty$). If $q = \infty$ then one has

$$\sup_{0<t<\varepsilon} \frac{\widetilde{\omega}(f,t)}{|\log t|} \leq c_0 \sup_{0<t<\varepsilon} \frac{|\nabla f|^*(t)}{|\log t|} \leq c \, \|f\,|B_{p\infty}^{1+\frac{n}{p}}(\mathbb{R}^n)\|. \qquad (14.27)$$

14. The super-critical case

Let $\varkappa(t)$ be an (arbitrary) positive function on $(0, \varepsilon]$ and let again $1 < q \le \infty$. Then

$$\sup_{0<t<\varepsilon} \frac{\varkappa(t)\,\widetilde{\omega}(f,t)}{|\log t|^{\frac{1}{q'}}} \le c\,\|f\,|B_{pq}^{1+\frac{n}{p}}(\mathbb{R}^n)\| \tag{14.28}$$

for some $c > 0$ and all $f \in B_{pq}^{1+\frac{n}{p}}(\mathbb{R}^n)$ if, and only if, $\varkappa$ is bounded (and the same assertion with $|\nabla f|^*(t)$ in place of $\widetilde{\omega}(f,t)$). But as discussed in 12.11 the above additional assumption that $\varkappa$ is monotone is rather immaterial.

14.3(ii) *The F-spaces* Let $\varkappa(t)$ be a positive monotonically decreasing function on $(0, \varepsilon]$. Let $0 < u \le \infty$. Let p and q be given by (14.6). Then: (ii$_1$)

$$\left(\int_0^\varepsilon \left(\frac{\varkappa(t)\,\widetilde{\omega}(f,t)}{|\log t|^{\frac{1}{p'}}} \right)^u \frac{dt}{t|\log t|} \right)^{\frac{1}{u}} \le c\,\|f\,|F_{pq}^{1+\frac{n}{p}}(\mathbb{R}^n)\| \tag{14.29}$$

for some $c > 0$ and all $f \in F_{pq}^{1+\frac{n}{p}}(\mathbb{R}^n)$ if, and only if, $\varkappa$ is bounded and $p \le u \le \infty$, with the modification

$$\sup_{0<t<\varepsilon} \frac{\varkappa(t)\,\widetilde{\omega}(f,t)}{|\log t|^{\frac{1}{p'}}} \le c\,\|f\,|F_{pq}^{1+\frac{n}{p}}(\mathbb{R}^n)\| \quad \text{if} \quad u = \infty; \tag{14.30}$$

and (ii$_2$)

$$\left(\int_0^\varepsilon \left(\frac{\varkappa(t)|\nabla f|^*(t)}{|\log t|^{\frac{1}{p'}}} \right)^u \frac{dt}{t|\log t|} \right)^{\frac{1}{u}} \le c\,\|f\,|F_{pq}^{1+\frac{n}{p}}(\mathbb{R}^n)\| \tag{14.31}$$

for some $c > 0$ and all $f \in F_{pq}^{1+\frac{n}{p}}(\mathbb{R}^n)$ if, and only if, $\varkappa$ is bounded and $p \le u \le \infty$ with a similar modification as in (14.30) if $u = \infty$. In particular,

$$\sup_{0<t<\varepsilon} \frac{\widetilde{\omega}(f,t)}{|\log t|^{\frac{1}{p'}}} \le c_0 \left(\int_0^\varepsilon \left(\frac{\widetilde{\omega}(f,t)}{|\log t|} \right)^p \frac{dt}{t} \right)^{\frac{1}{p}} \le c_1\,\|f\,|F_{pq}^{1+\frac{n}{p}}(\mathbb{R}^n)\| \tag{14.32}$$

(and similarly with $|\nabla f|^*(t)$ in place of $\widetilde{\omega}(f,t)$) are the two end-point cases according to (12.28). The two types of inequalities (14.29) and (14.31) are connected by

$$\left(\int_0^\varepsilon \left(\frac{\omega(f,t)}{|\log t|^{\frac{1}{p'}}} \right)^u \frac{dt}{t|\log t|} \right)^{\frac{1}{u}} \le c_0 \left(\int_0^\varepsilon \left(\frac{|\nabla f|^*(t)}{|\log t|^{\frac{1}{p'}}} \right)^u \frac{dt}{t|\log t|} \right)^{\frac{1}{u}}$$

$$\le c_1\,\|f\,|F_{pq}^{1+\frac{n}{p}}(\mathbb{R}^n)\| \tag{14.33}$$

for some $c_0 > 0$, $c_1 > 0$, and all $f \in F_{pq}^{1+\frac{n}{p}}(\mathbb{R}^n)$, where again $p \le u \le \infty$ (with the modification as in (14.30) if $u = \infty$). As in connection with (14.28) one does not need for the sharpness assertion in (14.30) that $\varkappa$ is monotone.

14.3(iii)   *Explanations*   The above inequalities with respect to $\widetilde{\omega}(f,t)$ follow from Theorem 14.2, Example 2 in 12.4, Definition 12.14 and the modified Proposition 12.10 with $\mathcal{E}_C A_{pq}^{1+\frac{n}{p}}(t)$ and $\widetilde{\omega}(f,t)$ in place of $\mathcal{E}_G A_{pq}^s(t)$ and $f^*(t)$, respectively. Or in other words, they simply describe what is meant by a continuity envelope. Furthermore, (14.26), (14.27), and (14.33) are covered by Step 1 of the proof of Theorem 14.2. The only point which is not immediately clear by the above theorem and its proof is the boundedness of $\varkappa$ in (14.24) and (14.31). By (12.14) this question can be reduced to

$$\sup_{0<t<\varepsilon} \frac{\varkappa(t)\,|\nabla f|^*(t)}{|\log t|^{\frac{1}{q'}}} \le c\,\|f\,|B_{pq}^{1+\frac{n}{p}}(\mathbb{R}^n)\| \tag{14.34}$$

and its F-counterpart. Hence we assume that we have (14.34) with $1 < q \le \infty$ for some positive function $\varkappa$ on $(0,\varepsilon]$. We wish to show that $\varkappa$ must be bounded. Let f^b be given by (14.17) with $b_j = 1$ if $j = 1, \ldots, J$ and $b_j = 0$ if $j > J$. Then by an argument similar to that in (14.22) we have

$$|\nabla f|(x) \ge c\,J \quad \text{in a cube} \quad [-\eta 2^{-J}, 0]^n, \quad J \in \mathbb{N}, \tag{14.35}$$

with $2^{-K-1} \le \eta \le 2^{-K}$ where $c > 0$ and $K \in \mathbb{N}$ are independent of J. Then $|\nabla f|^*(\eta^n\,2^{-Jn}) \ge c\,J$ and it follows by (14.34) and (14.18)

$$\frac{\varkappa(\eta^n\,2^{-Jn})\,J}{J^{\frac{1}{q'}}} \le c\,J^{\frac{1}{q}}. \tag{14.36}$$

Hence $\varkappa$ is bounded. This proves the $\varkappa$-sharpness also in (14.24) and (14.31).

14.4   More handsome inequalities

In 14.3 we tried to unwrap what is hidden in Theorem 14.2 and, with a switch from f^* to $|\nabla f|^*$, in Theorem 13.2. This may also be taken as an excuse for the undue length of 14.3 (compared with the lengths of the respective theorems). Nevertheless the formulations remain somewhat involved. But in case of $u = \infty$, this means (14.28) and (14.30) with $\varkappa = 1$, one can convert these assertions into more handsome inequalities which come also near to what is done in the literature. Let $v > 0$. By the definitions of the moduli of continuity in (12.69) and (12.72) we have with $0 < \varepsilon < 1$,

$$\sup_{0<t<\varepsilon} \frac{\widetilde{\omega}(f,t)}{|\log t|^v} = \sup_{0<t<\varepsilon} \frac{1}{t|\log t|^v} \sup_{x\in\mathbb{R}^n, |h|\leq t} |f(x+h) - f(x)|$$

$$= \sup_{|x-y|\leq \varepsilon} \frac{|f(x) - f(y)|}{|x-y| \, |\log|x-y||^v}. \qquad (14.37)$$

Then (14.28) and (14.30) with $\varkappa = 1$ can be reformulated as

$$|f(x) - f(y)| \leq c \, |x-y| \, |\log|x-y||^{\frac{1}{q'}} \, \|f \, | B_{pq}^{1+\frac{n}{p}}(\mathbb{R}^n)\|, \quad x \in \mathbb{R}^n, \, y \in \mathbb{R}^n, \qquad (14.38)$$

and $|x - y| < \varepsilon$, for some $c > 0$ and all $f \in B_{pq}^{1+\frac{n}{p}}(\mathbb{R}^n)$, where

$$0 < p < \infty, \quad 1 < q \leq \infty, \quad \frac{1}{q} + \frac{1}{q'} = 1,$$

and

$$|f(x) - f(y)| \leq c \, |x-y| \, |\log|x-y||^{\frac{1}{p'}} \, \|f \, | F_{pq}^{1+\frac{n}{p}}(\mathbb{R}^n)\|, \quad x \in \mathbb{R}^n, \, y \in \mathbb{R}^n, \qquad (14.39)$$

and $|x - y| < \varepsilon$, for some $c > 0$ and all $f \in F_{pq}^{1+\frac{n}{p}}(\mathbb{R}^n)$, where

$$1 < p < \infty, \quad 0 < q \leq \infty, \quad \frac{1}{p} + \frac{1}{p'} = 1;$$

with the special case

$$|f(x) - f(y)| \leq c \, |x-y| \, |\log|x-y||^{\frac{1}{p'}} \, \|f \, | H_p^{1+\frac{n}{p}}(\mathbb{R}^n)\|, \quad x \in \mathbb{R}^n, \, y \in \mathbb{R}^n, \qquad (14.40)$$

and $|x - y| < \varepsilon$, where $H_p^{1+\frac{n}{p}}(\mathbb{R}^n) = F_{p,2}^{1+\frac{n}{p}}(\mathbb{R}^n)$ are the Sobolev spaces, $1 < p < \infty$.

14.5 Borderline cases

This means in our context here $p = \infty$ and $s = 1$, hence the Besov spaces $B_{\infty q}^1(\mathbb{R}^n)$, where $0 < q \leq \infty$, with the Zygmund class $\mathcal{C}^1(\mathbb{R}^n) = B_{\infty\infty}^1(\mathbb{R}^n)$ as a special case. We refer to 1.2(iv), (v), especially (1.11). The extension of (14.38) to $p = q = \infty$ is given by

$$|f(x) - f(y)| \leq c \, |x-y| \, |\log|x-y|| \, \|f \, | \mathcal{C}^1(\mathbb{R}^n)\|, \quad x \in \mathbb{R}^n, \, y \in \mathbb{R}^n, \qquad (14.41)$$

and $|x - y| < \varepsilon < 1$, for some $c > 0$, and all $f \in \mathcal{C}^1(\mathbb{R}^n)$. It is due to A. Zygmund, [Zyg45], and may also be found in [Zyg77], Chapter II, Theorem 3.4, p. 44. It was apparently A. Zygmund who coined the word *smooth functions* in this context in his paper [Zyg45]. In [Zyg77], Notes, p. 375, he mentioned that B. Riemann was the first who considered smooth functions. B. Riemann discussed in his Habilitationsschrift [Rie'54] the possibility to represent a continuous periodic function on the interval $[0, 2\pi]$ in terms of trigonometric series: First he surveyed what had been done so far. Afterwards he studied in Sections 7–13 the indicated problem in detail based on the systematic use of second differences. This is just what A. Zygmund called almost 100 years later in [Zyg45] *smooth functions*. The extension of (14.25) to $p = \infty$ (and $1 < q \le \infty$) is given by

$$\sup_{0<t<\varepsilon} \frac{\widetilde{\omega}(f,t)}{|\log t|^{\frac{1}{q'}}} \le c_0 \left(\int_0^\varepsilon \left(\frac{\widetilde{\omega}(f,t)}{|\log t|} \right)^q \frac{dt}{t} \right)^{\frac{1}{q}} \le c_1 \|f \,|\, B^1_{\infty q}(\mathbb{R}^n)\| \qquad (14.42)$$

(with the obvious modification if $q = \infty$ which is essentially (14.41)). This has been proved very recently in [BoL00], Proposition 1. Furthermore, (14.25) with (14.4) and (14.32) with (14.6) follow from (14.42) and the embeddings

$$B^{1+\frac{n}{p}}_{pq}(\mathbb{R}^n) \subset B^1_{\infty q}(\mathbb{R}^n) \quad \text{and} \quad F^{1+\frac{n}{p}}_{pq}(\mathbb{R}^n) \subset B^1_{\infty p}(\mathbb{R}^n), \qquad (14.43)$$

[Triβ], Theorem 2.7.1, p. 129. (In the second embedding we used the first one and [Triβ], (15) on p. 131.) Our own approach is characterized by lifting the critical case, considered in Section 13, to the super-critical one considered here. This results in the sharper inequalities (14.26), (14.27), (14.33), where always $p < \infty$. But it is unclear whether there is something of this type if $p = \infty$. As discussed in connection with (13.3), based on Theorem 11.2, the question itself makes sense at least for the spaces $B^1_{\infty q}(\mathbb{R}^n)$ with $1 < q \le 2$, also for the space $bmo(\Omega)$, lifted by 1, according to 13.7. By (14.42) and (14.43) the sharpness assertions available for the spaces with $p < \infty$ can be carried over to the spaces $B^1_{\infty q}(\mathbb{R}^n)$. We get the complement

$$\mathfrak{E}_C B^1_{\infty q} = (|\log t|^{\frac{1}{q'}}, q), \quad 1 < q \le \infty, \qquad (14.44)$$

of (14.5). In particular, the f^b, given by (14.17), are extremal functions also for $B^1_{\infty q}(\mathbb{R}^n)$ and we have (14.18) with $p = \infty$. If $p = \infty$ then

$$f(x) = h(x)\, x_1 \log |x|, \quad x \in \mathbb{R}^n, \qquad (14.45)$$

with h given by (13.63), for example, is an extremal function in $\mathcal{C}^1(\mathbb{R}^n)$. This follows from

$$\frac{\partial f}{\partial x_1} \sim \log |x| \in bmo(\mathbb{R}^n) \subset \mathcal{C}^0(\mathbb{R}^n), \quad |x| \le \frac{\delta}{2}, \qquad (14.46)$$

the boundedness of the other first derivatives, and elementary calculations. The embedding mentioned is well known. We refer to [RuS96], p. 33. Other extremal functions may be found in 17.1.

14.6 Envelope functions and non-compactness

This remark applies equally to growth envelope functions and continuity envelope functions and to all cases (critical, super-critical, sub-critical). But it will be clear what is meant by looking at an example connected with the above considerations. Let Ω be a bounded C^∞ domain in $\mathbb{R}^n$ (one might think of the unit ball). Then $B^s_{pq}(\Omega)$ has the usual meaning according to Definition 5.3. Let $\alpha \geq 0$. Let $Lip^{(1,-\alpha)}(\Omega)$ be, by definition, the Banach space of all (complex-valued) continuous functions in Ω, such that

$$\|f\,|Lip^{(1,-\alpha)}(\Omega)\| = \sup_{x\in\Omega} |f(x)| + \sup_{\substack{x,y\in\Omega \\ 0<|x-y|\leq \frac{1}{2}}} \frac{|f(x)-f(y)|}{|x-y|\,|\log|x-y|\,|^\alpha} < \infty. \tag{14.47}$$

We use here the notation introduced in [EdH99]. Then (14.38), restricted to Ω, is equivalent to the continuous embedding

$$B^{1+\frac{n}{p}}_{pq}(\Omega) \subset Lip^{(1,-\frac{1}{q'})}(\Omega), \quad 0 < p < \infty, \quad 1 < q < \infty, \tag{14.48}$$

(where we excluded $q = \infty$). However this embedding is not compact. We prove this assertion by looking at the growth envelope function $|\log t|^{\frac{1}{q'}}$ for $B^{1+\frac{n}{p}}_{pq}(\Omega)$ (as for spaces on domains we refer also to 13.6). Since $q < \infty$ it follows that $C^\infty(\overline{\Omega})$, the restriction of $S(\mathbb{R}^n)$ on Ω, is dense in $B^{1+\frac{n}{p}}_{pq}(\Omega)$. Assume that the embedding (14.48) is compact. We fix a quasi-norm in $B^{1+\frac{n}{p}}_{pq}(\Omega)$. Then we find for any $\delta > 0$ finitely many functions

$$f_j \in C^\infty(\overline{\Omega}), \quad \|f_j\,|B^{1+\frac{n}{p}}_{pq}(\Omega)\| \leq 1 \quad \text{with} \quad j = 1, \ldots, M(\delta), \tag{14.49}$$

such that for any f with $\|f\,|B^{1+\frac{n}{p}}_{pq}(\Omega)\| \leq 1$,

$$\inf_j \frac{\widetilde{\omega}(f-f_j,t)}{|\log t|^{\frac{1}{q'}}} \leq \delta, \quad \text{uniformly for} \quad 0 < t < \frac{1}{2}. \tag{14.50}$$

Here $\{f_j\}$ is a δ-net. Furthermore we used (14.37) with $\varepsilon = \frac{1}{2}$. Since the functions f_j are smooth one obtains

$$\widetilde{\omega}(f,t) \leq C_\delta + \delta |\log t|^{\frac{1}{q'}} \quad \text{where} \quad 0 < t < \frac{1}{2}, \tag{14.51}$$

and hence by (12.77),

$$\mathcal{E}_C B_{pq}^{1+\frac{n}{p}}(t) \leq C_\delta + \delta |\log t|^{\frac{1}{q'}}, \quad 0 < t < \frac{1}{2}. \tag{14.52}$$

If $\delta > 0$ is small one gets a contradiction to

$$\mathcal{E}_C B_{pq}^{1+\frac{n}{p}}(t) \sim |\log t|^{\frac{1}{q'}}$$

when t is tending to zero. Hence the embedding (14.48) is not compact. But it was not so much our aim to prove this specific assertion. We wanted to make clear what happens if both source and target space have the *same* envelope functions (growth or continuity): The respective embeddings are not compact (at least in those cases where smooth functions are dense in the source space).

14.7 References

First we recall that (11.59) coincides with (14.40) if $1 + \frac{n}{p} = k \in \mathbb{N}$. This inequality in this version is due to [BrW80]. We refer also to our remarks in 11.8(vi). The extension (14.40) from the classical Sobolev spaces $W_p^k(\mathbb{R}^n)$ to $H_p^{1+\frac{n}{p}}(\mathbb{R}^n)$ may be found in [EdK95]. As mentioned in 14.5 the borderline case, in our notation used here, (14.41) goes back to [Zyg45] (and may be found with a new proof in [Zyg77], Chapter II, Theorem 3.4, p. 44). A Fourier-analytical proof of (14.38), at least in some cases, has been given in [Vis98]. The additional point of interest here is the use of spaces of type $Lip^{(1,-\alpha)}$ (on domains and in $\mathbb{R}^n$) according to (14.47) in connection with problems from physics. We refer in this context also to [Lio98], pp. 146, 152. The first full proof of (14.38) and (14.39), including sharpness assertions, was given in [EdH99]. In the context of this paper sharpness means that the exponents $\frac{1}{q'}$ and $\frac{1}{p'}$ in the log-terms in (14.38) and (14.39), respectively, cannot be replaced by a smaller exponent. The proofs are based on atomic decompositions. The borderline inequality (14.42) (without the middle term) has been derived in [KrS98] using extrapolation techniques. By (14.43), and as has also been mentioned in [KrS98] explicitly, this results in new proofs of (14.38), (14.39). As remarked in 14.5 the decisive improvement concerning the middle term in (14.42) is due to [BoL00]. Our own approach which resulted not only in Theorem 14.2, including the $\varkappa$-sharpness as described in 14.3, but also in the sharp assertions in 14.3 concerning $|\nabla f|^*(t)$, especially (14.26), (14.27), (14.33), is published here for the first time. Especially the concept of envelope functions and envelopes (here in connection with the super-critical case in the understanding of Theorem 14.2) came out in recent discussions with D. D. Haroske. A more systematic treatment will be given in [Har01]. In a different context lifting arguments have also been used in [EdK95] with a reference to [Adm75], Theorem

8.36, pp. 254–255. Somewhat different types of envelopes appear in [Net87b] in connection with optimal embeddings of F_{pq}^s-spaces in rearrangement-invariant spaces, preferably in sub-critical situations which will be treated in Section 15 below. Finally we add a remark in connection with 14.6. As mentioned, there is no hope that the continuous embeddings (14.48) are compact. However the situation is different if one replaces the target space in (14.48) by $Lip^{(1,-\alpha)}(\Omega)$ according to (14.47) with $\alpha > \frac{1}{q'}$. Then one has compact embeddings. The adequate notation to measure the degree of compactness are entropy numbers and approximation numbers. As for the general background we refer to [ET96], Chapter 1. But later on in connection with the spectral theory for fractal elliptic operators we repeat in 19.16 what is needed. A detailed study of entropy numbers and approximation numbers for problems treated in the present section (in the modification indicated above) has been given in [EdH99] and [EdH00]. This has been complemented in [Har00a]. The small survey [Har00b] summarizes these results.

15 The sub-critical case

15.1 Introduction

By the terminology of (10.5) the sub-critical case covers the spaces

$$B_{pq}^s(\mathbb{R}^n) \quad \text{and} \quad F_{pq}^s(\mathbb{R}^n) \quad \text{with } 0 < p < \infty, \, 0 < q \leq \infty, \, \sigma_p < s < \frac{n}{p}. \tag{15.1}$$

Recall our standard abbreviations

$$\sigma_p = n\left(\frac{1}{p} - 1\right)_+ \quad \text{and} \quad \sigma_{pq} = n\left(\frac{1}{\min(p,q)} - 1\right)_+ \tag{15.2}$$

where $0 < p < \infty$, $0 < q \leq \infty$. We are interested in sharp limiting embeddings (or better related inequalities) corresponding to the foot-point of the dashed line in Fig. 10.1 and given by

$$1 < r < \infty, \quad s > 0, \quad s - \frac{n}{p} = -\frac{n}{r}, \quad \text{and} \quad 0 < q \leq \infty. \tag{15.3}$$

We characterized in Theorem 11.4(i) those spaces (15.1) which are embedded in $L_r(\mathbb{R}^n)$. In Theorem 11.7(i) and (ii) we collected the classical more refined inequalities in the sub-critical context and we described their rather rich history in the points 11.8(i)–(iv). Again we are interested only in spaces which consist entirely of regular distributions; this means that they are embedded in $L_1^{loc}(\mathbb{R}^n)$. A final description has been given in Theorem 11.2. Compared with

(15.1) we exclude as previously borderline cases, we mean here those spaces $F_{pq}^s(\mathbb{R}^n)$ and $B_{pq}^s(\mathbb{R}^n)$ with $0 < p < \infty$ and $s = \sigma_p$ which are covered by (11.9) and (11.14), respectively. Otherwise we are very much in the same general situation as in 13.1. Again, as outlined in the introductory Section 10 we wish to measure the singularity behaviour of functions belonging to the spaces (15.1) in terms of the growth envelope as introduced in Definition 12.8. Instead of $\mathfrak{E}_G A_{pq}^s$ in (12.56) we use the more handsome version (12.60). Similarly as in Section 13, in the theorem below we first calculate explicitly the growth envelopes for all spaces in (15.1). Afterwards we describe what this means in terms of inequalities. As explained in detail in 11.8(i), (ii) the Lorentz spaces and their (quasi-)norms come in naturally, whereas as described in 11.8(iii), (iv) the Zygmund spaces and their (quasi-)norms are distinguished but (from the above point of view) not so natural target spaces. Nevertheless we collect in a corollary below sharp assertions concerning related inequalities also in these cases. Finally we complement the references given so far.

15.2 Theorem

Let

$$0 < q \leq \infty, \quad s > 0, \quad \text{and} \quad s - \frac{n}{p} = -\frac{n}{r} \quad \text{with} \quad 1 < r < \infty, \tag{15.4}$$

(the dashed line in Fig. 10.1). Then

$$\mathfrak{E}_G B_{pq}^s = (t^{-\frac{1}{r}}, q) \tag{15.5}$$

and

$$\mathfrak{E}_G F_{pq}^s = (t^{-\frac{1}{r}}, p). \tag{15.6}$$

Proof *Step 1* Let $0 < \varepsilon < 1$. In 11.8(ii) we proved (11.52). The interpolation argument used there applies to all cases covered by (15.4), [Triβ], Theorem 2.4.2, p. 64. Hence, together with (12.26), we obtain always

$$\sup_{0<t<\varepsilon} t^{\frac{1}{r}} f^*(t) \leq c_0 \left(\int_0^\varepsilon \left(t^{\frac{1}{r}} f^*(t) \right)^q \frac{dt}{t} \right)^{\frac{1}{q}} \leq c_1 \|f \,|\, B_{pq}^s(\mathbb{R}^n)\|. \tag{15.7}$$

If $q = \infty$ then one has only the first and the last term. By (12.37) it follows that

$$\mathcal{E}_G B_{pq}^s(t) \leq c\, t^{-\frac{1}{r}}, \quad 0 < t < \varepsilon. \tag{15.8}$$

15. The sub-critical case

As for the F-spaces we use Netrusov's observation described in (13.29). Similarly as in (13.30) it follows that

$$\sup_{0<t<\varepsilon} t^{\frac{1}{r}} f^*(t) \le c_0 \left(\int_0^\varepsilon \left(t^{\frac{1}{r}} f^*(t) \right)^p \frac{dt}{t} \right)^{\frac{1}{p}} \le c_1 \|f\,|F^s_{pq}(\mathbb{R}^n)\|. \qquad (15.9)$$

Hence we have also

$$\mathcal{E}_G F^s_{pq}(t) \le c t^{-\frac{1}{r}}, \quad 0 < t < \varepsilon. \qquad (15.10)$$

Step 2 Let ψ be given by (13.35). By [Triδ], Theorem 13.8, p. 75,

$$f_j(x) = 2^{j\frac{n}{r}} \psi(2^j x), \quad j \in \mathbb{N}, \qquad (15.11)$$

are atoms in all spaces $B^s_{pq}(\mathbb{R}^n)$ and at least in those spaces $F^s_{pq}(\mathbb{R}^n)$ where no moment conditions are needed, say $q \ge 1$ (ignoring constants, which may be chosen independent of j). Again by (12.37),

$$\mathcal{E}_G B^s_{pq}(d 2^{-jn}) \ge c f_j^*(d 2^{-jn}) \sim 2^{\frac{jn}{r}}, \quad j \in \mathbb{N}, \qquad (15.12)$$

for some $d > 0$ and $c > 0$. Together with (15.8) we obtain

$$\mathcal{E}_G B^s_{pq}(t) = t^{-\frac{1}{r}}, \quad 0 < t < \varepsilon. \qquad (15.13)$$

Similarly for the F-spaces as far as they are covered. If $q > 0$ is small then the moment conditions needed for the atoms can be incorporated in the same way as in Step 5 of the proof of Theorem 13.2. Then we have also

$$\mathcal{E}_G F^s_{pq}(t) = t^{-\frac{1}{r}}, \quad 0 < t < \varepsilon, \qquad (15.14)$$

without any restriction for q.

Step 3 It remains to prove that q and p in (15.5) and (15.6), respectively, are the correct numbers. Since we have already (15.7) and (15.9) we must show that q and p, respectively, cannot be improved by smaller numbers. Let $v < q$ and let

$$\left(\int_0^\varepsilon \left(t^{\frac{1}{r}} f^*(t) \right)^v \frac{dt}{t} \right)^{\frac{1}{v}} \le c \|f\,|B^s_{pq}(\mathbb{R}^n)\| \qquad (15.15)$$

for some $c > 0$ and all $f \in B^s_{pq}(\mathbb{R}^n)$. Let

$$f(x) = \sum_{j=1}^J 2^{j\frac{n}{r}} \psi(2^j x - x^0) \qquad (15.16)$$

with $x^0 \in \mathbb{R}^n$. If $|x^0|$ is large then the supports of the atoms in (15.16) are disjoint. It follows for some $d > 0$,

$$f^*(d2^{-jn}) \sim 2^{j\frac{n}{r}} \quad \text{where} \quad j = 1, \ldots, J. \tag{15.17}$$

We insert (15.17) in (15.15). Since (15.16) is an atomic decomposition we get

$$\left(\sum_{j=1}^{J} 1\right)^{\frac{1}{v}} \leq c_0 \, \|f \,|B_{pq}^s(\mathbb{R}^n)\| \leq c_1 \left(\sum_{j=1}^{J} 1\right)^{\frac{1}{q}}, \tag{15.18}$$

where $c_0 > 0$ and $c_1 > 0$ are independent of J. But this is a contradiction. In case of the F-spaces we assume $v < p$ and that we have (15.15) with F_{pq}^s in place of B_{pq}^s. Let first q be large, say $q \geq 1$, such that no moment conditions in the atomic decomposition (15.16) are needed. We apply the considerations in connection with the proof of (13.40). Then we get (15.18) with p in place of q on the right-hand side. We have again a contradiction. Finally if moment conditions are needed, then one has to modify the above constructions as indicated in Step 5 of the proof of Theorem 13.2.

15.3 Inequalities

The above theorem covers all cases (15.3). It excludes borderline situations as described in 15.1. Parallel to 13.3 we explain also in the sub-critical case considered now, which is hidden in the above theorem. We use Example 1 in 12.4, Definition 12.8 and Proposition 12.10. Let $0 < \varepsilon < 1$.

15.3(i) *The B-spaces* Let $\varkappa(t)$ be a positive monotonically decreasing function on $(0, \varepsilon]$. Let $0 < u \leq \infty$. Let p, q, s be given by (15.4). Then

$$\left(\int_0^\varepsilon \left(\varkappa(t) \, t^{\frac{1}{r}} \, f^*(t)\right)^u \frac{dt}{t}\right)^{\frac{1}{u}} \leq c \, \|f \,|B_{pq}^s(\mathbb{R}^n)\| \tag{15.19}$$

for some $c > 0$ and all $f \in B_{pq}^s(\mathbb{R}^n)$ if, and only if, $\varkappa$ is bounded and $q \leq u \leq \infty$, with the modification

$$\sup_{0 < t < \varepsilon} \varkappa(t) \, t^{\frac{1}{r}} \, f^*(t) \leq c \, \|f \,|B_{pq}^s(\mathbb{R}^n)\| \tag{15.20}$$

if $u = \infty$. Furthermore, (15.7) deals with the two end-point cases according to (12.26). Let $\varkappa$ be an arbitrary positive function on $(0, \varepsilon]$. Then (15.20) holds if, and only if, $\varkappa$ is bounded.

15.3(ii) *The F-spaces* Let $\varkappa(t)$ be a positive monotonically decreasing function on $(0, \varepsilon]$. Let $0 < u \leq \infty$. Let p, q, s be given by (15.4). Then

$$\left(\int_0^\varepsilon \left(\varkappa(t) \, t^{\frac{1}{r}} \, f^*(t) \right)^u \frac{dt}{t} \right)^{\frac{1}{u}} \leq c \, \|f \,|F_{pq}^s(\mathbb{R}^n)\| \tag{15.21}$$

for some $c > 0$ and all $f \in F_{pq}^s(\mathbb{R}^n)$ if, and only if, $\varkappa$ is bounded and $p \leq u \leq \infty$ (modified by (15.20) with F_{pq}^s in place of B_{pq}^s when $u = \infty$). Also the other assertions for the B-spaces after (15.20) have obvious counterparts, in particular the two end-point cases (15.9) according to (12.26).

The Lorentz spaces $L_{ru}(I_\varepsilon)$ were introduced in 11.6(i). The above theorem and the explanations just given can be reformulated in terms of natural and sharp embeddings of the spaces $B_{pq}^s(\mathbb{R}^n)$ and $F_{pq}^s(\mathbb{R}^n)$ with (15.4) into $L_{ru}(I_\varepsilon)$. We complement these assertions by looking at corresponding *optimal* embeddings into Zygmund spaces $L_r(\log L)_a(I_\varepsilon)$ according to 11.6(ii). By (11.46) the original definition (11.44) can be reformulated in terms of rearrangement. *Optimal* means here that for given r in (15.4) and in (11.46), (11.54), one asks for all numbers a for which we have the desired embedding, again formulated in terms of inequalities.

15.4 Corollary

Let p, q, s be given by (15.4) and let $0 < \varepsilon < 1$.

(i) *Then*

$$\left(\int_0^\varepsilon (|\log t|^a \, f^*(t))^r \, dt \right)^{\frac{1}{r}} \leq c \, \|f \,|F_{pq}^s(\mathbb{R}^n)\| \tag{15.22}$$

for some $c > 0$ and all $f \in F_{pq}^s(\mathbb{R}^n)$ if, and only if, $a \leq 0$.

(ii) *Let, in addition, $0 < q \leq r$. Then*

$$\left(\int_0^\varepsilon (|\log t|^a \, f^*(t))^r \, dt \right)^{\frac{1}{r}} \leq c \, \|f \,|B_{pq}^s(\mathbb{R}^n)\| \tag{15.23}$$

for some $c > 0$ and all $f \in B_{pq}^s(\mathbb{R}^n)$ if, and only if, $a \leq 0$.

(iii) *Let, in addition, $r < q \leq \infty$. Then (15.23) holds if, and only if, $a < \frac{1}{q} - \frac{1}{r}$.*

Proof *Step 1* If $a = 0$, then (15.22) and (15.23) with $q \leq r$ follow from Theorem 11.4(i). Let $r < q$. Then (15.23) with $a < \frac{1}{q} - \frac{1}{r}$ is a consequence of (11.66) and (15.7). This covers all if-parts.

Step 2 It remains to prove the only-if-parts of the corollary. First we insert f_j, given by (15.11) in (15.22) and (15.23). By the equivalence in (15.12) we get

$$j^a \leq c \quad \text{for all} \quad j \in \mathbb{N} \quad \text{and some} \quad c > 0. \tag{15.24}$$

Hence $a \leq 0$. This completes the proof of (i) and (ii). As for (iii) we modify (15.16) by

$$f(x) = \sum_{j=2}^{\infty} b_j \, 2^{j\frac{n}{r}} \, \psi(2^j x - x^0) \quad \text{with} \quad b_j = j^{-\frac{1}{q}} (\log j)^{-\frac{1}{r}}. \tag{15.25}$$

This is again an atomic decomposition and we have

$$\|f \,|\, B^s_{pq}(\mathbb{R}^n)\| \leq c \left(\sum_{j=2}^{\infty} b_j^q \right)^{\frac{1}{q}} < \infty. \tag{15.26}$$

On the other hand, (15.17) must be modified by

$$f^*(d \, 2^{-jn}) \sim j^{-\frac{1}{q}} (\log j)^{-\frac{1}{r}} 2^{j\frac{n}{r}}, \quad j = 2, 3, \ldots. \tag{15.27}$$

Inserted in the left-hand side of (15.23) with $a = \frac{1}{q} - \frac{1}{r}$ we obtain

$$\int_0^\varepsilon (|\log t|^a f^*(t))^r \, dt \sim \sum_{j=2}^{\infty} j^{-1} (\log j)^{-1} = \infty. \tag{15.28}$$

This proves the only-if-part of (iii).

15.5 Further references

Embeddings and related inequalities in sub-critical cases have a long and rich history. We tried to collect the relevant papers in 11.8(i), with respect to the Lorentz spaces L_{rq}, and in 11.8(iii), with respect to the Zygmund spaces $L_r(\log L)_a$. This will not be repeated here. In connection with the critical case we gave some additional references in 13.5, which apply at least partly also to the sub-critical case considered here. In particular, Ju. V. Netrusov anticipated in [Net87a], and also in [Net89a], in a somewhat different context, the concept of envelope functions and optimal embeddings in rearrangement-invariant spaces. More recent (and independent of each other and of Netrusov's work) treatments have been given in [CwP98] and in [EKP00] (restricted to Sobolev spaces, in contrast to Netrusov, who considered F^s_{pq} spaces). As for related capacity estimates we refer again to [Maz85] and to the recent paper [Sic99], where one finds also further references. This section is based on [Tri99d] and might be considered as an improved and extended version.

16 Hardy inequalities

16.1 Introduction

In this book we dealt so far several times with Hardy inequalities. But first we wish to mention that the whole story began with Hardy's note [Had28] and the famous Theorem 330 in [HLP52], p. 245 (in small print). As a consequence (ignoring constants) one gets the following assertion: *Let $1 < p < \infty$ and $m \in \mathbb{N}$. There is a number $c > 0$ such that*

$$\int_\mathbb{R} |t|^{-mp} |u(t)|^p \, dt \le c \int_\mathbb{R} \left| \frac{d^m u(t)}{dt^m} \right|^p dt \qquad (16.1)$$

for all

$$u \in S(\mathbb{R}) \quad \text{with} \quad \frac{d^j u}{dt^j}(0) = 0 \quad \text{for} \quad j = 0, \ldots, m-1 \, .$$

In the years after, and especially in the last decades, hundreds of papers and dozens of books have appeared dealing with numerous variations of inequalities of this type. The reader may consult [OpK90] and the references given there. As far as this book is concerned we refer to 5.7–5.12, making clear how different natural inequalities for F-spaces and B-spaces might be. Of special interest in this section is the following consequence of the previous results. Let Ω be a bounded C^∞ domain in $\mathbb{R}^n$. Let

$$\Gamma = \partial\Omega, \quad D(x) = dist\,(x, \Gamma) = \inf_{y \in \Gamma} |x - y|, \quad x \in \mathbb{R}^n, \qquad (16.2)$$

be the distance to Γ and, for $\varepsilon > 0$,

$$\Gamma_\varepsilon = \{x \in \mathbb{R}^n \,:\, D(x) < \varepsilon\} \qquad (16.3)$$

be a neighbourhood of Γ. Let

$$0 < p < \infty, \quad 0 < q \le \infty, \quad n\left(\frac{1}{p} - 1\right)_+ = \sigma_p < s < \frac{1}{p}. \qquad (16.4)$$

There is a number $c > 0$ such that

$$\int_{\Gamma_\varepsilon} D^{-sp}(x) |f(x)|^p \, dx \le c \, \|f \,|\, F^s_{pq}(\mathbb{R}^n)\|^p \qquad (16.5)$$

for all $f \in F_{pq}^s(\mathbb{R}^n)$. This follows from (5.104), (5.105). There one finds also the necessary explanations and further assertions of this type. This measures how singular a function f belonging to $F_{pq}^s(\mathbb{R}^n)$ near $\Gamma = \partial\Omega$ can be. Let Γ be an arbitrary, say, compact set on $\mathbb{R}^n$. Of interest is the behaviour of functions f belonging to a given space $F_{pq}^s(\mathbb{R}^n)$ or $B_{pq}^s(\mathbb{R}^n)$ near or at Γ. There are two different, but closely related aspects: traces on Γ and Hardy inequalities of type (16.5). If Γ is smooth (maybe $\Gamma = \partial\Omega$ as above) then the trace problem is more or less settled and treated in detail in most of the books mentioned in 1.1. Specific references and rather final formulations and proofs (excluding borderline cases) may be found in [Triβ], 2.7.2, 3.3.3, pp. 132, 200, and [Triγ], 4.4.2, 4.4.3, pp. 213–221. Sophisticated borderline cases have been treated recently in [Joh00] and [FJS00]. If Γ is an irregular, say, compact, set in $\mathbb{R}^n$ then the situation is different. We considered this problem in some detail in Section 9 and refer in particular to Theorems 9.3, 9.9, 9.21, and 9.33. There we quoted also the relevant literature. Special attention has been paid to d-sets. Of interest here is Proposition 9.13. One aim in the present section is to complement these trace assertions by a discussion about related Hardy inequalities. We outlined our intentions at the end of Section 10 and added also a warning concerning the outcome. As stated there we are interested with some preference in $\Gamma = \{0\}$, where we get sharp results.

But we look also at more general sets. In principle the method to get, for example (10.26) or (10.28), is quite simple. We use (10.23) as a vehicle to reduce Hardy inequalities to Theorems 13.2 and 15.2, and the related inequalities in 13.3 and 15.3, respectively. A few points should be mentioned.

First, in case of $\Gamma = \{0\}$ we deal both with F_{pq}^s-spaces and B_{pq}^s-spaces, although really satisfactory inequalities for the B-spaces look somewhat different. We refer to (5.77). This may justify that we later on concentrate on the F-spaces.

Secondly, if Γ is the boundary of a C^∞ domain or (part of) a hyper-plane and if one deals with the full spaces $F_{pq}^s(\mathbb{R}^n)$ or $B_{pq}^s(\mathbb{R}^n)$ (and not with appropriate subspaces) then there seems to be a clear distinction between those spaces having traces on Γ and those spaces with substantial Hardy inequalities. But if Γ is irregular the situation might be different. As examples, (16.5) may serve on the one hand and (10.26) on the other hand. In case of irregular compact sets Γ we have no final assertions, and the later parts of this section might be considered as a discussion of how to shed light on the possibly tricky interplay between Hardy inequalities, the geometry of irregular sets Γ and related measures. This justifies our restriction to examples, mostly $F_{pq}^{\frac{n}{p}}(\mathbb{R}^n)$.

We complement (16.3) by

$$K_\varepsilon = \{x \in \mathbb{R}^n : |x| < \varepsilon\}. \tag{16.6}$$

16.2 Theorem

(Critical case)

Let $0 < \varepsilon < 1$ and let $\varkappa(t)$ be a positive monotonically decreasing function on $(0, \varepsilon]$.

(i) Let
$$1 < p < \infty \quad \text{and} \quad 0 < q \leq \infty. \tag{16.7}$$

Then
$$\int_{K_\varepsilon} \left| \frac{\varkappa(|x|) f(x)}{\log |x|} \right|^p \frac{dx}{|x|^n} \leq c \left\| f \, | F_{pq}^{\frac{n}{p}}(\mathbb{R}^n) \right\|^p \tag{16.8}$$

for some $c > 0$ and all $f \in F_{pq}^{\frac{n}{p}}(\mathbb{R}^n)$ if, and only if, $\varkappa$ is bounded.

(ii) Let
$$0 < p < \infty \quad \text{and} \quad 1 < q < \infty. \tag{16.9}$$

Then
$$\int_{K_\varepsilon} \left| \frac{\varkappa(|x|) f(x)}{\log |x|} \right|^q \frac{dx}{|x|^n} \leq c \left\| f \, | B_{pq}^{\frac{n}{p}}(\mathbb{R}^n) \right\|^q \tag{16.10}$$

for some $c > 0$ and all $f \in B_{pq}^{\frac{n}{p}}(\mathbb{R}^n)$ if, and only if, $\varkappa$ is bounded.

Proof *Step 1* We prove (16.8) with $\varkappa = 1$. We may assume that the $\varepsilon > 0$ is so small that
$$a(x) = \left| |x|^{\frac{n}{p}} \log |x| \right|^{-1} \quad \text{is monotone in } K_\varepsilon$$

and, hence,
$$a^*(t) \sim t^{-\frac{1}{p}} |\log t|^{-1} \tag{16.11}$$

if $0 < t < c\varepsilon$ for some $c > 0$. Recall that $a^*(t)$ is the measure-preserving rearrangement of $a(x)$. Then (16.11) follows from the behaviour of $a(x)$ and $a^*(t)$ at $|x| \sim 2^{-\frac{j}{n}}$ and $t \sim 2^{-j}$, respectively, where $j \in \mathbb{N}$ is sufficiently large. We obtain

$$\int_{K_\varepsilon} \left| \frac{f(x)}{\log |x|} \right|^p \frac{dx}{|x|^n} = \int_{K_\varepsilon} a(x)^p |f(x)|^p \, dx = \int_0^{c\varepsilon^n} (af)^{*p}(t) \, dt$$

$$\leq \int_0^{c\varepsilon^n} a^{*p}(t) f^{*p}(t) \, dt \sim \int_0^{c\varepsilon^n} \left| \frac{f^*(t)}{\log t} \right|^p \frac{dt}{t} \leq c' \left\| f \, | F_{pq}^{\frac{n}{p}}(\mathbb{R}^n) \right\|^p. \tag{16.12}$$

The first inequality is a well-known property of rearrangement and may be found in [BeS88], p. 44. It goes back to [HLP52] (first edition 1934), Theorems 368 and 378. The last inequality comes from Theorem 13.2 or, more explicitly, from (13.62). Similarly one proves (16.10) with $\varkappa = 1$, where one has to use (13.57).

Step 2 We prove that $\varkappa$ in (16.8) must be bounded. Let $f(x)$ be a positive monotonically decreasing function in K_ε in radial directions. Since $\varkappa$ is also assumed to be monotone it follows in analogy to (13.62) by (12.28) and (16.8),

$$\sup_{0<t<c\varepsilon} \frac{\varkappa(t) \, f^*(t)}{|\log t|^{\frac{1}{p'}}} \leq c_1 \left(\int_0^{c\varepsilon} \left(\frac{\varkappa(t) \, f^*(t)}{|\log t|} \right)^p \frac{dt}{t} \right)^{\frac{1}{p}} \tag{16.13}$$

$$= c_1 \left(\int_{K_\varepsilon} \left| \frac{\varkappa(|x|) \, f(x)}{\log |x|} \right|^p \frac{dx}{|x|^n} \right)^{\frac{1}{p}} \leq c_2 \left\| f \, | F_{pq}^{\frac{n}{p}}(\mathbb{R}^n) \right\|.$$

Let $q \geq 1$. We insert f_J with (13.50), (13.51). This proves that $\varkappa$ must be bounded. If $q > 0$ is small, then one has to modify f_J as indicated in Steps 5 and 6 of the proof of Theorem 13.2. But this does not influence the above argument. Similarly one proves that $\varkappa$ in (16.10) must be bounded.

16.3 Theorem

(Sub-critical case)
Let $\varepsilon > 0$ and let $\varkappa(t)$ be a positive monotonically decreasing function on $(0, \varepsilon]$. Let

$$s > 0 \quad \text{and} \quad s - \frac{n}{p} = -\frac{n}{r} \quad \text{with} \quad 1 < r < \infty \tag{16.14}$$

(the dashed line in Fig. 10.1).
(i)   Let $0 < q \leq \infty$. Then

$$\int_{K_\varepsilon} \left| \varkappa(|x|) \, |x|^{\frac{n}{r}} f(x) \right|^p \frac{dx}{|x|^n} \leq c \, \| f \, | F_{pq}^s(\mathbb{R}^n) \|^p \tag{16.15}$$

for some $c > 0$ and all $f \in F_{pq}^s(\mathbb{R}^n)$ if, and only if, $\varkappa$ is bounded.
(ii)   Let $0 < q \leq r$. Then

$$\int_{K_\varepsilon} \left| \varkappa(|x|) \, |x|^{\frac{n}{r}} f(x) \right|^q \frac{dx}{|x|^n} \leq c \, \| f \, | B_{pq}^s(\mathbb{R}^n) \|^q \tag{16.16}$$

for some $c > 0$ and all $f \in B_{pq}^s(\mathbb{R}^n)$ if, and only if, $\varkappa$ is bounded.

Proof Let $\alpha \geq 0$. Then

$$\left(|x|^{-\alpha n}\right)^*(t) \sim t^{-\alpha} \quad \text{where} \quad t > 0. \tag{16.17}$$

This can be applied to the left-hand sides of (16.15) and (16.16) with $\alpha = 1 - \frac{p}{r} > 0$ and $\alpha = 1 - \frac{q}{r} \geq 0$, respectively. Then (16.15) and (16.16) with $\varkappa = 1$ follow from the counterpart of (16.12) on the one hand, and (15.21) with $u = p$ and (15.19) with $u = q$, respectively, on the other hand. If one inserts $f_j(x)$ given by (15.11) (with the indicated modification for the F-spaces when $q > 0$ is small) in (16.15) and (16.16), then it follows that $\varkappa$ must be bounded.

16.4 Comments and references

First we look at the sub-critical case. Using (16.14), the inequality (16.15) can be reformulated as

$$\int_{K_\varepsilon} \frac{|f(x)|^p}{|x|^{sp}}\, dx \leq c\, \|f\,|F_{pq}^s(\mathbb{R}^n)\|^p, \tag{16.18}$$

where again $0 < q \leq \infty$. As mentioned in 1.2 if $q = 2$ and $1 < p < \infty$ then $F_{pq}^s(\mathbb{R}^n)$ are the Sobolev spaces $H_p^s(\mathbb{R}^n)$ with the classical Sobolev spaces $W_p^s(\mathbb{R}^n)$ as a subclass if, in addition, $s \in \mathbb{N}$. Then inequalities of type (16.18) are known although explicit formulations are rare in the literature (especially in higher dimensions). But everything is included in the extensively treated problem of embeddings of Sobolev spaces in weighted L_p spaces, or more generally in L_p spaces with respect to Radon measures in $\mathbb{R}^n$. We dealt in Section 9 with questions of this type in the different context of traces. But the references given there apply also to the above case, in particular [Maz85], [AdH96], [Ver99]. Switching to general spaces $F_{pq}^s(\mathbb{R}^n)$ and $B_{pq}^s(\mathbb{R}^n)$ the situation is different. The first explicit inequality of type (16.18) for the general spaces $F_{pq}^s(\mathbb{R}^n)$ may be found in [Triβ], 2.8.6, p. 155, which covers also the one-dimensional version of (16.5). Such inequalities also have anisotropic counterparts, at least for anisotropic spaces of type $B_{pp}^s(\mathbb{R}^n)$ and $H_p^s(\mathbb{R}^n)$. We refer to [ST87], 4.3, pp. 202–209, and the literature mentioned there. If $p \neq q$ then inequalities of type (16.16) are not optimally adapted to the spaces $B_{pq}^s(\mathbb{R}^n)$. More natural inequalities may be found in [Triα], p. 319, and more general ones in [Tri99b]. But they do not fit in our scheme here. The above Theorem 16.3 is a modification of [Tri99d]. There one finds also additional discussions concerning the interrelation of rearrangement and Hardy inequalities. In the critical case as considered in Theorem 16.2 there are only very few papers. Restricted to classical Sobolev spaces $W_2^{\frac{n}{2}}(\mathbb{R}^n)$ inequalities of type (16.8) with log-terms may be found in [EgK90], Lemma 8, p. 155, and in [Sol94]. Restricted to $H_p^s(\mathbb{R}^n)$ and $B_{pp}^s(\mathbb{R}^n)$, Theorem 16.2 has been proved in [EdT99b].

We reduced the inequalities in the Theorems 16.2 and 16.3 to 13.2, 13.3 and 15.2, 15.3, respectively. It is clear that all the other inequalities mentioned there in 13.3 and 15.3 produce also sharp Hardy inequalities: One has to modify (16.12). Another possibility is to replace $\Gamma = \{0\}$ by more general sets. In principle this does not cause much trouble. But it is unclear to what extent or for which Γ one gets sharp and natural inequalities. We formulate a few results and complement them by some discussions. It comes out that under some additional geometrical restrictions the outcome is far from being optimal. In other words, the main aim of the rest of this section is to shed light on these problems. This may also justify that we restrict our attention to the critical case and in particular to $F_{pq}^{\frac{n}{p}}(\mathbb{R}^n)$. The first candidates beyond $\Gamma = \{0\}$ and, maybe, compact smooth surfaces are d-sets. Let $0 < d < n$. Then a compact set Γ in $\mathbb{R}^n$ is called a d-set if there are a Borel measure μ in $\mathbb{R}^n$ and two positive numbers c_1 and c_2 such that $supp\,\mu = \Gamma$ and

$$c_1 t^d \leq \mu(B(\gamma, t)) \leq c_2 t^d \quad \text{for all} \quad 0 < t < 1, \tag{16.19}$$

and $\gamma \in \Gamma$, where $B(\gamma, t)$ is a ball centred at $\gamma \in \Gamma$ and of radius t. Further details and references may be found in 9.12.

16.5 Proposition

Let $0 < d < n$ and let Γ be a compact d-set in $\mathbb{R}^n$. Let

$$D(x) = dist(x, \Gamma), \quad x \in \mathbb{R}^n, \tag{16.20}$$

be the distance of $x \in \mathbb{R}^n$ to Γ. Let p, q be given by (16.7). Let $0 < \varepsilon < 1$ and let Γ_ε be an ε-neighbourhood of Γ as in (16.3). Then

$$\int_{\Gamma_\varepsilon} \left| \frac{f(x)}{\log D(x)} \right|^p \frac{dx}{D^{n-d}(x)} \leq c \left\| f \,|\, F_{pq}^{\frac{n}{p}}(\mathbb{R}^n) \right\|^p \tag{16.21}$$

for some $c > 0$ and all $f \in F_{pq}^{\frac{n}{p}}(\mathbb{R}^n)$.

Proof Let

$$\Gamma_j = \left\{ x \in \mathbb{R}^n \,:\, 2^{-\frac{j+1}{n-d}} < D(x) \leq 2^{-\frac{j}{n-d}} \right\}, \quad j \geq J. \tag{16.22}$$

Then $vol\,\Gamma_j \sim 2^{-j}$. With

$$a(x) = |\log D(x)|^{-p} D^{d-n}(x)$$

one gets

$$a^*(t) \sim t^{-1} |\log t|^{-p}, \quad 0 < t < \delta < 1. \tag{16.23}$$

Now we obtain (16.21) in the same way as in (16.12).

16.6 Discussion

Let Γ be a hyper-plane in $\mathbb{R}^n$, say,

$$\Gamma = \mathbb{R}^{n-1} = \{x = (x', x_n) \in \mathbb{R}^n \,,\; x_n = 0\}\,, \qquad (16.24)$$

with $x' \in \mathbb{R}^{n-1}$ and $n \geq 2$. Let p, q be given by (16.7). For fixed $x' \in \mathbb{R}^{n-1}$ we use the one-dimensional version of (16.8) and obtain for $0 < \varepsilon < 1$,

$$\int_{-\varepsilon}^{\varepsilon} \left| \frac{f(x', x_n)}{\log |x_n|} \right|^p \frac{dx}{|x_n|} \leq c \left\| f(x', \cdot) \,|\, F_{pq}^{\frac{1}{p}}(\mathbb{R}) \right\|^p, \qquad x' \in \mathbb{R}^{n-1}. \qquad (16.25)$$

If $1 < p < \infty$ and $1 \leq q \leq \infty$, then by Theorem 4.4, the spaces $F_{pq}^{\frac{1}{p}}(\mathbb{R}^n)$ have the Fubini property. Together with (16.25) one obtains

$$\int_{\mathbb{R}_\varepsilon^{n-1}} \left| \frac{f(x)}{\log |x_n|} \right|^p \frac{dx}{|x_n|} \leq c \left\| f \,|\, F_{pq}^{\frac{1}{p}}(\mathbb{R}^n) \right\|^p \qquad (16.26)$$

for some $c > 0$ and all $f \in F_{pq}^{\frac{1}{p}}(\mathbb{R}^n)$, where $\mathbb{R}_\varepsilon^{n-1}$ is an ε-neighbourhood of $\mathbb{R}^{n-1}$ given by (16.24) according to (16.3). Since for fixed p with $1 < p < \infty$, the spaces $F_{pq}^{\frac{1}{p}}(\mathbb{R}^n)$ are monotone with respect to q, the inequality (16.26) holds for all p, q with (16.7). Even the $\varkappa$-sharpness of Theorem 16.2 extends from the one-dimensional case to the above situation:

Let $\varkappa(t)$ be a positive monotonically decreasing function on $(0, \varepsilon]$, where again $0 < \varepsilon < 1$, let p, q be given by (16.7). Then

$$\int_{\mathbb{R}_\varepsilon^{n-1}} \left| \frac{\varkappa(|x_n|) f(x)}{\log |x_n|} \right|^p \frac{dx}{|x_n|} \leq c \left\| f \,|\, F_{pq}^{\frac{1}{p}}(\mathbb{R}^n) \right\|^p \qquad (16.27)$$

for some $c > 0$ and all $f \in F_{pq}^{\frac{1}{p}}(\mathbb{R}^n)$ if, and only if, $\varkappa$ is bounded.

The if-part is covered by (16.26). We outline how the only-if-part can be proved by modification of previous arguments. Let

$$S_j = \{x \in \mathbb{R}^n \,:\, |x'| < 1,\, |x_n| < 2^{-j}\}, \qquad j \in \mathbb{N}. \qquad (16.28)$$

We modify (13.50) by

$$f_J(x) = J^{-\frac{1}{p}} \sum_{j=1}^{J} \sum_{l=1}^{2^{j(n-1)}} 2^{-j\frac{n-1}{p}} \left[2^{j\frac{n-1}{p}} \psi \left(2^{j-1}(x - x^{j,l}) \right) \right], \qquad (16.29)$$

where $[\cdots]$ are correctly normalized atoms or quarks in $F_{pq}^{\frac{1}{p}}(\mathbb{R}^n)$ (we refer to (2.16)) and where $x^{j,l}$ stands for suitable lattice-points. (We assume, say,

$q \geq 1$, such that no moment conditions are needed. The necessary additional modifications if $q > 0$ is small have been indicated in Step 5 of the proof of Theorem 13.2.) We have a counterpart of (13.51) with $n = 1$ in the first equivalence and with $F_{pq}^{\frac{1}{p}}(\mathbb{R}^n)$ in place of $F_{pq}^{\frac{n}{p}}(\mathbb{R}^n)$ in the second equivalence: We note that the arguments in connection with and after (13.40) with a reference to 2.15 apply also to (16.29). Then the desired $\varkappa$-sharpness follows as in Step 2 of the proof of Theorem 16.2. If one compares the sharp assertion (16.26) with (16.21) where now $d = n - 1$, then it is quite clear that in this special case, (16.21) does not say very much. Even worse: Since for any $\delta > 0$ the space $F_{pq}^{\frac{1}{p}+\delta}(-\varepsilon, \varepsilon)$ is continuously embedded in $C(-\varepsilon, \varepsilon)$ (in obvious notation and with a reference to, say, [Triβ], 2.7.1) one has an immediate and rather obvious counterpart of (16.26) with $F_{pq}^{\frac{1}{p}+\delta}(\mathbb{R}^n)$ in place of $F_{pq}^{\frac{1}{p}}(\mathbb{R}^n)$ and with an arbitrary positive integrable function $a(x_n)$ in place of $|x_n|^{-1}|\log|x_n||^{-p}$. Then, in this special case, (16.21) with $n \geq 2$ is obvious. On the other hand, the above arguments depend on the special structure of Γ in (16.24) and on the possibility to apply the Fubini Theorem 4.4. But this is not the case if Γ is a general d-set or an arbitrary fractal. In other words, the problem arises under which geometrical conditions for Γ the inequality (16.21) is substantial and sharp. Finally one can use (16.26) to complement our considerations in 5.23 and also of (16.5). We formulate the outcome.

16.7 Corollary

Let Ω be a bounded C^∞ domain in $\mathbb{R}^n$ and let Γ, Γ_ε, and $D(x)$ be given by (16.2), (16.3). Let $0 < \varepsilon < 1$ and let $\varkappa(t)$ be a positive monotonically decreasing function on $(0, \varepsilon]$. Let p, q be given by (16.7). Then

$$\int_{\Gamma_\varepsilon} \left| \frac{\varkappa(D(x)) f(x)}{\log D(x)} \right|^p \frac{dx}{D(x)} \leq c \left\| f \, | F_{pq}^{\frac{1}{p}}(\mathbb{R}^n) \right\|^p \tag{16.30}$$

for some $c > 0$ and all $f \in F_{pq}^{\frac{1}{p}}(\mathbb{R}^n)$ if, and only if, $\varkappa$ is bounded.

Proof This follows from (16.27) and standard localization arguments.

16.8 Remark

If p, q, s are given by (5.104), then we have the sharp Hardy inequality (5.105). If now p, q are restricted by (16.7) and $s = \frac{1}{p}$, then

$$\int_\Omega \left| \frac{f(x)}{1 + |\log D(x)|} \right|^p \frac{dx}{D(x)} \leq c \left\| f \, | F_{pq}^{\frac{1}{p}}(\Omega) \right\|^p. \tag{16.31}$$

This is an immediate consequence of (16.30).

16.9 Proposition

Let μ be a finite Radon measure in $\mathbb{R}^n$ and let $\Gamma = \operatorname{supp}\mu$ be compact. Let p, q be given by (16.7), $0 < \varepsilon < 1$, and

$$I_{p,\varepsilon}(x) = \int_{B(x,\varepsilon)} \frac{\mu(d\gamma)}{|x-\gamma|^n |\log|x-y||^p} \leq \infty, \quad x \in \mathbb{R}^n, \tag{16.32}$$

where $B(x,\varepsilon)$ is a ball centred at $x \in \mathbb{R}^n$ and of radius ε. Then

$$\int_{\mathbb{R}^n} I_{p,\varepsilon}(x)\, |f(x)|^p\, dx \leq c\, \left\| f\, |F_{pq}^{\frac{n}{p}}(\mathbb{R}^n) \right\|^p \tag{16.33}$$

for some $c > 0$ and all $f \in F_{pq}^{\frac{n}{p}}(\mathbb{R}^n)$.

Proof Let χ_ε be the characteristic function of K_ε given by (16.6). Let $\gamma \in \Gamma$. Then it follows by (16.8) that

$$\int_{\mathbb{R}^n} \left| \frac{f(x)}{\log|x-\gamma|} \right|^p \chi_\varepsilon(x-\gamma) \frac{dx}{|x-\gamma|^n} \leq c\, \left\| f\, |F_{pq}^{\frac{n}{p}}(\mathbb{R}^n) \right\|^p. \tag{16.34}$$

Integration with respect to μ and application of Fubini's theorem results in (16.33).

16.10 Remark

As mentioned at the end of 16.4 the Propositions 16.5 and 16.9 are far from final. This is also clear from the discussion in 16.6 and the more satisfactory assertions in 16.7 and 16.8. We mainly wanted to make clear that there might be a sophisticated interplay between the geometry of irregular fractal sets Γ and the singularity behaviour of functions belonging to spaces $F_{pq}^s(\mathbb{R}^n)$ and $B_{pq}^s(\mathbb{R}^n)$ near Γ. We restricted ourselves in the course of this discussion to the critical case extending Theorem 16.2. But of course one can deal in the same way with the sub-critical case as considered in Theorem 16.3.

17 Complements

17.1 Green's functions as envelope functions

Looking at (13.7) or (13.9) one may ask whether there are functions f belonging to $B_{pq}^{\frac{n}{p}}(\mathbb{R}^n)$ or $F_{pq}^{\frac{n}{p}}(\mathbb{R}^n)$ such that $f^*(t)$ is equivalent to $|\log t|^{\frac{1}{q'}}$ or $|\log t|^{\frac{1}{r'}}$, respectively. If $q < \infty$ in (13.7), then it follows from (13.57) that this is impossible since in such a case the middle term diverges. Because always $p < \infty$, one has by (13.62) a corresponding argument for the spaces $F_{pq}^{\frac{n}{p}}(\mathbb{R}^n)$. Similarly for the sub-critical case according to Theorem 15.2 and (15.7). Corresponding questions can also be asked for the super-critical case considered in Theorem 14.2. If $q = \infty$ then the situation is different. We deal first with the critical case as covered by Theorem 13.2 and by 13.3. Let δ be the usual δ-distribution in $\mathbb{R}^n$ with the origin as the off-point. Then

$$\delta \in B_{p\infty}^{n(\frac{1}{p}-1)}(\mathbb{R}^n) \quad \text{where} \quad 0 < p \leq \infty. \tag{17.1}$$

This is well known and also an easy consequence of (2.37). Let again $-\Delta$ be the Laplacian in $\mathbb{R}^n$. By well-known lifting properties of $-\Delta + id$ it follows that

$$G = (id - \Delta)^{-\frac{n}{2}} \delta \in B_{p\infty}^{\frac{n}{p}}(\mathbb{R}^n), \quad 0 < p \leq \infty, \tag{17.2}$$

where G might be considered as the Green's function of the fractional power $(id - \Delta)^{\frac{n}{2}}$ of $id - \Delta$. We claim that $G(x)$ is a C^∞ function in $\mathbb{R}^n \setminus \{0\}$ which decays exponentially if $|x| \to \infty$ and

$$G(x) \sim |\log |x|| \quad \text{if} \quad |x| < \varepsilon \quad \text{and hence} \quad G^*(t) \sim |\log t| \tag{17.3}$$

if $0 < t < \varepsilon < 1$. Hence $G(x)$ is an extremal function for $B_{p\infty}^{\frac{n}{p}}(\mathbb{R}^n)$. By Definition 12.8 and (12.60), and in agreement with (13.7), (13.58) we have

$$\mathcal{E}_G B_{p\infty}^{\frac{n}{p}}(t) = G^*(t) \sim |\log t|, \quad 0 < t < \varepsilon < 1. \tag{17.4}$$

We outline a proof. Let

$$g(x) = \int_0^\infty e^{-t-\frac{|x|^2}{4t}} \frac{dt}{t}, \quad x \in \mathbb{R}^n, \quad x \neq 0. \tag{17.5}$$

By well-known properties of the Fourier transform of $e^{-\frac{|x|^2}{2}}$ and with respect to dilations $x \to cx$, $c \neq 0$, e.g., [Tri92], pp. 100/101, it follows that

$$\begin{aligned}(Fg)(\xi) &= \int_0^\infty e^{-t} F\left(e^{-\frac{|x|^2}{4t}}\right)(\xi) \frac{dt}{t} \\ &= \int_0^\infty (2t)^{\frac{n}{2}} e^{-t} e^{-t|\xi|^2} \frac{dt}{t} = c(1+|\xi|^2)^{-\frac{n}{2}}\end{aligned} \tag{17.6}$$

for some $c > 0$. Here we used the Fubini theorem. This is possible since an integration over $\mathbb{R}^n$ in (17.5) results in a convergent integral over $\mathbb{R}^n \times [0, \infty)$. Application of the Fourier transform to G, introduced in (17.2), gives

$$G(x) = c \int_0^\infty e^{-t - \frac{|x|^2}{4t}} \frac{dt}{t}, \qquad x \in \mathbb{R}^n, \quad x \neq 0, \tag{17.7}$$

for some $c > 0$. We estimate $G(x)$. Let $|x| \geq 1$. We split the integral in (17.7) in

$$\int_0^{|x|} e^{-t - \frac{|x|^2}{4t}} \frac{dt}{t} \leq \int_0^{|x|} e^{-\frac{|x|^2}{4t}} \frac{dt}{t} = c \int_{\frac{|x|}{4}}^\infty e^{-\tau} \frac{d\tau}{\tau} \leq c' e^{-\frac{|x|}{4}} \tag{17.8}$$

and in

$$\int_{|x|}^\infty e^{-|x|} e^{-(\sqrt{t} - \frac{|x|}{2\sqrt{t}})^2} \frac{dt}{t} \leq \int_{|x|}^\infty e^{-|x|} e^{-tc} \frac{dt}{t} \leq c' e^{-c''|x|}. \tag{17.9}$$

This proves the exponential decay of $G(x)$ if $|x| \to \infty$. (Of course all constants in the above estimate are positive.) Let $|x| > 0$ be small. Then by (17.7) and $t = |x|^2 \tau$,

$$G(x) \sim 1 + \int_0^1 e^{-t - \frac{|x|^2}{4t}} \frac{dt}{t} \sim 1 + \int_0^1 e^{-\frac{|x|^2}{4t}} \frac{dt}{t}$$

$$\sim 1 + \int_0^{|x|^{-2}} e^{-\frac{1}{4\tau}} \frac{d\tau}{\tau} \sim |\log |x||. \tag{17.10}$$

Hence by (17.2), the decay assertions, and (17.3) it follows that $G(x)$ materializes the envelope function in (13.7) with $q = \infty$. Furthermore by (17.2) and (13.87), the Green's function G belongs also to $F^0_{\infty q}(\mathbb{R}^n)$ for all $0 < q \leq \infty$, in particular

$$G \in bmo(\mathbb{R}^n).$$

Finally we mention that (17.3) in case of $n = 2$ is essentially the well-known behaviour of the Green's function of the Laplacian in the plane.

The super-critical case can be reduced to the critical one as follows. Let g be given by (17.5). Then it comes out that

$$h(x) = x_1 g(x) \in B_{p\infty}^{1+\frac{n}{p}}(\mathbb{R}^n), \quad 0 < p \leq \infty, \tag{17.11}$$

and
$$|\nabla h|^*(t) \sim \widetilde{w}(h,t) \sim |\log t|, \quad 0 < t < \varepsilon, \tag{17.12}$$

where $\widetilde{w}(h,t)$ are the divided differences introduced in (12.72). In particular, h is an extremal function according to (14.27) and we have by Definition 12.14, (12.85) and (14.5),

$$\mathcal{E}_C B_{p\infty}^{1+\frac{n}{p}}(t) = \widetilde{w}(h,t) \sim |\log t|. \tag{17.13}$$

We outline a proof. By similar arguments as in (17.6) it follows that

$$h(x) = c \frac{\partial}{\partial x_1} \int_0^\infty e^{-t-\frac{|x|^2}{4t}} dt = c' \frac{\partial}{\partial x_1} (id - \Delta)^{-\frac{n}{2}-1} \delta \in B_{p\infty}^{1+\frac{n}{p}}(\mathbb{R}^n). \tag{17.14}$$

Furthermore, as in (17.10) we obtain

$$\frac{\partial h}{\partial x_1} = g(x) + c x_1^2 \int_0^\infty e^{-t-\frac{|x|^2}{4t}} \frac{dt}{t^2} \sim |\log |x|| + \frac{x_1^2}{|x|^2}. \tag{17.15}$$

This proves (17.12) (one needs only an estimate from below, since the estimate from above is covered by (14.27)).

Finally, formulas like (17.7) originate from heat kernels and their relations to Green's functions. We refer to [Dav89], 3.4, pp. 99–105, for details.

17.2 Further limiting embeddings

In all three cases, critical, super-critical, sub-critical, treated in Sections 13, 14, 15, respectively, we avoided borderline situations. This means in the critical and sub-critical case spaces with parameters as described in (13.2), (13.3) as far as they are covered by Theorem 11.2. In the super-critical case we excluded $p = \infty$ in the source spaces and concentrated on the target spaces exclusively on $s = 1$, $p = \infty$ (the dotted line in Fig. 10.1). This might be justified by the history of the topic which we tried to collect in Theorem 11.7 and on which we commented in 11.8. It would be of interest to have a closer look at these omitted spaces. In the context of a more systematic study it might be even reasonable to modify the sub-division of the spaces covered by Theorem 11.2 in the above three distinguished cases as follows:

(i) to extend the *sub-critical* case as described in (10.5) to those spaces with $s = \sigma_p = n(\frac{1}{p} - 1)_+$ covered by Theorem 11.2,

(ii) to extend the *critical* case as described in (10.6) to those spaces with $p = \infty$ covered by Theorem 11.2, and

(iii) to call all other spaces covered by Theorem 11.2 *super-critical*.

Any subdivision of the spaces covered by Theorem 11.2 depends on the admitted target spaces. This means in the sub-critical and critical case spaces with $s = 0$ according to Fig. 10.1 and in the super-critical case $s = 1$, $p = \infty$. A somewhat more general case of interest in connection with target spaces is given by

$$s = 1, \quad 1 < p \leq \infty, \tag{17.16}$$

(again in the understanding of Fig. 10.1). We give a brief description of the set-up in a slightly more general context. We use standard notation. Let $m \in \mathbb{N}$ and $1 < p \leq \infty$. Then

$$\omega_m(f,t)_p = \sup_{|h| \leq t} \|\Delta_h^m f \,|L_p(\mathbb{R}^n)\|, \quad 0 < t < \infty, \tag{17.17}$$

is the usual mth order modulus of continuity, [BeS88], 5.3, p. 332 or [DeL93], 2.7, p. 44. Here $\Delta_h^m f$ is given by (1.12). The classical Besov spaces

$$B_{pq}^1(\mathbb{R}^n) \quad \text{with} \quad 1 \leq p \leq \infty, \quad 1 \leq q \leq \infty,$$

described in 1.2(v), can be normed by

$$\|f\,|B_{pq}^1(\mathbb{R}^n)\| = \|f\,|L_p(\mathbb{R}^n)\| + \left(\int_0^\varepsilon \left(\frac{\omega_2(f,t)_p}{t}\right)^q \frac{dt}{t}\right)^{\frac{1}{q}} \tag{17.18}$$

if $q < \infty$ and by

$$\|f\,|B_{p\infty}^1(\mathbb{R}^n)\| = \|f|L_p(\mathbb{R}^n)\| + \sup_{0 < t < \varepsilon} \frac{\omega_2(f,t)_p}{t} \tag{17.19}$$

if $q = \infty$. Here $0 < \varepsilon < 1$, [Triβ], 2.5.12, p. 110. Now we incorporate a log-term in (17.18), (17.19). Let $b \in \mathbb{R}$. Then

$$B_{pq}^{(1,-b)}(\mathbb{R}^n) \quad \text{with} \quad 1 \leq p \leq \infty, \quad 1 \leq q \leq \infty,$$

is the collection of all $f \in L_p(\mathbb{R}^n)$ with

$$\|f\,|B_{pq}^{(1,-b)}(\mathbb{R}^n)\| = \|f\,|L_p(\mathbb{R}^n)\| + \left(\int_0^\varepsilon \left(\frac{\omega_2(f,t)_p}{t|\log t|^b}\right)^q \frac{dt}{t}\right)^{\frac{1}{q}} < \infty \tag{17.20}$$

(obviously modified if $q = \infty$). These spaces can be characterized in Fourier-analytical terms. Let φ_k be the same functions as in (2.33)–(2.35). In generalization of (2.37),

$$\left(\sum_{j=0}^{\infty} 2^{jq}(1+j)^{-bq} \, \|(\varphi_j \widehat{f})^{\vee} \, | L_p(\mathbb{R}^n)\| \right)^{\frac{1}{q}} \tag{17.21}$$

(obviously modified if $q = \infty$) is an equivalent norm in $B_{pq}^{(1,-b)}(\mathbb{R}^n)$. These spaces, in their general version of $B_{pq}^{(s,-b)}(\mathbb{R}^n)$ with (2.36) and $b \in \mathbb{R}$ go back to H.-G. Leopold in 1998 and may be found in [Leo98] and [Leo00a]. The point of interest in our context of distinguished target spaces is to replace the second differences $\omega_2(f,t)_p$ in (17.20) by the first differences $\omega(f,t)_p = \omega_1(f,t)_p$ and to introduce in this way spaces $Lip_{pq}^{(1,-\alpha)}(\mathbb{R}^n)$ of Lipschitz type, consisting of all $f \in L_p(\mathbb{R}^n)$ such that

$$\|f \, | Lip_{pq}^{(1,-\alpha)}(\mathbb{R}^n)\| = \|f \, | L_p(\mathbb{R}^n)\| + \left(\int_0^\varepsilon \left(\frac{\omega(f,t)_p}{t |\log t|^\alpha} \right)^q \frac{dt}{t} \right)^{\frac{1}{q}} \tag{17.22}$$

is finite (with the usual modification if $q = \infty$). Here

$$1 \leq p \leq \infty, \quad 0 < q \leq \infty, \quad \alpha > \frac{1}{q},$$

(with $\alpha \geq 0$ if $q = \infty$). The restriction on α is natural. This follows from the considerations in 12.12: If $\alpha \leq \frac{1}{q}$ (with $\alpha < 0$ in case of $q = \infty$) then, with exception of $f = 0$, there are no functions f such that (17.22) is finite. These spaces were introduced in [Har00a], Definition 1. If $p = \infty$ in (17.22) then the inequalities in 14.3 for the super-critical case can be reformulated (at least locally) in terms of these target spaces. Hence it is reasonable to extend these considerations from $p = \infty$ to, say, $1 < p < \infty$. But there is a decisive difference between these two cases. Recall that the classical Sobolev space $W_p^1(\mathbb{R}^n)$ with $1 < p < \infty$ can be equivalently normed by

$$\|f \, | L_p(\mathbb{R}^n)\| + \sup_{0<t<\varepsilon} \frac{\omega(f,t)_p}{t}, \quad f \in W_p^1(\mathbb{R}^n). \tag{17.23}$$

We refer to [Ste70], Proposition 3, p. 139, [Nik77], 4.8, p. 213 (first edition 1969) and [DeL93], p. 53. In other words, $W_p^1(\mathbb{R}^n)$ with $1 < p < \infty$ coincides with $Lip_{p\infty}^{(1,0)}(\mathbb{R}^n)$. Replacing the first differences in (17.23) by second or higher differences, one gets (17.19) and hence the larger spaces $B_{p\infty}^1(\mathbb{R}^n)$. We do not go into detail. A thorough investigation of all these spaces, especially their mutual embeddings, may be found in [Har00a] with [EdH99] and [EdH00] as forerunners. We refer also to the small survey [Har00b]. Finally we mention that limiting embeddings especially in the super-critical case for spaces with dominating mixed derivatives have been considered in [KrS98].

17.3 Logarithmic spaces

Let Ω be a bounded C^∞ domain in $\mathbb{R}^n$. We assume that $|\Omega| = \varepsilon < 1$. Let $1 < p < \infty$ and $a \in \mathbb{R}$. Then the spaces $L_p(\log L)_a(\Omega)$ can be introduced much as in 11.6(ii) as the collection of all $f \in L_1(\Omega)$ such that

$$\int_\Omega |f(x)|^p \log^{ap}(2 + |f(x)|) \, dx < \infty. \tag{17.24}$$

As in (11.46), these spaces can also be characterized as the collection of all $f \in L_1(\Omega)$ such that

$$\left(\int_0^\varepsilon (|\log t|^a \, f^*(t))^p \, dt \right)^{\frac{1}{p}} < \infty \tag{17.25}$$

(equivalent quasi-norms). Details and references are given in 11.6(ii). Based on [EdT95] we proved in [ET96], 2.6.2, Theorem 1, pp. 69/70, another characterization of these spaces with the consequence that they have the same mapping properties with respect to pseudodifferential operators and fractional powers of elliptic operators as the space $L_p(\Omega)$ with $1 < p < \infty$, itself. In particular one can define logarithmic Sobolev spaces $H_p^s(\log H)_a(\Omega)$ by lifting of $L_p(\log L)_a(\Omega)$ in the following way. Let

$$A_m f = (\mathrm{id} - \Delta)^m f, \quad m \in \mathbb{N}, \tag{17.26}$$

and let $A_{m,N} f = A_m f$ be the corresponding Neumann operator with the domain of definition

$$\mathrm{dom}\, A_{m,N} = \left\{ f \in H_p^{2m}(\Omega) : \frac{\partial^{j+m} f}{\partial \nu^{j+m}} \Big|_{\partial\Omega} = 0 \text{ if } j = 0, \ldots, m-1 \right\}, \tag{17.27}$$

where ν is the outer normal with respect to $\partial\Omega$. Let

$$a \in \mathbb{R}, \quad 1 < p < \infty, \quad 0 < \tau \le \frac{1}{2} \quad \text{and} \quad s = 2m\tau.$$

Then one can define

$$H_p^s(\log H)_a(\Omega) = A_{m,N}^{-\tau} L_p(\log L)_a(\Omega). \tag{17.28}$$

We refer for details and explanations to [ET96], 2.6.3, pp. 75–81. In particular, (17.28) imitates the lifting (1.8). If $s \in \mathbb{N}$, then one obtains, as should be the case,

$$H_p^s(\log H)_a(\Omega) = \{f \in L_1(\Omega) : D^\alpha f \in L_p(\log L)_a(\Omega) , |\alpha| \leq s\} , \qquad (17.29)$$

with the equivalent norms

$$\sum_{|\alpha| \leq s} \|D^\alpha f \,|L_p(\log L)_a(\Omega)\| . \qquad (17.30)$$

If $s \in \mathbb{N}$ and $s = \frac{n}{p}$, then one is in the critical case with logarithmically modified classical Sobolev spaces (the dotted line in Fig. 10.1). One may ask for counterparts of Theorem 13.2 and related inequalities in 13.3. Some results of Trudinger type as in (11.56) and 11.8(v) may be found in [FLS96]. Extensions to the fractional case, including Sobolev-Orlicz spaces, have been given in [EdK95]. The interest in these logarithmic Sobolev spaces comes also from the regularity properties of the Jacobian. References can be found in [FLS96]. This may justify having a closer look at these logarithmic spaces from the point of view of sharp inequalities as treated in this chapter.

17.4 Compact embeddings

We proved in 14.6 by geometrical reasoning that the sharp embeddings (14.48) cannot be compact. If one replaces $-\frac{1}{q'}$ in (14.48) by $-\alpha$ with $-\alpha < -\frac{1}{q'}$, then one gets a larger space and it turns out that the corresponding embedding is compact. The degree of compactness can be measured in terms of entropy numbers and approximation numbers. Definitive results in this direction have been obtained in [EdH99], [EdH00], [Har00b] and recently in [CoK00]. This covers also the more general spaces normed by (17.22), and the delicate interplay between these spaces and also in relation to the spaces introduced in (17.18), (17.19), (17.20). As for the latter spaces we refer also to [Leo98] and [Leo00a]. The general background may be found in [ET96] and, in connection with weighted and logarithmic spaces as discussed in 17.3, in [Har97], [Har98], and [Har00c].

Chapter III
Fractal Elliptic Operators

18 Introduction

Let Γ be a compact d-set in $\mathbb{R}^n$, where $n \in \mathbb{N}$ and $0 < d < n$, according to (9.1) or (9.67). By our previous considerations, d-sets are especially well adapted to the function spaces $B_{pq}^s(\mathbb{R}^n)$ and $F_{pq}^s(\mathbb{R}^n)$ as treated in this book. In particular, by 9.19 and 9.18 such sets satisfy the ball condition with the consequences for traces as described in Theorem 9.21. Furthermore, there is a natural way of introducing function spaces on Γ as mentioned in (9.125) with a reference to [Triδ], Definition 20.2, p. 159. Quarkonial representations of such spaces, in the larger context of more general fractals, have been given in 9.29–9.33. These results pave the way to a substantial theory for fractal elliptic (pseudo)differential operators in continuation of [Triδ], Chapter V. We refer especially to [Triδ], Section 26, where we discussed in detail our point of view, compared with what has been done in the literature. In particular, there are different interpretations of what is called a fractal drum. The physical background of our approach in [Triδ] and also here, as far as spectral theory is concerned, may be found in [Triδ], 30.1, and will be briefly repeated in 19.1. As indicated one could continue the studies started in [Triδ], Chapters IV and V, based now on the theory developed in Chapter I of the present book on the level of (degenerate) fractal elliptic pseudodifferential operators, including fractional powers of elliptic differential operators. But this is not our aim. As far as elliptic operators are concerned we stick to the Laplacian as an outstanding example.

We give a brief description of what follows. Let Γ be a compact d-set in $\mathbb{R}^n$ equipped with the Radon measure μ such that

$$supp\,\mu = \Gamma, \quad \mu(B(\gamma,r)) \sim r^d \quad \text{where} \quad \gamma \in \Gamma \quad \text{and} \quad 0 < r < 1, \qquad (18.1)$$

according to the beginning of 9.12, in particular (9.67). Let Ω be a bounded C^∞ domain in $\mathbb{R}^n$ with $\Gamma \subset \Omega$ and let $-\Delta$ be the Dirichlet Laplacian in Ω. Let $n - 2 < d < n$. Then, roughly speaking,

$$tr^\Gamma : \quad \varphi \mapsto \varphi\mu, \tag{18.2}$$

sending $\varphi \in D(\Omega)$ into $D'(\Omega)$, can be extended by continuity to a bounded map from $\overset{\circ}{H}{}^1(\Omega)$ into $H^{-1}(\Omega)$ and, as a consequence,

$$B = (-\Delta)^{-1} \circ tr^\Gamma \tag{18.3}$$

turns out to be a self-adjoint operator in $\overset{\circ}{H}{}^1(\Omega)$. At least in case of $n = 2$, this operator has physical relevance, describing the vibrations of a drum with the fractal membrane Γ. A detailed discussion may be found in [Triδ], Section 30. In particular,

$$\varrho_k \sim k^{-1 + \frac{n-2}{d}}, \quad k \in \mathbb{N}, \tag{18.4}$$

where ϱ_k are the (ordered by magnitude) positive eigenvalues of B. One aim of Section 19 is to complement these results. In particular we discuss the smoothness of the eigenfunctions and prove that the largest eigenvalue ϱ_1 is simple. If $n = 2$ then by (18.4) we have the classical Weyl behaviour $\varrho_k \sim k^{-1}$ for all d-sets Γ with $0 < d < 2$ and the related measures μ. It is the second aim of Section 19 to discuss under what circumstances for general compact Radon measures μ in the plane this remarkable property remains valid. Section 20 deals with the Dirichlet problem for $-\Delta$ in $\Omega \setminus \Gamma$ with given boundary data at Γ, where Γ is a d-set with $\Gamma \subset \Omega$ and $n - 2 < d < n$. In Section 21 we return to the Riemannian manifolds of hyperbolic type as studied in Section 7, especially to the (fractally bounded) d-domains according to the end of 7.5. We are interested in the so-called negative spectrum of the (self-adjoint) operator

$$H_\beta = -\Delta_g + \varrho\, id - \beta\, g^{-\varkappa} \quad \text{with} \quad \varkappa > 0 \quad \text{and} \quad \beta > 0, \tag{18.5}$$

where g has the same meaning as in (7.9), $\varrho > 0$ is sufficiently large, and Δ_g is the related Laplace-Beltrami operator according to (7.16). Of particuliar interest, and in slight modification of (18.5), is the behaviour of the hydrogen atom in this infinite hyperbolic world. We rely on the techniques developed in Section 7, especially on the quarkonial decompositions from Theorem 7.22. With exception of the end of Section 19 (Weyl measures in the plane) all fractals involved in Sections 19–21 are d-sets. On the other hand we developed in Section 9 a theory for more general fractals and their relations to function spaces. One may ask to what extent the d-sets in Sections 19–21 can be replaced by more general fractals. In principle this should be possible for some classes of fractals. But one must have in mind that d-sets are tailored to the spaces $B^s_{pq}(\mathbb{R}^n)$ and $F^s_{pq}(\mathbb{R}^n)$ considered so far in this book. Changing the class of admitted fractals then, despite a few rather satisfactory assertions obtained in Section 9, one needs apparently some new optimally adapted function spaces. We describe this method in the remaining two Sections 22 and 23 in case of so-called (d, Ψ)-sets and their related function spaces. In generalization of (18.1) a compact set Γ in $\mathbb{R}^n$ is called a (d, Ψ)-set if there is a Radon measure μ with

$$\operatorname{supp}\mu = \Gamma, \quad \mu(B(\gamma,r)) \sim r^d \Psi(r) \quad \text{where} \quad \gamma \in \Gamma \text{ and } 0 < r < \varepsilon, \qquad (18.6)$$

for some $0 < \varepsilon < 1$. Here $r^d \Psi(r)$ might be considered as a perturbation of r^d, typically $0 < d < n$ and $\Psi(r) = |\log r|^b$ for some $b \in \mathbb{R}$. Inspired by some previous work of H.-G. Leopold in [Leo98] we formulated this theory in [EdT98] and outlined the proofs in [EdT99a] parallel to the relevant arguments in [Triδ]. In the meantime there are several papers, preprints and PhD-theses dealing with the subject in detail. This might justify our being very brief here and giving the necessary references. In Section 22 we describe the theory of the underlying spaces $B^{(s,\Psi)}_{pq}(\mathbb{R}^n)$, including quarkonial decompositions, entropy numbers and spaces on (d, Ψ)-fractals Γ. The related spectral theory for operators of type (18.3) will be outlined in Section 23, including the music of a rusty drum.

19 Spectral theory for the fractal Laplacian

19.1 The physical background

We modify the explanations given in [Triδ], 30.1, pp. 233–234. Let Ω be a bounded domain in the plane $\mathbb{R}^2$ with C^∞ boundary $\partial\Omega$, interpreted as a membrane fixed at its boundary. Vibrations of such a membrane in $\mathbb{R}^3$ are measured by the deflection $v(x,t)$, where $x = (x_1, x_2) \in \Omega$, and $t \geq 0$ stands for the time. In other words, the point $(x_1, x_2, 0)$ in $\mathbb{R}^3$ with $(x_1, x_2) \in \Omega$ of the membrane at rest, is deflected to $(x_1, x_2, v(x,t))$. Up to constants the usual physical description is given by

$$\Delta v(x,t) = m(x) \frac{\partial^2 v(x,t)}{\partial t^2}, \quad x \in \Omega, \quad t \geq 0, \qquad (19.1)$$

and

$$v(y,t) = 0 \quad \text{if} \quad y \in \partial\Omega, \quad t \geq 0, \qquad (19.2)$$

where $\Delta = \frac{\partial^2}{\partial x_1^2} + \frac{\partial^2}{\partial x_2^2}$ and the right-hand side of (19.1) is Newton's law with the mass density $m(x)$. To find the eigenfrequencies one has to insert $v(x,t) =$

$u(x)e^{i\lambda t}$ with $\lambda \in \mathbb{R}$ in (19.1) and obtains

$$-\Delta u(x) = \lambda^2 m(x)\, u(x)\,, \quad x \in \Omega\,; \quad u(y) = 0 \quad \text{if} \quad y \in \partial\Omega\,, \qquad (19.3)$$

where one is interested in non-trivial solutions $u(x)$. Hence one asks for eigenfunctions and eigenvalues of the operator

$$B = (-\Delta)^{-1} \circ m(\cdot)\,, \qquad (19.4)$$

where $(-\Delta)^{-1}$ is the inverse of the *Dirichlet Laplacian* $-\Delta$. We use the notation *Dirichlet Laplacian* always with the understanding that vanishing boundary data at $\partial\Omega$ are incorporated into the domains of definition for $-\Delta$ in the function spaces considered, preferably $B^s_{pq}(\Omega)$ and $H^s_p(\Omega)$ with $1 < p \leq \infty$ and $s > \frac{1}{p}$ (this will be specified in greater detail in the next subsection). If ϱ is a positive eigenvalue of B then $\lambda = \varrho^{-\frac{1}{2}}$ is the related eigenfrequency. We are interested in the problem of what happens when the mass density $m(x)$ shrinks to a fractal set Γ and a related Radon measure μ with

$$supp\,\mu = \Gamma \subset \Omega\,. \qquad (19.5)$$

Hence we ask for eigenfrequencies and eigenfunctions of drums with a fractal membrane. This is what we call fractal drums and fractal Laplacians (extending this notation to $n \in \mathbb{N}$). Otherwise the term *fractal drums* has several meanings in the literature. We gave in [Triδ], 26.1, 26.2, pp. 199–201, a short description which will not be repeated here.

19.2 The mathematical background

Let Ω be a bounded C^∞ domain in $\mathbb{R}^n$. By Definition 5.3 the spaces $B^s_{pq}(\Omega)$ and $F^s_{pq}(\Omega)$ are the restrictions of the spaces $B^s_{pq}(\mathbb{R}^n)$ and $F^s_{pq}(\mathbb{R}^n)$, respectively, to Ω. Of interest for us are especially the Besov spaces

$$B^s_{pq}(\Omega)\,, \quad 1 \leq p \leq \infty\,, \quad 1 \leq q \leq \infty\,, \quad s \in \mathbb{R}\,, \qquad (19.6)$$

with the Hölder-Zygmund spaces

$$\mathcal{C}^s(\Omega) = B^s_{\infty\infty}(\Omega)\,, \quad s \in \mathbb{R}\,, \qquad (19.7)$$

as a special case, and the Sobolev spaces

$$H^s_p(\Omega) = F^s_{p2}(\Omega)\,, \quad 1 < p < \infty\,, \quad s \in \mathbb{R}\,, \qquad (19.8)$$

with the abbreviation

$$H^s(\Omega) = H^s_2(\Omega)\,, \quad s \in \mathbb{R}\,. \qquad (19.9)$$

Recall that
$$H^1(\Omega) = W_2^1(\Omega) \tag{19.10}$$
is the very classical Sobolev space which will play a decisive role in what follows. We use standard notation as introduced in 1.2 and 2.1. Let
$$1 \le p \le \infty, \quad 1 \le q \le \infty \quad \text{and} \quad s > \frac{1}{p}. \tag{19.11}$$
Then we have the well-known trace property
$$tr_{\partial\Omega} B_{pq}^s(\Omega) = B_{pq}^{s-\frac{1}{p}}(\partial\Omega). \tag{19.12}$$
This may be found in [Triα], 4.7.1, pp. 329–330, or [Triβ], 3.3.3, p. 200. Since $\partial\Omega$ is an $(n-1)$-set, (19.12) is also a very special case of (9.143) with $t_p = \frac{n-1}{p}$ by (9.120) (complemented now by $p = 1$ and $p - \infty$). As always
$$-\Delta = -\sum_{j=1}^{n} \frac{\partial^2}{\partial x_j^2} \tag{19.13}$$
is the Laplacian in $\mathbb{R}^n$ and in Ω. If p, q, s are given by (19.11), then $-\Delta$ maps
$$\{g \in B_{pq}^s(\Omega) : tr_{\partial\Omega} g = 0\} \quad \text{isomorphically onto} \quad B_{pq}^{s-2}(\Omega). \tag{19.14}$$
This is a well-known assertion extending mapping properties for classical Sobolev spaces and may be found in [Triα], 5.7.1, Remark 1, p. 402 (with a correction in the 1995 edition), complemented by [Triβ], 4.3.3, 4.3.4. A systematic treatment has been given in [RuS96], 3.5.2, p. 130. If the domain of definition is given by one of the spaces on the left-hand part of (19.14), then we call $-\Delta$ the *Dirichlet Laplacian*. In particular, if p, q, s are given by (19.11) and if $(-\Delta)^{-1}$ stands for the inverse of the Dirichlet Laplacian, then
$$(-\Delta)^{-1} : B_{pq}^{s-2}(\Omega) \mapsto B_{pq}^s(\Omega), \tag{19.15}$$
is a bounded map. Let $\overset{\circ}{H}{}^1(\Omega)$ be the completion of $D(\Omega)$ in $H^1(\Omega)$. This coincides with the spaces on the left-hand part of (19.14) with $s = 1$ and $p = q = 2$. In particular,
$$(-\Delta)^{-1} : H^{-1}(\Omega) \mapsto \overset{\circ}{H}{}^1(\Omega) \tag{19.16}$$
is an isomorphic map.

Next we clarify what is meant by tr^Γ in (18.2) and, in the following subsection, by B in (18.3).

We use the equivalence $\sim$ in
$$a_k \sim b_k \quad \text{or} \quad \varphi(x) \sim \psi(x) \tag{19.17}$$
always to mean that there are two positive numbers c_1 and c_2 such that
$$c_1 a_k \le b_k \le c_2 a_k \quad \text{or} \quad c_1 \varphi(x) \le \psi(x) \le c_2 \varphi(x) \tag{19.18}$$
for all admitted values of the discrete variable k or the continuous variable x. Here a_k, b_k are positive numbers and $\varphi(x)$, $\psi(x)$ are positive functions. Let $0 < d \le n$. Then, as always in this book, a compact set Γ in $\mathbb{R}^n$ is called a d-set if there is a Radon measure μ with (18.1). Recall that such a measure μ is uniquely determined up to equivalences and one might think about $\mu = \mathcal{H}^d|\Gamma$, the restriction of the Hausdorff measure $\mathcal{H}^d$ in $\mathbb{R}^n$ to Γ. A short proof and a few basic facts concerning d-sets may be found in [Triδ], Section 3, pp. 5–7, including some references. The trace operator tr_Γ was introduced in 9.2 and studied in detail in 9.32 and 9.33 in connection with rather general compact sets Γ and related Radon measures μ. As mentioned in 9.34(i) in case of d-sets with $0 < d < n$ we have $t_p = \frac{d}{p}$ in (9.143) and in addition (9.149). We wish to incorporate $p = \infty$ in (9.149). However instead of referring to the rather general assertions in 9.32–9.34 one can rely on the simpler and more specific results in [Triδ], Theorems 18.2 and 18.6 on pp. 136 and 139. On this basis we describe what we need in the sequel. Let
$$1 < p \le \infty \quad \text{and} \quad 0 < d < n. \tag{19.19}$$
Then by [Triδ], Theorem 18.6,
$$tr_\Gamma B_{p1}^{\frac{n-d}{p}}(\mathbb{R}^n) = L_p(\Gamma) \quad \text{if} \quad 1 < p < \infty, \tag{19.20}$$
complemented by
$$B_{\infty 1}^0(\mathbb{R}^n) \subset C(\mathbb{R}^n) \quad \text{if} \quad p = \infty, \tag{19.21}$$
where the latter is a well-known embedding which may be found in [Triβ], 2.7.1, Remark 2, p. 130. It complements (11.22). Let id_Γ be the identification operator introduced in (9.16). By (9.167), (9.168) we have
$$id_\Gamma L_p(\Gamma) = B_{p\infty}^{-\frac{n-d}{p'},\Gamma}(\mathbb{R}^n), \quad 1 < p \le \infty, \quad \frac{1}{p} + \frac{1}{p'} = 1. \tag{19.22}$$
This coincides with [Triδ], Theorem 18.2, p. 136 and applies also to $p = \infty$. Let
$$tr^\Gamma = id_\Gamma \circ tr_\Gamma \tag{19.23}$$
and let $\Gamma \subset \Omega$. Then one can replace $\mathbb{R}^n$ in (19.20)–(19.22) by Ω and obtains
$$tr^\Gamma : B_{p1}^{\frac{n-d}{p}}(\Omega) \mapsto B_{p\infty}^{-\frac{n-d}{p'}}(\Omega), \quad 1 < p \le \infty. \tag{19.24}$$

19.3 The operator B

We clip together the mapping properties of the Dirichlet Laplacian, given by (19.15), (19.11), and of tr^Γ given by (19.24), (19.19), and put

$$B = (-\Delta)^{-1} \circ tr^\Gamma. \qquad (19.25)$$

We wish to look at B as a bounded operator acting in an admitted space $B^s_{pq}(\Omega)$. The smoothness one loses in (19.24) is $n-d$, the gain in (19.15) is 2. Hence

$$n - 2 < d < n \quad (\text{with } 0 < d < 1 \text{ if } n = 1) \qquad (19.26)$$

is a natural restriction. Then we have also

$$2 - \frac{n-d}{p'} > \frac{2}{p} \geq \frac{n-d}{p} \quad \text{with} \quad 1 < p \leq \infty \quad \text{and} \quad 2 - \frac{n-d}{2} > 1. \qquad (19.27)$$

We refer also to the Figures 19.1 and 19.2 below. In particular, B is continuous in

$$B^{2-\frac{n-d}{p'}}_{p\infty}(\Omega), \quad 1 < p \leq \infty, \quad \text{and compact in} \quad \overset{\circ}{H}{}^1(\Omega), \qquad (19.28)$$

where we used (19.16) in addition. Let $f \in \overset{\circ}{H}{}^1(\Omega)$ and $g \in \overset{\circ}{H}{}^1(\Omega)$. We choose as the scalar product in $\overset{\circ}{H}{}^1(\Omega)$,

$$(f, g)_{\overset{\circ}{H}{}^1(\Omega)} = \sum_{j=1}^{n} \int_\Omega \frac{\partial f}{\partial x_j}(x) \frac{\partial \overline{g}}{\partial x_j}(x)\, dx. \qquad (19.29)$$

Then we have, by completion of functions belonging to $D(\Omega)$, for functions $f \in \overset{\circ}{H}{}^1(\Omega)$ and $g \in \overset{\circ}{H}{}^1(\Omega)$,

$$\begin{aligned}
(Bf, g)_{\overset{\circ}{H}{}^1(\Omega)} &= \sum_{j=1}^{n} \int_\Omega \frac{\partial Bf}{\partial x_j}(x) \frac{\partial \overline{g}}{\partial x_j}(x)\, dx \\
&= \int_\Omega -\Delta\left((-\Delta)^{-1} \circ tr^\Gamma\right) f(x) \overline{g}(x)\, dx \\
&= \int_\Gamma f(\gamma) \overline{g}(\gamma)\, \mu(d\gamma). \qquad (19.30)
\end{aligned}$$

We refer for a more detailed justification to [Triδ], 28.6, 30.2, pp. 226, 234. In particular, B is a non-negative self-adjoint compact operator in $\overset{\circ}{H}{}^1(\Omega)$, generated by the quadratic form on the right-hand side of (19.30) with

$$\|\sqrt{B}f \,|\overset{\circ}{H}{}^1(\Omega)\| = \|f\,|L_2(\Gamma)\|, \quad f \in \overset{\circ}{H}{}^1(\Omega), \qquad (19.31)$$

and, consequently, null-space

$$N(B) = \left\{ f \in \overset{\circ}{H}{}^{1}(\Omega) \; : \; tr_\Gamma f = 0 \right\}. \tag{19.32}$$

Furthermore, if ϱ_k are the positive eigenvalues of B, repeated according to multiplicity and ordered by their magnitude, then

$$\varrho_k \sim k^{-1+\frac{n-2}{d}}, \quad k \in \mathbb{N}, \tag{19.33}$$

in the sense of (19.17), (19.18), and the explanations given there. Proofs may be found in [Triδ] at the places indicated above. We complement these assertions now by a closer look at the eigenfunctions belonging to ϱ_k, the simplicity of the largest eigenvalue ϱ_1 and by the observation that $D(\Omega\backslash\Gamma)$ is dense in $N(B)$ given by (19.32). We prove the latter assertion separately in a larger context in 19.5 which might be of self-contained interest and which is also helpful for the Dirichlet problem in a fractal setting considered in Section 20. Otherwise we refer to Theorem 19.7 where we give a full formulation of all results, known and new.

19.4 Generalizations, comments

We concentrate in Theorem 19.7 below on the operator B given by (19.25) where Γ is a d-set, equipped with a Radon measure μ according to (18.1), for example $\mu = \mathcal{H}^d|\Gamma$. Then B can also be written as

$$B = (-\Delta)^{-1} \circ \mu, \tag{19.34}$$

provided one interprets μ as tr^Γ with (19.23). One may ask for generalizations of $(-\Delta)^{-1}$ on the one hand and/or μ with $supp\,\mu = \Gamma$ and tr^Γ according to (19.23) on the other hand. As far as the inverse $(-\Delta)^{-1}$ of the Dirichlet Laplacian is concerned we remark that almost all of our arguments in [Triδ], Sections 27–30, and also many arguments below rely only on qualitative mapping properties as described in (19.15), (19.16), and embedding theorems for B-spaces on Ω and on Γ. But this can be generalized replacing $(-\Delta)^{-1}$ by

$$b_1 \circ b(x,D) \circ b_2, \quad b_l \in L_{r_l}(\Gamma), \quad 1 < r_l \leq \infty, \tag{19.35}$$

where $b(x,D)$ are suitable pseudodifferential operators (of negative order of smoothness) with the special case

$$b(x,D) = A^{-\frac{\varkappa}{m}}, \quad 0 < \varkappa < 2m, \tag{19.36}$$

where A^σ are fractional powers of regular elliptic differential operators A of order $2m$ with suitable boundary conditions. The functions b_1 and b_2 in (19.35) indicate that this is a degenerate operator. We refer to [ET96], Chapter 5, where we dealt with operators of this type in a non-fractal setting and to [Triδ], Chapter 5, for a fractal setting. In this book we restrict ourselves to the non-degenerate operator $(-\Delta)^{-1}$ in the above context, with the only exception being at the end of this chapter in Section 23 in connection with what we wish to call the sintered drum.

As far as the fractal part, so far d-sets, is concerned, the situation is different. We have now an elaborated theory of fractals and their function spaces developed in Section 9 at hand. Subsections 9.24–9.33 are especially well adapted to our approach to a spectral theory for operators of type B in (19.34), now with more general compact Radon measures μ in $\mathbb{R}^n$. This will not be done in a systematic way. But there are two, as we hope, interesting exceptions. First, in Subsections 19.12–19.18 below we take a closer look at fractal drums in the plane $\mathbb{R}^2$. Then

$$\varrho_k \sim k^{-1}, \quad k \in \mathbb{N}, \tag{19.37}$$

in (19.33) for all compact d-sets in $\mathbb{R}^2$ with $0 < d < 2$. We discuss in some detail for which more general compact Radon measures μ in the plane and related operators B in (19.34) the Weyl behaviour (19.37) of the eigenvalues remains valid. Secondly, as indicated in Section 18, we describe at the end of this chapter in Section 23 (with some preparation in Section 22) what happens when one generalizes d-sets and their measures in (19.34) by (d, Ψ)-sets and their measures according to (18.6).

All the operators and all the measures μ, including the indicated generalizations connected with (d, Ψ)-sets and the measures in the plane treated at the end of this section, are isotropic in some sense: there are no distinguished directions in $\mathbb{R}^n$, or all directions in $\mathbb{R}^n$ are equal. However in connection with fractal geometry, iterated function systems etc. it is quite natural to ask what happens if μ in (19.34) is a related anisotropic or nonisotropic Radon measure in the plane. We studied these problems in some detail in [Triδ], Chapter 5, especially Section 30. More recent results may be found in [FaT99]. But there are no final results comparable with (19.33), (19.37). The situation is somewhat better if one looks at semi-elliptic operators of prototype

$$Au = \frac{\partial^4 u}{\partial x_1^4} - \frac{\partial^2 u}{\partial x_2^2}, \quad x = (x_1, x_2) \in \mathbb{R}^2, \tag{19.38}$$

and anisotropic fractals with adapted anisotropies. We refer to [Far98], [Far99], [She99] and [NaS00]. We do not deal with problems of this type here. But it would be of interest to check what can be said in this direction using the results from Section 9.

As indicated at the end of 19.3 we prove first a density assertion in a somewhat larger context. Let Γ be a compact d-set in $\mathbb{R}^n$ with $0 < d < n$. By (19.20),

$$H^\sigma_{p,\Gamma}(\mathbb{R}^n) = \{f \in H^\sigma_p(\mathbb{R}^n) : tr_\Gamma f = 0\} \tag{19.39}$$

and

$$B^\sigma_{pq,\Gamma}(\mathbb{R}^n) = \{f \in B^\sigma_{pq}(\mathbb{R}^n) : tr_\Gamma f = 0\} \tag{19.40}$$

with

$$1 < p < \infty, \quad 0 < q \le \infty, \quad \sigma > \frac{n-d}{p}, \tag{19.41}$$

are well defined and closed subspaces of $H^\sigma_p(\mathbb{R}^n) = F^\sigma_{p2}(\mathbb{R}^n)$ and $B^\sigma_{pq}(\mathbb{R}^n)$, respectively. We used this type of notation before in (9.164) and (9.171).

19.5 Proposition

Let Γ be a compact d-set in $\mathbb{R}^n$ with $0 < d < n$. Let

$$1 < p < \infty, \quad 1 \le q < \infty \quad \text{and} \quad 0 < s < 1. \tag{19.42}$$

Then $D(\mathbb{R}^n \backslash \Gamma)$ is dense in

$$H^{s+\frac{n-d}{p}}_{p,\Gamma}(\mathbb{R}^n) \quad \text{and in} \quad B^{s+\frac{n-d}{p}}_{pq,\Gamma}(\mathbb{R}^n). \tag{19.43}$$

Proof *Step 1* The main point of the proof is to use the deep Theorem 10.1.1 in [AdH96], p. 281, due to Yu. V. Netrusov. For this purpose we need some preparation. Let K be a compact set in $\mathbb{R}^n$. Let $1 < p < \infty$ and $\alpha > 0$. By [AdH96], Definition 2.2.6, p. 20, complemented by Corollary 2.6.8, p. 44,

$$C_{\alpha,p}(K) = \inf\{\|\varphi \,|\, H^\alpha_p(\mathbb{R}^n)\|^p : \varphi \in S(\mathbb{R}^n), \varphi \ge 1 \text{ on } K\} \tag{19.44}$$

is called the (α, p)-capacity of K. Here the admitted functions φ are real. By [AdH96], Definition 2.2.5, p. 20, and the explanations given on that page a property is said to hold (α, p)-quasi-everywhere, (α, p)-q.e. for short, if it is true for all $x \in \mathbb{R}^n$ with exception of a set E with $C_{\alpha,p}(E) = 0$. Recall that a point $x \in \mathbb{R}^n$ is called a Lebesgue point for a locally integrable function $f(x)$ in $\mathbb{R}^n$ if

$$f(x) = \lim_{r \to 0} |B(x,r)|^{-1} \int_{B(x,r)} f(y)\,dy, \tag{19.45}$$

19. Spectral theory for the fractal Laplacian

where again $B(x,r)$ stands for a ball in $\mathbb{R}^n$ centred at $x \in \mathbb{R}^n$ and of radius $0 < r < 1$. By [AdH96], 6.1, pp. 157–158, it follows that in each equivalence class $[f] \in H_p^\alpha(\mathbb{R}^n)$ there is an (α,p)-q.e. uniquely determined representative such that (19.45) holds (α,p)-q.e. Such a representative f is taken there to define traces on sets. Let now Γ be the above d-set and let $\alpha = s + \frac{n-d}{p}$ with $1 < p < \infty$ and $0 < s < 1$. Then we have

$$\|tr_\Gamma \varphi \,|L_p(\Gamma)\| \leq c\, \|\varphi\,|H_p^\alpha(\mathbb{R}^n)\|, \quad \varphi \in S(\mathbb{R}^n). \tag{19.46}$$

Assume $\Gamma_0 \subset \Gamma$ and $C_{\alpha,p}(\Gamma_0) = 0$. Let, in addition, φ be real and $\varphi \geq 1$ on Γ_0. Then

$$\mu(\Gamma_0) \leq \int_{\Gamma_0} \varphi(\gamma)^p\, \mu(d\gamma) \leq c\, \|\varphi\,|H_p^\alpha(\mathbb{R}^n)\|^p, \tag{19.47}$$

and by (19.44) we get $\mu(\Gamma_0) = 0$. In particular we have (19.45) for the above representative μ-a.e. On the other hand this is also the definition of a trace according to [JoW84], 2.1, p. 15. This coincides also with our way of introducing traces via inequalities, first for smooth functions and afterwards by completion. Hence trace assertions from [AdH96] can be used in our context. We finish our preparation by the following observation. Let Γ be the above d-set. Then

$$C_{\sigma,p}(\Gamma) = 0, \quad \text{where} \quad 1 < p < \infty \quad \text{and} \quad 0 \leq \sigma < \frac{n-d}{p}. \tag{19.48}$$

This follows essentially from an atomic (or quarkonial) argument as used several times in [Triδ], for example on p. 129. We outline the procedure. We cover Γ by N_j balls B_k of radius 2^{-j} and construct a subordinated resolution of unity $\{\varphi_k^j\}_{k=1}^\infty \subset D(\mathbb{R}^n)$ with

$$N_j \sim 2^{jd}, \quad supp\, \varphi_k^j \subset B_k, \quad \sum_{k=1}^{N_j} \varphi_k^j(x) = 1, \tag{19.49}$$

in a neighbourhood of Γ. Here $j \in \mathbb{N}$ and the equivalence constants in (19.49) are independent of j. Then we may assume that

$$\varphi^j(x) = 2^{j(\sigma - \frac{n}{p})} \sum_{k=1}^{N_j} \varphi_k^j(x)\, 2^{-j(\sigma - \frac{n}{p})}, \quad x \in \mathbb{R}^n, \tag{19.50}$$

is an atomic decomposition of φ^j in, say, $B_{pp}^\sigma(\mathbb{R}^n)$ (ignoring immaterial constants). We refer to [Triδ], 13.3 and 13.8, pp. 73, 75. We have to check the

ℓ_p-sum of the N_j terms, each $2^{j(\sigma-\frac{n}{p})}$. But

$$\|\varphi^j \,|B_{pp}^\sigma(\mathbb{R}^n)\| \leq c\, 2^{j(\sigma-\frac{n}{p})} \left(\sum_{k=1}^{N_j} 1\right)^{\frac{1}{p}}$$

$$\leq c'\, 2^{j(\sigma-\frac{n}{p}+\frac{d}{p})} \to 0 \quad \text{if} \quad j \to \infty. \tag{19.51}$$

By embedding we have also

$$\|\varphi^j \,|H_p^\sigma(\mathbb{R}^n)\| \to 0 \quad \text{if} \quad j \to \infty. \tag{19.52}$$

Now (19.48) follows from (19.44).

Step 2 After these preparations we can apply [AdH96], Theorem 10.1.1, p. 281, with $\alpha = s + \frac{n-p}{d}$ to $H_p^\alpha(\mathbb{R}^n)$. Since $s < 1$, we have (19.48) for all $\sigma = \alpha - k$ with $k \in \mathbb{N}$ and $\sigma \geq 0$. Hence by Step 1 and the explanations given in [AdH96], p. 234, part (a) of Theorem 10.1.1 in [AdH96] reduces to $\mathrm{tr}_\Gamma f = 0$ and, as a consequence, $D(\mathbb{R}^n \setminus \Gamma)$ is dense in $H_{p,\Gamma}^\alpha(\mathbb{R}^n)$.

Step 3 We prove the corresponding assertion for the B-spaces by real interpolation. Let

$$1 < p < \infty, \quad 1 \leq q < \infty, \quad -\infty < \sigma_0 < \sigma_1 < \infty. \tag{19.53}$$

Let $0 < \theta < 1$. Then

$$\left(H_p^{\sigma_0}(\mathbb{R}^n), H_p^{\sigma_1}(\mathbb{R}^n)\right)_{\theta,q} = B_{pq}^\sigma(\mathbb{R}^n), \quad \sigma = (1-\theta)s_0 + \theta s_1. \tag{19.54}$$

We refer for details to [Triβ], 2.4.1, 2.4.2, pp. 62–64. In particular,

$$\|f \,|B_{pq}^\sigma(\mathbb{R}^n)\| \sim \left(\int_0^\infty t^{-\theta q} K(t,f)^q \,\frac{dt}{t}\right)^{\frac{1}{q}}, \tag{19.55}$$

where $K(t,f)$ is Peetre's K-functional

$$K(t,f) = \inf\left(\|f_0 \,|H_p^{\sigma_0}(\mathbb{R}^n)\| + t\,\|f_1 \,|H_p^{\sigma_1}(\mathbb{R}^n)\|\right), \tag{19.56}$$

and the infimum is taken over all representations

$$f = f_0 + f_1 \quad \text{with} \quad f_0 \in H_p^{\sigma_0}(\mathbb{R}^n) \quad \text{and} \quad f_1 \in H_p^{\sigma_1}(\mathbb{R}^n). \tag{19.57}$$

Now let

$$\sigma_0 = \frac{n-d}{p} + s_0 \quad \text{and} \quad \sigma_1 = \frac{n-d}{p} + s_1 \quad \text{with} \quad 0 < s_0 < s_1 < 1. \tag{19.58}$$

We wish to prove that

$$\left(H_{p,\Gamma}^{\sigma_0}(\mathbb{R}^n), H_{p,\Gamma}^{\sigma_1}(\mathbb{R}^n)\right)_{\theta,q} = B_{pq,\Gamma}^\sigma(\mathbb{R}^n) \tag{19.59}$$

where
$$\sigma = \frac{n-d}{p} + s \quad \text{with} \quad s = (1-\theta)s_0 + \theta s_1. \tag{19.60}$$

By [JoW84], Chapter VII, 2.1, Theorem 3, p. 197, there is a common linear extension operator ext from the trace spaces
$$B_{pp}^{s_j}(\Gamma) \quad \text{into} \quad H_p^{\sigma_j}(\mathbb{R}^n), \quad \text{where} \quad j=0,1. \tag{19.61}$$

Then
$$P = id - ext \circ tr_\Gamma \tag{19.62}$$

is a common projection of
$$H_p^{\sigma_j}(\mathbb{R}^n) \quad \text{onto} \quad H_{p,\Gamma}^{\sigma_j}(\mathbb{R}^n), \quad \text{where} \quad j = 0,1. \tag{19.63}$$

Now by [Tri α], 1.17.1, p. 118,
$$\left(PH_p^{\sigma_0}(\mathbb{R}^n), PH_p^{\sigma_1}(\mathbb{R}^n)\right)_{\theta,q} = P\left(H_p^{\sigma_0}(\mathbb{R}^n), H_p^{\sigma_1}(\mathbb{R}^n)\right)_{\theta,q}. \tag{19.64}$$

This coincides with (19.59). One can give also a direct argument to prove (19.64). Let $f \in H_{p,\Gamma}^{\sigma_0}(\mathbb{R}^n)$ be optimally decomposed according to (19.57) and (19.56) with respect to the interpolation couple $H_p^{\sigma_0}(\mathbb{R}^n)$, $H_p^{\sigma_1}(\mathbb{R}^n)$. Then
$$f = Pf = Pf_0 + Pf_1 \tag{19.65}$$

shows that the K-functionals for the couples $H_p^{\sigma_0}(\mathbb{R}^n)$, $H_p^{\sigma_1}(\mathbb{R}^n)$ and $H_{p,\Gamma}^{\sigma_0}(\mathbb{R}^n)$, $H_{p,\Gamma}^{\sigma_1}(\mathbb{R}^n)$ are equivalent (independently of t). Now (19.59) with (19.58) follows from (19.54). By the density assertion for interpolation spaces, [Triα], 1.6.2, p. 39, $H_{p,\Gamma}^{\sigma_1}(\mathbb{R}^n)$ is dense in $B_{pq,\Gamma}^{\sigma}(\mathbb{R}^n)$. Since, by Step 2, $D(\mathbb{R}^n \backslash \Gamma)$ is dense in $H_{p,\Gamma}^{\sigma_1}(\mathbb{R}^n)$ it follows that $D(\mathbb{R}^n \backslash \Gamma)$ is also dense in $B_{pq,\Gamma}^{\sigma}(\mathbb{R}^n)$.

19.6 Complements and references

The crucial point of the proof of the above proposition is the use of a special case of the deep Theorem 10.1.1 in [AdH96], p. 281, attributed to Yu. V. Netrusov. This results in the density assertion for the H-spaces. As for the B-spaces one relies afterwards on the density properties of interpolation spaces and the substantial assertion concerning the existence of a common linear extension operator under the circumstances of the proposition proved in [JoW84], p. 197. Density problems in function spaces have a long and rich history going back to the very beginning of the theory of function spaces, [Sob50] (and

Sobolev's original papers from 1936–1938). If $\Gamma = \partial\Omega$ is the boundary of a bounded C^∞ domain Ω in $\mathbb{R}^n$, then we have had final answers for a long time: see [Triα], 4.7.1, p. 330, together with references and historical remarks. If Γ is not smooth (maybe an arbitrary set in $\mathbb{R}^n$ or the non-smooth boundary of a bounded domain) then the situation is more complicated and closely connected with the so-called problem of *spectral synthesis*. What is meant by spectral synthesis may be found in [AdH96], 9.13, p. 279, and in [Hed84]. In the latter paper the problem was posed (but as mentioned there, partial results had been obtained before, including assertions by the author of this paper and by the author of this book). The final solution of this problem for classical Sobolev spaces $W_p^m(\mathbb{R}^n)$ with $m \in \mathbb{N}$ and $1 < p < \infty$ was given in [HeW83]. This coincides essentially with [AdH96], Theorem 9.1.3, p. 234. Further explanations and references may be found in [AdH96], 9.1 and 9.13 (Note section). As said above, the extension of this theory to the (fractional) Sobolev spaces

$$H_p^s(\mathbb{R}^n), \quad 1 < p < \infty, \quad s > 0, \tag{19.66}$$

is the subject of [AdH96], Section 10. The version given there was especially prepared after discussions with Yu. V. Netrusov (as acknowledged there). Further references are given in [AdH96], 10.4, p. 303. The considerations are based on [Net92], where a solution of this problem was announced for the larger scale of the spaces

$$F_{pq}^s(\mathbb{R}^n), \quad s > 0, \quad 1 < p < \infty, \quad 1 < q < \infty. \tag{19.67}$$

Some special cases, based on the techniques developed in [JoW84], were considered before in [Mar87a] and [Wal91] (where $\Gamma = \partial\Omega$ is the boundary of a non-smooth domain, Lipschitz or d-set). We refer also to [FaJ00].

We return to the main subject of this section, the study of the operator B, given by (19.25). We complement the notation introduced in 19.2 and 19.3. As there, let Γ be a compact set in $\mathbb{R}^n$ with $\Gamma \subset \Omega$, where Ω is a bounded C^∞ domain in $\mathbb{R}^n$. Then as usual, $\overset{\circ}{H}{}^1(\Omega)$ and $\overset{\circ}{H}{}^1(\Omega\setminus\Gamma)$ is the completion of $D(\Omega)$ and $D(\Omega\setminus\Gamma)$ in $H^1(\Omega)$ and $H^1(\Omega\setminus\Gamma)$, respectively. Recall that we use $\sim$ according to (19.17), (19.18).

19.7 Theorem

Let Ω be a bounded C^∞ domain in $\mathbb{R}^n$ (where $n \in \mathbb{N}$) and let Γ be a compact d-set according to (18.1) such that $\Gamma \subset \Omega$ and $n - 2 < d < n$ (with $0 < d < 1$ when $n = 1$). Then B, given by

$$B = (-\Delta)^{-1} \circ tr^\Gamma, \tag{19.68}$$

as introduced in 19.2, 19.3, is a non-negative compact self-adjoint operator in $\overset{\circ}{H}{}^1(\Omega)$ with null-space

$$N(B) = \overset{\circ}{H}{}^1(\Omega\setminus\Gamma). \tag{19.69}$$

Furthermore, B is generated by the quadratic form

$$(Bf,g)_{\overset{\circ}{H}{}^1(\Omega)} = \int_\Gamma f(\gamma)\,\overline{g(\gamma)}\,\mu(d\gamma), \quad f \in \overset{\circ}{H}{}^1(\Omega),\ g \in \overset{\circ}{H}{}^1(\Omega), \tag{19.70}$$

with (19.29) as the scalar product in $\overset{\circ}{H}{}^1(\Omega)$. Let ϱ_k be the positive eigenvalues of B, repeated according to multiplicity and ordered by their magnitude, and let $u_k(x)$ be the related eigenfunctions,

$$Bu_k = \varrho_k u_k, \quad k \in \mathbb{N}. \tag{19.71}$$

(i) The largest eigenvalue is simple,

$$\varrho_1 > \varrho_2 \geq \varrho_3 \cdots. \tag{19.72}$$

Furthermore,

$$\varrho_k \sim k^{-1+\frac{n-2}{d}}, \quad k \in \mathbb{N}. \tag{19.73}$$

(ii) The eigenfunctions $u_k(x)$ are (classical) harmonic functions in $\Omega\setminus\Gamma$,

$$\Delta u_k(x) = 0 \quad \text{if} \quad x \in \Omega\setminus\Gamma. \tag{19.74}$$

(iii) Let $1 < p \leq \infty$, $\frac{1}{p} + \frac{1}{p'} = 1$, and $\varepsilon \in \mathbb{R}$. Then

$$u_k \in B_{p\infty}^{2-\frac{n-d}{p'}+\varepsilon}(\Omega) \quad \text{if, and only if,} \quad \varepsilon \leq 0, \tag{19.75}$$

in particular,

$$u_k \in C^{2-n+d+\varepsilon}(\Omega) \quad \text{if, and only if,} \quad \varepsilon \leq 0. \tag{19.76}$$

(iv) The eigenfunctions $u_1(x)$ have no zeroes in Ω,

$$u_1(x) = c\,u(x) \quad \text{with} \quad c \in \mathbb{C} \quad \text{and} \quad u(x) > 0 \quad \text{if} \quad x \in \Omega. \tag{19.77}$$

Proof *Step 1* The necessary explanations of what is meant by the operator B have been given in 19.2, 19.3 with a reference for further details to [Triδ], especially Sections 28 and 30. In particular, the above theorem might be considered as an extension of Theorem 30.2 in [Triδ], p. 234. By this theorem, B is a non-negative compact self-adjoint operator in $\overset{\circ}{H}{}^1(\Omega)$ with null-space

$$N(B) = \left\{ f \in \overset{\circ}{H}{}^1(\Omega) : tr_\Gamma f = 0 \right\}, \qquad (19.78)$$

generated by the quadratic form (19.70) and with (19.73). By (19.27) or Fig. 19.1, and Proposition 19.5, (19.78) coincides with (19.69). Hence it remains to prove that the largest eigenvalue ϱ_1 is simple and all assertions concerning the eigenfunctions, including (19.77).

Step 2 First we prove

$$u_k \in B_{p\infty}^{2-\frac{n-d}{p'}}(\Omega), \quad \text{where} \quad 1 < p \le \infty, \quad k \in \mathbb{N}, \qquad (19.79)$$

with the special case

$$u_k \in \mathcal{C}^{2-n+d}(\Omega), \quad k \in \mathbb{N}. \qquad (19.80)$$

This is essentially a matter of spectral invariance. Let

$$1 < p \le \infty, \quad \frac{1}{p}\max(1, n-d) < s < 2 - \frac{n-d}{p'}. \qquad (19.81)$$

Then $(\frac{1}{p}, s)$ belongs to the shaded regions in Figures 19.1 and 19.2.

By the explanations given in 19.3 and by (19.24) the operator B might also be considered as a compact operator in $B_{pq}^s(\Omega)$ with $1 \le q \le \infty$. We denote this restriction of B to $B_{pq}^s(\Omega)$ temporarily by $B(p,s)$ (the index q is unimportant and we may treat it as fixed in what follows). By the well-known spectral theory for compact operators in Banach spaces (Riesz's theorem) it follows that any eigenvalue $\varrho \ne 0$ of $B(p,s)$ has finite algebraic multiplicity. By (19.28) we have

$$u \in B_{p\infty}^{2-\frac{n-d}{p'}}(\Omega) \qquad (19.82)$$

for any (associated) eigenfunction (the upper line in the Figures 19.1 and 19.2). Now let q be fixed and let (p_0, s_0) and (p_1, s_1) be two couples with (19.81) and

$$1 < p_0 < p_1 \le \infty, \quad s_1 - \frac{n}{p_1} = s_0 - \frac{n}{p_0}, \qquad (19.83)$$

such that not only the couples $(\frac{1}{p_0}, s_0)$ and $(\frac{1}{p_1}, s_1)$ belong to the shaded regions in the figures, but also $(\frac{1}{p_0}, s_1)$, as indicated by the triangles in the figures. Recall

$$B_{p_0 q}^{s_0}(\Omega) \subset B_{p_1 q}^{s_1}(\Omega) \subset B_{p_0 q}^{s_1}(\Omega). \qquad (19.84)$$

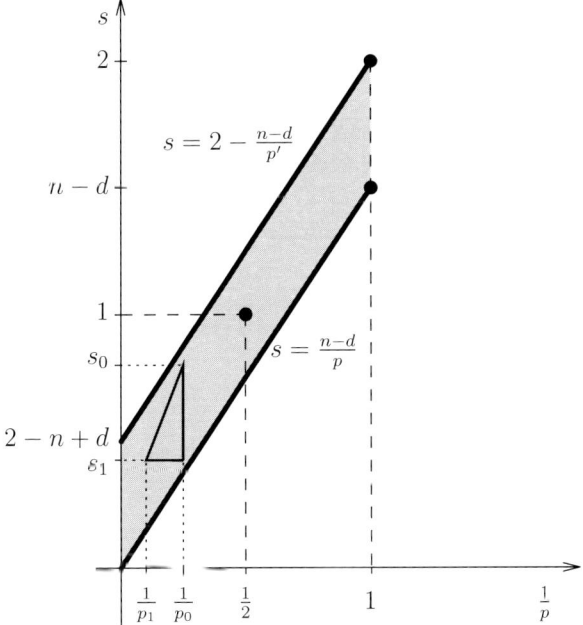

Fig. 19.1 $n-2 < d < n-1$

The first inclusion is a well-known (sharp) embedding, [Triβ], 2.7.1, p. 129. The second one comes from the assumption that Ω is bounded. It can be proved by using equivalent norms for B^s_{pq}-spaces based on local means, [Triγ], 2.4.6, p. 122, and Hölder's inequality for $L_p(\Omega)$-spaces. But as will be indicated in 19.8 below, (19.84) is also an immediate consequence of quarkonial representations as considered in Chapter I of this book. Now it follows easily that any eigenvalue $\varrho \neq 0$ and any related associated eigenfunction $u(x)$ of $B(p_0, s_0)$ is also an eigenvalue and an associated eigenfunction of $B(p_1, s_1)$ and we have (19.82) both with respect to (p_0, s_0) and (p_1, s_1) in place of (p, s). Then one obtains that the eigenvalues $\varrho \neq 0$, including their algebraic multiplicities and their root systems (associated eigenfunctions), for $B(p_0, s_0)$, $B(p_1, s_1)$, and $B(p_0, s_1)$ coincide. Starting from (p_0, s_0) any other couple in the shaded regions in the Fig. 19.1 and 19.2 can be reached by finitely many steps, including $H^1(\Omega) = B^1_{2,2}(\Omega)$. One can replace $H^1(\Omega)$ by $\overset{\circ}{H}{}^1(\Omega)$, which does not influence the above arguments. This proves (19.79) with the special case (19.80).

Step 3 We prove (ii). We apply the Dirichlet Laplacian to

$$(-\Delta)^{-1} \circ tr^\Gamma u_k = \varrho_k u_k, \quad k \in \mathbb{N}, \tag{19.85}$$

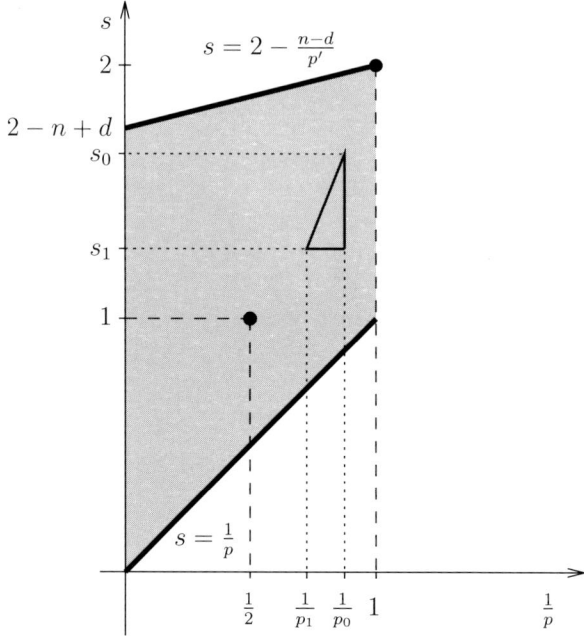

Fig. 19.2 $n-1 \leq d < n$

and obtain by (19.79),

$$\Delta u_k = -\varrho_k^{-1} tr^\Gamma u_k \in B_{p\infty}^{-\frac{n-d}{p'}}(\Omega), \quad 1 < p \leq \infty. \tag{19.86}$$

By (19.23) we have $supp\, tr^\Gamma u_k \subset \Gamma$ and hence

$$\Delta u_k = 0 \quad \text{in} \quad D'(\Omega \backslash \Gamma). \tag{19.87}$$

It is one of the very basic assertions of the theory of distributions that a distributional solution of the Laplace equation is also a classical solution.

Step 4   We prove (iii). By Step 2 it remains to disprove

$$u_k \in B_{p\infty}^{2-\frac{n-d}{p'}+\varepsilon}(\Omega) \quad \text{for some} \quad 1 < p \leq \infty \quad \text{and} \quad \varepsilon > 0. \tag{19.88}$$

Assume that we have (19.88). Let $\psi \in D(\Omega)$. By (19.86) and (19.23) we get

$$(\Delta u_k, \psi) = -\varrho_k^{-1} \int_\Gamma \psi(\gamma) \, (tr_\Gamma u_k)(\gamma) \, \mu(d\gamma). \tag{19.89}$$

Using the same notation as in (19.22), introduced in (9.168), we obtain

$$\Delta u_k \in B_{p\infty}^{-\frac{n-d}{p'}+\varepsilon, \Gamma}(\Omega). \tag{19.90}$$

The Hausdorff dimension of Γ is d. Then it follows from the distributional characterization of the Hausdorff dimension in [Triδ], Theorem 17.8, p. 130, that the space on the right-hand side of (19.90) is trivial (consists of the null distribution only). Hence $\Delta u_k = 0$ in $D'(\Omega)$ and

$$\Delta u_k(x) = 0 \text{ in } \Omega \quad \text{with} \quad u_k(y) = 0 \text{ if } y \in \partial\Omega, \tag{19.91}$$

in the classical sense. We obtain $u_k = 0$. This is a contradiction.

Step 5 Let $\varrho = \varrho_1$ be the largest eigenvalue of B. We prove that there is an eigenfunction $u = u_1$ with

$$Bu = \varrho u, \quad u(x) > 0 \quad \text{if} \quad x \in \Omega. \tag{19.92}$$

Recall that the self-adjoint non-negative compact operator B in $\overset{\circ}{H}{}^1(\Omega)$ is generated by the quadratic form (19.70). By a classical assertion of Hilbert space theory which goes back to Hilbert and Courant, [CoH24], a non-trivial function $v \in \overset{\circ}{H}{}^1(\Omega)$ is an eigenfunction of B with respect to the eigenvalue ϱ if, and only if,

$$\int_\Gamma |v(\gamma)|^2 \mu(d\gamma) = \varrho \sum_{j=1}^n \int_\Omega \left|\frac{\partial v}{\partial x_j}(x)\right|^2 dx. \tag{19.93}$$

In particular, $\bar{v}$, $\operatorname{Re} v$, and $\operatorname{Im} v$ are also eigenfunctions (or identically zero). Hence we may assume that v is real. Then

$$\varrho \sum_{j=1}^n \int_\Omega \left|\frac{\partial |v|}{\partial x_j}(x)\right|^2 dx = \varrho \sum_{j=1}^n \int_\Omega \left|\frac{\partial v}{\partial x_j}(x)\right|^2 dx = \int_\Gamma |v(\gamma)|^2 \mu(d\gamma). \tag{19.94}$$

As for the first equality we refer to [GiT77], Lemma 7.6, p. 145 or to [Zie89], 2.1.8, p. 47. Hence, $|v| \in \overset{\circ}{H}{}^1(\Omega)$ is also an eigenfunction with respect to the largest eigenvalue ϱ. Then we have

$$Bw = \varrho w \quad \text{with} \quad w(x) = |v|(x) - v(x) \geq 0. \tag{19.95}$$

First we assume that w does not vanish identically. Then w is an eigenfunction. By (19.80) it follows that w is continuous in $\overline{\Omega}$. Let $G(x,y)$ be the (symmetric) Green's function for the Dirichlet Laplacian Δ in Ω. We have

$$\varrho w(x) = (-\Delta)^{-1} \circ tr^\Gamma w(x) = \int_\Gamma G(x,\gamma)\, (tr_\Gamma w)(\gamma)\, \mu(d\gamma), \quad x \in \Omega. \tag{19.96}$$

(In 19.8 below we add a comment about this construction.) Since $w(x) \geq 0$ is non-trivial and continuous, and $G(x, \gamma) > 0$ if $x \in \Omega$ and $\gamma \in \Gamma$, it follows that there is a neighbourhood of some point $\gamma^0 \in \Gamma$ where w is strictly positive. Then, by (19.96) we have $w(x) > 0$ for all $x \in \Omega$. Hence $v(x) < 0$ for all $x \in \Omega$ and $u = -v$ satisfies (19.92). Next we assume that $w(x)$ in (19.95) is identically zero. Then we have $v(x) \geq 0$. By the same arguments as above we get $v(x) > 0$ in Ω and, hence, $u = v$ satisfies (19.92). We proved a little bit more than stated: If v is a real eigenfunction with respect to the largest eigenvalue ϱ, then v is in Ω either strictly positive or strictly negative.

Step 6 We prove (i) and (iv). In Step 1 we mentioned that (19.73) is known. We prove that $\varrho = \varrho_1$ is simple. Let us assume that ϱ is not simple. Then by Step 5 there are two real $\overset{\circ}{H}{}^1(\Omega)$-orthogonal eigenfunctions $v^1(x)$ and $v^2(x)$ with

$$\int_\Gamma v^1(\gamma) \, v^2(\gamma) \, \mu(d\gamma) = \left(Bv^1, v^2\right)_{\overset{\circ}{H}{}^1(\Omega)} = \varrho \left(v^1, v^2\right)_{\overset{\circ}{H}{}^1(\Omega)} = 0. \qquad (19.97)$$

By the end of Step 5 we may assume that v^1 and v^2 are strictly positive continuous functions in Ω and hence on Γ. But this contradicts (19.97). Hence ϱ is simple and we have by Step 5 also (19.77). The proof is complete.

19.8 Two comments

We comment on two points of the above proof. First, as we have noted, (19.84) is known. But the two embeddings follow also from the quarkonial representations for B-spaces. By (19.83) the normalizing factors for the (s_0, p_0)-β-quarks and the (s_1, p_1)-β-quarks in (2.16) are the same. Then the first embedding in (19.84) follows from (2.7), Definition 2.6(i) and the monotonicity of the ℓ_p-spaces. As for the second embedding we remark that one needs $N_\nu \sim 2^{\nu n}$ balls of radius $c 2^{-\nu}$ to cover Ω in connection with the (s_1, p_1)-β-quarks in (2.16). Furthermore, by Hölder's inequality applied to $\lambda_l \in \mathbb{C}$,

$$\left(\sum_{l=1}^{N_\nu} |\lambda_l|^{p_0}\right)^{\frac{1}{p_0}} \leq c' \, 2^{\nu(\frac{1}{p_0} - \frac{1}{p_1})n} \left(\sum_{l=1}^{N_\nu} |\lambda_l|^{p_1}\right)^{\frac{1}{p_1}} \qquad (19.98)$$

for some $c' > 0$. By Definition 5.3 quarkonial decompositions for B-spaces on Ω can be restricted to the above-indicated N_ν balls. The first factor on the right-hand side of (19.98) compensates the normalizing factors for (s_1, p_1)-β-quarks and (s_1, p_0)-β-quarks in (2.16). Then one gets the second embedding in (19.84) as above as a consequence of (2.7) and Definition 2.6. Secondly, we add a comment on (19.96). If $x \in \Omega \backslash \Gamma$ then the right-hand side of (19.96) is well defined, since both $G(x,\gamma)$ and $(tr_\Gamma w)(\gamma)$ are continuous and the outcome coincides with the left-hand side: The latter follows by definition for the Sobolev mollifications $(tr_\Gamma w)_h(x)$. The rest is a matter of completion. However these observations remain also valid if $x \in \Gamma$. Let $n \geq 3$. Then

$$G(x,\gamma) \sim |x-\gamma|^{2-n} \quad \text{near } x.$$

This singularity is well compensated by (18.1), since $d > n-2$. Similarly if $n = 2$. We return to this point in greater detail in Section 20 in connection with single layer potentials.

19.9 Discussion

As said above, Theorem 19.7 is the continuation of Theorem 30.2 in [Triδ], p. 234. We described in 19.3 what was known so far. We repeated the corresponding assertions in Theorem 19.7 with references to [Triδ]. Detailed discussions and interpretations connected with the physical background described in 19.1 may be found in [Triδ], 30.3, 30.4, pp. 235–236, including the very few references dealing with problems of this type. We took over the crucial equivalence (19.73). Its proof in [Triδ] is based on quarkonial representations, entropy numbers and approximation numbers. At that time the attempt to prove assertions of type (19.73) was the decisive impetus to develop the theory of quarkonial decompositions for function spaces. We return to this technique in the later parts of this section and also in some other sections of this chapter. In comparison with the spectral theory for, say, the (Dirichlet) Laplacian in bounded domains with smooth or fractal boundary one may ask whether (19.73) can be strengthened by

$$\varrho_k = c\, k^{-1+\frac{n-2}{d}}(1+o(1)), \quad k \in \mathbb{N}, \tag{19.99}$$

where $c > 0$ is a suitable constant and $o(1)$ is a remainder term tending to zero if k tends to infinity. But this cannot be expected. There is even a counterexample. We refer to [Triδ], 30.4, pp. 235–236, and the literature mentioned there.

19.10 Nullstellenfreiheit

Part (iv) of Theorem 19.7, including the simplicity of the largest eigenvalue is the fractal version of Courant's classical assertion for the Dirichlet Laplacian in Ω (1924). Courant's strikingly short elegant proof on less than one page may be found in [CoH24], pp. 398–399, where the title '*Charakterisierung der ersten Eigenfunktion durch ihre Nullstellenfreiheit*' indicates what follows in a few lines. Based on quadratic forms Courant relies (in recent language) on

H^1-arguments. But he did not bother very much about the technical rigour of his few-lines-proof. Problems of this type have a long and rich history at least since that time. More recent versions may be found in [Tay96], pp. 315–316. We refer also to [Tai96] for generalizations and to [ReS78], Theorem XIII.43, for an abstract version.

19.11 Singular perturbations; the case $n = 1$

First we deal with $n = 1$. Then we have by Theorem 19.7 for B given by (19.68) and d-sets Γ with $0 < d < 1$,

$$\varrho_k \sim k^{-1-\frac{1}{d}} \quad \text{and} \quad u_k \in \mathcal{C}^{1+d}(\Omega), \quad k \in \mathbb{N}. \tag{19.100}$$

Of course, we have $-\Delta u(x) = -u''(x)$ in this case and (19.74) means that the eigenfunctions $u_k(x)$ are linear in $\Omega \setminus \Gamma$. In this case, B makes sense for any finite Radon measure μ with, say,

$$supp\, \mu = \Gamma \subset \Omega = (-1, 1). \tag{19.101}$$

First we remark that

$$L_p(\Gamma) \subset L_1(\Gamma) \subset B^0_{1,\infty}(\Omega) \subset B^{-\frac{1}{p'}}_{p,\infty}(\Omega), \quad 1 \leq p \leq \infty, \tag{19.102}$$

where again $\frac{1}{p} + \frac{1}{p'} = 1$. The first embedding is Hölder's inequality, the last embedding follows again from [Triβ], 2.7.1, p. 129. As for the middle embedding we refer to (9.9)–(9.11), extended to $L_1(\Gamma)$ (complex measures). All spaces in the shaded region in Fig. 19.2, now applied to $n = 1$, are continuously embedded in $C(\Omega)$, the space of continuous functions on the interval $(-1, 1)$. Hence there are pointwise traces for all spaces of interest in Theorem 19.7 and its proof. We have for the eigenfunctions u_k,

$$u_k \in B^{2-\frac{1}{p'}}_{p\infty}(\Omega) = B^{1+\frac{1}{p}}_{p\infty}(\Omega) \subset \mathcal{C}^1(\Omega), \quad 1 < p \leq \infty. \tag{19.103}$$

We have a quick look at the case

$$\Gamma = \bigcup_{j=1}^{N} \{a_j\} \quad \text{with} \quad -1 < a_1 < \cdots < a_N < 1. \tag{19.104}$$

Then $d = \dim_H \Gamma = 0$. Let δ_{a_j} be the δ-distribution with the off-points a_j. In this case we have

$$Bu(x) = (-\Delta)^{-1} \circ tr^\Gamma u(x) \qquad (19.105)$$

$$= \int_{-1}^{1} G(x,y) \sum_{j=1}^{N} u(a_j) \delta_{a_j}(y) \, dy$$

$$= \sum_{j=1}^{N} u(a_j) G(x, a_j), \quad x \in (-1, 1),$$

where $G(x,y)$ is the Green's function for the Dirichlet Laplacian $-\Delta u = -u''$ on $(-1,1)$ with respect to the off-point $y \in (-1,1)$. Recall that $G(x,y)$ for fixed y is continuous on $[-1,1]$, linear in $[-1,y)$ and in $(y,1]$, with $G(-1,y) = G(1,y) = 0$ and $G''(\cdot, y) = \delta_y$. Hence Bu is a polygonal line, the dimension of the range of B is N, and the N eigenvalues of B are the N real eigenvalues of the symmetric matrix $(G(a_j, a_k))_{j,k=1}^N$. The physical interpretation is a vibrating string where the whole mass is evenly distributed in the N points a_j.

We add some comments. As mentioned, in the one-dimensional case B makes sense for any finite Radon measure μ with (19.101). In case of Lebesgue measure, say, restricted to Ω, we have $d = 1$ in (19.100), which means $\varrho_k \sim k^{-2}$. One can extend this assertion to

$$\varrho_k \sim k^{-2} \quad \text{if, and only if,} \quad |\Gamma| > 0,$$

and

$$k^2 \varrho_k \to 0 \quad \text{when} \quad k \to \infty \quad \text{if, and only if,} \quad |\Gamma| = 0.$$

We do not go into detail. We only mention that these observations can be proved using entropy numbers for estimates from above and approximation numbers for estimates from below. The abstract background will be described in 19.16 below and used later extensively. However this phenomenon is not new. It can be found in a larger context in [BiS74] and in [Bor70].

If $n \geq 2$ then it is not possible by our method to deal with arbitrary finite Radon measures μ with compact support. On the other hand, in the slightly different but nearby context of quantum mechanics, it is of interest to study, say, $-\Delta + \delta$ in $\mathbb{R}^n$, or related operators with strongly singular measures. This attracted a lot of attention since the 1960s. The state of the art may be found in [AlK00] with almost a thousand references. The methods there and here are different. Nevertheless the question arises of whether they can complement each other.

19.12 Fractal drums in the plane

In 19.11 we discussed fractal strings, where $n = 1$. Also the cases $n = 2$ and $n = 3$ are of physical interest. As explained in 19.1, if $n = 2$ then one might think of vibrations of a drum $\Omega \subset \mathbb{R}^2$ with a fractal membrane $\Gamma \subset \Omega$. Let Ω be a bounded C^∞ domain in the plane $\mathbb{R}^2$, let Γ be a compact d-set with $\Gamma \subset \Omega$ and $0 < d < 2$ and let B be given by (19.68). Then we have by Theorem 19.7,

$$\varrho_k \sim k^{-1} \quad \text{and} \quad u_k \in \mathcal{C}^d(\Omega), \quad k \in \mathbb{N}. \tag{19.106}$$

In the plane, -1 is the classical Weyl exponent concerning the distribution of eigenvalues of the inverse of the Dirichlet Laplacian $(-\Delta)^{-1}$ in bounded smooth domains. We discussed this problem in greater detail in [Triδ], 26.1–26.3 and 28.9–28.11, pp. 199–202 and 230, respectively. In particular, the exponent -1 in (19.106) is independent of d. This observation can be immediately extended (based on the techniques developed in [Triδ], explained and used later on in this section and in the following sections of this chapter) to the following situation: Let Ω be a bounded C^∞ domain in the plane, let $(-\Delta)^{-1}$ be the inverse of the related Dirichlet Laplacian, let Γ_j where $j = 1, \ldots, N$ be pairwise disjoint compact d_j-sets with $0 < d_j < 2$ and

$$\Gamma = \bigcup_{j=1}^{N} \Gamma_j \subset \Omega. \tag{19.107}$$

Naturally, Γ is equipped with the Radon measure $\mu = \sum_{j=1}^{N} \mu_j$, where $\mu_j = \mathcal{H}^{d_j}|\Gamma_j$ is the restriction of the Hausdorff measure $\mathcal{H}^{d_j}$ to Γ_j. Then B, given by (19.68), is well defined and the distribution of the positive eigenvalues ϱ_k is again equivalent to k^{-1}. The question arises for which more general Radon measures μ in the plane, B makes sense and the positive eigenvalues ϱ_k are equivalent to k^{-1} (Weylian behaviour).

19.13 Definition

Let μ be a finite Radon measure in the plane $\mathbb{R}^2$ with compact support. Then μ is called a *Weyl measure* if for any bounded C^∞ domain Ω in $\mathbb{R}^2$ with

$$\operatorname{supp} \mu = \Gamma \subset \Omega \tag{19.108}$$

the quadratic form (19.70) generates a non-negative self-adjoint compact operator B in $\overset{\circ}{H}{}^1(\Omega)$ with

$$\varrho_k \sim k^{-1}, \quad k \in \mathbb{N}, \tag{19.109}$$

where ϱ_k are the positive eigenvalues of B, repeated according to multiplicity and ordered by their magnitude.

19.14 Discussion

Let $n = 2$ and let μ, Γ, Ω be as in the above definition. We ask under what conditions μ is a Weyl measure. By (19.70) and 9.2, especially (9.14), one has first to check whether there is a constant $c > 0$ such that

$$\|tr_\Gamma \, \varphi \,|L_2(\Gamma)\| \le c \,\|\varphi\,|\overset{\circ}{H}{}^1(\Omega)\| = c \left(\sum_{j=1}^n \int_\Omega \left| \frac{\partial \varphi}{\partial x_j}(x) \right|^2 dx \right)^{\frac{1}{2}} \quad (19.110)$$

for all $\varphi \in D(\Omega)$. Then, by completion, the trace operator tr_Γ,

$$tr_\Gamma \,:\, \overset{\circ}{H}{}^1(\Omega) \mapsto L_2(\Gamma), \quad (19.111)$$

exists, where $L_2(\Gamma)$ must be understood with respect to the given measure μ according to the notation (9.14). A refined version of what is meant by traces has been given in Step 1 of the proof of Proposition 19.5. Recall the classical duality assertion

$$\left(\overset{\circ}{H}{}^1(\Omega) \right)' = H^{-1}(\Omega) \quad (19.112)$$

in the understanding of the dual pairing $(D(\Omega), D'(\Omega))$, [Tria], 4.8.2, p. 332. Then we have by (9.19)–(9.21) for the identification operator id_Γ,

$$id_\Gamma \,:\, L_2(\Gamma) \mapsto H^{-1}(\Omega). \quad (19.113)$$

Let again tr^Γ be given by (19.23). Then we obtain by (19.111), (19.113) and (19.16) that

$$B = (-\Delta)^{-1} \circ tr^\Gamma \,:\, \overset{\circ}{H}{}^1(\Omega) \mapsto \overset{\circ}{H}{}^1(\Omega), \quad (19.114)$$

is a bounded operator and as indicated in 19.3 the generator of the quadratic form (19.70). To make clear what is going on we remark that we have for any finite compactly supported Radon measure μ in the plane

$$id_\Gamma \, L_2(\Gamma) \subset B^0_{1,\infty}(\Omega) \subset B^{-1}_{2,\infty}(\Omega). \quad (19.115)$$

This follows from (9.9), (9.12), extended to complex measures μ (in particular $f \in L_2(\Gamma)$). Since $H^{-1}(\Omega) = B^{-1}_{2,2}(\Omega)$ the spaces in (19.113) and (19.115) differ only by the third index. This makes clear that we have a rather delicate limiting situation. In particular by 13.1 or Theorem 11.4(ii), especially (11.23), always with $n = 2$, there is no continuous embedding of $H^1(\Omega)$ in $C(\Omega)$. As a consequence a measure μ with (19.111) must be diffuse (or non-atomic; as for notation we refer to 9.17). By Theorem 13.2 with $n = 2$ the measure μ with (19.111) must compensate the singularity behaviour of functions belonging to $H^1(\Omega)$ expressed by the growth envelope

$$\mathfrak{E}_G H^1 = \left(\sqrt{|\log t|}, 2 \right). \tag{19.116}$$

On the other hand, by Theorem 9.3, we have a necessary and sufficient criterion under which circumstances the trace operator (19.111) exists. Let $f_{\nu m}$ be given by (9.26). Then tr_Γ in (19.111) exists if, and only if,

$$\sup \sum_{\nu \in \mathbb{N}_0} \sum_{m \in \mathbb{Z}^2} f_{\nu m}^2 < \infty, \tag{19.117}$$

where the supremum is taken over all

$$f \in L_2(\Gamma) \quad \text{with} \quad f \geq 0 \quad \text{and} \quad \|f \,|L_2(\Gamma)\| \leq 1.$$

This follows from (9.29) with $s = 1$ and $n = p = 2$. Inserting $f = 1$ on $\Gamma = \operatorname{supp} \mu$ in (19.117) one gets that

$$\sum_{\nu \in \mathbb{N}_0} \sum_{m \in \mathbb{Z}^2} \mu(2Q_{\nu m})^2 < \infty \tag{19.118}$$

is necessary for (19.111), whereas by Corollary 9.8,

$$\sum_{\nu=0}^{\infty} \sup_{m \in \mathbb{Z}^2} \mu(2Q_{\nu m}) < \infty \tag{19.119}$$

is sufficient for (19.111). It might be of interest to have a closer look at the interplay of the growth envelope (19.116), the conditions (19.117), (19.118), (19.119) and the the operator B in (19.114). This will not be done here in a systematic way. We wish to find out sufficient conditions ensuring that μ is a Weyl measure according to Definition 19.13.

19.15 Strongly diffuse measures

Let μ be a finite compactly supported Radon measure in the plane $\mathbb{R}^2$. If μ has an atom then (19.118) is violated. Hence any Weyl measure must be necessarily diffuse. As for notation we refer to [Bou56], §5.10, p. 61 or [Die75], 13.18, p. 215, where diffuse means non-atomic. But one easily finds diffuse measures which do not satisfy (19.118). We describe an example. Let μ be concentrated on the line segment

$$I = \left\{ (x,y) : 0 \le x \le \frac{1}{2},\ y = 0 \right\} \tag{19.120}$$

in $\mathbb{R}^2$ and let the restriction $\mu|I$ be given by

$$\frac{d\mu}{dt} = \left(t |\log t|^{b+1} \right)^{-1}, \quad b > 0,\ 0 < t < \frac{1}{2}. \tag{19.121}$$

Let Q_ν be the square

$$Q_\nu = [0, 2^{-\nu+1}] \times [-2^{-\nu}, 2^{-\nu}], \quad \nu \in \mathbb{N}. \tag{19.122}$$

Then we have, by definition,

$$\mu(Q_\nu) = \int_0^{2^{-\nu+1}} \frac{d\mu}{dt}(t)\, dt \sim \nu^{-b}, \quad \nu \in \mathbb{N}. \tag{19.123}$$

If $0 < b \le \frac{1}{2}$, then (19.118) is wrong; if $b > 1$, then the sufficient condition (19.119) is satisfied. In any case if μ is a Weyl measure, then it must be diffuse in a qualified way. We describe the class of measures we have in mind.

Let μ be a finite compactly supported Radon measure in the plane. Then, by definition:

(i) μ satisfies the doubling condition if there is a number $c \ge 1$ such that

$$\mu(B(\gamma, 2r)) \le c\,\mu(B(\gamma, r)), \tag{19.124}$$

for all $\gamma \in \Gamma = \operatorname{supp}\mu$, all r, $0 < r < 1$, and all circles $B(\gamma, r)$ centred at γ and of radius r.

(ii) μ is strongly diffuse if it satisfies the doubling condition and if there is a number $\varkappa$, $0 < \varkappa < 1$, such that

$$\mu(Q_1) \le \frac{1}{2} \mu(Q_0) \tag{19.125}$$

for any square Q_0 centred at some point $\gamma_0 \in \Gamma = \operatorname{supp}\mu$ and of side-length r with $0 < r < 1$ and any sub-square Q_1 with $Q_1 \subset Q_0$ centred at some point $\gamma_1 \in \Gamma$ and of side-length $\varkappa r$.

Of course, in (ii) all squares Q_0 and Q_1 for all points $\gamma_0 \subset \Gamma$, $\gamma_1 \subset \Gamma$, and all r, $0 < r < 1$, with the indicated properties must be checked. Incorporating the doubling condition in the definition of strongly diffuse measures is convenient for us. One may call μ (not necessarily satisfying the doubling condition) to be *directionally strongly diffuse* if one has (19.125) for the squares Q_0, Q_1 as indicated, now, in addition, with sides parallel to the axes of coordinates. Then the two conditions (doubling and directionally strongly diffuse) are independent. The Dirac measure satisfies (19.124), but not (19.125). Conversely, let $\mu = \mu_1 + \mu_2$, where μ_1 is generated by the restriction of the Lebesgue line measure to I in (19.120) and μ_2 is the restriction of the Lebesgue measure on $\mathbb{R}^2$ to a square with I as a base-line. Then μ is directionally strongly diffuse, but does not satisfy the doubling condition. If one admits in the definition of directionally strongly diffuse measures not only squares with sides parallel to the axes, but arbitrary squares, then there is no $\varkappa > 0$ such that (19.125) is always satisfied with respect to the above measure $\mu = \mu_0 + \mu_1$. Hence the definition of directionally strongly diffuse measures depends on the axes of coordinates. This might be of some interest in connection with anisotropic problems. But for our (isotropic) problems the above definition of strongly diffuse measures which includes the doubling condition is at least convenient.

19.16 Entropy numbers and approximation numbers

The second main result of this section is to prove that strongly diffuse measures in the plane are Weyl measures according to 19.13. This will be done in Theorem 19.17 below. We rely on entropy numbers and approximation numbers and their relations to the spectral theory of compact operators in quasi-Banach spaces. We developed this theory in [ET96], 1.3, pp. 7–22, in detail with complete proofs and with references to the literature. In [Triδ], Section 6, pp. 33–35, we summarized the main assertions needed, mostly following [ET96], and complemented the references. The proof of (19.73) given in [Triδ], Sections 28 and 30, is (as we hope) an outstanding example of this technique. We give here a very brief description of the bare minimum needed to provide an understanding of the relations between

entropy numbers, approximation numbers, and eigenvalues,

on an abstract level (restricting ourselves to Banach spaces, in contrast to [ET96] and [Triδ], where we developed this theory in quasi-Banach spaces). The interplay of this abstract theory with quarkonial decompositions as developed in Chapter I of this book produces equivalences of type (19.73) and (19.109). But this will be explained in the course of the proof of Theorem 19.17 below and in the following sections of this chapter.

Let A, B be complex Banach spaces. The family of all linear and bounded operators $T : A \mapsto B$ will be denoted by $L(A, B)$ or $L(A)$ if $A = B$. Let

$$U_B = \{ b \in B : \|b \,|\, B\| \leq 1 \}$$

be the unit ball in the Banach space B. Then the entropy numbers are defined as follows.

Let A, B be complex Banach spaces and let $T \in L(A, B)$. Then for all $k \in \mathbb{N}$ the kth entropy number $e_k(T)$ of T is defined by

$$e_k(T) = \inf \left\{ \varepsilon > 0 \,:\, T(U_A) \subset \bigcup_{j=1}^{2^{k-1}} (b_j + \varepsilon U_B) \text{ for some } b_1, \ldots, b_{2^{k-1}} \in B \right\}. \tag{19.126}$$

Of course $e_1(T) = \|T\|$. Recall that

$$T \in L(B) \text{ is compact if, and only if, } e_k(T) \to 0 \text{ for } k \to \infty. \tag{19.127}$$

Furthermore, if $T \in L(B)$ is compact, then the spectrum of T, apart from the point 0, consists solely of eigenvalues of finite algebraic multiplicity. This is Riesz's theory. Let $\{\varrho_k(T)\}_{k \in \mathbb{N}}$ be the sequence of all non-zero eigenvalues of the compact operator $T \in L(B)$, repeated according to algebraic multiplicity and ordered so that

$$|\varrho_1(T)| \geq |\varrho_2(T)| \geq \cdots \to 0. \tag{19.128}$$

If T has only m $(< \infty)$ distinct eigenvalues and M is the sum of their algebraic multiplicities we put $\varrho_n(T) = 0$ for all $n > M$. The crucial observation in our context is *Carl's inequality*

$$|\varrho_k(T)| \leq \sqrt{2}\, e_k(T), \quad k \in \mathbb{N}, \tag{19.129}$$

proved in [Carl81] and [CaT80] for Banach spaces and extended to quasi-Banach spaces in [ET96]. We followed [Triδ], Section 6 and [ET96], 1.3. There one finds further results and references to the extensive literature. Furthermore we need the approximation numbers which are defined as follows.

Let A, B be complex Banach spaces and let $T \in L(A, B)$. Then for all $k \in \mathbb{N}$ the kth approximation number $a_k(T)$ of T is defined by

$$a_k(T) = \inf \{ \|T - L\| \,:\, L \in L(A, B), \operatorname{rank} L < k \}, \tag{19.130}$$

where $\operatorname{rank} L$ is the dimension of the range of L.

Of course $a_1(T) = \|T\|$. Let now H be a Hilbert space and $T \in L(H)$ be a compact self-adjoint operator.

Let $\{\lambda_k\}_{k \in \mathbb{N}}$ be the sequence of all eigenvalues of T, repeated according to their geometric multiplicity and ordered so that

$$|\lambda_1(T)| \geq |\lambda_2(T)| \geq \cdots. \tag{19.131}$$

Then

$$|\lambda_k(T)| = a_k(T), \quad k \in \mathbb{N}. \tag{19.132}$$

This is a well-known classical assertion. References, further information, and also comments about the relations between $e_k(T)$ and $a_k(T)$ have been given in [ET96], 1.3, p. 7–22, and [Triδ], 24.3–24.7, p. 191–192. We do not repeat these assertions with exception of some rather elementary inequalities which are needed later on.

Let A, B, C be complex Banach spaces, let $S \in L(A,B)$, $T \in L(A,B)$ and $R \in L(B,C)$. Let h_k be either the entropy numbers e_k or the approximation numbers a_k.

(i) $\|T\| = h_1(T) \geq h_2(T) \geq \cdots$.

(ii) *For all $k \in \mathbb{N}$, $l \in \mathbb{N}$,*

$$h_{k+l-1}(R \circ S) \leq h_k(R)\, h_l(S), \tag{19.133}$$

and

$$h_{k+l-1}(S + T) \leq h_k(S) + h_l(T). \tag{19.134}$$

To avoid a misunderstanding we remark that above either all h_j are entropy numbers or all h_j are approximation numbers (no mixed inequalities).

19.17 Theorem

A finite, compactly supported, strongly diffuse Radon measure in the plane $\mathbb{R}^2$ as defined in 19.15 (ii), is a Weyl measure according to 19.13.

Proof *Step 1* We begin with some preliminaries. Let Ω be a bounded C^∞ domain in the plane $\mathbb{R}^2$ with

$$\operatorname{supp}\mu = \Gamma \subset \Omega, \tag{19.135}$$

where μ is a finite Radon measure with (19.124), (19.125). We may assume $\mu(\Gamma) = 1$. We must prove that the right-hand side of (19.70) is a bounded quadratic form in $\overset{\circ}{H}{}^1(\Omega)$. Then, by the previous considerations, the generator B, defined by (19.70), is non-negative and self-adjoint and can be represented by (19.68) in the interpretation of (19.114), based on (19.113), and the explanations given in 19.3. Hence first one has to prove that

$$\operatorname{tr}_\Gamma : \overset{\circ}{H}{}^1(\Omega) \mapsto L_2(\Gamma) \tag{19.136}$$

is bounded. As before, $L_2(\Gamma)$ is the L_2-space with respect to the given measure μ. We begin with a closer look at the interplay of μ, the geometry of its support $\Gamma = supp\,\mu$ and the quarkonial set-up in 9.24.

Step 2 Let μ be the above measure with $\mu(\Gamma) = 1$ and the compact support $\Gamma = supp\,\mu$. We use the quarkonial set-up in 9.24, where we now replace the balls

$$B(\gamma^{k,m}, c_2\, 2^{-k}) \quad \text{by the open cubes} \quad Q_{k,m} \tag{19.137}$$

with sides parallel to the axes, centred at $\gamma^{k,m} \in \Gamma$ and with side-length $c_2\, 2^{-k}$. This is immaterial for the quarkonial approach described there. Otherwise we use the same notation as in 9.24. We may assume in addition that

$$\{\gamma^{k,m} : m = 1, \ldots, M_k\} \subset \{\gamma^{k+1,m} : m = 1, \ldots, M_{k+1}\}, \quad k \in \mathbb{N}_0, \tag{19.138}$$

and that for any cube $Q_{k+1,m}$ there is a cube $Q_{k,l}$ with

$$Q_{k+1,m} \subset Q_{k,l}, \quad k \in \mathbb{N}_0. \tag{19.139}$$

The additional assumption (19.138) does not cause any problems. If (19.139) is not satisfied one can replace all squares $Q_{k,m}$ for all admitted k and m by $2Q_{k,m}$, centred at $\gamma^{k,m}$ and of side-length $c_2\, 2^{-k+1}$. It follows by geometrical reasoning that these modified cubes have the desired property. Hence, we assume that also (19.139) holds from the very beginning. Again this modification is immaterial for the quarkonial approach in 9.24 and what follows afterwards. We may also assume that only one cube with $k = 0$ is needed to cover Γ. Let K_j be the number of all cubes $Q_{k,m}$ with

$$2^{-j} < \mu(Q_{k,m}) \leq 2^{-j+1}, \quad j \in \mathbb{N}_0. \tag{19.140}$$

We wish to estimate the number of these cubes. Let $\varkappa = 2^{-L}$ with $L \in \mathbb{N}$ in 19.15(ii). Let $Q_{k,m}$ be a cube with (19.140) and let $Q_{k+L,l}$ be a sub-cube of $Q_{k,m}$ in iteration of (19.139). Then by (19.125) and (19.140),

$$\mu(Q_{k+L,l}) \leq 2^{-j}. \tag{19.141}$$

Let $K_{j,0}$ be the number of cubes $Q_{kL,m}$ with $k \in \mathbb{N}_0$ and (19.140). By the above construction (19.139) and (19.141) these cubes are disjoint for different values of k. Together with the controlled overlapping of cubes with the same k it follows that

$$2^{-j} K_{j,0} \leq \sum \mu(Q_{kL,m}) \leq c, \tag{19.142}$$

where $c > 0$ is independent of j, the sum is taken over all k and m where the respective cubes have the property (19.140). Similarly one can estimate $K_{j,l}$ with $l = 0, 1, \ldots, L-1$, being the number of cubes $Q_{kL+l,m}$ with the property (19.140). We get

$$K_j = \sum_{l=0}^{L-1} K_{j,l} \leq c\, 2^j = N_j \tag{19.143}$$

for some $c > 0$ which is independent of j.

Step 3 After these preparations we first observe that there is a number $D > 0$ (diffusion number) with

$$\mu(Q_{j,m}) \leq c\, 2^{-jD}, \quad j \in \mathbb{N}, \tag{19.144}$$

for some $c > 0$: If L has the above meaning and

$$kL \leq j < (k+1)L, \tag{19.145}$$

then, using (19.139) and iteratively (19.125),

$$\mu(Q_{j,m}) \leq \mu(Q_{kL,l}) \leq c\, 2^{-k} \leq c'\, 2^{-jD} \tag{19.146}$$

with $D = L^{-1}$. We apply Corollary 9.8(ii) with

$$p = r = n = 2, \quad s = 1 \quad \text{and} \quad d = D > 0, \tag{19.147}$$

and obtain by (9.28) that

$$\operatorname{tr}_\Gamma : \quad H^1(\mathbb{R}^2) \hookrightarrow L_2(\Gamma) \tag{19.148}$$

is a bounded operator. Since $\Gamma \subset \Omega$ this is the same as (19.136). Let $t = t_2 \geq \frac{D}{2} > 0$ be the typical number introduced in Definition 9.25, where the balls $\tilde{B}_{km}$ in (9.109) can be replaced by the above cubes $Q_{k,m}$ (since we have the doubling condition it does not matter whether we choose $Q_{k,m}$ or $2Q_{k,m}$). Here D has the same meaning as in (19.146). We apply Theorem 9.33 to $H^t(\Gamma) = B^t_{2,2}(\Gamma)$ and $p = q = n = 2$ and obtain

$$\operatorname{tr}_\Gamma \overset{\circ}{H}{}^1(\Omega) = \operatorname{tr}_\Gamma H^1(\mathbb{R}^2) = H^t(\Gamma). \tag{19.149}$$

We denote temporarily the embedding operator from $H^t(\Gamma)$ in $L_2(\Gamma)$ by

$$\mathrm{id}_t : \quad H^t(\Gamma) \hookrightarrow L_2(\Gamma). \tag{19.150}$$

We wish to prove that the entropy numbers of id_t can be estimated by

$$e_k(id_t) \leq c\, k^{-\frac{1}{2}}, \quad k \in \mathbb{N}, \tag{19.151}$$

for some $c > 0$. We rely on the quarkonial representations in Definition 9.29. By Definition 9.27 the $(t, 2)$-β-quarks, responsible for $H^t(\Gamma)$, are given by

$$(\beta qu)_{km}(\gamma) = 2^{k|\beta|}\,(\gamma - \gamma^{k,m})^\beta\, \psi^{k,m}(\gamma), \quad \gamma \in \Gamma, \tag{19.152}$$

now with respect to the above cubes $Q_{k,m}$ in place of the balls B_{km}. By Definition 9.29,

$$g \in H^t(\Gamma) \quad \text{with} \quad \|g\,|H^t(\Gamma)\| \leq 1 \tag{19.153}$$

can be represented as

$$g(\gamma) = \sum_{\beta \in \mathbb{N}_0^2} \sum_{k=0}^{\infty} \sum_{m=1}^{M_k} \lambda_{km}^\beta\, (\beta qu)_{km}(\gamma) \tag{19.154}$$

where the β-quarks are given by (19.152), and $\lambda_{km}^\beta \in \mathbb{C}$ with

$$\sup_{\beta \in \mathbb{N}_0^2} 2^{\varrho|\beta|} \left(\sum_{k=0}^{\infty} \sum_{m=1}^{M_k} |\lambda_{km}^\beta|^2 \right)^{\frac{1}{2}} \leq 2. \tag{19.155}$$

Here $\varrho > 0$ has the same meaning as in 9.29 and can be chosen arbitrarily large. By Remark 9.30 and Proposition 9.31 all representations of type (19.154), (19.155) converge absolutely and unconditionally and can be rearranged as one wishes. In particular, one can re-organize the collection of all cubes by

$$\{Q_{k,m} : k \in \mathbb{N}_0;\, m = 1, \ldots, M_k\} = \{\widetilde{Q}_{j,m} : j \in \mathbb{N}_0;\, m = 1, \ldots, K_j\}, \tag{19.156}$$

where $\widetilde{Q}_{j,m}$ are the cubes with (19.140), and K_j can be estimated by (19.143). Hence, g can be represented as

$$g(\gamma) = \sum_{\beta \in \mathbb{N}_0^2} \sum_{k=0}^{\infty} \sum_{m=1}^{K_k} \widetilde{\lambda}_{km}^\beta\, \widetilde{(\beta qu)}_{km}(\gamma), \tag{19.157}$$

where $\widetilde{(\beta qu)}_{km}$ are the β-quarks related to $\widetilde{Q}_{k,m}$ and $\widetilde{\lambda}_{km}^\beta$ are corresponding coefficients with

$$\sup_{\beta \in \mathbb{N}_0^2} 2^{\varrho|\beta|} \left(\sum_{k=0}^{\infty} \sum_{m=1}^{K_k} |\widetilde{\lambda}_{km}^\beta|^2 \right)^{\frac{1}{2}} \leq 2. \tag{19.158}$$

We prove (19.151) by factorization through sequence spaces. For this purpose we introduce some notation in modification of [Triδ], Sections 8 and 9. Let

$$\delta \in \mathbb{R}, \quad \varkappa > 0, \quad \text{and} \quad L_k \in \mathbb{N} \text{ where } k \in \mathbb{N}_0. \tag{19.159}$$

Then

$$\ell_{\infty,\varkappa}\left[\ell_q\left(2^{k\delta}\ell_p^{L_k}\right)\right], \quad 1 \le p < \infty, \quad 1 \le q < \infty, \tag{19.160}$$

is the Banach space of all complex-valued sequences

$$x = \left\{x_{k,m}^{\beta} : \beta \in \mathbb{N}_0^2; \; k \in \mathbb{N}_0; \; m = 1, \ldots, L_k\right\} \tag{19.161}$$

with

$$\left\|x \mid \ell_{\infty,\varkappa}\left[\ell_q\left(2^{k\delta}\ell_p^{L_k}\right)\right]\right\|$$

$$= \sup_{\beta \in \mathbb{N}_0^2} 2^{\varkappa|\beta|} \left(\sum_{k=0}^{\infty}\left(\sum_{m=1}^{L_k} 2^{k\delta p}|x_{k,m}^{\beta}|^p\right)^{\frac{q}{p}}\right)^{\frac{1}{q}} < \infty. \tag{19.162}$$

Now we factorize id_t in (19.150) by

$$id_t = T \circ id^t \circ S \tag{19.163}$$

with

$$S : H^t(\Gamma) \mapsto \ell_{\infty,\varkappa_1}\left[\ell_2\left(\ell_2^{N_k}\right)\right], \tag{19.164}$$

$$id^t : \ell_{\infty,\varkappa_1}\left[\ell_2\left(\ell_2^{N_k}\right)\right] \mapsto \ell_{\infty,\varkappa_2}\left[\ell_1\left(2^{-\frac{k}{2}}\ell_2^{N_k}\right)\right], \tag{19.165}$$

$$T : \ell_{\infty,\varkappa_2}\left[\ell_1\left(2^{-\frac{k}{2}}\ell_2^{N_k}\right)\right] \mapsto L_2(\Gamma). \tag{19.166}$$

Here S maps the unit ball in $H^t(\Gamma)$ by (19.153), (19.157), (19.158) in the indicated sequence space now with $\varkappa_1 > 0$ sufficiently large and N_k, given by (19.143). Of course, if $K_k < N_k$, then the remaining components under the map S are zero, by definition. Furthermore, id^t in (19.165) is the identity between the indicated sequence spaces with $\varkappa_1 > \varkappa_2 > 0$. Finally, T in (19.166) is given by

$$Tx = \sum_{\beta \in \mathbb{N}_0^2} \sum_{k=0}^{\infty} \sum_{m=1}^{K_k} x_{k,m}^{\beta} (\beta qu)\widetilde{{}_{km}}(\gamma), \quad \gamma \in \Gamma, \tag{19.167}$$

where x is the sequence (19.161) with $L_k = N_k$, terms with $K_k < m \le N_k$ are simply neglected, and $(\beta qu)_{km}^\sim$ are the β-quarks in (19.157). First we have a look at T. For fixed $\beta \in \mathbb{N}_0^2$ we have

$$\left\| \sum_{k=0}^{\infty} \sum_{m=1}^{K_k} x_{k,m}^\beta (\beta qu)_{km}^\sim \,|L_2(\Gamma)\right\|$$

$$\le \sum_{k=0}^{\infty} \left(\int_\Gamma \left| \sum_{m=1}^{K_k} x_{k,m}^\beta (\beta qu)_{km}^\sim(\gamma) \right|^2 \mu(d\gamma) \right)^{\frac{1}{2}}$$

$$\le c\, 2^{r|\beta|} \sum_{k=0}^{\infty} \left(\sum_{m=1}^{K_k} \left| x_{k,m}^\beta \right|^2 2^{-k} \right)^{\frac{1}{2}}. \tag{19.168}$$

The first estimate is simply the triangle inequality. The second estimate comes from (19.140), the support properties of the involved cube $\widetilde{Q}_{k,m}$ described in Step 2 and $r > 0$ in $2^{r|\beta|}$ is the same constant as in 2.4 and 2.5 adapted to the above more general situation (19.152), and $c > 0$ is independent of β. We always assume

$$\varkappa_1 > \varkappa_2 > r > 0 \tag{19.169}$$

in (19.164)–(19.166). In particular the operator T in (19.166) is linear and bounded. We have (19.163), where S is bounded (not necessarily linear) and T is linear and bounded. As for id^t we can apply [Triδ], Theorem 9.2, p. 47, and obtain

$$c_k(id^t) \sim k^{-\frac{1}{2}}, \quad k \in \mathbb{N}. \tag{19.170}$$

Here we use (19.143). Now (19.151) follows from (19.170) and (19.163).

Step 4 We summarize what is known so far. We always assume that Γ and Ω are related by (19.135). By (19.149) and the compactness of id_t, (19.150), (19.151), it follows that

$$tr_\Gamma : \overset{\circ}{H}{}^1(\Omega) \mapsto L_2(\Gamma) \tag{19.171}$$

is compact. We have (19.113) for the identification operator id_Γ and hence,

$$tr^\Gamma = id_\Gamma \circ tr_\Gamma : \overset{\circ}{H}{}^1(\Omega) \mapsto H^{-1}(\Omega) \tag{19.172}$$

is compact. Together with (19.16) it follows that

$$B = (-\Delta)^{-1} \circ tr^\Gamma \quad \text{is compact in} \quad \overset{\circ}{H}{}^1(\Omega). \tag{19.173}$$

As discussed in 19.14 we have as in (19.70), based on (19.29),

$$(Bf, g)_{\overset{\circ}{H}^1(\Omega)} = \int_\Gamma f(\gamma)\, \overline{g(\gamma)}\, \mu(d\gamma)\,, \tag{19.174}$$

and the counterpart of (19.78),

$$N(B) = \left\{ f \in \overset{\circ}{H}^1(\Omega) \,:\, tr_\Gamma f = 0 \right\} \tag{19.175}$$

for null-space. Hence B is a non-negative, compact, self-adjoint operator in $\overset{\circ}{H}^1(\Omega)$. By the usual Weyl decomposition and (19.149) we have

$$\overset{\circ}{H}^1(\Omega) = N(B) \oplus tr_\Gamma \overset{\circ}{H}^1(\Omega) = N(B) \oplus H^t(\Gamma)\,. \tag{19.176}$$

We identify the orthogonal complement of $N(B)$ in $\overset{\circ}{H}^1(\Omega)$ with $H^t(\Gamma)$ and denote the restriction of B to $H^t(\Gamma)$ by B^Γ. Then B^Γ is a positive self-adjoint compact operator with the same eigenvalues,

$$\varrho_1 \geq \varrho_2 \geq \cdots > 0, \quad \varrho_k \to 0 \quad \text{if} \quad k \to \infty\,, \tag{19.177}$$

as B, and the related orthonormal eigenfunctions u_k,

$$B^\Gamma u_k = \varrho_k u_k\,, \quad k \in \mathbb{N}, \quad u_k \in H^t(\Gamma)\,, \tag{19.178}$$

span $H^t(\Gamma)$. Furthermore, by (19.174),

$$\left\| \sqrt{B^\Gamma}\, f \,|\, H^t(\Gamma) \right\| = \| f \,|\, L_2(\Gamma) \| \tag{19.179}$$

and $\{u_k\}$ is also an orthogonal system in $L_2(\Gamma)$. By construction, $D|\Gamma$, the restriction of $D(\mathbb{R}^2)$ on Γ, is dense in $H^t(\Gamma)$. Since μ is a Radon measure, $D|\Gamma$ is also dense in $L_2(\Gamma)$. This follows from the proof of Theorem 3.8 in [Triδ], p. 7 (with a reference to [Mat95], Theorem 1.10, p. 11). Hence any $g \in D|\Gamma$ can be approximated in $H^t(\Gamma)$ and consequently also in $L_2(\Gamma)$ by linear combinations of $\{u_k\}$. In particular, $\{u_k\}$ is a complete orthogonal system in $L_2(\Gamma)$. Let B_Γ be the extension of B^Γ to $L_2(\Gamma)$. It has the same eigenvalues and eigenfunctions. By (19.179) with $\sqrt{B_\Gamma}$ in place of $\sqrt{B^\Gamma}$ and (19.150) we have in obvious notation,

$$\sqrt{B_\Gamma} = id_t \circ \sqrt{B_\Gamma}\, (L_2 \mapsto H^t)\,. \tag{19.180}$$

By (19.129) and (19.151) we get

$$\sqrt{\varrho_k} \leq c\, k^{-\frac{1}{2}}, \quad k \in \mathbb{N}\,. \tag{19.181}$$

According to Definition 19.13 it remains to prove the converse inequality.

Step 5 We use again the coverings of Γ by cubes as constructed in Step 2. By the doubling condition (19.124) there is a number $C > 0$ such that for any two cubes with (19.139),

$$\mu(Q_{k,l}) \leq C\,\mu(Q_{k+1,m}), \quad k \in \mathbb{N}_0. \tag{19.182}$$

Let $j \in \mathbb{N}_0$. We ask for all cubes $Q_{k,l}$ with

$$2^{-j} < \mu(Q_{k,l}) \leq C\,2^{-j}, \quad j \in \mathbb{N}_0, \tag{19.183}$$

such that there is no larger cube in which $Q_{k,l}$ is contained according to the hierarchy (19.139) having this property. We denote these largest cubes by $Q^*_{j,m}$. We claim

$$\Gamma \subset \bigcup_m Q^*_{j,m}. \tag{19.184}$$

Let $\gamma \in \Gamma$ and $\gamma \in Q_{k,m}$. If k is large then $\mu(Q_{k,m}) < 2^{-j}$. This follows from (19.144). Stepping iteratively from k to $k-1$, then by (19.182) there must a number k with (19.183). There might be even smaller k's with (19.183). But in any case there is a largest (or several largest) cubes with this property. This proves (19.184). Let P_j be the number of cubes $Q^*_{j,m}$. By (19.184) we have $P_j \geq c\,2^j$ for some $c > 0$. Two such cubes $Q^*_{j,l} = Q_{k,m}$ with different values of k are disjoint by construction. For the same level k we have the above controlled overlapping. In any case there are C^∞ functions $\varphi^{*j,l}$ with

$$\mathrm{supp}\,\varphi^{*j,l} \subset Q^*_{j,l}, \quad \mathrm{supp}\,\varphi^{*j,l} \cap \mathrm{supp}\,\varphi^{*j,m} = \emptyset \quad \text{if} \quad l \neq m, \tag{19.185}$$

$$\|\varphi^{*j,l}\,|\overset{\circ}{H}{}^1(\Omega)\| \sim 1 \quad \text{and} \quad \|\varphi^{*j,l}\,|L_2(\Gamma)\| \sim 2^{-\frac{j}{2}}, \tag{19.186}$$

where all equivalence constants are independent of j. In the last equivalence we used again the doubling condition. We obtain for linear combinations of these functions,

$$\left\| \sum_{l=1}^{P_j} \lambda_{j,l}\,\varphi^{*j,l} \,\Big|\overset{\circ}{H}{}^1(\Omega)\right\| \sim \left(\sum_{l=1}^{P_j} |\lambda_{j,l}|^2 \right)^{\frac{1}{2}}, \tag{19.187}$$

and, by (19.174),

$$\left\| \sqrt{B}\left(\sum_{l=1}^{P_j} \lambda_{j,l}\,\varphi^{*j,l} \right) \Big|\overset{\circ}{H}{}^1(\Omega)\right\| \sim \left\| \sum_{l=1}^{P_j} \lambda_{j,l}\,\varphi^{*j,l} \,\Big|L_2(\Gamma)\right\|$$

$$\sim 2^{-\frac{j}{2}} \left(\sum_{l=1}^{P_j} |\lambda_{j,l}|^2 \right)^{\frac{1}{2}}. \tag{19.188}$$

We wish to estimate the approximation numbers $a_{P_j}(\sqrt{B})$, defined by (19.130). Let L be a corresponding operator in (19.130) with $rank\, L < P_j$. Then one finds a non-trivial linear combination φ as used in (19.187), (19.188) with $L\varphi = 0$. Inserted in (19.130) one gets

$$a_{P_j}(\sqrt{B}) \geq c\, 2^{-\frac{j}{2}}, \quad j \in \mathbb{N}. \tag{19.189}$$

Using $P_j \geq c'\, 2^j$ and (19.132) we obtain for some $c > 0$,

$$\sqrt{\varrho_k} \geq c\, k^{-\frac{1}{2}}, \quad k \in \mathbb{N}. \tag{19.190}$$

The proof is complete.

19.18 Problems and comments: Weyl measures

19.18(i) We proved that

$$B = (-\Delta)^{-1} \circ tr^{\Gamma} \quad \text{exists and that} \quad \varrho_k \sim k^{-1}, \quad k \in \mathbb{N}, \tag{19.191}$$

for its positive eigenvalues ϱ_k according to Definition 19.13 if μ is strongly diffuse. This assumption is isotropic (there are no distinguished directions). In [Triδ], Section 30 and in [FaT99] we dealt also with anisotropic and non-isotropic measures μ in the plane and related fractals (ferns, grasses etc.). For corresponding fractal drums and related operators B we obtained only estimates,

$$c_1\, k^{-a} \leq \varrho_k \leq c_2\, k^{-b}, \quad k \in \mathbb{N}, \quad 0 < b \leq 1 \leq a < \infty, \tag{19.192}$$

for the (ordered) positive eigenvalues, $c_1 > 0$, $c_2 > 0$. It is unclear whether these estimates are sharp. The main problem in this context is to

characterize all Weyl measures

in the plane. In particular, the question arises to what extent isotropy assumptions (doubling condition, strongly diffuse) are really necessary. With respect to the just indicated anisotropic fractal beauties, monsters, grasses and ferns one can rephrase (in a somewhat vague Hamletian spirit):

Is the music of the ferns Weylian or alien: that is the question.

The situation can be described as follows: Let Γ (and the related measure) be a genuine anisotropic, but otherwise quite regular fractal, for example the PXT-fractal from [Triδ], pp. 14/15 and 20. The question is of whether the positive eigenvalues ϱ_k behave like k^{-1} (Weyl measure) or at least like $\psi(k)^{-1}$, where

ψ is a tame function; or whether the clash of anisotropic fractals and measures on the one hand with isotropic elliptic operators on the other hand creates a chaotic distribution of eigenvalues. Here a distribution function $\psi(k)$ might be called *tame*, when
$$\psi(2^j) \sim \psi(2^{j+1}), \quad j \in \mathbb{N}.$$
In this case one has the desirable equivalence
$$\varrho_k \sim \psi(k)^{-1} \sim e_k(B_\Gamma), \quad k \in \mathbb{N}.$$
Some related details may be found in 19.18(vii) below. Finally we refer to the end of 9.34(viii), where we described a characterization under which circumstances a compact set Γ in $\mathbb{R}^n$ preserves the Markov inequality. These sets play a significant role in the theory of (isotropic) function spaces on Γ as has been developed by A. Jonsson and H. Wallin, [JoW84], [JoW97]. It is quite clear that such sets Γ are isotropic in the indicated sense. This sheds additional light on the above problem when genuine anisotropic fractals collide with isotropic function spaces and isotropic elliptic operators.

19.18(ii) We refer in this context also to [NaS94], [NaS95]. These papers deal with eigenvalue distributions for operators of type (18.3), (18.2), where μ is a self-similar measure defined by probabilistic IFS (Iterated Function Systems) in $\mathbb{R}^n$. This set-up goes back to [Hut81] and may also be found in [Fal90], Chapter 17, especially p. 263, [Fal97], 11.2, p. 192, and is connected with *multifractal* measures. Restricted to $n = 2$ one has always (19.191).

Hence, all the considered multifractal measures in the plane $\mathbb{R}^2$ are Weyl measures according to Definition 19.13.

There are always some separation conditions. But then it follows from the construction in [Fal97], pp. 192, and 36, 37, that all these measures fit in the scheme of 19.15 or at least in the reasoning of the proof of Theorem 19.17. One has always (19.125). Hence, Step 2 and Step 5 of the proof of Theorem 19.17 can be applied ensuring the estimate $\varrho_k \geq c k^{-1}$, $c > 0$, from below. The doubling condition need not be satisfied in any of these cases. But this is also not necessary in this very regular construction. One can follow the arguments in Step 3 of the proof of Theorem 19.17 directly and gets finally the estimate $\varrho_k \leq c k^{-1}$ from above. On the other hand on random isotropic (but otherwise special) fractals the situation is less favourable. We refer to [Ham00]. There one finds spectral assertions for Laplacians on some types of fractals. Although the set-up is different from our approach it makes clear that spectral theory of generators of quadratic forms on fractals seems to be rather complicated.

19.18(iii) There are further examples of Weyl measures where the doubling condition is not necessarily satisfied. We refer to Theorem 23.2 and in particular, to the end of the discussion in 23.3. In any case the roles of the two assumptions (19.124) (doubling condition) and (19.125) seem to be different. We used (19.125) in a decisive way. On the other hand the doubling condition was convenient in the general case but apparently it is not needed if one has additional information about the nature of the underlying measure (probabilistic IFS, or $b\mu$ as at the end of 23.3).

19.18(iv) Furthermore, the estimate $\varrho_k \geq ck^{-1}$, $c > 0$, is a local matter (it is sufficient to have such an estimate in a neighbourhood of some point $\gamma \in \Gamma$), whereas the estimate $\varrho_k \leq ck^{-1}$ might be a global matter. This effect is not well reflected by the hypotheses of Theorem 19.17 and the other examples mentioned above.

19.18(v) The example of a measure $\mu = \mu_1 + \mu_2$ at the end of 19.15 which is not strongly diffuse and does not satisfy the doubling condition is also a Weyl measure. Obviously, μ_1 and μ_2 are Weyl measures. Let B_1 and B_2 be the related operators according to (19.191). Then we have by (19.134),

$$e_{2k}(B_1 + B_2) \leq e_k(B_1) + e_k(B_2) \leq ck^{-1} \tag{19.193}$$

as the desired estimate from above. Hence $\varrho_k \leq ck^{-1}$. As mentioned in 19.18(iv) the converse is a local matter which applies to the case considered.

19.18(vi) As for the estimate from above one has the following rather satisfactory assertion:

$$\sup_k k\,\varrho_k < \infty \quad \text{if, and only if,} \quad \sup_k k\,e_k(B) < \infty. \tag{19.194}$$

This is a consequence of (19.129) on the one hand and of $\varrho_k = a_k(B)$ in (19.132) and the relations between approximation numbers and entropy numbers as described in [ET96], Theorem 1.3.3(ii), p. 15, on the other hand.

19.18(vii) The assertion of the previous point can be complemented as follows. Let id_t be given by (19.150).

If μ is a Weyl measure according to Definition 19.13 then

$$e_k(id_t) \sim k^{-\frac{1}{2}}, \quad k \in \mathbb{N}. \tag{19.195}$$

As for justification we first remark that by (19.180),

$$e_k(id_t) \sim e_k(\sqrt{B_\Gamma}), \quad k \in \mathbb{N}. \tag{19.196}$$

Recall that $\sqrt{\varrho_k}$ are the eigenvalues of $\sqrt{B_\Gamma}$. We have

$$\sqrt{\varrho_k} \leq \sqrt{2}\,e_k(\sqrt{B_\Gamma}) \leq c\,a_k(\sqrt{B_\Gamma}) = c\,\sqrt{\varrho_k}, \quad k \in \mathbb{N}, \tag{19.197}$$

where the first inequality and the equality come from (19.129) and (19.132), respectively. The crucial second inequality is covered by [ET96], Theorem 1.3.3(i), p. 15, where one needs that

$$\varrho_{2j-1} \sim \varrho_{2j}, \quad j \in \mathbb{N}, \tag{19.198}$$

(the equivalence constants are independent of j). Together with (19.196) and $\varrho_k \sim k^{-1}$ one gets (19.195). There is the following converse assertion:

Let B be compact and assume that we have in addition (19.198) for its positive eigenvalues. Then μ is a Weyl measure if, and only if, (19.195) holds.

This claim is a consequence of (19.197) which, in turn, follows from the additional hypothesis (19.198) and the quoted literature.

19.18(viii) Further discussions about the Weylian behaviour of positive eigenvalues of more general fractal elliptic operators in $\mathbb{R}^n$ with $n \in \mathbb{N}$ may be found in [Triδ], 28.9–28.11, 30.3, pp. 230, 235.

19.19 The degenerate case

We illustrate Theorem 19.17 and also the problems in 19.18 by glancing at the degenerate (non-fractal) case in the plane. Let Ω be a bounded C^∞ domain in the plane $\mathbb{R}^2$ and let

$$\left\{ x \in \mathbb{R}^2 : |x| \le \frac{1}{2} \right\} = \Gamma \subset \Omega. \tag{19.199}$$

Let μ_L be the Lebesgue measure in $\mathbb{R}^2$ and let b be a weight function with

$$b \in L_1(\mathbb{R}^2), \quad b(x) > 0 \text{ a.e. in } \Gamma, \quad b(x) = 0 \text{ if } |x| > \frac{1}{2}. \tag{19.200}$$

We equip Γ with the Radon measure $\mu = b(x)\,\mu_L$ and ask, in our previous notation, for existence, boundedness, compactness, spectral properties, of

$$B = (-\Delta)^{-1} \circ tr^\Gamma \quad \text{as an operator in } \overset{\circ}{H}{}^1(\Omega). \tag{19.201}$$

Recall that $(-\Delta)^{-1}$ is the inverse of the Dirichlet Laplacian in $\overset{\circ}{H}{}^1(\Omega)$. Specializing 19.14 to the above situation, (19.110) reduces to the Hardy inequality

$$\int_\Gamma b(x)\,|\varphi(x)|^2\,dx \le c \sum_{j=1}^2 \int_\Omega \left|\frac{\partial \varphi}{\partial x_j}(x)\right|^2 dx, \quad \varphi \in D(\Omega). \tag{19.202}$$

If there is a number $c > 0$ with (19.202) for all $\varphi \in D(\Omega)$, then B is a bounded operator in $\overset{\circ}{H}{}^1(\Omega)$ which, in our context, and with an obvious interpretation can be written as

$$B = (-\Delta)^{-1} \circ b, \qquad (19.203)$$

justified by

$$(B\varphi, \psi)_{\overset{\circ}{H}{}^1(\Omega)} = \int_\Omega -\Delta \left((-\Delta)^{-1} \circ b\varphi\right) \overline{\psi} \, dx = \int_\Gamma b \varphi \overline{\psi} \, dx, \qquad (19.204)$$

where $\varphi \in D(\Omega)$, $\psi \in D(\Omega)$. The boundedness of B is naturally and intimately related to the sharp inequalities in the critical case as described in Theorem 13.2 and detailed in 13.3. Let b^* be the rearrangement of b, and let for some $c > 0$,

$$b^*(t) \leq \frac{c}{t |\log t|^2}, \qquad 0 < t < \frac{\pi}{4}. \qquad (19.205)$$

Then it follows by (13.62) with $n = p = q = 2$,

$$\int_\Gamma b(x) |\varphi(x)|^2 \, dx \leq \int_0^{\frac{\pi}{4}} b^*(t) \varphi^*(t)^2 \, dt \leq c \, \|\varphi \, |H^1(\mathbb{R}^2)\|^2. \qquad (19.206)$$

As for the first inequality we refer to (16.12) and the comments and references afterwards.

We specify b in (19.200) by

$$b(x) = \frac{\varkappa(|x|)}{|x|^2 |\log |x||^2}, \qquad |x| \leq \frac{1}{2}, \qquad (19.207)$$

where $\varkappa(t)$ is a positive monotone (decreasing or increasing) function in the interval $(0, \frac{1}{2})$. If $\varkappa$ is increasing we put $\varkappa(0) = \lim_{t \downarrow 0} \varkappa(t)$. With respect to this special choice of b we have for B, given by (19.203), in the above interpretation as an operator in $\overset{\circ}{H}{}^1(\Omega)$, the following assertions:

$$B \quad \text{is bounded if, and only if,} \quad \varkappa \text{ is bounded,} \qquad (19.208)$$

and

$$B \quad \text{is compact if, and only if,} \quad \varkappa(0) = 0. \qquad (19.209)$$

Here (19.208) is a special case of the sharp Hardy inequality in Theorem 16.2(i) with $n = p = q = 2$.

We prove that B with $\varkappa = 1$ is not compact. Let $\varphi(x) \geq 0$ be monotonically decreasing in radial directions from the origin. Then we get by (13.62), Theorem 16.2(i) with $n = p = q = 2$, and (16.11),

$$\sup_{0<t<\frac{\pi}{4}} \frac{\varphi^*(t)}{|\log t|^{\frac{1}{2}}} \leq c \left(\int_0^{\frac{\pi}{4}} \frac{\varphi^*(t)^2}{t|\log t|^2} dt \right)^{\frac{1}{2}} \tag{19.210}$$

$$\sim \left(\int_\Gamma \frac{\varphi(x)^2}{|x|^2 |\log |x||^2} dx \right)^{\frac{1}{2}} \leq c' \|\varphi \, | \overset{\circ}{H}{}^1(\Omega)\|.$$

This applies in particular to the extremal functions in (13.50)–(13.52) approximating the growth envelope function $\mathcal{E}_G H^1(t)$. But now we are in the same situation as in 14.6. If the embedding related to the last inequality in (19.210) is compact, then the set of the above extremal functions in the space quasinormed by the left-hand side of (19.210) is pre-compact. This is a contradiction by the same arguments as in 14.6.

Let $\varkappa(0) = 0$. Then $\varkappa(\delta) \downarrow 0$ if $\delta \downarrow 0$. We decompose Γ in $|x| \leq \delta$ and in $\delta < |x| \leq \frac{1}{2}$ and get

$$\int_\Gamma \frac{\varkappa(|x|)}{|x|^2 |\log |x||^2} |f(x)|^2 \, dx \leq \varkappa(\delta) \int_\Gamma \frac{|f(x)|^2}{|x|^2 |\log |x||^2} dx + c_\delta \int_\Gamma |f(x)|^2 \, dx \tag{19.211}$$

for some $c_\delta > 0$. Now it is clear that the embedding of $\overset{\circ}{H}{}^1(\Omega)$ in the space related to the left-hand side of (19.211) is compact: the first term on the right-hand side creates a $\varkappa(\delta)$-net and for the second term one has the classical compact embedding of $H^1(\Omega)$ in $L_2(\Omega)$. This completes the proof of (19.209).

There remains the problem under which circumstances for b in general, and for b given by (19.207) with $\varkappa(0) = 0$ in particular, $\mu = b(x) \mu_L$ is a Weyl measure according to Definition 19.13, and hence

$$\varrho_k \sim k^{-1}, \quad k \in \mathbb{N}, \tag{19.212}$$

for the positive eigenvalues of B, given by (19.203). This is not so clear so far (to the author). If $\delta > 0$ and $\varkappa(t) = |\log t|^{-\delta}$, then we have (19.209), but

$$\mu = |x|^{-2} |\log |x||^{-2-\delta} \mu_L, \quad |x| \leq \frac{1}{2}, \tag{19.213}$$

is not strongly diffuse according to (19.125), and hence we cannot apply Theorem 19.17. On the other hand,

$$\mu = |x|^{-2+\varepsilon} \mu_L, \quad \varepsilon > 0, \quad |x| \leq \frac{1}{2}, \tag{19.214}$$

is strongly diffuse. Hence, by Theorem 19.17, it is a Weyl measure, and we have (19.212).

In [ET96], Chapter 5, we developed a spectral theory for degenerate elliptic operators, especially the examples on p. 211 of [ET96] are related to the above considerations. There one finds also further references.

20 The fractal Dirichlet problem

20.1 Introduction

Let Ω be a bounded C^∞ domain in $\mathbb{R}^n$, where, temporarily, $n \geq 3$, and let $\Gamma = \partial\Omega$ be equipped naturally with a Radon measure μ equivalent (or equal) to $\mathcal{H}^{n-1}|\Gamma$ (the restriction of the Hausdorff measure $\mathcal{H}^{n-1}$ in $\mathbb{R}^n$ to Γ). The single layer potential G,

$$(Gh)(x) = \int_\Gamma \frac{h(\gamma)}{|x-\gamma|^{n-2}} \mu(d\gamma), \quad x \in \mathbb{R}^n, \tag{20.1}$$

makes sense both in $\mathbb{R}^n$ and on Γ (using the same letter G) if, for example, h is bounded. Since Γ is a compact C^∞ manifold,

$$H^s(\Gamma) = B^s_{2,2}(\Gamma), \quad s \in \mathbb{R}, \tag{20.2}$$

can be introduced in a canonical way via local charts. It turns out that G (restricted to Γ) makes sense for some spaces $H^s(\Gamma)$. In particular,

$$G H^{-\frac{1}{2}}(\Gamma) = H^{\frac{1}{2}}(\Gamma) \tag{20.3}$$

is an isomorphic mapping. This has the consequence that the uniquely determined solution $u(x)$ of the (almost) classical Dirichlet problem

$$\Delta u(x) = 0, \quad x \in \Omega, \quad u \in H^1(\Omega), \tag{20.4}$$

$$tr_\Gamma u = g \in H^{\frac{1}{2}}(\Gamma), \tag{20.5}$$

for given g, can be uniquely represented by (20.1) as $u = Gh$ with some $h \in H^{-\frac{1}{2}}(\Gamma)$. It is the main aim of this section to extend these observations to fractals. We rely on the techniques developed in Section 19. We describe what can be expected. First we remark that the boundary $\Gamma = \partial\Omega$ of the above C^∞ domain Ω is an $(n-1)$-set according to (18.1). In particular, the singularity $|x-\gamma|^{2-n}$ with $x \in \Gamma$ and $\gamma \in \Gamma$, is well compensated by (18.1) with $d = n-1$. In addition it is quite clear that this argument applies to any compactly supported d-set Γ with $d > n-2$. One obtains the generalization

$$G H^{-1+\frac{n-d}{2}}(\Gamma) = H^{1-\frac{n-d}{2}}(\Gamma), \quad n-2 < d < n, \qquad (20.6)$$

of (20.3). Then we are precisely in the context of Section 19, especially Theorem 19.7. Let the fractal Γ, the domain Ω and the related inverse of the Dirichlet Laplacian $(-\Delta)^{-1}$ be as there, and let the trace operator tr_Γ and the identification operator id_Γ as in 19.2. Then

$$tr_\Gamma \circ (-\Delta)^{-1} \circ id_\Gamma \, H^{-1+\frac{n-d}{2}}(\Gamma) = H^{1-\frac{n-d}{2}}(\Gamma), \qquad (20.7)$$

combined with (20.6) (where G stand now for the Green's operator with respect to Ω), is the concise version of the uniquely determined solution $u \in H^1(\Omega)$ of the Dirichlet problem,

$$\Delta u(x) = 0 \quad \text{in} \quad \Omega\backslash\Gamma, \quad tr_{\partial\Omega}\, u = 0, \qquad (20.8)$$

$$tr_\Gamma\, u = g \in H^{1-\frac{n-d}{2}}(\Gamma), \qquad (20.9)$$

including its (uniquely determined) representation as a single layer potential.

The plan of the section is the following. First we need some preparation: some duality assertions; how to introduce the spaces $H^s(\Gamma)$ with $s \in \mathbb{R}$; mapping properties of the operator B, given by (19.68), restricted to Γ. This will be done in 20.2–20.6 The main result of this section, concerning the Dirichlet problem (20.6)–(20.9), may be found in Theorem 20.7. Afterwards we return in Corollary 20.10 to the Dirichlet problem as described at the beginning of this subsection, but now under the assumption that the boundary $\partial\Omega = \Gamma$ of a bounded domain Ω in $\mathbb{R}^n$ is a d-set with $n-1 \le d < n$.

20.2 Some notation

We collect some notation partly repeating earlier definitions. Let $0 < d < n$. A compact set Γ in $\mathbb{R}^n$ is called a d-set if there is a Radon measure μ in $\mathbb{R}^n$ and two positive numbers c_1 and c_2 such that $supp\, \mu = \Gamma$ and

$$c_1\, r^d \le \mu(B(\gamma,r)) \le c_2\, r^d \quad \text{for all} \quad 0 < r < 1, \qquad (20.10)$$

and all $\gamma \in \Gamma$, where $B(\gamma, r)$ is the ball centred at $\gamma \in \Gamma$ and of radius r. Up to equivalences, μ is uniquely determined and may be identified with the restriction $\mathcal{H}^d|\Gamma$ of the Hausdorff measure $\mathcal{H}^d$ in $\mathbb{R}^n$ on Γ. Comments and references may be found at the beginning of 9.12.

Let

$$H^s(\mathbb{R}^n) = F_{2,2}^s(\mathbb{R}^n) = B_{2,2}^s(\mathbb{R}^n), \quad s \in \mathbb{R}, \tag{20.11}$$

be the (special) Sobolev spaces as introduced in (1.9). If Ω is a bounded domain in $\mathbb{R}^n$, then by Definition 5.3.,

$$H^s(\Omega) = F_{2,2}^s(\Omega) = B_{2,2}^s(\Omega), \quad s \in \mathbb{R}, \tag{20.12}$$

are the restrictions of $H^s(\mathbb{R}^n)$ on Ω.

Let $\Gamma \subset \Omega$, where Γ is a compact d-set in $\mathbb{R}^n$ with $0 < d < n$ and Ω is a bounded domain in $\mathbb{R}^n$. The trace operator tr_Γ has been introduced in 9.2 and studied in detail in 9.32 and 9.34 in connection with rather general compact sets Γ and related Radon measures μ. As mentioned in 9.34(i) for d-sets with $0 < d < n$ we have $t_p = \frac{d}{p}$ in (9.143). We are interested here only in the special case (20.11), (20.12) and we extend this notation to Γ by

$$H^s(\Gamma) = B_{2,2}^s(\Gamma), \quad s > 0. \tag{20.13}$$

Then, by (9.143),

$$H^s(\Gamma) = tr_\Gamma\, H^{s+\frac{n-d}{2}}(\mathbb{R}^n) = tr_\Gamma\, H^{s+\frac{n-d}{2}}(\Omega), \quad s > 0. \tag{20.14}$$

Some further information is given in 19.2, including references to [Triδ], Theorems 18.2 and 18.6 on pp. 136 and 139. The identification operator id_Γ has been introduced in (9.16). We rely in particular on the duality assertions (9.20), (9.21).

Let ω be an arbitrary bounded domain in $\mathbb{R}^n$. Of special interest for us are

$$H^1(\omega), \quad H^{-1}(\omega), \quad \text{and} \quad \overset{\circ}{H}{}^1(\omega). \tag{20.15}$$

Recall that $H^1(\omega)$ and $H^{-1}(\omega)$ are defined by restriction of $H^1(\mathbb{R}^n)$ and $H^{-1}(\mathbb{R}^n)$, respectively, on ω. As usual, $\overset{\circ}{H}{}^1(\omega)$ is the completion of $D(\omega)$ in $H^1(\omega)$. We have in $H^1(\mathbb{R}^n)$ the explicit norm (1.4) with $s = 1$ and $p = 2$. Then it follows by standard arguments that for any bounded domain ω the space $\overset{\circ}{H}{}^1(\omega)$ can be equivalently normed by

$$\|f\,|\overset{\circ}{H}{}^1(\omega)\| = \left(\sum_{j=1}^n \left\| \frac{\partial f}{\partial x_j}\,|L_2(\omega) \right\|^2 \right)^{\frac{1}{2}} \tag{20.16}$$

(with a corresponding scalar product). As usual, $D'(\omega)$ is the collection of all (complex-valued) distributions in ω, the dual of $D(\omega)$.

20.3 Proposition

Let ω be an arbitrary bounded domain in $\mathbb{R}^n$. Then

$$\left(\overset{\circ}{H}{}^1(\omega)\right)' = H^{-1}(\omega) \quad \text{(equivalent norms)} \tag{20.17}$$

with respect to the dual pairing $(D(\omega), D'(\omega))$.

Proof This assertion is well known if ω is smooth and also if ω is replaced by $\mathbb{R}^n$,

$$\left(H^1(\mathbb{R}^n)\right)' = H^{-1}(\mathbb{R}^n), \tag{20.18}$$

with respect to the dual pairing $(D(\mathbb{R}^n), D'(\mathbb{R}^n))$. Now let ω be the above arbitrary domain. Let $g \in \left(\overset{\circ}{H}{}^1(\omega)\right)'$ according to the dual pairing $(D(\omega), D'(\omega))$. By the explicit norm (1.4) of $H^1(\mathbb{R}^n)$ we have

$$|g(\varphi)| \leq \|g\| \, \|\varphi \,|H^1(\omega)\| = \|g\| \, \|\varphi \,|H^1(\mathbb{R}^n)\|, \quad \varphi \in D(\omega). \tag{20.19}$$

Interpreting $\overset{\circ}{H}{}^1(\omega)$ as a closed subspace of $H^1(\mathbb{R}^n)$, we find by (20.18) an element $h \in H^{-1}(\mathbb{R}^n)$ with

$$g(\varphi) = (h, \varphi), \quad \|h\,|H^{-1}(\mathbb{R}^n)\| \sim \|g\|, \quad \varphi \in D(\omega). \tag{20.20}$$

With $G = h|\omega$ (restriction to ω) we have

$$g(\varphi) = (G, \varphi), \quad \|G\,|H^{-1}(\omega)\| \leq c\|g\|, \tag{20.21}$$

where $c > 0$ is independent of g. Conversely, let $G \in H^{-1}(\omega)$. There is an element $h \in H^{-1}(\mathbb{R}^n)$ with $G = h|\omega$ and

$$\|h\,|H^{-1}(\mathbb{R}^n)\| \sim \|G\,|H^{-1}(\omega)\|. \tag{20.22}$$

Let $g(\varphi) = (G, \varphi)$ if $\varphi \in D(\omega)$. Then we have

$$\begin{aligned}|g(\varphi)| &= |(h, \varphi)| \leq \|h\,|H^{-1}(\mathbb{R}^n)\| \, \|\varphi\,|H^1(\mathbb{R}^n)\| \\ &\leq c\|G\,|H^{-1}(\omega)\| \, \|\varphi\,|H^1(\omega)\|, \quad \varphi \in D(\omega). \end{aligned} \tag{20.23}$$

Hence,

$$g \in \left(\overset{\circ}{H}{}^1(\omega)\right)' \quad \text{and} \quad \|g\| \leq c\|G\,|H^{-1}(\omega)\|. \tag{20.24}$$

Together with (20.21) and the usual interpretation we have (20.17).

20.4 Remark

We did not use in the above proof that ω is bounded. In other words, in extending Definition 5.3 to arbitrary domains in $\mathbb{R}^n$, (20.17) remains valid. But in connection with (20.16) we used that ω is bounded. Checking the proof it is clear that it depends on the special nature of the norm (1.4). An extension to more general spaces is not clear. The right way to look at duality in (smooth or non-smooth) domains is indicated in (5.157). We refer also to [Triα], 4.8.1, p. 332 (in the second edition, 1995), which works also for non-smooth domains.

20.5 Spaces on d-sets

Let Γ be again a compact d-set in $\mathbb{R}^n$ with $0 < d < n$ and let

$$D(\Gamma) = \bigcap_{s>0} H^s(\Gamma). \tag{20.25}$$

Since $S(\mathbb{R}^n)$ is dense in any space $H^\sigma(\mathbb{R}^n)$ with $\sigma \in \mathbb{R}$, it follows by (20.14) that

$$D(\Gamma) = tr_\Gamma \, S(\mathbb{R}^n). \tag{20.26}$$

In particular, $D(\Gamma)$ is a locally convex space, naturally equipped with norms, for example $\| \cdot \, |H^k(\Gamma)\|$, $k \in \mathbb{N}$. It is dense in $H^s(\Gamma)$ with $s > 0$. The latter assertion is also an immediate consequence of the quarkonial representation in Theorem 9.33 and Definition 9.29. Recall that $D(\Gamma)$ is also dense in $L_2(\Gamma)$. This follows from the properties of Radon measures and has been mentioned with some references in Step 4 of the proof of Theorem 19.17. But in case of d-sets we have also

$$L_2(\Gamma) = tr_\Gamma \, B_{2,1}^{\frac{n-d}{2}}(\mathbb{R}^n), \quad 0 < d < n, \tag{20.27}$$

as remarked in (9.124) with a reference to [Triδ], Theorem 18.6, p. 139. Hence, the density of $D(\Gamma)$ in $L_2(\Gamma)$ is also a consequence of Theorem 9.33, Definition 9.29 and (9.135). Imitating the definition of $S'(\mathbb{R}^n)$ as the dual of $S(\mathbb{R}^n)$, we introduce in an obvious way $D'(\Gamma)$ by

$$D'(\Gamma) = (D(\Gamma))'. \tag{20.28}$$

As usual we identify the dual of $L_2(\Gamma)$ with $L_2(\Gamma)$ itself and we use the scalar product of $L_2(\Gamma)$ as the dual pairing of $D(\Gamma)$ and $D'(\Gamma)$. In this way we can introduce

$$H^s(\Gamma) = \left(H^{-s}(\Gamma)\right)', \quad -\infty < s < 0, \tag{20.29}$$

and one obtains

$$(H^s(\Gamma))' = H^{-s}(\Gamma), \quad s \in \mathbb{R}, \quad \text{with} \quad H^0(\Gamma) = L_2(\Gamma), \tag{20.30}$$

and

$$D'(\Gamma) = \bigcup_{s \in \mathbb{R}} H^s(\Gamma).$$

If $\Gamma = \partial\Omega$ is, say, the boundary of a bounded C^∞ domain in $\mathbb{R}^n$, then as mentioned in 20.1, the spaces $H^s(\Gamma)$ have a direct meaning. In this case, (20.30) is an assertion which can be checked easily.

20.6 The operator B and single layer potentials

We collect what has been said before for the operator B and complement these assertions. Let again Ω be a bounded C^∞ domain in $\mathbb{R}^n$ and let $(-\Delta)^{-1}$ be the inverse of the Dirichlet Laplacian $-\Delta$ with

$$(-\Delta)^{-1}: \quad H^{-1}(\Omega) \mapsto \overset{\circ}{H}{}^1(\Omega) \tag{20.31}$$

according to (19.16), including the explanations and references given there. Let $G(x,y)$ with $x \in \overline{\Omega}$ and $y \in \Omega$ be the classical Green's function for the Dirichlet Laplacian. In particular,

$$G(x,y) = 0 \quad \text{if} \quad x \in \partial\Omega, \, y \in \Omega \quad \text{and} \quad G(x,y) \sim |x-y|^{2-n} \tag{20.32}$$

if $n \geq 3$ (and $x \in \Omega$, $y \in \Omega$), with the usual modifications in case of $n = 1$ or $n = 2$. Furthermore,

$$(-\Delta)^{-1} f(x) = \int_\Omega G(x,y) f(y) \, dy, \quad f \in H^{-1}(\Omega), \tag{20.33}$$

in the usual interpretation according to (20.31). Let Γ be a compact d-set in $\mathbb{R}^n$ with

$$n - 2 < d < n \quad (0 < d < 1 \text{ if } n = 1) \quad \text{and} \quad \Gamma \subset \Omega. \tag{20.34}$$

Then, in the usual notation, G,

$$(Gf)(x) = \int_\Gamma G(x,\gamma) f(\gamma) \mu(d\gamma), \quad x \in \overline{\Omega}, \tag{20.35}$$

is called a single layer potential. If $n \geq 3$ then we have the singularity (20.32). Since $d > n - 2$, it follows by (20.10) that the restriction of G to Γ, denoted by G^Γ,

$$(G^\Gamma f)(\lambda) = \int_\Gamma G(\lambda, \gamma) f(\gamma) \mu(d\gamma), \quad \lambda \in \Gamma, \tag{20.36}$$

makes sense if, say, $f \in D(\Gamma)$. Of interest are the mapping properties of the operator G^Γ between spaces $H^s(\Gamma)$ and the connection to the operator B, given by

$$B = (-\Delta)^{-1} \circ tr^\Gamma, \quad \text{with} \quad tr^\Gamma = id_\Gamma \circ tr_\Gamma, \tag{20.37}$$

where tr_Γ is the above trace operator and id_Γ is again the identification operator as mentioned in 20.2 (with a reference to (9.16) and (9.20)). As for B we have Theorem 19.7 and Theorem 19.17 (where $n = 2$). We collect what we need. By (19.69) and (20.14) the Weyl decomposition (19.176) is given by

$$\mathring{H}^1(\Omega) = \mathring{H}^1(\Omega \setminus \Gamma) \oplus tr_\Gamma \mathring{H}^1(\Omega) = \mathring{H}^1(\Omega \setminus \Gamma) \oplus H^{1-\frac{n-d}{2}}(\Gamma). \tag{20.38}$$

As in Step 4 of the proof of Theorem 19.17 we denote the restriction of B to $H^{1-\frac{n-d}{2}}(\Gamma)$ by B^Γ. In analogy to (19.179) we have

$$\left\| \sqrt{B^\Gamma} f \,|\, H^{1-\frac{n-d}{2}}(\Gamma) \right\| = \| f \,|\, L_2(\Gamma) \|. \tag{20.39}$$

Repeating the arguments given there, $\sqrt{B^\Gamma}$ can be extended by continuity (using now the same letter) to an isomorphic map from $L_2(\Gamma)$ onto $H^{1-\frac{n-d}{2}}(\Gamma)$. Both on $L_2(\Gamma)$ and on $H^{1-\frac{n-d}{2}}(\Gamma)$ the operator $\sqrt{B^\Gamma}$ is positive, compact, and self-adjoint with the same eigenvalues and eigenfunctions. Now by (20.29) and the usual arguments, $\sqrt{B^\Gamma}$ (identified with its dual) is also an isomorphic map from

$$H^{-1+\frac{n-d}{2}}(\Gamma) \quad \text{onto} \quad L_2(\Gamma). \tag{20.40}$$

Hence,

$$B^\Gamma = \sqrt{B^\Gamma} \circ \sqrt{B^\Gamma} : \quad H^{-1+\frac{n-d}{2}}(\Gamma) \mapsto H^{1-\frac{n-d}{2}}(\Gamma) \tag{20.41}$$

is an isomorphic map. As for the connection with the above Green's function we first remark that

$$id_\Gamma : \quad H^{-1+\frac{n-d}{2}}(\Gamma) \mapsto H^{-1}(\Omega). \tag{20.42}$$

This follows from (20.38) (or (20.14) with $s = 1 - \frac{n-d}{2}$) and (9.20), (20.17). By (20.31) we have

$$(-\Delta)^{-1} \circ id_\Gamma : \quad H^{-1+\frac{n-d}{2}}(\Gamma) \mapsto H^1(\Omega). \tag{20.43}$$

Interpreting (20.33) as a dual pairing now with f replaced by $id_\Gamma h$ with $h \in H^{-1+\frac{n-d}{2}}(\Gamma)$ we get

$$(-\Delta)^{-1} \circ id_\Gamma h = \int_\Gamma G(x, \gamma) \, h(\gamma) \, \mu(d\gamma), \quad x \in \Omega. \tag{20.44}$$

Restricting this H^1-function to Γ, hence $x = \lambda \in \Gamma$, we obtain

$$tr_\Gamma (-\Delta)^{-1} \circ id_\Gamma h(\lambda) = \int_\Gamma G(\lambda, \gamma) \, h(\gamma) \, \mu(d\gamma), \quad \lambda \in \Gamma, \tag{20.45}$$

and by (20.36), (20.41),

$$B^\Gamma = tr_\Gamma \circ (-\Delta)^{-1} \circ id_\Gamma = G^\Gamma : \quad H^{-1+\frac{n-d}{2}}(\Gamma) \mapsto H^{1-\frac{n-d}{2}}(\Gamma), \tag{20.46}$$

as an isomorphic map. After all these preparations we can now prove rather easily the following assertions.

20.7 Theorem

Let Ω be a bounded C^∞ domain in $\mathbb{R}^n$ and let $G(x,y)$ with $x \in \overline{\Omega}$, $y \in \Omega$, be the classical Green's function for the Dirichlet Laplacian in Ω. Let Γ be a compact d-set in $\mathbb{R}^n$ with $\Gamma \subset \Omega$ and $n - 2 < d < n$ (interpreted as $0 < d < 1$ in case of $n = 1$). Let $g \in H^{1-\frac{n-d}{2}}(\Gamma)$. Then the Dirichlet problem

$$u \in H^1(\Omega), \quad \Delta u(x) = 0 \quad \text{in} \quad \Omega \backslash \Gamma, \tag{20.47}$$

$$tr_{\partial\Omega} u = 0, \quad tr_\Gamma u = g, \tag{20.48}$$

has a unique solution u. Furthermore this solution u can be represented as a single layer potential

$$u(x) = \int_\Gamma G(x, \gamma) \, h(\gamma) \, \mu(d\gamma), \quad x \in \Omega, \tag{20.49}$$

with a uniquely determined distribution $h \in H^{-1+\frac{n-d}{2}}(\Gamma)$.

Proof *Step 1* The existence of a solution u and the representation (20.49), including its interpretation, can be obtained from 20.6 as follows. By (20.46) we find for given

$$g \in H^{1-\frac{n-d}{2}}(\Gamma) \quad \text{a distribution} \quad h \in H^{-1+\frac{n-d}{2}}(\Gamma)$$

with

$$u(x) = (-\Delta)^{-1} \circ id_\Gamma \, h \in \overset{\circ}{H}{}^{1}(\Omega) \qquad (20.50)$$

and

$$tr_\Gamma \, u = G^\Gamma h = g \,. \qquad (20.51)$$

Since $supp \, \Delta u \subset \Gamma$, we have

$$\Delta u(x) = 0 \quad \text{if} \quad x \in \Omega \backslash \Gamma \quad \text{(classical harmonic function)}\,. \qquad (20.52)$$

Hence, u satisfies (20.47), (20.48). Furthermore, (20.49) follows from (20.50) and (20.44). Conversely if u with (20.47), (20.48) is given by (20.49) with some $h \in H^{-1+\frac{n-d}{2}}(\Gamma)$, then we have by (20.46) also (20.51) and hence $h = (G^\Gamma)^{-1} g$ is uniquely determined.

Step 2 To prove the uniqueness of u with (20.47), (20.48), it is sufficient to show $u = 0$ if $g = 0$. Hence, let u be a solution of (20.47), (20.48) with $g = 0$. Let $\omega = \Omega \backslash \Gamma$. Then $\partial \omega = \partial \Omega \cup \Gamma$. We apply Proposition 19.5 with $p = 2$ and $1 > s = 1 - \frac{n-d}{2} > 0$ and get

$$u \in \{v \in H^1(\omega) : tr_{\partial \omega} = 0\} = \overset{\circ}{H}{}^{1}(\omega)\,, \qquad (20.53)$$

where we used the notation introduced in 20.2. Then we have for any $\varphi \in D(\omega)$,

$$\int_\omega \sum_{j=1}^{n} \frac{\partial u}{\partial x_j} \frac{\partial \overline{\varphi}}{\partial x_j} \, dx = (-\Delta u, \varphi) = 0\,. \qquad (20.54)$$

Since $u \in \overset{\circ}{H}{}^{1}(\omega)$ it follows that $\frac{\partial u}{\partial x_j} = 0$ with $j = 1, \ldots, n$ and finally $u = 0$.

20.8 Comments and the variational approach

20.8(i) The above proof of uniqueness is based on Proposition 19.5, which, in turn, used the substantial Theorems 9.1.3 and 10.1.1 in [AdH96], pp. 234, 281 (here Theorem 9.1.3 dealing with Sobolev spaces is sufficient). We refer to 19.6 for further comments. The proof of the existence and the representation of a solution u with (20.47)–(20.49) is based on the techniques developed in connection with Theorems 19.7 (with a reference to [Triδ], Sections 28 and 30) and 19.17. Basically one shifts the problem from Ω to Γ and proves that the resulting operator B^Γ is an isomorphic map as indicated in (20.41). The rest is interpretation of this observation. If Γ is a smooth or Lipschitz-continuous surface in $\mathbb{R}^n$ (maybe the boundary of a respective domain), then $d = n - 1$. Hence we have by (20.41)

$$B^\Gamma : \quad H^{-\frac{1}{2}}(\Gamma) \mapsto H^{\frac{1}{2}}(\Gamma), \tag{20.55}$$

as indicated in (20.3). As will be seen below in Corollary 20.10 and its proof, the single layer potential on the right-hand side of (20.49) can be replaced by the single layer potential according to (20.1). Furthermore, after this modification (20.6) corresponds to (20.41).

20.8(ii) If one looks only for the existence of a solution u with (20.47), (20.48), and not for its representation, one can use the very classical variational approach as follows. Let again $g \in H^{1-\frac{n-d}{2}}(\Gamma)$. Then by definition there is a function

$$v \in \overset{\circ}{H}{}^1(\Omega) \quad \text{with} \quad tr_\Gamma v = g. \tag{20.56}$$

Let again $\omega = \Omega \backslash \Gamma$. Then we ask for a function h with

$$h \in \overset{\circ}{H}{}^1(\omega) \quad \text{and} \quad -\Delta h = \Delta v \in H^{-1}(\omega). \tag{20.57}$$

This is the point where one needs Proposition 20.3. It follows that the right-hand side of

$$\sum_{j=1}^n \int_\omega \frac{\partial h}{\partial x_j} \frac{\partial \varphi}{\partial x_j} \, dx = (\Delta v)(\varphi), \quad \varphi \in D(\omega), \tag{20.58}$$

is a linear and continuous functional on $\overset{\circ}{H}{}^1(\omega)$ and, hence, can be represented by the left-hand side with some $h \in \overset{\circ}{H}{}^1(\omega)$. Then we have (20.57) and $u = v + h$ is a solution of (20.47), (20.48).

20.8(iii) We concentrate in this section as in Section 19 on the Dirichlet Laplacian $-\Delta$. But it is quite clear that many of our considerations can be generalized as indicated in 19.4 with a reference to [Triδ], Sections 27–30. This applies also to Theorem 20.7. Similarly the variational approach outlined in 20.8(ii) can be used for wider classes of elliptic operators to prove the existence of weak solutions of the Dirichlet problem for elliptic equations in non-smooth domains. The corresponding uniqueness assertions can be based on density properties as mentioned at the beginning of 20.8(i). Both together (variational approach and density assertions) result in existence and uniqueness assertions for Dirichlet problems in non-smooth domains. This has a long history and goes back to [Sob50] (and Sobolev's original papers in 1936–1938). These problems have also been considered in some detail in [AdH96], Corollary 9.1.8, Theorem 9.1.9, pp. 236, 237, including Sobolev's original problems connected with the poly-harmonic operator.

20.9 The Dirichlet problem in domains with a fractal boundary

Let Ω be a bounded domain in $\mathbb{R}^n$ and let its boundary $\Gamma = \partial \Omega$ be a d-set. Then we have necessarily $d \geq n - 1$. We assume $n - 1 \leq d < n$ (with $d > 0$ in case of $n = 1$; then Ω is a disconnected bounded open set). By (20.14) and in modification of Theorem 20.7 we ask for solutions u of

$$u \in H^1(\Omega), \quad \Delta u(x) = 0 \text{ in } \Omega, \quad \operatorname{tr}_\Gamma u = g \in H^{1-\frac{n-d}{2}}(\Gamma), \qquad (20.59)$$

where g is given. Let K be an open ball with $\overline{\Omega} \subset K$ and let $G^K(x,y)$ be the Green's function for the Dirichlet Laplacian with respect to K. We apply Theorem 20.7 to K in place of Ω and to $\Gamma = \partial \Omega$. Then one finds an element $h \in H^{-1+\frac{n-d}{2}}(\Gamma)$ such that

$$u(x) = \int_\Gamma G^K(x,\gamma) h(\gamma) \mu(d\gamma) \qquad (20.60)$$

solves (20.47), (20.48), with K in place of Ω. We have in particular (20.59) as a solution. As for the uniqueness we recall first our point of view: By (20.12) all spaces on Ω are defined by restriction of the corresponding spaces on $\mathbb{R}^n$. This avoids problems of extendability and traces on $\Gamma = \partial \Omega$ for intrinsically defined spaces. As for traces we discussed this problem in some detail in 9.34(vi) (but this is hardly needed here from our point of view). Hence the uniqueness problem can be formulated as follows: Let $u \in H^1(\mathbb{R}^n)$ and let

$$\Delta u(x) = 0 \text{ in } \Omega \quad \text{and} \quad \operatorname{tr}_\Gamma u = 0. \qquad (20.61)$$

Then one has to prove that $u(x) = 0$ in Ω. By Proposition 19.5 it follows that $D(\mathbb{R}^n \setminus \Gamma)$ is dense in

$$\{v \in H^1(\mathbb{R}^n) : \operatorname{tr}_\Gamma v = 0\}. \qquad (20.62)$$

Let $\varphi \in D(\mathbb{R}^n \setminus \Gamma)$. Then by (20.61),

$$\int_\Omega \sum_{n=1}^n \frac{\partial u}{\partial x_j} \frac{\partial \overline{\varphi}}{\partial x_j} \, dx = 0, \quad \varphi \in D(\mathbb{R}^n \setminus \Gamma). \qquad (20.63)$$

Since u belongs to the spaces in (20.62) it can be approximated in $H^1(\mathbb{R}^n)$ by functions belonging to $D(\mathbb{R}^n\backslash\Gamma)$. Then it follows from (20.63) by standard arguments, first $u(x) = c$ in Ω and afterwards $u(x) = 0$ in Ω. Hence we have existence and uniqueness for the Dirichlet problem (20.59). But the representation (20.60) is unsatisfactory since it depends on K. However at least in case of $n \geq 3$ this awkward description can be replaced by a more natural one.

20.10 Corollary

Let Ω be a bounded domain in $\mathbb{R}^n$ where $n \in \mathbb{N}$, and let $\Gamma = \partial\Omega$ be a d-set with $n - 1 \leq d < n$ (interpreted as $0 < d < 1$ in case of $n = 1$ as described at the beginning of 20.9). Let $g \in H^{1-\frac{n-d}{2}}(\Gamma)$.

(i)   Then the Dirichlet problem
$$u \in H^1(\Omega), \quad \Delta u(x) = 0 \text{ in } \Omega, \quad \text{tr}_\Gamma u = g, \qquad (20.64)$$
has a uniquely determined solution.

(ii)   Let, in addition, $n \geq 3$. Then there is a uniquely determined distribution $h \in H^{-1+\frac{n-d}{2}}(\Gamma)$ such that the solution u of (20.64) can be represented as
$$u(x) = \int_\Gamma \frac{h(\gamma)}{|x-\gamma|^{n-2}} \mu(d\gamma), \quad x \in \Omega. \qquad (20.65)$$

Proof   Part (i) is covered by 20.9. In addition, we have in any case the representation (20.60). It remains to prove (20.65). Let now $n \geq 3$. Let K be a ball centred at the origin with radius R and $\overline{\Omega} \subset K$. Then the representation (20.60) can be written as
$$u(x) = \int_\Gamma \left(\frac{c}{|x-\gamma|^{n-2}} + d_K(x,\gamma) \right) h(\gamma)\, \mu(d\gamma), \quad x \in K, \qquad (20.66)$$
where $c^{-1} = (n-2)|\omega_n|$ is independent of K (here $|\omega_n|$ is the volume of the unit sphere). Furthermore $d_K(x,\gamma)$ is the harmonic correction of the singularity function $c|x-\gamma|^{2-n}$ such that $G^K(x,\gamma) = 0$ if $x \in \partial K$. Recall
$$d_K(x,\gamma) = -\frac{1}{(n-2)|\omega_n|} \left(\frac{R}{|\gamma|}\right)^{n-2} \frac{1}{|x-\gamma^M|^{n-2}}, \quad \gamma^M = \frac{R^2}{|\gamma|^2}\gamma, \qquad (20.67)$$
where γ^M is the inverse point of $\gamma \neq 0$ with respect to K. This may be found in any relevant textbook, for example [GiT77], p. 19. Denoting the operator with respect to the right-hand side of (20.60) by B_K we obtain
$$(B_K h)(x) = (Ph)(x) + (D_K h)(x), \quad x \in K, \qquad (20.68)$$

with
$$(Ph)(x) = \int_\Gamma \frac{c}{|x-\gamma|^{n-2}} h(\gamma)\, \mu(d\gamma), \quad c^{-1} = (n-2)|\omega_n|, \tag{20.69}$$

and
$$(D_K h)(x) = \int_\Gamma d_K(x,\gamma)\, h(\gamma)\, \mu(d\gamma). \tag{20.70}$$

Let B_K^Γ, P^Γ, D_K^Γ be the restrictions of B_K, P, D_K on Γ. We have (20.39) with B_K^Γ in place of B^Γ (uniformly with respect to K). Then by the arguments given there, in particular (20.41),
$$B_K^\Gamma : \quad H^{-1+\frac{n-d}{2}}(\Gamma) \mapsto H^{1-\frac{n-d}{2}}(\Gamma) \tag{20.71}$$

is for all K an isometric map. Let Γ_1 be a neighbourhood of Γ (maybe collecting all points in $\mathbb{R}^n$ with distance smaller than 1 to Γ). Let R be large. Now it follows from (20.70) in obvious notation (dual pairings)

$$\begin{aligned}
|D_x^\alpha D_K h(x)| &= \left| \int_\Gamma D_x^\alpha d_K(x,\gamma)\, h(\gamma)\, \mu(d\gamma) \right| \\
&= |(D_x^\alpha d_K(x,\cdot), id_\Gamma h)| \\
&\leq \left\| D_x^\alpha d_K(x,\cdot) \,|\, H^1(\Gamma_1) \right\| \left\| id_\Gamma h \,|\, H^{-1}(\Gamma_1) \right\| \\
&\leq \frac{c_\alpha}{R^{n-2}} \left\| h \,|\, H^{-1+\frac{n-d}{2}}(\Gamma) \right\|, \quad x \in \Gamma_1, \alpha \in \mathbb{N}_0^n.
\end{aligned} \tag{20.72}$$

We use (20.67) and also (20.42) adapted to our situation. The estimate is uniform with respect to $x \in \Gamma_1$ and α, restricted to $|\alpha| = 0$ and $|\alpha| = 1$. In particular we have an estimate in $H^1(\Gamma_1)$ and by application of tr_Γ as in (20.38) we obtain

$$\left\| D_K^\Gamma h \,|\, H^{1-\frac{n-d}{2}}(\Gamma) \right\| \leq \frac{c}{R^{n-2}} \left\| h \,|\, H^{-1+\frac{n-d}{2}}(\Gamma) \right\|, \tag{20.73}$$

where c is independent of K (and, hence, of R). By (20.68) we have
$$P^\Gamma = B_K^\Gamma - D_K^\Gamma = B_K^\Gamma \circ (id - (B_K^\Gamma)^{-1} \circ D_K^\Gamma). \tag{20.74}$$

By (20.71) and (20.73) with R large it follows that the last operator on the right-hand side of (20.74) is invertible and hence P^Γ is an isomorphic map between the spaces in (20.71). Then any $g \in H^{1-\frac{n-d}{2}}(\Gamma)$ can be uniquely represented as

$$g = P^\Gamma h, \quad h \in H^{-1+\frac{n-d}{2}}(\Gamma). \tag{20.75}$$

Now we are in the same position as in 20.6 and Theorem 20.7. In particular, u in (20.65) is a solution of (20.64).

20.11 Spaces on d-sets, revisited

Let Γ be again a compact d-set in $\mathbb{R}^n$ with $0 < d < n$. We considered in 20.5 the spaces $H^s(\Gamma)$. If $s > 0$ then we have the quarkonial description from Definition 9.29 and Theorem 9.33. But especially in the cases of interest for us in connection with Theorem 20.7 and Corollary 20.10 one has rather simple explicit descriptions. Let $0 < s < 1$. Then by [JoW84], p. 103,

$$\|u\,|H^s(\Gamma)\| = \left(\int_\Gamma |u(\gamma)|^2\,\mu(d\gamma)\right)^{\frac{1}{2}} + \left(\int_{\Gamma \times \Gamma} \frac{|u(\gamma) - u(\varrho)|^2}{|\gamma - \varrho|^{2s+d}}\,\mu(d\gamma)\,\mu(d\varrho)\right)^{\frac{1}{2}} \tag{20.76}$$

is an equivalent norm in $H^s(\Gamma)$. This applies in particular to the spaces in Theorem 20.7 and Corollary 20.10 where $s = 1 - \frac{n-d}{2}$.

20.12 Formalizations: the smooth case

Let Γ and Ω be as in Theorem 20.7. Then the assertion that the operator G^Γ in (20.46) is an isomorphic map is a concise version of (20.47)–(20.49): existence and representation. This coincides with (20.7). The counterpart in case of Corollary 20.10(ii) is given by

$$tr_\Gamma \circ P \circ id_\Gamma : \quad H^{-1+\frac{n-d}{2}}(\Gamma) \mapsto H^{1-\frac{n-d}{2}}(\Gamma) \tag{20.77}$$

as an isomorphic map, where P is given by (20.69). Let in addition $\Gamma = \partial\Omega$ be smooth (say, Ω is a bounded C^∞ domain in $\mathbb{R}^n$ with $n \geq 3$). Then we have $d = n - 1$ both in (20.46) and (20.77), in particular

$$tr_\Gamma \circ P \circ id_\Gamma : \quad H^{-\frac{1}{2}}(\Gamma) \mapsto H^{\frac{1}{2}}(\Gamma) \tag{20.78}$$

as an isomorphic map. This is the precise version of (20.3). As mentioned there, one might consider $\Gamma = \partial\Omega$ as a compact Riemannian manifold. Then the spaces $H^s(\Gamma)$ can be defined directly for all $s \in \mathbb{R}$ via local charts. Let $-\Delta_g$ be the related Laplace-Beltrami operator. If $\varrho > 0$ is sufficiently large then

$$-\Delta_g + \varrho\,id : \quad H^s(\Gamma) \mapsto H^{s-2}(\Gamma), \quad s \in \mathbb{R}, \tag{20.79}$$

is an isomorphic map. Some information concerning spaces on Riemannian manifolds may be found in 7.2 (although in a somewhat different context); otherwise we refer to [Triδ], 1.11 and Chapter 7. As in the euclidean case (or the n-torus) the fractional powers of the positive-definite self-adjoint operator

$-\Delta_g + \varrho\, id$ with pure point spectrum have the expected mapping properties. In particular,

$$\sqrt{-\Delta_g + \varrho\, id}\, : \quad H^s(\Gamma) \mapsto H^{s-1}(\Gamma), \quad s \in \mathbb{R}. \tag{20.80}$$

Combined with (20.78) one gets

$$tr_\Gamma \circ P \circ id_\Gamma \circ \sqrt{-\Delta_g + \varrho\, id}\, : \quad H^{\frac{1}{2}}(\Gamma) \mapsto H^{\frac{1}{2}}(\Gamma) \tag{20.81}$$

as an isomorphic map. This sheds some light on the relations between the Laplacians on Ω and $\Gamma = \partial\Omega$.

20.13 L_2-L_p-theory

Again let Ω be a bounded domain in $\mathbb{R}^n$ with boundary $\Gamma = \partial\Omega$. The quality of the boundary Γ has a strong influence on the Dirichlet problem as treated above. If Γ is C^∞ then one has a complete theory in all reasonable spaces B^s_{pq} and F^s_{pq} even with $p < 1$, where the final results go back to [FrR95], which may also be found in [RuS96], 3.5. Looking for an L_p-theory for non-smooth boundaries there is apparently a big difference between $\Gamma \in C^1$ and $\Gamma \in Lip^1$. We refer to [Ken94], [JeK95], Theorem 5.1, p. 191, and in particular to [FMM98]. The last paper deals also with mapping properties of single layer potentials of type (20.69) on $\Gamma \in Lip^1$ from $B^{-s}_{pp}(\Gamma)$ onto $B^{1-s}_{pp}(\Gamma)$ for some s with $0 < s < 1$ and p with $\left|\frac{1}{p} - \frac{1}{2}\right| \leq \delta$ for some $\delta > 0$, where δ depends on the Lipschitz constant of Γ, [FMM98], Theorems 3.1 and 8.1. We refer also to [Zan00]. This seems to suggest that in the fractal case as treated in Corollary 20.10 there might be little hope to step from L_2 to L_p (or B^s_{pp}). Further information and additional references concerning layer potentials, boundary integrals and Dirichlet problems for Laplacians (on Riemannian manifolds) in Lipschitz domains may be found in [MeM00] and in Chapter 4 of [Tay00]. We refer in this context also to the survey [JoW97], where the authors summarize their contributions to boundary value problems of the above type.

Let again P^Γ be the operator P given by (20.69), restricted to the d-set Γ. We have the mapping properties as described at the end of 20.10. One may extend these considerations to Riesz potentials on d-sets Γ in $\mathbb{R}^n$ given by

$$(P^\Gamma_s h)(\lambda) = \int_\Gamma \frac{h(\gamma)}{|\lambda - \gamma|^{d-s}} \mu(d\gamma), \quad \lambda \in \Gamma,$$

where $d > s > 0$. Mapping properties of these operators and related spectral problems have been studied recently by M. Zähle, [Zah00], in the framework of an L_2-theory.

20.14 Classical solutions

Let Ω be a bounded domain in $\mathbb{R}^n$ where $n \geq 2$ and let $\Gamma = \partial \Omega$ be its boundary. Let $C(\Gamma)$ be the space of all continuous functions on Γ, obviously normed, and let $Lip(\Gamma)$ be the space of all Lipschitz continuous functions on Γ, normed in analogy to (10.17). Let $g \in C(\Gamma)$ or, more restrictive, $g \in Lip(\Gamma)$. The classical Dirichlet problem for the Laplacian asks for harmonic functions u in Ω with

$$\Delta u(x) = 0 \text{ in } \Omega \quad \text{and} \quad u(x) \to g(\gamma) \text{ if } x \in \Omega, \ x \to \gamma \in \Gamma. \tag{20.82}$$

The first decisive step for arbitrary bounded domains Ω goes back to Perron, [Per23], and is known as the method of subharmonic and superharmonic functions. One gets always a harmonic function u in Ω. The main problem is to clarify for which $\gamma \in \Gamma$ or for which boundaries Γ one has the desired pointwise boundary behaviour as indicated in (20.82). The final solution goes back to Wiener and can be described as follows. Let $\Omega^c = \mathbb{R}^n \setminus \Omega$ be the complement of Ω in $\mathbb{R}^n$ and let $C_{1,2}(K)$ be the $(1,2)$-capacity according to (19.44), this means with respect to the above classical Sobolev space $H^1(\mathbb{R}^n)$. Then one has (20.82) for a given point $\gamma \in \Gamma$ and $g \in C(\Gamma)$ if, and only if,

$$\int_0^1 \frac{C_{1,2}(B(\gamma, r) \cap \Omega^c)}{r^{n-1}} dr = \infty, \tag{20.83}$$

where $B(\gamma, r)$ is again a ball centred at $\gamma \in \Gamma$ and of radius r. In particular there is a natural connection to the theory of weak solutions, this means H^1-solutions. We do not go into detail. A description of the method of subharmonic functions may be found in [GiT77], 2.8, pp. 23–27. As for the Wiener criterion (20.83) we refer to [AdH96], Theorem 6.3.3, p. 165, and to [Ken94], p. 5. In our context it is of interest under what additional conditions for Γ the solutions u from Corollary 20.10 are classical solutions with respect to Perron's method. By [Ken94], p. 5 (restricted to our situation) one has the following assertion:

The Wiener criterion (20.83) is necessary and sufficient such that for every $g \in Lip(\Gamma)$ the (weak) solution u in (20.64) is continuous in $\overline{\Omega}$ (and hence a classical solution).

We discuss (20.83) in connection with the domains in Corollary 20.10. First we remark that always

$$C_{1,2}(B(\gamma, r)) \sim r^{n-2}, \quad \gamma \subset \mathbb{R}^n, \quad 0 < r < 1. \tag{20.84}$$

This can be checked easily; one can use the norm in (20.16). Secondly, by Remark 9.19 any d-set in $\mathbb{R}^n$ with $d < n$ satisfies the ball condition according to 9.16. This suggests the following modification of Definition 9.16:

A bounded domain Ω in $\mathbb{R}^n$ is said to satisfy the outer ball condition if there is a number $0 < \eta < 1$ such that for any ball $B(\gamma, r)$ centred at $\gamma \in \Gamma = \partial \Omega$ and of radius $0 < r < 1$ there is a ball $B(y, \eta r)$ (centred at $y \in \mathbb{R}^n$ and of radius ηr) with

$$B(y, \eta r) \subset \Omega^c \cap B(\gamma, r). \tag{20.85}$$

Now we get the following assertion:

Let Ω be a domain as in Corollary 20.10 satisfying in addition the outer ball condition and let $g \in Lip(\Gamma)$. Then the solution u in (20.64) is a classical solution.

As for the proof we first remark that

$$Lip(\Gamma) \subset H^s(\Gamma), \quad 0 < s < 1, \tag{20.86}$$

as a consequence of (20.76). Secondly, by (20.84) we have (20.83) for all $\gamma \in \Gamma$. Outer ball conditions and modifications of them have been used by several authors on different occasions (also in [Ken94], p. 4, called class S). In [TrW96] we called such a domain exterior regular in connection with intrinsic atomic characterizations of spaces in domains. A short description and further references may also be found in [ET96], 2.5.1, p. 59.

21 Spectral theory on manifolds

21.1 Introduction

The aim of this section is twofold. First we continue our studies from Section 7 on function spaces on manifolds. Let M be the Riemannian manifold as introduced in 7.2 and let $F_{pq}^s(M, g^\varkappa)$ be the spaces defined in 7.8. We characterized in Proposition 7.17 the compact embeddings

$$F_{p_1 q_1}^{s_1}(M, g^{\varkappa_1}) \subset F_{p_2 q_2}^{s_2}(M, g^{\varkappa_2}). \tag{21.1}$$

The first aim is to find out the degree of compactness expressed in terms of entropy numbers as introduced in 19.16. The motivation comes from the close connection between entropy numbers and eigenvalues of compact operators described by (19.129). Under restriction to d-domains as introduced in 7.5 and discussed in 7.6 we get in Theorem 21.3 a definitive result. We rely on the quarkonial decomposition from Theorem 7.22 and the techniques developed in [Triδ] and used so far in this book in connection with Theorem 19.17. The second (and main) aim of this section is to use this result in the indicated way as a starting point for a spectral theory of elliptic operators in d-domains. We developed in [ET96], Chapter 5, in a rather systematic way a spectral theory for degenerate pseudodifferential operators based on estimates of entropy numbers. The fractal counterpart has been studied in [Triδ], Chapter 5. There are always two distinguished problems. The first is the question of the distribution of eigenvalues. Theorem 19.7, in particular (19.73), and Theorem 19.17 with Definition 19.13, are typical examples in the present book. The second type of problems is connected with the so-called *negative spectrum*. Based on the techniques used in [Triδ] and Theorem 21.3 one can develop a spectral theory of weighted (fractal) pseudodifferential operators in d-domains. But this will not be done here in detail. We restrict ourselves to an example, concentrating now on the negative spectrum for suitable operators. Let $-\Delta_g$ be the Laplace-Beltrami operator from (7.54). If ϱ is sufficiently large then we have the lifting property (7.57). In particular, $-\Delta_g$ with its domain of definition,

$$dom\,(-\Delta_g) = H^2(M) = F^2_{2,2}(M), \tag{21.2}$$

is self-adjoint in $L_2(M)$ ($= F^0_{2,2}(M)$) and bounded from below. Let $\varrho \in \mathbb{R}$ be such that

$$Spec\,(-\Delta_g + \varrho\,id) \subset [1, \infty). \tag{21.3}$$

Then we are interested in the negative spectrum of the relatively compact perturbation

$$H_\beta = -\Delta_g + \varrho\,id - \beta\,g^{-\varkappa}, \quad \varkappa > 0, \quad \beta > 0, \tag{21.4}$$

of $-\Delta_g + \varrho\,id$. In other words, we ask for the behaviour of

$$N_\beta = \#\,\{Spec\,(H_\beta) \cap (-\infty, 0)\} \tag{21.5}$$

as $\beta \to \infty$. Problems of this type attracted a lot of attention in the euclidean setting (i.e., with $\mathbb{R}^n$ in place of M). They originate from (euclidean) quantum mechanics and the semi-classical limit $\hbar \to 0$ (Planck's constant tending to zero) and $\beta \sim \hbar^{-2}$, considered here. In Theorem 21.7 we prove

$$N_\beta \sim \beta^{\frac{d}{\varkappa}} \quad \text{if} \quad 0 < \varkappa < \frac{2d}{n} \quad (\beta \to \infty), \tag{21.6}$$

$$N_\beta \sim \beta^{\frac{n}{2}} \quad \text{if} \quad \varkappa > \frac{2d}{n} \quad (\beta \to \infty), \tag{21.7}$$

where M is the above d-domain. This might be considered as the main assertion of this section. We complement these results in Corollary 21.11 and in 21.12 by a look at hydrogen-like operators of type

$$H^\beta = -\Delta_g + \varrho\,id - \beta g^{-1}\left(1 + \left(\frac{1}{|x|_g} - 1\right)_+\right) \tag{21.8}$$

in this hyperbolic world (M, g), where $|x|_g$ is the Riemannian distance of $x \subset M$ to a fixed off-point.

21.2 Preliminaries

We recall what we need in the sequel. We always assume that Ω is a d-domain in $\mathbb{R}^n$ according to 7.5. This means in particular that we have the covering (7.23) with (7.24). Equivalences are used as in (7.10). We denote Ω also by M or (M, g) when converted in a Riemannian manifold as in 7.2. Also, though not really necessary, we assume $n \geq 2$ and that Ω is connected. Typical examples of d-domains, called thorny star-like d-domains, have been discussed in 7.6. In these distinguished but otherwise characteristic cases of infinite Riemannian manifolds of hyperbolic type we comment briefly on the relations between Riemannian distances, the generating function g given by (7.9) and the Riemannian volume of the slices Ω_j in (7.21). Let $|x|_g$ be the Riemannian distance of $x \in M$ to, say, $0 \in M$. By (7.11), (7.21), the Riemannian width of each slice is approximately 1. If one starts from a point $x \in \Omega_j$, then $g(x) \sim 2^j$, and one needs approximately j steps of Riemannian length 1 to reach 0. Hence,

$$g(x) \sim 2^j \sim 2^{m(x)\,|x|_g}, \quad x \in \Omega_j, \tag{21.9}$$

with $m(x) \sim 1$. Furthermore we have for the Riemannian volume $\mathrm{vol}_g\, \Omega_j$ of Ω_j and of the balls $\Omega^j = \bigcup_{l=0}^{j} \Omega_l$,

$$\mathrm{vol}_g\, \Omega_j \sim \mathrm{vol}_g\, \Omega^j \sim 2^{jd} \sim 2^{d\,m(x)\,|x|_g}, \quad x \in \Omega_j. \tag{21.10}$$

This reflects the well-known fact that the volume of balls in Riemannian manifolds of hyperbolic type grows exponentially with the radius. In the general case of the above Riemannian manifolds the argument of stepping down from slice to slice might be more complicated in dependence on the geometry of Ω. But nevertheless (21.9), (21.10) in the above distinguished star-like d-domains reflect typical behaviour.

In Definition 7.8 we introduced the spaces $F^s_{pq}(M)$ and $F^s_{pq}(M, g^\varkappa)$, where

$$0 < p \leq \infty, \quad 0 < q \leq \infty, \quad s \in \mathbb{R}, \quad \varkappa \in \mathbb{R}, \tag{21.11}$$

(with $q = \infty$ if $p = \infty$), where we now always assume that M is a d-domain in $\mathbb{R}^n$. For our later purposes we abbreviate

$$H^s(M) = F^s_{2,2}(M), \quad H^s(M, g^\varkappa) = F^s_{2,2}(M, g^\varkappa), \quad s \in \mathbb{R},\, \varkappa \in \mathbb{R}, \tag{21.12}$$

in particular,

$$L_2(M) = H^0(M) \quad \text{and} \quad L_2(M, g^\varkappa) = H^0(M, g^\varkappa), \quad \varkappa \in \mathbb{R}. \tag{21.13}$$

Of special interest for us is the Laplace-Beltrami operator in M,

$$-\Delta_g = -g^{-n} \sum_{j=1}^{n} \frac{\partial}{\partial x_j} \left(g^{n-2} \frac{\partial}{\partial x_j} \right). \tag{21.14}$$

This operator has been studied in great detail on manifolds of the above type and more general Riemannian manifolds.
In particular, $-\Delta_g$ with $dom\,(-\Delta_g) = D(M)$ is essentially self-adjoint and with $dom\,(-\Delta_g) = H^2(M)$ self-adjoint and positive in $L_2(M)$. We refer to [Str83], [Dav89], 5.2.3, p. 151, [Shu92], [Skr98], [Tri88], [Triγ], Chapter 7. Furthermore if $\varrho \in \mathbb{R}$ is sufficiently large, then

$$-\Delta_g + \varrho\,id\;:\quad F_{pq}^{s+2}(M, g^{\varkappa}) \mapsto F_{pq}^s(M, g^{\varkappa}) \tag{21.15}$$

with (21.11) and, in particular,

$$-\Delta_g + \varrho\,id\;:\quad H^2(M) \mapsto L_2(M) \tag{21.16}$$

are isomorphic mappings. We refer to Theorem 7.15.

Recall that we describe the degree of compactness in terms of entropy numbers according to 19.16.

21.3 Theorem

Let $n \in \mathbb{N}$ with $n > 2$ and let $n - 1 \le d < n$. Let Ω be a connected bounded d-domain in $\mathbb{R}^n$. Let (M, g) (or M for short) be the related non-compact Riemannian manifold with the function spaces according to Definition 7.8. Let $\varkappa_1 \in \mathbb{R}$, $\varkappa_2 \in \mathbb{R}$,

$$-\infty < s_2 < s_1 < \infty, \quad 0 < p_1 \le p_2 \le \infty, \quad 0 < q_1 \le \infty,\; 0 < q_2 \le \infty, \tag{21.17}$$

($q_1 = \infty$ if $p_1 = \infty$, and $q_2 = \infty$ if $p_2 = \infty$) with

$$\delta = \left(s_1 - \frac{n}{p_1}\right) - \left(s_2 - \frac{n}{p_2}\right) > 0, \quad \varkappa = \varkappa_1 - \varkappa_2 > 0. \tag{21.18}$$

Then

$$id\;:\quad F_{p_1 q_1}^{s_1}(M, g^{\varkappa_1}) \mapsto F_{p_2 q_2}^{s_2}(M, g^{\varkappa_2}) \tag{21.19}$$

is compact and for the related entropy numbers $e_k = e_k(id)$ we have

$$e_k \sim k^{-\frac{s_1 - s_2}{n}}, \quad k \in \mathbb{N}, \quad \text{if}\quad \varkappa > \frac{\delta d}{n}, \tag{21.20}$$

$$e_k \sim k^{-\frac{\varkappa}{d} + \frac{1}{p_2} - \frac{1}{p_1}}, \quad k \in \mathbb{N}, \quad \text{if}\quad \varkappa < \frac{\delta d}{n}. \tag{21.21}$$

Proof *Step 1* First we wish to prove that the entropy numbers e_k in (21.20), (21.21) can be estimated from above by the respective right-hand sides. We begin with some preparation. By the lifting property (21.15) we may assume that s_1 and s_2 are sufficiently large that Theorem 7.22 can be applied to the spaces in (21.19). By Theorem 7.15 we may also assume $\varkappa_2 = 0$. Recall that the quasi-norms in (7.99) are defined by (7.87), (7.83). Using (2.10) we wish to replace the somewhat complicated sequence spaces f_{pq}^Ω by their simpler b-counterparts. We use the notation introduced in [Triδ], Sections 8 and 9 (slightly modified). Let $(E_j)_{j \in \mathbb{N}_0}$ be a sequence of natural numbers, let

$$\sigma \in \mathbb{R}, \quad 0 < p \leq \infty, \quad 0 < q \leq \infty.$$

Then by $\ell_q\left(2^{j\sigma}\ell_p^{E_j}\right)$ we shall mean the linear space of all complex sequences λ with (7.82) (and E_j in place of L_j), endowed with the quasi-norm

$$\left\| \lambda \,|\, \ell_q\left(2^{j\sigma}\ell_p^{E_j}\right) \right\| = \left(\sum_{j=0}^{\infty} \left(\sum_{l=1}^{E_j} 2^{j\sigma p} |\lambda_{jl}|^p \right)^{\frac{q}{p}} \right)^{\frac{1}{q}} < \infty \tag{21.22}$$

with obvious modification if $p = \infty$ and/or $q = \infty$. In case of $\sigma = 0$ we write $\ell_q\left(\ell_p^{E_j}\right)$. Let in addition $\varrho \in \mathbb{R}$ and let now λ be given by (7.85), (7.86) (again with E_j in place of L_j). Then we put

$$\left\| \lambda \,|\, \ell_{\infty,\varrho}\left[\ell_q\left(2^{j\sigma}\ell_p^{E_j}\right)\right] \right\| = \sup_{\beta \in \mathbb{N}_0^n} 2^{\varrho|\beta|} \left\| \lambda^\beta \,|\, \ell_q\left(2^{j\sigma}\ell_p^{E_j}\right) \right\| \tag{21.23}$$

as quasi-norms of corresponding spaces. In analogy to [Triδ], pp. 163–165, we wish to reduce the estimate from above of the entropy numbers e_k in (21.20), (21.21) to

$$e_k \leq c\, e_k(Id), \quad k \in \mathbb{N}, \tag{21.24}$$

where

$$Id: \quad \ell_{\infty,\varrho_1}\left[\ell_{q_1}\left(2^{r\varkappa}\ell_{p_1}^{E_r}\right)\right] \hookrightarrow \ell_{\infty,\varrho_2}\left[\ell_{q_2}\left(\ell_{p_2}^{E_r}\right)\right] \tag{21.25}$$

is the identity. We always assume that

$$\varrho_1 > \varrho_2 > 0 \quad \text{large}, \tag{21.26}$$

such that Theorem 7.22 can be applied. Furthermore $\varkappa = \varkappa_1 > 0$ (recall that $\varkappa_2 = 0$) has the above meaning. The numbers q_1 and q_2 are unimportant. We may assume $q_1 \geq p_1$ and $q_2 \leq p_2$ (the F-spaces are monotone with respect to the q-index). Crucial for the estimate of $e_k(Id)$ is the knowledge of E_r. Afterwards one gets (21.24) factorizing id in (21.19) via Id in (21.25) in a similar way as in Step 3 of the proof of Theorem 19.17. First we estimate E_r. Let $f \in F^{s_1}_{p_1 q_1}(M, g^\varkappa)$ be given by (7.98), where $(\beta qu)_{jl}(x)$ are $(s_1, p_1, \varkappa)$-β-quarks according to (7.88). Comparing this expansion with a corresponding representation in $F^{s_2}_{p_2 q_2}(M)$ (recall $\varkappa_2 = 0$) one gets

$$f = \sum_{\beta \in \mathbb{N}_0^n} \sum_{j=0}^{\infty} \sum_{l=1}^{L_j} \lambda^\beta_{jl} (\beta qu)_{jl}(x) = \sum_{\beta \in \mathbb{N}_0^n} \sum_{j=0}^{\infty} \sum_{l=1}^{L_j} \lambda^\beta_{jl} 2^{-k\varkappa} 2^{-(j-k)\delta} (\beta qu)^*_{jl}(x), \tag{21.27}$$

where $(\beta qu)^*_{jl}(x)$ denote temporarily the corresponding $(s_2, p_2, 0)$-β-quarks. Here δ is given by (21.18) and k by (7.89). If $r \in \mathbb{N}_0$ then we have to estimate the number E_r of balls K_{jl} given by (7.78) with

$$\operatorname{dist}(K_{jl}, \partial\Omega) \sim 2^{-m} \quad \text{for} \quad m = 0, \ldots, j+c \tag{21.28}$$

and

$$r\varkappa = m\varkappa + (j-m)\delta. \tag{21.29}$$

Here we used $m \leq j+c$ according to (7.90). For fixed m with $m = 0, \ldots, r+c'$ we have to estimate the number E^m_r of balls K_{jl} of radius $\sim 2^{-j}$ which, say, intersect Ω_m given by (7.21) with (7.24), and where j is calculated by (21.29). Hence

$$E^m_r \sim (\operatorname{vol}\Omega_m) 2^{jn} \sim 2^{-m(n-d)} 2^{jn} \sim 2^{md} 2^{(r-m)\frac{\varkappa n}{\delta}}, \tag{21.30}$$

where we used again (7.24). Summation over m results in

$$E_r \sim 2^{\frac{r\varkappa n}{\delta}} \sum_{m=0}^{r+c'} 2^{m(d-\frac{\varkappa n}{\delta})}. \tag{21.31}$$

We obtain

$$E_r \sim 2^{rd} \quad \text{if} \quad d > \frac{n\varkappa}{\delta}, \tag{21.32}$$

$$E_r \sim 2^{\frac{rn\varkappa}{\delta}} \quad \text{if} \quad d < \frac{n\varkappa}{\delta}. \tag{21.33}$$

Let Λ_r be the collection of those E_r couples (j, l) with the above property. Recall that the series (21.27) converge absolutely and can be rearranged by

$$f = \sum_{\beta \in \mathbb{N}_0^n} \sum_{r=0}^{\infty} \sum_{(j,l) \in \Lambda_r} \lambda^\beta_{jl} (\beta qu)_{jl}(x) \tag{21.34}$$

and

$$f = \sum_{\beta \in \mathbb{N}_0^n} \sum_{r=0}^{\infty} 2^{-r\varkappa} \sum_{(j,l) \in \Lambda_r} \lambda^\beta_{jl} (\beta qu)^*_{jl}(x). \tag{21.35}$$

For fixed $r \in \mathbb{N}_0$ the balls K_{jl} with $(j,l) \in \Lambda_r$ have only a controlled overlapping such that we have an obvious counterpart of Proposition 2.3. Furthermore by 7.23 we may assume that the coefficients $\lambda_{jl}^\beta(f)$ depend linearly on f. We construct the linear operator S,

$$S : F_{p_1 q_1}^{s_1}(M, g^{\varkappa_1}) \mapsto \ell_{\infty, \varrho_1}\left[\ell_{q_1}\left(2^{r\varkappa}\ell_{p_1}^{E_r}\right)\right] \tag{21.36}$$

by

$$Sf = \eta = \{\eta^\beta : \beta \in \mathbb{N}_0^n\}, \tag{21.37}$$

$$\eta^\beta = \left\{2^{-r\varkappa}\lambda_{jl}^\beta : r \in \mathbb{N}_0, (j,l) \in \Lambda_r\right\}, \tag{21.38}$$

where f is given by (21.34) with $\lambda_{jl}^\beta = \lambda_{jl}^\beta(f)$. To justify that S is continuous we claim

$$\left\|Sf \,|\ell_{\infty, \varrho_1}\left[\ell_{q_1}\left(2^{r\varkappa}\ell_{p_1}^{E_r}\right)\right]\right\| = \left\|\lambda \,|\ell_{\infty, \varrho_1}\left[\ell_{q_1}\left(\ell_{p_1}^{E_r}\right)\right]\right\|$$

$$\leq c \left\|\lambda \,|f_{p_1 q_1}^\Omega\right\|_{\varrho_1} \sim \left\|f \,|F_{p_1 q_1}^{s_1}(M, g^{\varkappa_1})\right\|. \tag{21.39}$$

The first equality comes from the construction (in obvious notation) and the notational agreement after (21.22). As mentioned above we can apply Proposition 2.3 now with $p = p_1$ and $q = q_1 \geq p_1$. Together with (7.87) we obtain the inequality in (21.39). The final equivalence follows from Theorem 7.22 and the choice of $\lambda_{jl}^\beta(f)$ according to 7.23. Next we construct the linear operator T,

$$T : \ell_{\infty, \varrho_2}\left[\ell_{q_2}\left(\ell_{p_2}^{E_r}\right)\right] \mapsto F_{p_2 q_2}^{s_2}(M), \tag{21.40}$$

by

$$T\zeta = \sum_{\beta \in \mathbb{N}_0^n} \sum_{r=0}^\infty \sum_{(j,l) \in \Lambda_r} \zeta_{jl}^\beta (\beta q u)_{jl}^*(x) \tag{21.41}$$

where again $(\beta q u)_{jl}^*(x)$ are the above $(s_2, p_2, 0)$-β-quarks and

$$\zeta = \{\zeta^\beta : \beta \in \mathbb{N}_0^n\}, \quad \zeta^\beta = \left\{\zeta_{jl}^\beta : r \in \mathbb{N}_0, (j,l) \in \Lambda_r\right\}. \tag{21.42}$$

Obviously, T is linear. But it is also continuous, since we can again apply Proposition 2.3 now with $p = p_2$ and $q = q_2 \leq p_2$, and afterwards Theorem 7.22. Let id and Id be given by (21.19) and (21.25) with (21.26). Then by (21.34), (21.35),

$$id = T \circ Id \circ S. \tag{21.43}$$

By [Triδ], Theorem 9.2, p. 47 and (21.32) we have

$$e_k(Id) \sim k^{-\frac{\varkappa}{d}+\frac{1}{p_2}-\frac{1}{p_1}} \quad \text{if} \quad d > \frac{n\varkappa}{\delta} \tag{21.44}$$

and by (21.33)

$$e_k(Id) \sim k^{-\frac{\delta}{n}+\frac{1}{p_2}-\frac{1}{p_1}} = k^{-\frac{s_1-s_2}{n}} \quad \text{if} \quad d < \frac{n\varkappa}{\delta}. \tag{21.45}$$

Together with (21.43) (and (19.133)) we get

$$e_k(id) \leq c k^{-\frac{s_1-s_2}{n}}, \quad k \in \mathbb{N}, \quad \text{if} \quad \varkappa > \frac{\delta d}{n}, \tag{21.46}$$

$$e_k(id) \leq c k^{-\frac{\varkappa}{d}+\frac{1}{p_2}-\frac{1}{p_1}}, \quad k \in \mathbb{N}, \quad \text{if} \quad \varkappa < \frac{\delta d}{n}, \tag{21.47}$$

for some $c > 0$.

Step 2 We prove the converse of (21.46), (21.47). In case of (21.20) we must show that there is a number $c > 0$ such that

$$e_k\left(id : F^{s_1}_{p_1 q_1}(M, g^{\varkappa_1}) \hookrightarrow F^{s_2}_{p_2 q_2}(M, g^{\varkappa_2})\right) k^{\frac{s_1-s_2}{n}} \geq c \tag{21.48}$$

for all $k \in \mathbb{N}$. By the lifting property (21.15) we may assume that s_2 (and hence also s_1) is large. Assume that there is no $c > 0$ with (21.48) for all $k \in \mathbb{N}$. Then we find a sequence $k_v \to \infty$ with

$$e_{k_v}\left(id : F^{s_1}_{p_1 q_1}(M, g^{\varkappa_1}) \hookrightarrow F^{s_2}_{p_2 q_2}(M, g^{\varkappa_2})\right) k_v^{\frac{s_1-s_2}{n}} \to 0 \quad \text{if} \quad v \to \infty. \tag{21.49}$$

By Theorem 7.15(ii), (21.46) and (19.133) one can even specify

$$F^{s_2}_{p_2 q_2}(M, g^{\varkappa_2}) = F^0_{p_2, 2}(M, g^{-\frac{n}{p_2}}) = L_{p_2}(\Omega), \quad 1 < p_2 < \infty, \tag{21.50}$$

where we used Theorem 7.10, obviously extended to $s_2 = 0$, $1 < p_2 < \infty$, and, by (1.9), (1.10),

$$F^0_{p_2, 2}(\mathbb{R}^n) = H^0_{p_2}(\mathbb{R}^n) = L_{p_2}(\mathbb{R}^n). \tag{21.51}$$

Hence we must disprove (21.49) with the specification (21.50). As for the excluded case $p_2 = \infty$ we add a remark in 21.4 below. Similarly in case of (21.21). Let $r \in \mathbb{N}_0$ and let $\widetilde{\Lambda}_r$ be a subset of the above set Λ_r, introduced after (21.33), such that corresponding balls K_{jl} with $(j, l) \in \widetilde{\Lambda}_r$ are pairwise disjoint. By the above construction one may select $\widetilde{E}_r$ balls with this property such that

$$\widetilde{E}_r \sim E_r, \quad r \in \mathbb{N}_0, \tag{21.52}$$

where E_r is given by (21.32), (21.33). By (7.79) the corresponding quarks in (7.88) have disjoint supports. We put $(qu)_{jl} = (\beta qu)_{jl}$ if $\beta = 0$. Let

$$s > 0, \quad 0 < p_1 < \infty, \quad 1 < p_2 < \infty, \quad p_1 \leq p_2 < \infty, \quad 0 < q \leq \infty, \tag{21.53}$$

$$\delta = s - \frac{n}{p_1} + \frac{n}{p_2} > 0, \quad \varkappa = \varkappa_1 + \frac{n}{p_2} > 0. \tag{21.54}$$

Let $\lambda = \left\{ \lambda_{jl} : (j,l) \in \widetilde{\Lambda}_r \right\}$ for fixed $r \in \mathbb{N}_0$. Let A,

$$A: \ell_{p_1}^{\widetilde{E}_r} \mapsto F_{p_1 q}^s(M, g^{\varkappa_1}) \tag{21.55}$$

be given by

$$A\lambda = \sum_{(j,l) \in \widetilde{\Lambda}_r} \lambda_{jl} (qu)_{jl}(x), \quad x \in \Omega, \tag{21.56}$$

where

$$(qu)_{jl}(x) = 2^{-(j-m)(s-\frac{n}{p_1}) - m\varkappa_1} \psi_{j,l}(x) \tag{21.57}$$

are $(s, p_1, \varkappa_1)$-0-quarks according to (7.88) and $m = m(j,l)$ has the same meaning as in (21.28). By Theorem 7.22 and the support properties of the balls K_{jl} involved it follows that

$$\|A\lambda \,|\, F_{p_1 q}^s(M, g^{\varkappa_1})\| \leq c \,\|\lambda \,|\, \ell_{p_1}^{\widetilde{E}_r}\| \tag{21.58}$$

(where c is independent of r). By (21.29) and (21.54) we have

$$(qu)_{jl}(x) = 2^{-r\varkappa + j\frac{n}{p_2}} \psi_{j,l}(x), \quad x \in \Omega, \tag{21.59}$$

and

$$A\lambda = 2^{-r\varkappa} \sum_{(j,l) \in \widetilde{\Lambda}_r} \lambda_{jl} \, 2^{j\frac{n}{p_2}} \psi_{j,l}(x). \tag{21.60}$$

Let again $(j,l) \in \widetilde{\Lambda}_r$. We may assume that there are functions

$$\chi_{j,l}(x) \in D(K_{j,l}), \quad |\chi_{j,l}(x)| \leq c, \tag{21.61}$$

and

$$\int_\Omega \psi_{j,l}(x) \chi_{j,l}(x) \, dx = 2^{-jn}, \quad (j,l) \in \widetilde{\Lambda}_r. \tag{21.62}$$

Let B,
$$B : L_{p_2}(\Omega) \mapsto \ell_{p_2}, \qquad (21.63)$$
be given by
$$Bf = \left\{ 2^{jn-j\frac{n}{p_2}} 2^{r\varkappa} \int_\Omega f(x) \chi_{j,l}(x)\, dx \ : \ (j,l) \in \widetilde{\Lambda}_r \right\}. \qquad (21.64)$$

By Hölder's inequality we have
$$\| Bf \,|\ell_{p_2}\|^{p_2} \le \sum_{(j,l)\in\widetilde{\Lambda}_r} 2^{jn\frac{p_2}{p_2'}} 2^{r\varkappa p_2} \int_{K_{j,l}} |f(x)|^{p_2}\, dx\, 2^{-jn\frac{p_2}{p_2'}}$$
$$\le 2^{r\varkappa p_2} \| f\, |L_{p_2}(\Omega)\|^{p_2}. \qquad (21.65)$$

Hence,
$$\|B\| \le 2^{r\varkappa}. \qquad (21.66)$$

Let id be given by (21.19) with the specification (21.50). Let id' be the identity from $\ell_{p_1}^{\widetilde{E}_r}$ into ℓ_{p_2}. Then we have by (21.60), (21.62), (21.64),
$$id^r = B \circ id \circ A. \qquad (21.67)$$

In particular one gets by (21.58), (21.66) for the related entropy numbers
$$e_k(id^r) \le c\, 2^{r\varkappa} e_k(id), \quad k \in \mathbb{N}, \qquad (21.68)$$
where c is independent of $r \in \mathbb{N}_0$. By [Triδ], Theorem 7.3, p. 37,
$$e_{\widetilde{E}_r}(id) \ge c\, 2^{-r\varkappa} \widetilde{E}_r^{\frac{1}{p_2} - \frac{1}{p_1}}, \quad r \in \mathbb{N}_0, \qquad (21.69)$$
$c > 0$, independent of r. In case of (21.52), (21.32) we have
$$e_{c 2^{rd}}(id) \ge c'\, 2^{-rd(\frac{\varkappa}{d} - \frac{1}{p_2} + \frac{1}{p_1})}, \quad r \in \mathbb{N}_0. \qquad (21.70)$$

In case of (21.52), (21.33) we put $a = \frac{n\varkappa}{\delta}$ and get
$$e_{c 2^{ra}}(id) \ge c'\, 2^{-ra(\frac{\varkappa}{a} - \frac{1}{r_2} + \frac{1}{r_1})} = c'\, 2^{-ra\frac{s}{n}}, \quad r \in \mathbb{N}_0, \qquad (21.71)$$
where we used (21.54). Here $c > 0$, $c' > 0$, are independent of $r \in \mathbb{N}$. But this disproves (21.49) with (21.50). As said above, this is sufficient to prove the converse of (21.46), (21.47).

21.4 The case $p = \infty$

In Step 2 of the above proof we excluded so far the case $p_2 = \infty$ (and hence also $p_1 = \infty$). Let now $p_2 = \infty$ and let $\mathcal{C}^s(\Omega) = B^s_{\infty\infty}(\Omega)$ with $0 < s < 1$ be the Hölder-Zygmund spaces, normed by first differences as in (1.11)–(1.13) and according to Definition 5.3. By Theorem 7.10 and direct arguments one gets

$$F^\sigma_{\infty\infty}(M, g^\sigma) \subset \mathcal{C}^\sigma(\Omega), \quad 0 < \sigma < 1, \tag{21.72}$$

where by Proposition 7.12 and (5.103) these two spaces coincide if Ω is smooth. By the same arguments as at the beginning of Step 2 it is sufficient to disprove (21.49) or the related counterpart based on (21.21) by the specification (21.50) with $\mathcal{C}^{s_2}(\Omega)$ in place of $L_{p_2}(\Omega)$. Combining this with the entropy numbers of

$$id : \quad \mathcal{C}^{s_2}(\Omega) \to L_\infty(\Omega) \tag{21.73}$$

according to [Triδ], Theorem 23.2, p. 186 (which works for any bounded domain), it turns out that one can take also $L_\infty(\Omega)$ as a target space to disprove (21.49). The rest is the same as in Step 2 which works (including the references) also for $p_2 = \infty$.

21.5 Remark

Theorem 21.3 coincides with [Tri99c], Theorem 2.4.2, where we outlined also a proof. This theorem might be considered as the hyperbolic counterpart of the weighted euclidean situation with the replacement of

$$F^s_{pq}(M, g^\varkappa) \quad \text{by} \quad F^s_{pq}(\mathbb{R}^n, w_\varkappa), \quad w_\varkappa(x) = (1 + |x|^2)^{\frac{\varkappa}{2}}. \tag{21.74}$$

The outcome is similar. The splitting point $\varkappa = \frac{\delta d}{n}$ in (21.20), (21.21) is then $\varkappa = \delta$. First results in this weighted euclidean case have been obtained in [HaT94a]. An improved version may be found in [ET96], 4.3.2, pp. 170–171. The state of art, including estimates for approximation numbers may be found in [Har95], [Har97], [Har98], [Har00c]. In the papers by D. D. Haroske special attention is paid to limiting case $\varkappa = \delta$. We excluded here in the hyperbolic situation the corresponding limiting case $\varkappa = \frac{\delta d}{n}$. Furthermore in the euclidean setting there is a rather satisfactory discussion of cases with $p_2 < p_1$. It should be possible to combine the techniques developed in the euclidean case with the above hyperbolic arguments and to develop a more complete theory including $\varkappa = \frac{\delta d}{n}$ and also spaces with $p_2 < p_1$.

21.6 Spectral theory

Let $\lambda > 0$, $0 \le \gamma \le 1$, and

$$b(\cdot, D) \in \Psi_{1,\gamma}^{-\lambda}(\mathbb{R}^n) \tag{21.75}$$

be a pseudodifferential operator in $\mathbb{R}^n$ in the Hörmander class indicated. Let $1 \le r_1 \le \infty$, $1 \le r_2 \le \infty$, and

$$b_1 \in L_{r_1}(\mathbb{R}^n, w_{\varkappa_1}), \quad b_2 \in L_{r_2}(\mathbb{R}^n, w_{\varkappa_2}), \quad \varkappa_1 \in \mathbb{R}, \quad \varkappa_2 \in \mathbb{R}, \tag{21.76}$$

$\varkappa = \varkappa_1 + \varkappa_2 > 0$ (in obvious notation according to (21.74)). Based on [HaT94b] we developed in [ET96], 5.4, a spectral theory of the degenerate pseudodifferential operators

$$B = b_2\, b(\cdot, D)\, b_1 \tag{21.77}$$

with the weighted euclidean counterpart of Theorem 21.3 as indicated in 21.5 as starting point. More recent results in this direction may be found in [Har98]. In [Triδ], Chapter 5, we discussed a corresponding spectral theory for fractal pseudodifferential operators. Armed with Theorem 21.3 one can try to extend this theory from the euclidean case to the above manifolds of hyperbolic type. Pseudodifferential operators on manifolds with bounded geometry and positive injectivity radius, especially the Laplace-Beltrami operator (21.14) have been considered with great intensity and may also be found in the references given in 21.2. In particular the mapping properties proved there can be taken as the starting point for a spectral theory of hyperbolic counterparts of (21.77). But this will not be done here.

As described in the Introduction 21.1 we concentrate ourselves on the special but interesting problem of the negative spectrum as explained in (21.3)–(21.5). We commented there also on the physical background (at least in the euclidean case). As for the mathematical background we complement our previous assertions. Recall that the Laplace-Beltrami operator $-\Delta_g$, given by (21.14) and considered as an operator in $L_2(M)$, is essentially self-adjoint on $D(M) = D(\Omega)$, and self-adjoint with $dom(-\Delta_g) = H^2(M)$. If $\varrho \in \mathbb{R}$ is sufficiently large then not only (21.16), but also (21.15) are isomorphic maps. Of course for any $\varkappa > 0$ the multiplication operator $f \mapsto g^{-\varkappa} f$ is symmetric and bounded on $L_2(M)$. Furthermore, by Theorem 7.15,

$$I - (-\Delta_g + \varrho\, id)^{-1} g^{-\varkappa} : \quad L_2(M) \mapsto H^2(M, g^\varkappa), \tag{21.78}$$

is an isomorphic map and by Proposition 7.17 (or Theorem 21.3),

$$B = (-\Delta_g + \varrho\, id)^{-1} g^{-\varkappa} : \quad L_2(M) \mapsto L_2(M), \tag{21.79}$$

is compact. Hence, H_β, given by (21.4), is for any $\beta > 0$ a relatively compact perturbation of $-\Delta_g + \varrho\,id$. By Kato's criterion, [Tri92], Theorem 3, p. 208, or [Dav95], Theorem 1.4.2, p. 18 (attributed there to Rellich, 1939), which applies in particular to relatively compact perturbations, it follows that

$$H_\beta = -\Delta_g + \varrho\,id - \beta g^{-\varkappa}, \quad \varkappa > 0, \quad dom\,(H_\beta) = H^2(M), \qquad (21.80)$$

is a self-adjoint operator in $L_2(M)$. Furthermore, since the operator in (21.79) is compact one gets by [Dav95], Theorem 8.4.3, p. 167, or in a more general context of Banach spaces by [EdE87], Theorem 2.1 on p. 418, that the essential spectra of these two operators coincide,

$$EssSpec\,(H_\beta) = EssSpec\,(-\Delta_g + \varrho\,id) \subset [1, \infty), \qquad (21.81)$$

where the latter inclusion comes from our assumption (21.3) concerning the choice of ϱ. Let $Spec\,(H_\beta)$ be the spectrum of the self-adjoint operator H_β in $L_2(M)$, given by (21.80). Then N_β in (21.5) counts the finite number of negative eigenvalues of H_β, called the negative spectrum. As usual, $\#A$ denotes the number of elements of the finite set A. Recall again that equivalence, expressed by $\sim$, is used as explained in (7.10) and (19.17), (19.18).

21.7 Theorem

Let $n \in \mathbb{N}$ with $n \geq 2$ and let $n-1 \leq d < n$. Let Ω be a connected d-domain in $\mathbb{R}^n$ according to 7.5. Let M be the related non-compact Riemannian manifold. Let H_β be the self-adjoint operator in $L_2(M)$ given by (21.80) with $\beta > 0$. Let

$$N_\beta = \#\{Spec\,(H_\beta) \cap (-\infty, 0)\} \qquad (21.82)$$

be the number of the negative eigenvalues of H_β. Then

$$N_\beta \sim \beta^{\frac{d}{\varkappa}} \quad \text{if} \quad 0 < \varkappa < \frac{2d}{n}, \quad \beta \to \infty, \qquad (21.83)$$

$$N_\beta \sim \beta^{\frac{n}{2}} \quad \text{if} \quad \varkappa > \frac{2d}{n}, \quad \beta \to \infty. \qquad (21.84)$$

Proof *Step 1* Let the compact operator B and the isomorphic operator I be given by (21.79) and (21.78), respectively. Let

$$id: H^2(M, g^\varkappa) \mapsto L_2(M) \qquad (21.85)$$

be the identity. Then we have

$$B = id \circ I \quad \text{and} \quad id = B \circ I^{-1}. \qquad (21.86)$$

Let $e_k(B)$ and $e_k(id)$ be the corresponding entropy numbers according to 19.16. It follows by (21.86),

$$e_k(B) \sim e_k(id), \quad k \in \mathbb{N}. \tag{21.87}$$

In order to estimate N_β from above we use the entropy version of the Birman-Schwinger principle as described in [Triδ], 31.1, p. 243, with references to [HaT94b] and [ET96], 5.4.1, p. 223. In the latter book one finds also further discussions about the Birman-Schwinger principle including the relevant literature. We have

$$N_\beta \le \# \left\{ k \in \mathbb{N} : \sqrt{2}\,\beta\,e_k(B) \ge 1 \right\}. \tag{21.88}$$

We specify id in (21.19) by (21.85) and obtain

$$e_k(id) \sim k^{-\frac{2}{n}}, \quad k \in \mathbb{N}, \quad \varkappa > \frac{2d}{n}, \tag{21.89}$$

$$e_k(id) \sim k^{-\frac{\varkappa}{d}}, \quad k \in \mathbb{N}, \quad 0 < \varkappa < \frac{2d}{n}. \tag{21.90}$$

Now (21.87)–(21.90) result in

$$N_\beta \lesssim c\beta^{\frac{n}{2}} \quad \text{if} \quad \varkappa > \frac{2d}{n}, \quad \beta \to \infty, \tag{21.91}$$

$$N_\beta \le c\beta^{\frac{d}{\varkappa}} \quad \text{if} \quad 0 < \varkappa < \frac{2d}{n}, \quad \beta \to \infty, \tag{21.92}$$

for some $c > 0$ which is independent of β.

Step 2 We prove the converse inequalities. Recall that H_β is self-adjoint in $L_2(M)$. We shift the problem to quadratic forms and the Max-Min principle. Hence, by (21.80) we have

$$(H_\beta f, f)_{L_2(M)} \sim \|f\,|H^1(M)\|^2 - \beta \|g^{-\frac{\varkappa}{2}} f\,|L_2(M)\|^2 \tag{21.93}$$
$$= \|f\,|H^1(M)\|^2 - \beta \|f\,|L_2(M, g^{-\frac{\varkappa}{2}})\|^2, \quad f \in H^2(M).$$

Let $r \in \mathbb{N}$ and let $\widetilde{\Lambda}_r$ be the same set of couples (j,l) as in Step 2 of the proof of Theorem 21.3. Recall that balls K_{jl} with $(j,l) \in \widetilde{\Lambda}_r$ are pairwise disjoint. Otherwise they are located by (21.28) with (21.29). We apply this construction to

$$H^1(M) \quad \text{and} \quad L_2(M, g^{-\frac{\varkappa}{2}}). \tag{21.94}$$

In particular we have $\delta = 1$ and $\varkappa$ must be replaced $\frac{\varkappa}{2}$. Hence, (21.29) specifies to

$$r\frac{\varkappa}{2} = m\frac{\varkappa}{2} + (j - m). \tag{21.95}$$

Furthermore, $\widetilde{E}_r$, the number of balls involved, can be estimated by (21.52) with (21.32), (21.33) as

$$\widetilde{E}_r \sim 2^{rd} \quad \text{if} \quad 0 < \varkappa < \frac{2d}{n}, \tag{21.96}$$

$$\widetilde{E}_r \sim 2^{\frac{rn\varkappa}{2}} \quad \text{if} \quad \varkappa > \frac{2d}{n}. \tag{21.97}$$

Now one could follow the quarkonial arguments in Step 2 of the proof of Theorem 21.3 starting with (21.55), (21.57). But maybe it is more transparent to do the calculations directly. Recall that the radius of K_{jl} is $\sim 2^{-j}$. Let $\psi_{jl}(x)$ be the same functions as in (21.57), given by (7.79), (7.80). In particular they have disjoint supports. To estimate the $H^1(M)$-norms we use Theorem 7.10. By (21.28) and the necessary notational modifications in (7.38) we may assume

$$\|\psi_{jl}\,|H^1(M)\| \sim 2^{-m(1-\frac{n}{2})} \|\psi_{jl}\,|H^1(\mathbb{R}^n)\| \sim 2^{(j-m)(1-\frac{n}{2})}. \tag{21.98}$$

On the other hand, again by (7.38),

$$\|\psi_{jl}\,|L_2(M, g^{-\frac{\varkappa}{2}})\| \sim 2^{-\frac{m}{2}\varkappa} 2^{-(j-m)\frac{n}{2}}. \tag{21.99}$$

Comparing (21.98) with (21.99) it follows from (21.95) that there is an orthonormal system

$$\left\{ \chi_t^r(x) \,:\, t = 1, \ldots, \widetilde{E}_r \right\}, \quad r \in \mathbb{N}, \tag{21.100}$$

in $H^1(M)$ which is orthogonal in $L_2(M, g^{-\frac{\varkappa}{2}})$ and

$$\|\chi_t^r\,|L_2(M, g^{-\frac{\varkappa}{2}})\| \sim 2^{-r\frac{\varkappa}{2}}, \quad r \in \mathbb{N}, \tag{21.101}$$

where the equivalence constants are independent of r. We choose $\beta = c\,2^{r\varkappa}$ with $c > 0$ small (but independent of $r \in \mathbb{N}$) and insert finite linear combinations of these functions χ_t^r in (21.93). We find that for these functions the quadratic form is always negative. Hence,

$$(H_\beta f, f)_{L_2(M)} \quad \text{with} \quad \beta = c\,2^{r\varkappa} \tag{21.102}$$

is negative on a subspace of $H^1(M)$ of dimension $\widetilde{E}_r$. By the Max-Min principle, which may be found in [EdE87], pp. 489–492, it follows that

$$N_\beta \geq \widetilde{E}_r \quad \text{with} \quad \beta = c\,2^{r\varkappa}. \tag{21.103}$$

By (21.96), (21.97) we get

$$N_\beta \geq c_1\,2^{rd} = c_2\,\beta^{\frac{d}{\varkappa}} \quad \text{if} \quad 0 < \varkappa < \frac{2d}{n}, \tag{21.104}$$

$$N_\beta \geq c_1\,2^{\frac{rn\varkappa}{2}} = c_2\,\beta^{\frac{n}{2}} \quad \text{if} \quad \varkappa > \frac{2d}{n}, \tag{21.105}$$

for some positive numbers c_1 and c_2. This is the desired converse of (21.91), (21.92).

21.8 Local singularities and hydrogen-like operators

As said in the introduction 21.1, in a euclidean setting problems as treated in Theorem 21.7 originate from semi-classical limits $\hbar \to 0$ (Planck's constant tending to zero) and $\beta \sim \hbar^{-2}$. Related potentials, with the Coulomb potential as an outstanding example, have natural local singularities. But this is not the case with the potential $g^{-\varkappa}$ in (21.80). However there are no serious problems in incorporating local singularities in our context. This will be done in the remaining subsections of this Section 21.

To provide a better understanding we have first a glance at the hydrogen atom in $\mathbb{R}^3$. The corresponding Hamiltonian operator $\mathcal{H}_H$ is given by

$$\mathcal{H}_H f = -\frac{\hbar^2}{2m}\Delta f - \frac{e^2}{|x|} f, \quad \mathrm{dom}\,(\mathcal{H}_H) = H^2(\mathbb{R}^3), \tag{21.106}$$

where e is the charge and m is the mass of the electron (spinning around the origin in $\mathbb{R}^3$, where the nucleus is located). It is a self-adjoint operator in $L_2(\mathbb{R}^3)$, its essential spectrum coincides with $[0,\infty)$, and the pairwise different eigenvalues E_j are given by

$$E_j = -\frac{me^4}{2\hbar^2 j^2} \quad \text{with multiplicity } j^2, \tag{21.107}$$

where $j \in \mathbb{N}$. These assertions may be found in many books, for example [Tri92], 7.3.3, 7.3.4, pp. 434, 438. We adapt $\mathcal{H}_H$ to (21.80), (21.81) and put with $\beta = \hbar^{-2}$,

$$H_\beta f = 2m\beta\, \mathcal{H}_H f + f = (-\Delta + \mathrm{id})f - \beta \frac{a}{|x|} f, \quad \mathrm{dom}\,(H_\beta) = H^2(\mathbb{R}^3), \tag{21.108}$$

where $a > 0$ is the resulting constant. We have

$$\mathrm{EssSpec}\,(H_\beta) = [1, \infty), \quad \text{eigenvalues } \lambda_j = 1 - c\frac{\beta^2}{j^2} \tag{21.109}$$

with multiplicity j^2 for some $c > 0$ and $j \in \mathbb{N}$. Hence we get in this case for N_β given by (21.82), in obvious interpretation,

$$N_\beta \sim \sum_{j=1}^{\beta} j^2 \sim \beta^3, \quad \beta \to \infty. \tag{21.110}$$

We preferred a direct approach to obtain (21.110). But this has little to do with the explicit knowledge of the eigenvalues. It follows also from qualitative arguments using the technique developed in [ET96], 5.4.8, 5.4.9, pp. 238–242.

Now we switch from the euclidean situation to the hyperbolic one. Hence we replace $\mathbb{R}^3$ by the d-domain Ω in $\mathbb{R}^n$ according to 7.5 and denote again the related non-compact Riemannian manifold by M. We have Theorem 21.7. On the other hand, the Coulomb potential in (21.108) and its first derivatives describing the resulting forces are comparable with the volume of spheres in $\mathbb{R}^3$ centred at the origin. The situation in spaces M of the above type is different. We discussed this point in 21.2. In particular, (21.9), (21.10) suggest that something like Coulomb potentials in M should have an exponential decay when measured in terms of the Riemannian distance $|x|_g$ from, say, the origin. Hence the far-distance behaviour might be reasonably expressed by $g^{-\varkappa}$ with $\varkappa > 0$ as in (21.80), which coincides with (21.4). On the other hand locally, near the nucleus, the Coulomb potential in M (say with $n = 3$) should be similar as in $\mathbb{R}^3$. Hence something like $|x|_g^{-1}$ looks reasonable. Then one gets, with the additional specification $\varkappa = 1$, the operator H^β in (21.8). We call such operators hydrogen-like. But we could not find in the literature anything like a quantum mechanics in hyperbolic spaces providing a better justification of operators like H^β in (21.8) than the above vague arguments.

We look at operators of type (21.8) from a slightly more general point of view expressing local singularities by $b \in L_r(M)$. Then we have first to complement Theorem 21.3 in the operator version of (21.87). This will be done in 21.9. Afterwards we extend Theorem 21.7 to potentials with local singularities.

We introduced the weighted Sobolev spaces $H_p^s(M, g^\varkappa)$ in (7.42) with the weighted Lebesgue spaces

$$L_p(M, g^\varkappa) = L_p(\Omega, g^{\frac{n}{p}+\varkappa}), \quad 1 < p < \infty, \quad \varkappa \in \mathbb{R}, \tag{21.111}$$

as special cases. The latter follows from (7.33), (7.39), (7.34) with $s = 0$, $q = 2$. As always, $-\Delta_g$ stands for the Laplace-Beltrami operator (21.14) and we assume again that $\varrho \in \mathbb{R}$ is chosen so large that, in specification of (21.15),

$$-\Delta_g + \varrho\, id : \quad L_p(M) \mapsto H_p^{-2}(M), \quad 1 < p < \infty, \tag{21.112}$$

is an isomorphic map.

21.9 Corollary (to Theorem 21.3)

Let Ω be the same d-domain as in Theorem 21.3 and let M be the related Riemannian manifold. Let

$$1 < q \leq p < \infty, \quad \varkappa > 0, \tag{21.113}$$

and
$$b \in L_r(M) \quad \text{with} \quad \frac{1}{q} - \frac{1}{p} = \frac{1}{r} < \frac{2}{n}. \tag{21.114}$$

Then the operator
$$B^b = (-\Delta_g + \varrho\, id)^{-1}\, g^{-\varkappa} b : \quad L_p(M) \mapsto L_p(M), \tag{21.115}$$

is compact. There is a number $c > 0$ such that for the related entropy numbers $e_k = e_k(B^b)$,

$$e_k \leq c k^{-\frac{2}{n}}, \quad k \in \mathbb{N}, \quad \text{if} \quad \varkappa > \left(2 - \frac{n}{r}\right)\frac{d}{n}, \tag{21.116}$$

$$e_k \leq c k^{-\frac{\varkappa}{d} - \frac{1}{r}}, \quad k \in \mathbb{N}, \quad \text{if} \quad 0 < \varkappa < \left(2 - \frac{n}{r}\right)\frac{d}{n}. \tag{21.117}$$

Proof We factorize B^b by
$$B^b = (-\Delta_g + \varrho\, id)^{-1} \circ id \circ g^{-\varkappa} b \tag{21.118}$$

with
$$g^{-\varkappa} b \; : \; L_p(M) \mapsto L_q(M, g^\varkappa), \tag{21.119}$$
$$id \; : \; L_q(M, g^\varkappa) \mapsto H_p^{-2}(M), \tag{21.120}$$
$$(-\Delta_g + \varrho\, id)^{-1} \; : \; H_p^{-2}(M) \mapsto L_p(M). \tag{21.121}$$

Here $g^{-\varkappa} b$ is interpreted as a multiplication operator; the assertion itself follows from (21.111) and Hölder's inequality based on (21.114). By (21.112) we have (21.121). As for id we can apply Theorem 21.3 with
$$\delta = 2 + \frac{n}{p} - \frac{n}{q} = 2 - \frac{n}{r} > 0. \tag{21.122}$$

Then (21.116), (21.117) follow from (21.20), (21.21), respectively.

21.10 Discussion

We are not so much interested in the operator B^b itself. Our aim is to perturb the operator H_β in (21.80) and in Theorem 21.7 so that local singularities are included. Using again the Birman-Schwinger principle as explained in Step 1 of the proof of Theorem 21.7, we must estimate the entropy numbers $e_k(B_b)$ of the operator B_b,
$$B_b = B + B^b = (-\Delta_g + \varrho\, id)^{-1}\, g^{-\varkappa}(1 + b), \tag{21.123}$$

where B is given by (21.79) and B^b by (21.114), (21.115). We claim that

$$e_{2k}(B_b) \le e_k(B) + e_k(B^b) \le c\, e_k(B), \quad k \in \mathbb{N}. \tag{21.124}$$

The first inequality comes from (19.134). As for the second inequality we consider B in $L_p(M)$. We have again (21.87), where id in (21.85) is modified in an obvious way, and by (21.89), (21.90), equivalence in (21.116), (21.117) with $r = \infty$. This covers also the case $r = \infty$ in (21.124). Now let $r < \infty$ with respect to B^b. Then it follows that $e_k(B^b)$ can always be estimated from above by $ce_k(B)$. If $\varkappa = (2 - \frac{n}{r})\frac{d}{n}$ one may replace $\varkappa$ by $\varkappa - \varepsilon$ at the expense of r.

21.11 Corollary (to Theorem 21.7)

Let Ω be the same d-domain as in Theorem 21.7 and let M be the related Riemannian manifold. Let

$$\infty \ge r > \max\left(2, \frac{n}{2}\right), \tag{21.125}$$

$$b \in L_r(M) \text{ real}, \quad \operatorname{supp} b \text{ compact in } \Omega. \tag{21.126}$$

Let

$$H_\beta^b = -\Delta_g + \varrho\, id - \beta\, g^{-\varkappa}(1+b), \quad \varkappa > 0, \quad \operatorname{dom}(H_\beta^b) = H^2(M). \tag{21.127}$$

Then for $\beta > 0$ and $\varrho \in \mathbb{R}$ large, H_β^b is self-adjoint in $L_2(M)$ with

$$\operatorname{EssSpec}(H_\beta^b) = \operatorname{EssSpec}(-\Delta_g + \varrho\, id) \subset [1, \infty). \tag{21.128}$$

Let

$$N_\beta^b = \#\left\{\operatorname{Spec}(H_\beta^b) \cap (-\infty, 0)\right\} \tag{21.129}$$

be the number of negative eigenvalues of H_β^b. Then

$$N_\beta^b \sim \beta^{\frac{d}{\varkappa}} \quad \text{if } 0 < \varkappa < \frac{2d}{n}, \quad \beta \to \infty, \tag{21.130}$$

$$N_\beta^b \sim \beta^{\frac{n}{2}} \quad \text{if } \varkappa > \frac{2d}{n}, \quad \beta \to \infty. \tag{21.131}$$

Proof By Corollary 21.9 with $p = 2$ the remarks in 21.6 apply also to H_β^b in place of H_β. Hence H_β^b is self-adjoint, $g^{-\varkappa}(1+b)$ is a relatively compact perturbation of $-\Delta_g + \varrho\, id$ and we have (21.128). By Step 1 of the proof of Theorem 21.7, the estimates (21.124), combined with (21.87) and (21.89), (21.90), we get the counterpart of (21.91), (21.92). Since we assumed that $supp\, b$ is compact in Ω we can apply the arguments from Step 2 of the proof of Theorem 21.7, restricted to those values of m such that the balls with (21.28) have an empty intersection with $supp\, b$. Then we obtain the counterpart of (21.104), (21.105).

21.12 Hydrogen-like operators

Let
$$2 \leq n-1 \leq d < n \tag{21.132}$$
and let H^β be the operator from (21.8),
$$H^\beta = -\Delta_g + \varrho\, id - \beta\, g^{-1}\left(1 + \left(\frac{1}{|x|_g} - 1\right)_+\right), \tag{21.133}$$

in specification of H^b_β in (21.127) with $\varkappa = 1$ and $b(x) = \left(\frac{1}{|x|_g} - 1\right)_+$. Recall that $a_+ = \max(a, 0)$ if $a \in \mathbb{R}$. By (21.132) we have $n \geq 3$. If ϱ is large then we get (21.128). Now it follows by Corollary 21.11 and in particular by (21.130) that

$$N^\beta = \#\left\{Spec\,(H^\beta) \cap (-\infty, 0)\right\} \sim \beta^d, \quad \beta \to \infty. \tag{21.134}$$

In 21.8 we called operators of type (21.133) hydrogen-like. At least in case of $n = 3$ this might be justified as far as the Coulomb potential $b(x) \sim |x|_g^{-1}$ near the off-point is concerned. Whether there is an argument in favour of $\varkappa = 1$ or any other value of $\varkappa > 0$ is not so clear. Let us take this for granted. Then the physicists in such a hypothetical hyperbolic world are in a remarkable position. They claim that they can find out by experiments in a laboratory, calculating the exponent d in (21.134), how fractal the invisible boundary of their infinite world might be (explaining to their astonished country(wo)men that this originates from tiny irregularities in the first millionth of a second after the Big Bang). Recall that we have (21.110) in our euclidean world $\mathbb{R}^3$.

22 Isotropic fractals and related function spaces

22.1 Analysis versus (fractal) geometry

Entropy numbers and eigenvalues may be considered as two sides of the same coin. Entropy numbers are the geometrical twins of eigenvalues. This opinion is based on Carl's observation (19.129) at an abstract level. It is also supported by the equivalences in (19.73), Theorem 19.17 combined with (19.109), and the assertions about the negative spectrum in Theorem 21.7 and Corollary 21.11 where estimates for entropy numbers play an important role. In 19.18(vii) we remarked that in case of Hilbert spaces, entropy numbers and eigenvalues might be even equivalent. The decisive link between the abstract inequality (19.129) on the one hand, and the more concrete sharp estimates from above of related eigenvalue distributions on the other hand, is the possibility to calculate the entropy numbers of compact embeddings between function spaces and the consequences for respective operators. Theorem 21.3 and Corollary 21.9 may serve as typical examples. We followed this path in [ET96], Chapter 5, based on the earlier papers mentioned there, for regular and degenerate elliptic pseudodifferential operators in $\mathbb{R}^n$ and on domains Ω in $\mathbb{R}^n$. At that time estimates of entropy numbers of compact embeddings between function spaces were based on the original Fourier-analytical definitions of $B^s_{pq}(\mathbb{R}^n)$ and $F^s_{pq}(\mathbb{R}^n)$ as described in (2.37), (2.38), and Theorem 2.9. These tools (including atomic decompositions) proved to be inadequate when switching from $\mathbb{R}^n$ and domains Ω in $\mathbb{R}^n$ to fractals, preferably d-sets according to (9.148). Extending this definition to $d = 0$ and $d = n$ we call a compact set Γ in $\mathbb{R}^n$ a d-set with $0 \leq d \leq n$ if there is a Radon measure μ with

$$\operatorname{supp}\mu = \Gamma, \quad \mu(B(\gamma, r)) \sim r^d, \quad 0 < r < 1, \tag{22.1}$$

where again $B(\gamma, r)$ is a ball in $\mathbb{R}^n$ centred at $\gamma \in \Gamma$ and of radius r. In particular, if Ω is a smooth bounded domain in $\mathbb{R}^n$, then $\Gamma = \overline{\Omega}$ is a (rather special) n-set. To handle related fractal elliptic (pseudo)differential operators we developed in [Triδ] the theory of quarkonial (subatomic) decompositions of the spaces $B^s_{pq}(\mathbb{R}^n)$ and $F^s_{pq}(\mathbb{R}^n)$. This has been extended in Chapter I of the present book in several directions. Seen from the point of view of applications to fractal elliptic operators, quarkonial decompositions compared with, say, atomic representations have two major advantages. First the building blocks, called β-quarks, are given in a constructive way as in Definition 2.4, and secondly they are highly flexible as indicated in 2.14. This paved the way for the development in [Triδ], Chapter IV, of a theory of function spaces on fractals, which in case of d-sets, looks satisfactory. On this basis we dealt in [Triδ], Theorem 30.2, p. 234, with isotropic fractal drums, described by operators B,

$$B = (-\Delta)^{-1} \circ tr^\Gamma, \tag{22.2}$$

where Γ is a compact d-set in $\mathbb{R}^n$ with $n - 2 < d < n$. We repeated the assertions obtained there at the beginning of Theorem 19.7, where one finds also the necessary explanations. In particular, if ϱ_k are the positive eigenvalues of B, then according to Theorem 19.7(i),

$$\varrho_k \sim k^{-1+\frac{n-2}{d}}, \quad k \in \mathbb{N}. \tag{22.3}$$

As said above, this part of Theorem 19.7 is not new, and (22.3) coincides with [Triδ], Theorem 30.2, (30.7), p. 234. One can take the equivalence (22.3) as a strong hint that also in case of d-sets Γ, the method outlined above is natural and well adapted and characterized by the key words:

Spaces of type

$$B^s_{pq}(\mathbb{R}^n), \quad F^s_{pq}(\mathbb{R}^n), \quad \textit{special cases} \quad H^s_p(\mathbb{R}^n), \quad W^k_p(\mathbb{R}^n), \tag{22.4}$$

on $\mathbb{R}^n$, their relations to compact d-sets Γ, estimates of entropy numbers of corresponding compact embeddings, and assertions of type (22.3) for respective operators B in (22.2).

The outcome is remarkable (in our opinion): *The spaces in (22.4), including all their special cases and their restrictions to domains Ω in $\mathbb{R}^n$, are natural, effective and well adapted not only when dealing with partial differential operators in $\mathbb{R}^n$, and on domains Ω, but also in connection with d-sets.*

One may ask for indicators providing an understanding of this success. The trace operator tr_Γ and the identification operator id_Γ have the same meaning as in 9.2. Furthermore by 9.19 any compact d-set in $\mathbb{R}^n$ with $d < n$ satisfies the ball condition. Let Γ be a compact d-set with $0 < d < n$. Then one gets:

(i) ([Triδ], Theorem 18.2, p. 136)

$$\mathrm{id}_\Gamma L_p(\Gamma) = B^{-\frac{n-d}{p'},\Gamma}_{p,\infty}(\mathbb{R}^n), \quad 1 < p \leq \infty, \quad \frac{1}{p} + \frac{1}{p'} = 1, \tag{22.5}$$

where we used the notation introduced in 9.34(viii).

(ii) ([Triδ], Corollary 18.12, p. 142, combined with 9.19 above)

$$\mathrm{tr}_\Gamma B^{\frac{n-d}{p}}_{pq}(\mathbb{R}^n) = L_p(\Gamma), \quad 0 < p < \infty, \quad 0 < q \leq \min(1,p). \tag{22.6}$$

On the other hand we discussed in Section 9 in some detail traces of spaces of type (22.4) on compact sets Γ which are supports of finite Radon measures. We refer to Theorem 9.3 and, as far as spaces on Γ are concerned, to Definition 9.29 and Theorem 9.33. The outcome is more or less satisfactory. But it is apparently a rather different question to ask for *tailored spaces* to meet the needs of operators B of type (22.2) with the desired outcome of type (22.3) for the distribution of the positive eigenvalues of B.

We dealt several times with operators of type B in (22.2) and their spectral theory when Γ is not a d-set. As for anisotropic and nonisotropic fractals in the plane $\mathbb{R}^2$, we refer to the discussion in 19.4 and the literature mentioned there. Special but very interesting cases are the Weyl measures from Definition 19.13 with Theorem 19.17 as the (isotropic) outcome. The rather direct proof of Theorem 19.17 might be considered as an interesting example of the intense interplay between fractal geometry and analysis in the above framework and in a particuliar situation. We refer also to the discussion in 19.18, and the literature mentioned there. Whereas the special *multifractal measures* μ in $\mathbb{R}^n$ with $n = 2$ mentioned there fit in the above scheme since they are Weyl measures, this is presumably not the case when $n \neq 2$. Hence it would be of interest to deal in greater detail with these multifractals, to find tailored function spaces, or arguments with geometry versus analysis of the above type.

We return to the original problem: analysis versus geometry. In Section 9 spaces of type (22.4) are given (analysis) and we ask for traces on given compact sets Γ (geometry). Conversely if Γ is given (geometry) and B in (22.2) is considered, then one may ask for *tailored spaces* (analysis) producing (22.3) (if possible). In case of d-sets these two ways of looking at function spaces coincide and we took (22.5), (22.6) as indicators of this fortunate outcome. In general this cannot be expected, although one has (in our opinion) the striking interplay between analysis and fractal geometry described in 19.18(vii). We discussed this situation in 9.34(i) and (ii) and we indicated there and also at the beginning of 9.1 and in Introduction 18 what intentions we have now: We replace d-sets by more general but isotropic (d, Ψ)-sets Γ, construct tailored spaces according to B in (22.2) and ask for counterparts of (22.3). Section 22 deals with (d, Ψ)-sets and related tailored spaces where the counterparts of (22.5), (22.6) serve as criteria. The spectral theory of the corresponding operators will be described in Section 23. As for the (d, Ψ)-fractals we give proofs. Otherwise we restrict ourselves to a report written in the style of a survey.

To justify this procedure we have a look at the state of the art and at the literature. Our inspiration to extend what has been done in [Triδ], Chapters IV and V, for d-sets and related operators B of type (22.2) to the more general (d, Ψ)-sets comes from the note [Leo98] by H.-G. Leopold, who studied embedding properties of spaces of type $B_{pq}^{(s,-b)}(\mathbb{R}^n)$ as briefly mentioned in 17.2 in delicate limiting situations, including estimates for entropy numbers. We summarized our own results in [EdT98] and in an extended version [EdT99a]. Obviously in [EdT98], but also in [EdT99a], we gave only outlines of proofs. We restricted all parameters as much as possible and claimed afterwards that everything is parallel to the respective arguments in [Triδ]. The situation is now much better. There are several (forthcoming) papers, preprints, PhD-theses dealing with all relevant aspects in detail. First one must ask for a generalization of [Triδ], Sections 8 and 9, dealing with entropy numbers of compact embeddings between weighted sequence spaces of a similar type as has been used before in Step 3 of the proof of Theorem 19.17 and in Step 1 of the proof of Theorem 21.3. All that one needs (and more) may be now found in [Leo00a], [Leo00b], [Mou01b], [Leo01] and [KuS00]. As for the related function spaces of type

$$B_{pq}^{(s,\Psi)}(\mathbb{R}^n) \quad \text{and} \quad F_{pq}^{(s,\Psi)}(\mathbb{R}^n) \tag{22.7}$$

all details are given in [Mou99], [Mou01a] and [Mou01b]. This applies also to the corresponding spaces on (d,Ψ)-sets. We refer in this context also to [Bri99], [Bri00], [Bri01].

22.2 Admissible functions

We outlined our intentions in Section 18: We wish to generalize compact d-sets, given by (18.1), to (d,Ψ)-sets with (18.6). Here $r^d \Psi(r)$ might be considered as a perturbation of r^d. Hence first we must clarify which type of functions Ψ we have in mind. As for notation we recall first our use of $\sim$, explained in (19.17), (19.18). Furthermore we adopt the following convention. A real function Ψ on the interval $(0,1]$ is said to be *monotone* if it is either monotonically decreasing or increasing, where monotone decreasing (increasing) means monotone not increasing (not decreasing). As always log is taken with respect to base 2.

A real function Ψ on the interval $(0,1]$ is called *admissible* if it is positive and monotone on $(0,1]$, and if

$$\Psi(2^{-j}) \sim \Psi(2^{-2j}), \quad j \in \mathbb{N}_0. \tag{22.8}$$

Let $0 < c < 1$ and $b \in \mathbb{R}$. Then

$$\Psi(x) = |\log cx|^b, \quad 0 < x \le 1, \tag{22.9}$$

is an example of an admissible function. We collect a few properties of admissible functions.

22.3 Proposition

Let Ψ be an admissible function on the interval $(0,1]$.

(i) Let $\varkappa \in \mathbb{R}$. Then $\Psi^\varkappa$ is also admissible.

(ii) Let $0 < \tau_1 < 1$ and $0 < \tau_2 < 1$. Then

$$\Psi(\tau_1^j) \sim \Psi(\tau_2^j), \quad j \in \mathbb{N}_0. \tag{22.10}$$

(iii) Furthermore,

$$\Psi(x) \sim \Psi(x^2), \quad 0 < x \le 1. \tag{22.11}$$

(iv) *There are positive numbers c_1, c_2, b and c with $0 < c < 1$ such that*

$$c_1 \,|\log cx|^{-b} \leq \Psi(x) \leq c_2 \,|\log cx|^b, \quad 0 < x \leq 1. \tag{22.12}$$

Proof *Step 1* Part (i) is obvious. In particular, to prove (ii) and (iii) we may assume that Ψ is decreasing. First we remark that for given $\varkappa$ with $0 < \varkappa < 1$, there is a number $d > 0$ such that

$$\Psi(\varkappa x) \leq d\,\Psi(x) \quad \text{for all} \quad 0 < x \leq 1. \tag{22.13}$$

Now (22.11) follows from (22.8) (it is the continuous version of (22.8)). Furthermore it is sufficient to prove (22.10) for τ_1^l and τ_2^l in place of τ_1 and τ_2, respectively, where $l \in \mathbb{N}$. In particular we assume that both τ_1 and τ_2 are small,

$$2^{-2^{k_1}} \leq \tau_1 < \tau_2 \leq 2^{-2^{k_2}} \quad \text{where} \quad k_1 \in \mathbb{N}_0 \quad \text{and} \quad k_2 \in \mathbb{N}_0. \tag{22.14}$$

Then it follows by iterative application of (22.8),

$$\begin{aligned}\Psi(\tau_2^j) \leq \Psi(\tau_1^j) &\leq \Psi\left(2^{-j2^{k_1}}\right) \leq 2^{b(k_1-k_2)} \Psi\left(2^{-j2^{k_2}}\right) \\ &\leq 2^{b(k_1-k_2)} \Psi(\tau_2^j), \quad j \in \mathbb{N}_0,\end{aligned} \tag{22.15}$$

for some $b > 0$. This proves (22.10).

Step 2 It remains to prove the right-hand side of (22.12) for an admissible decreasing function Ψ. Let

$$2^{-2^{k+1}} \leq x \leq 2^{-2^k} \quad \text{for some} \quad k \in \mathbb{N}. \tag{22.16}$$

As in (22.15) we obtain by iterative application of (22.8),

$$\Psi(x) \leq 2^{bk} \Psi(2^{-2}) \leq c\,|\log x|^b. \tag{22.17}$$

This proves (22.12). $\square$

22.4 Definition

Let Γ be a compact set in $\mathbb{R}^n$ where $n \in \mathbb{N}$.

(i) Let $0 < d < n$ and let Ψ be an admissible function according to 22.2. Then Γ is called a (d, Ψ)-set if there is a Radon measure μ in $\mathbb{R}^n$ such that for any ball $B(\gamma, r)$ in $\mathbb{R}^n$ centred at $\gamma \in \Gamma$ and of radius r with $0 < r < 1$,

$$\operatorname{supp} \mu = \Gamma, \quad \mu(B(\gamma, r)) \sim r^d \,\Psi(r). \tag{22.18}$$

(ii) Let Ψ be a decreasing admissible function according to 22.2 with $\Psi(x) \to \infty$ if $x \to 0$. Then Γ is called an (n, Ψ)-set if there is a Radon measure μ in $\mathbb{R}^n$ with the above properties and $d = n$ in (22.18).

22.5 Discussion

If $\Psi = 1$ in (22.18) then we have (22.1) and hence Γ is a d-set. So far d-sets played a decisive role at several places in this book. Some discussions may be found in 22.1. By the above definition and by (22.12) it is clear that (d, Ψ)-sets might be considered as isotropic perturbations of the (isotropic) d-sets. As for basic notation of fractal geometry we refer again to [Fal85], [Fal90], [Mat95] and the brief summary in [Triδ], Sections 1–3. Note that the requirement that μ in (22.18) be Radon is not an additional restriction but a natural outcome, see [Triδ], Section 2, pp. 2–4, with the references given there to [Fal85] and [Mat95]. In particular, [Mat95], p. 55, is now of relevance. Recall the following basic facts of fractal geometry: For any d with $0 \le d \le n$ there are compact d-sets Γ in $\mathbb{R}^n$. If $\dim_H \Gamma$ denotes the Hausdorff dimension of Γ, then $\dim_H \Gamma = d$. If in addition $d < n$, then $|\Gamma| = 0$, where $|\Gamma|$ is the Lebesgue measure of Γ. Furthermore for any d with $0 \le d \le n$ there are self-similar d-sets defined by IFS (Iterated Function Systems). Details may be found in the above references. We recall that any d-set Γ with $d < n$ satisfies the ball condition. We refer to 9.19 and 9.16. By Theorem 9.21 this property is of great interest in connection with traces of functions on Γ. The question arises whether for any admitted couple (d, Ψ) there is a (d, Ψ)-set according to Definition 22.4 and whether the above properties of d-sets can be extended to (d, Ψ)-sets. We describe the outcome in 22.6 and 22.8 and define in 22.7 what we wish to call pseudo self-similar sets.

22.6 Proposition

Let Γ be a compact (d, Ψ)-set according to Definition 22.4 in $\mathbb{R}^n$ and let μ be the related measure.

(i)   *Then*

$$|\Gamma| = 0 \quad \text{and} \quad \dim_H \Gamma = d. \tag{22.19}$$

(ii)   *Let, in addition, $n = 2$. Then μ is a Weyl measure according to Definition 19.13.*

(iii)   *Let $\widetilde{\mu}$ be a second Radon measure on Γ, satisfying (22.18) with $\widetilde{\mu}$ in place of μ. Then μ and $\widetilde{\mu}$ are equivalent.*

(iv)   *Furthermore, Γ satisfies the ball condition according to Definition 9.16 if, and only if, $d < n$.*

Proof   *Step 1*   We prove (i)–(iii). We again use covering arguments. Let $j \in \mathbb{N}$. There are N_j balls $B(\gamma_l^j, 2^{-j})$ centred at $\gamma_l^j \in \Gamma$ and of radius 2^{-j} such that

$$\Gamma \subset \bigcup_{l=1}^{N_j} B(\gamma_l^j, 2^{-j}), \quad \sum_{l=1}^{N_j} \mu\left(B(\gamma_l^j, 2^{-j})\right) \sim 1. \tag{22.20}$$

Hence, by (22.18),

$$N_j \sim 2^{jd} \Psi^{-1}(2^{-j}), \quad j \in \mathbb{N}, \tag{22.21}$$

where all equivalence constants are independent of j. Then we have

$$|\Gamma| \leq \sum_{l=1}^{N_j} \left|B(\gamma_l^j, 2^{-j})\right| \leq c\, 2^{-jn} N_j \sim 2^{-j(n-d)} \Psi^{-1}(2^{-j}). \tag{22.22}$$

If $j \to \infty$ then we obtain the first assertion in (22.19) in all cases. The second assertion in (22.19) is well known for d-sets. This can be extended to (d, Ψ)-sets by using standard arguments which can be found, for example, in [Triδ], pp. 3/4, and (22.12). Details and further information may be found in [Mou01b], 2.1. This completes the proof of (i). As for (ii) we note that μ satisfies the doubling condition and is strongly diffuse according to (19.124), (19.125). We obtain by Theorem 19.17 that μ is a Weyl measure (in the plane). Let μ and $\tilde{\mu}$ be two measures as in (iii). To compare μ and $\tilde{\mu}$ one can apply the same covering arguments as for d-sets as in, for example, [Triδ], proof of Theorem 3.4, pp. 5/6. It follows that μ and $\tilde{\mu}$ are equivalent. Details may be found in [Mou01b], Proposition 2.4.

Step 2 We prove (iv). By Proposition 9.18 and (9.83) we must check whether there are numbers $\lambda > 0$ and $c > 0$ such that for all $j \in \mathbb{N}_0$ and all $\varkappa \in \mathbb{N}_0$,

$$\Psi(2^{-j}) \leq c\, 2^{\varkappa(n-\lambda-d)} \Psi(2^{-j-\varkappa}). \tag{22.23}$$

If $d < n$ and if Ψ is decreasing then one may choose $\lambda = n - d$. Let $d < n$ and let Ψ be increasing. If $j = 0$ and

$$0 < \lambda < n - d, \tag{22.24}$$

then (22.23) follows from (22.12). Let $j \in \mathbb{N}$ and $\varkappa \leq 2j$. Then (22.23) with (22.24) is a consequence of (22.8). Let $j \in \mathbb{N}$ and

$$2^k j \leq \varkappa < 2^{k+1} j, \quad \text{with} \quad k \in \mathbb{N}. \tag{22.25}$$

By iterative application of (22.8) and (22.25) we obtain

$$\Psi(2^{-j}) \leq c\, 2^{bk} \Psi(2^{-2^{k+2}j}) \leq c\, 2^{bk} \Psi(2^{-j-\varkappa}) \leq c\, 2^{b\log \varkappa} \Psi(2^{-j-\varkappa}), \tag{22.26}$$

where c and b are suitable positive numbers. We again get (22.23) with (22.24). Hence Γ satisfies the ball condition if $d < n$. Let $d = n$. Assume that Γ satisfies the ball condition. Then we have (22.23) with $d = n$ for some $c > 0$ and $\lambda > 0$ and all $j \in \mathbb{N}_0$ and $\varkappa \in \mathbb{N}_0$. By (22.12) we obtain

$$\Psi(2^{-j}) \le c\, 2^{-\varkappa \lambda} (j+\varkappa)^b \qquad (22.27)$$

for some $c > 0$ and $b > 0$. For fixed $j \in \mathbb{N}_0$ and $\varkappa \to \infty$ we get a contradiction. Hence if $d = n$ then Γ does not satisfy the ball condition.

22.7 Pseudo self-similar sets and ψIFS

So far it is not clear whether there are (d, Ψ)-sets for each admitted couple (d, Ψ). As mentioned in 22.5 in case of d-sets we have distinguished related self-similar fractals created by IFS (Iterated Function Systems). The corresponding procedure may be found in the standard references quoted in 22.5 or in [Triδ], Section 4. Using the notation introduced there we describe now what we wish to call *pseudo iterated function systems*, abbreviated by ψIFS. But we do not study ψIFS for their own sake here. We outline the typical procedure of varying contractions and restrict otherwise ourselves to a simple, but for our purpose sufficient case. We follow [EdT99a]. Let $Q = [0,1]^n$ be the closed unit cube in $\mathbb{R}^n$. Let $N \in \mathbb{N}$ with $N \ge 2$. Let

$$A_{l_1 \cdots l_j} x = \varrho_j x + x^{l_1 \cdots l_j}, \quad j \in \mathbb{N}, \quad 1 \le l_k \le N, \qquad (22.28)$$

be N^j contractions in $\mathbb{R}^n$ with $x^{l_1 \cdots l_j} \in \mathbb{R}^n$ and

$$0 < \inf_j \varrho_j \le \sup_j \varrho_j < 1. \qquad (22.29)$$

The cubes

$$Q_{l_1 \cdots l_j} = A_{l_1 \cdots l_j} \circ A_{l_1 \cdots l_{j-1}} \circ \cdots \circ A_{l_1} Q \qquad (22.30)$$

have side-length $\prod_{m=1}^{j} \varrho_m$. We assume here that the *strong separation condition*

$$Q_{l_1 \cdots l_j} \cap Q_{m_1 \cdots m_j} = \emptyset, \quad (l_1, \ldots, l_j) \ne (m_1, \ldots, m_j), \qquad (22.31)$$

at each level $j \in \mathbb{N}$, and the *inclusion property*

$$Q_{l_1 \cdots l_j l} \subset Q_{l_1 \cdots l_j}, \quad l = 1, \ldots, N, \qquad (22.32)$$

at each level $j \in \mathbb{N}_0$ hold. Of course, $j = 0$ in (22.32) means $Q_l \subset Q$. If $\varrho_j = \varrho$ for some $0 < \varrho < 1$ and all $j \in \mathbb{N}$ and if the points $x^{l_1 \cdots l_j}$ are chosen appropriately, then one has the well-known procedure of IFS, creating self-similar (isotropic) fractals. As there, [Triδ], p. 11, we put also in the above more general case

$$(AQ)^j = \bigcup Q_{l_1 \cdots l_j}, \quad j \in \mathbb{N}, \tag{22.33}$$

where the union is taken for fixed j over all cubes in (22.30). By (22.32) the resulting sequence of sets is decreasing and its attractor

$$\Gamma = (AQ)^\infty = \bigcap_{j \in \mathbb{N}} (AQ)^j = \lim_{j \to \infty} (AQ)^j \tag{22.34}$$

is called a *pseudo self-similar set*. The strong separation condition (22.31) is not really necessary. It would be sufficient to have (22.31), (22.32) for the respective open interiors. But as said we do not study ψIFS and resulting pseudo self-similar sets for their own sake here.

22.8 Proposition

For any couple d and Ψ according to Definition 22.4 there is a pseudo self-similar (d, Ψ)-set.

Proof **Step 1** First we construct a sequence of cubes with the properties described in 22.7. Let $m \in \mathbb{N}$ be large,

$$N = m^n, \quad \varrho = m^{-\frac{n}{d}} \quad \text{and} \quad \varrho_j = \varrho \psi(j) \tag{22.35}$$

with

$$\psi(j) = \left(\frac{\Psi(\varrho^{j-1})}{\Psi(\varrho^j)}\right)^{\frac{1}{d}}, \quad \text{where} \quad j \in \mathbb{N}. \tag{22.36}$$

By (22.11) there are two positive numbers c and C such that

$$c \le \psi(j) \le C \quad \text{for} \quad j \in \mathbb{N}. \tag{22.37}$$

Let first $d = n$. Then Ψ is decreasing and we may choose $C = 1$ in (22.37). One may even assume $\psi(j) < 1$ for $j \in \mathbb{N}$ in this case where $d = n$. If not, one can replace ϱ^j (but not ϱ^{j-1}) on the right-hand side of (22.36) by ϱ^{j+k} with some $k = k(j)$ such that we have $\psi(j) < 1$ after this replacement. This does not influence the following constructions (after appropriate modifications). By (22.37) we have (22.29). Let $j = 1$. Then we divide Q naturally in N cubes with side length $\varrho_1 = m^{-1}\psi(1) < m^{-1}$ such that we have (22.31) with $j = 1$. By iteration one can construct a ψIFS and a related pseudo self-similar set Γ with (22.34).

Let now $d < n$. Let $j = 1$. Then we again divide Q naturally in N subcubes of side-length $\varrho_1 = m^{-\frac{n}{d}} \psi(1)$. By (22.35) and (22.37) we have

$$\varrho_1 \leq m^{-\frac{n}{d}} C < m^{-1} \quad \text{if } m \text{ is large} . \tag{22.38}$$

Hence (22.31) with $j = 1$ can be obtained. Iteration gives again a ψIFS and a related pseudo self-similar set Γ according to (22.34).

Step 2 We construct a measure μ on Γ with (22.18). We use the well-known mass distribution procedure, see [Fal90], pp. 13–14, and Proposition 1.7 there, distributing evenly the unit mass during the procedure described above. The resulting Radon measure μ has, by construction, the support property in (22.18). As for the second property in (22.18) we note that the cubes $Q_{l_1 \dots l_j}$ in (22.30) have side-length

$$r_j = c \, \varrho^j \, \Psi^{-\frac{1}{d}}(\varrho^j), \quad j \in \mathbb{N}, \tag{22.39}$$

where $c = \Psi(1)^{\frac{1}{d}} > 0$. By the mass distribution procedure and (22.35) we obtain

$$\mu(Q_{l_1 \dots l_j}) = N^{-j} = \varrho^{jd} = c' \, r_j^d \, \Psi(\varrho^j), \quad j \in \mathbb{N}. \tag{22.40}$$

Using (22.12) and (22.11) we have

$$c_1 \, \varrho^{2j} \leq r_j \leq c_2 \, \varrho^{\frac{j}{2}} \quad \text{and} \quad \Psi(\varrho^j) \sim \Psi(r_j), \tag{22.41}$$

where $c_1 > 0$ and $c_2 > 0$. Hence,

$$\mu(Q_{l_1 \dots l_j}) \sim r_j^d \, \Psi(r_j), \quad j \in \mathbb{N}. \tag{22.42}$$

By (22.39) and (22.11) we have $r_{j+1} \sim r_j$. Then (22.18) is a consequence of (22.42).

22.9 Function spaces: Preliminaries

As indicated in Section 18 and in greater detail in 22.1 we are now interested in spaces of type (22.7) tailored to (d, Ψ)-sets according to Definition 22.4. Since (d, Ψ)-sets might be considered as perturbations of d-sets, one can expect something similar for the spaces in (22.7) compared with $B^s_{pq}(\mathbb{R}^n)$ and $F^s_{pq}(\mathbb{R}^n)$. This is largely the case and one could follow the Weierstrassian approach to the latter spaces developed in Sections 2 and 3. But this will not be done here. We restrict ourselves to a summary. Then it might be better to begin with the Fourier-analytical definition and to say afterwards how atoms, β-quarks etc appear. The first steps have been carried out in [EdT98] and [EdT99a] under the restriction to the B-scale and the range of the admitted parameters. By the recent work of S. D. de Moura, we refer in particular to [Mou01b], the full theory for B-spaces, F-spaces and all parameters is now available. As a compromise we follow mainly [EdT99a], complemented by some assertions taken from [Mou01b], avoiding full generality. As said above, no proofs are given. Brief outlines may be found in [EdT99a] (under the indicated restrictions), full details are presented in [Mou01b] and [Mou01a].

We use the notation introduced in 2.8. In particular, the functions φ_k have the same meaning as in (2.33)–(2.35). First we define the counterparts of (2.37) and (2.38).

22.10 Definition

Let

$$s \in \mathbb{R}, \quad 0 < p \leq \infty, \quad 0 < q \leq \infty, \qquad (22.43)$$

and let Ψ be an admissible function according to 22.2.

(i)   *Then $B_{pq}^{(s,\Psi)}(\mathbb{R}^n)$ is the collection of all $f \in S'(\mathbb{R}^n)$ such that*

$$\|f \,|\, B_{pq}^{(s,\Psi)}(\mathbb{R}^n)\| = \left(\sum_{j=0}^{\infty} 2^{jsq} \Psi(2^{-j})^q \|(\varphi_j \widehat{f})^\vee \,|\, L_p(\mathbb{R}^n)\|^q \right)^{\frac{1}{q}} \qquad (22.44)$$

(with the usual modification if $q = \infty$) is finite.

(ii)   *Let in addition $p < \infty$. Then $F_{pq}^{(s,\Psi)}(\mathbb{R}^n)$ is the collection of all $f \in S'(\mathbb{R}^n)$ such that*

$$\|f \,|\, F_{pq}^{(s,\Psi)}(\mathbb{R}^n)\| = \left\| \left(\sum_{j=0}^{\infty} 2^{jsq} \Psi(2^{-j})^q |(\varphi_j \widehat{f})^\vee(\cdot)|^q \right)^{\frac{1}{q}} \,\Big|\, L_p(\mathbb{R}^n) \right\| \qquad (22.45)$$

(with the usual modification if $q = \infty$) is finite.

22.11 Comments

If $\Psi = 1$ then one has the usual Fourier-analytical definition of the spaces $B_{pq}^s(\mathbb{R}^n)$ and $F_{pq}^s(\mathbb{R}^n)$. On this basis we developed in [Triβ] and [Triγ] the theory of these spaces in detail. Many properties obtained there can be carried over to the above spaces $B_{pq}^{(s,\Psi)}(\mathbb{R}^n)$ and $F_{pq}^{(s,\Psi)}(\mathbb{R}^n)$. In particular, they are independent of the chosen resolution of unity $\{\varphi_k\}$. Furthermore they are quasi-Banach spaces (Banach spaces if $p \geq 1$ and $q \geq 1$). We refer to [Mou01b], 1.2 and 1.3, where one finds detailed proofs. In particular characterizations in terms of local means as developed in [Triγ], 2.4.6, 2.5.3, for the spaces $F_{pq}^s(\mathbb{R}^n)$ and $B_{pq}^s(\mathbb{R}^n)$ can be extended to the above spaces; Theorems 1.10 and 1.12 in [Mou01b] include also some improvements even in case of $\Psi = 1$. By (22.12) it is quite clear that s remains the main smoothness and Ψ stands for an additional finer tuning.

22.12 Atoms and β-quarks

In [Triδ], Sections 13 and 14, and in Sections 2 and 3 of this book we developed the theory of atomic and quarkonial decompositions for the spaces $B_{pq}^s(\mathbb{R}^n)$ and $F_{pq}^s(\mathbb{R}^n)$. There one finds also the necessary references to the literature. By [Mou01b], 1.4, there is a full counterpart of this theory for the spaces introduced in Definition 22.10. In particular the simultaneous proof of Theorems 1.18 and 1.23 in [Mou01b], covering atomic and quarkonial decompositions, fits in the scheme of the above Sections 2 and 3. We do not describe the full theory here. We wish to provide an understanding of the necessary modifications needed now, compared with the above Sections 2 and 3, where it is sufficient for our later purpose to restrict the considerations to those spaces where no moment conditions for atoms and β-quarks are necessary. Otherwise we use the notation introduced in Section 2.

Let ψ and ψ^β with $\beta \in \mathbb{N}_0^n$ be the same functions as in Definition 2.4. Let $s \in \mathbb{R}$ and $0 < p \leq \infty$. Let Ψ be an admissible function according to 22.2. Then, in generalization of (2.16),

$$(\beta qu)_{\nu m}(x) = 2^{-\nu(s-\frac{n}{p})} \Psi(2^{-\nu})^{-1} \psi^\beta(2^\nu x - m), \quad x \in \mathbb{R}^n, \qquad (22.46)$$

is called an (s, p, Ψ)-β-quark. Again $\nu \in \mathbb{N}_0$ and $m \in \mathbb{Z}^n$. We describe briefly the atomic counterpart of $(\beta qu)_{\nu m}$. The cubes $Q_{\nu m}$ and $cQ_{\nu m}$ with $c > 0$ have the same meaning as at the beginning of 2.2. Let $c > 1$. Then a K times differentiable complex-valued function $a(x)$ in $\mathbb{R}^n$ is called an $(s, p, \Psi)_{K,L}$-atom if for some $\nu \in \mathbb{N}_0$,

$$\operatorname{supp} a \subset c Q_{\nu m} \quad \text{for some} \quad m \in \mathbb{Z}^n, \qquad (22.47)$$

$$|D^\alpha a(x)| \leq 2^{-\nu(s-\frac{n}{p})+|\alpha|\nu} \Psi(2^{-\nu})^{-1} \quad \text{for} \quad |\alpha| \leq K, \qquad (22.48)$$

$$\int_{\mathbb{R}^n} x^\gamma a(x)\, dx = 0 \quad \text{if} \quad |\gamma| \leq L. \qquad (22.49)$$

Explanations may be found in [Triδ], pp. 73–74, and in [Mou01b], Definition 1.14. Let b_{pq} and f_{pq} be the sequence spaces introduced in (2.7) and (2.8), respectively. Let $0 < p \leq \infty$, $0 < q \leq \infty$. Then we put again

$$\sigma_p = n\left(\frac{1}{p} - 1\right)_+ \quad \text{and} \quad \sigma_{pq} = n\left(\frac{1}{\min(p,q)} - 1\right)_+ . \tag{22.50}$$

Here $a_+ = \max(a, 0)$ where $a \in \mathbb{R}$. We shall use the abbreviation (2.22). Let r be the same number as in (2.14). Then we have the following generalization of Definition 2.6 and Theorem 2.9 (now in the reverse order).

22.13 Theorem

Let Ψ be an admissible function according to 22.2.

(i) Let

$$0 < p \leq \infty, \quad 0 < q \leq \infty, \quad s > \sigma_p, \tag{22.51}$$

and let $\varrho > r$. Then $f \in S'(\mathbb{R}^n)$ belongs to $B_{pq}^{(s,\Psi)}(\mathbb{R}^n)$ if, and only if, it can be represented as

$$f = \sum_{\beta \in \mathbb{N}_0^n} \sum_{\nu=0}^{\infty} \sum_{m \in \mathbb{Z}^n} \lambda_{\nu m}^{\beta} (\beta qu)_{\nu m}(x), \tag{22.52}$$

unconditional convergence in $S'(\mathbb{R}^n)$, where $(\beta qu)_{\nu m}$ are (s, p, Ψ)-β-quarks and

$$\|\lambda \,|b_{pq}\|_\varrho = \sup_{\beta \in \mathbb{N}_0^n} 2^{\varrho|\beta|} \|\lambda^\beta \,|b_{pq}\| < \infty. \tag{22.53}$$

Furthermore, the infimum in (22.53) over all admissible representations (22.52) is an equivalent quasi-norm in $B_{pq}^{(s,\Psi)}(\mathbb{R}^n)$.

(ii) Let

$$0 < p < \infty, \quad 0 < q \leq \infty, \quad s > \sigma_{pq}, \tag{22.54}$$

and let $\varrho > r$. Then $f \in S'(\mathbb{R}^n)$ belongs to $F_{pq}^{(s,\Psi)}(\mathbb{R}^n)$ if, and only if, it can be represented by (22.52), unconditional convergence in $S'(\mathbb{R}^n)$, where $(\beta qu)_{\nu m}$ are (s, p, Ψ)-β-quarks and

$$\|\lambda \,|f_{pq}\|_\varrho = \sup_{\beta \in \mathbb{N}_0^n} 2^{\varrho|\beta|} \|\lambda^\beta \,|f_{pq}\| < \infty. \tag{22.55}$$

Furthermore, the infimum in (22.55) over all admissible representations (22.52) is an equivalent quasi-norm in $F_{pq}^{(s,\Psi)}(\mathbb{R}^n)$.

22.14 Remark

The above theorem coincides essentially with [Mou01b], Theorem 1.23. First steps, especially concerning the correct normalization factors in (22.46) and (22.48), may be found in [EdT99a], 3.3, p. 97. The discussions in 2.5 and 2.7 can be taken over without substantial changes. This explains the role of $\varrho > r$. Furthermore (22.52) converges unconditionally in $L_{\bar{p}}(\mathbb{R}^n)$ with $\bar{p} = \max(p, 1)$. The generalization of Definition 3.4 combined with Theorem 3.6 to spaces $B_{pq}^{(s,\Psi)}(\mathbb{R}^n)$ and $F_{pq}^{(s,\Psi)}(\mathbb{R}^n)$ for all $s \in \mathbb{R}$ (again in reverse order) may be found in [Mou01b], Corollary 1.27.

22.15 Tailored spaces: preliminaries

By our discussions in 22.1 the spaces $B_{pq}^s(\mathbb{R}^n)$ are optimally adapted not only to $\mathbb{R}^n$ and domains in $\mathbb{R}^n$ but also to compact d-sets. Apparently, (22.5) and (22.6) are the decisive criteria. Hence we ask for counterparts for the spaces $B_{pq}^{(s,\Psi)}(\mathbb{R}^n)$ with respect to (d, Ψ)-sets. The trace operator tr_Γ and the identification operator id_Γ have the same meaning as before. We refer to 9.2 and 9.32, obviously extended to $L_p(\Gamma)$ with $p < 1$ in case of tr_Γ, and also to [Triδ], 18.5, p. 138. Furthermore we extend (9.168) by

$$B_{pq}^{(s,\Psi),\Gamma}(\mathbb{R}^n)$$

$$= \left\{ f \in B_{pq}^{(s,\Psi)}(\mathbb{R}^n) : f(\varphi) = 0 \text{ if } \varphi \in S(\mathbb{R}^n) \text{ and } \varphi|\Gamma = 0 \right\} \quad (22.56)$$

where $s \in \mathbb{R}$, $0 < p \leq \infty$, $0 < q \leq \infty$, and where Γ is a compact set in $\mathbb{R}^n$. This also generalizes Definition 17.2 in [Triδ], where one finds on pp. 125–126 a discussion. We have $supp\, f \subset \Gamma$ if f belongs to the space in (22.56). If Γ is a (d, Ψ)-set then $|\Gamma| = 0$ by (22.19). Hence in this case, $B_{pq}^{(s,\Psi),\Gamma}(\mathbb{R}^n)$ is trivial if $B_{pq}^{(s,\Psi)}(\mathbb{R}^n)$ is a subset of $L_1^{loc}(\mathbb{R}^n)$. In particular $B_{pq}^{(s,\Psi),\Gamma}(\mathbb{R}^n)$ is trivial if $1 \leq p \leq \infty$ and $s > 0$. Recall that, as usual, $\frac{1}{p} + \frac{1}{p'} = 1$ if $1 \leq p \leq \infty$. As before, $L_p(\Gamma)$ and its quasi-norm must always be understood according to (9.13).

22.16 Theorem

Let $1 < p \leq \infty$.

(i) Let Γ be a compact (d, Ψ)-set according to Definition 22.4 (i). Then

$$id_\Gamma L_p(\Gamma) = B_{p,\infty}^{(-\frac{n-d}{p'}, \Psi^{-\frac{1}{p'}}),\Gamma}(\mathbb{R}^n). \quad (22.57)$$

(ii) Let Γ be a compact (n, Ψ)-set according to Definition 22.4(ii) and let in addition

$$\sum_{j=0}^{\infty} \Psi^{-\frac{1}{p'}}(2^{-j}) < \infty. \tag{22.58}$$

Then

$$id_\Gamma L_p(\Gamma) = B_{p,\infty}^{(0, \Psi^{-\frac{1}{p'}}), \Gamma}(\mathbb{R}^n). \tag{22.59}$$

22.17 Remark

This theorem coincides with Theorem 2.16 in [EdT99a]. It generalizes Theorem 18.2 in [Triδ], p. 136, from d-sets to (d, Ψ)-sets. In [EdT99a] we outlined also a proof claiming that one can follow the arguments in [Triδ] since the main ingredients have appropriate counterparts: local means and atoms. This is now available in detail. We refer to the comments in 22.11 and in 22.12. On this basis a detailed proof has been given in [Bri00], Theorem 3.12. Some arguments (in particular the application of local means) may also be found in [Mou01b], Proposition 2.12. Obviously, (22.57) extends (22.5). This, together with an appropriate counterpart of (22.6), is our criterion for whether spaces of type $B_{pq}^{(s, \Psi^b)}(\mathbb{R}^n)$ are tailored with respect to (d, Ψ)-sets. The restrictions $p > 1$ and, in case of $d = n$ the additional assumption (22.58), are needed to prove the sharp equalities in (22.57), (22.59). The inclusion

$$id_\Gamma L_p(\Gamma) \subset B_{p,\infty}^{(-\frac{n-d}{p'}, \Psi^{-\frac{1}{p'}}), \Gamma}(\mathbb{R}^n) \tag{22.60}$$

holds for all p with $1 \le p \le \infty$ and all (d, Ψ)-sets Γ according to Definition 22.4. The special case

$$id_\Gamma L_1(\Gamma) \subset B_{1,\infty}^{0, \Gamma}(\mathbb{R}^n) \tag{22.61}$$

is an immediate consequence of

$$M(\mathbb{R}^n) \subset B_{1,\infty}^{0}(\mathbb{R}^n), \tag{22.62}$$

where $M(\mathbb{R}^n)$ is the space of all complex-valued finite Radon measures in $\mathbb{R}^n$ normed in an obvious way. We refer to [Triδ], 18.3, p. 138. A direct proof has also been given in 9.2, formula (9.9).

22.18 Theorem

(i) Let Γ be a compact (d, Ψ)-set according to Definition 22.4(i). Let

$$0 < p < \infty, \quad 0 < q \leq \min(p, 1). \tag{22.63}$$

Then

$$tr_\Gamma B_{pq}^{(\frac{n-d}{p}, \Psi^{\frac{1}{p}})}(\mathbb{R}^n) = L_p(\Gamma). \tag{22.64}$$

(ii) Let Γ be a compact (n, Ψ)-set according to Definition 22.4(ii). Let

$$1 < p < \infty, \quad \sum_{j=0}^{\infty} \Psi^{-\frac{1}{p}}(2^{-j}) < \infty. \tag{22.65}$$

Then

$$tr_\Gamma B_{p,1}^{(0, \Psi^{\frac{1}{p}})}(\mathbb{R}^n) = L_p(\Gamma). \tag{22.66}$$

22.19 Remark

If $1 < p < \infty$ then the above theorem coincides with Theorem 2.19 in [EdT99a]. The proof outlined there is based on the duality of tr_Γ and id_Γ as described in 9.2, Theorem 22.16, and

$$\left(B_{p',1}^{(\frac{n-d}{p'}, \Psi^{\frac{1}{p'}})}(\mathbb{R}^n) \right)' = B_{p,\infty}^{(-\frac{n-d}{p'}, \Psi^{-\frac{1}{p'}})}(\mathbb{R}^n), \tag{22.67}$$

the latter in generalization of [Triβ], 2.11.2, pp. 178–179. As for $p \leq 1$ in part (i) we refer to [Bri00], Theorem 4.1. Here one needs that Γ satisfies the ball condition according to Proposition 22.6 (this makes also clear that at least the proof in case of $p \leq 1$ cannot be extended to $d = n$). If $\Psi = 1$ then (i) coincides with (22.6) and also with [Triδ], Corollary 18.12(i) on p. 142.

Next we use the above theorem as the starting point for the introduction of B-spaces on Γ. First we recall what has been done so far in this context. In case of arbitrary finite compactly supported Radon measures μ in $\mathbb{R}^n$ we defined in 9.29 some spaces $B_{pq}^s(\Gamma)$ on $\Gamma = \operatorname{supp} \mu$ which, according to Theorem 9.33 are traces of suitable spaces $B_{pq}^\sigma(\mathbb{R}^n)$. Now we specify Γ as a compact (d, Ψ)-set equipped with the respective Radon measure. One can follow Definition 9.29 and introduce spaces $B_{pq}^{(s, \Psi^a)}(\Gamma)$ via (9.132), (9.133), where now $(\beta qu)_{km}(\gamma)$ are β-quarks of type (22.46). Afterwards one asks for counterparts of (9.143). We prefer here the converse procedure taking the appropriate modification of

(9.143) as starting point. This approach has the advantage that it works for all values of p with $0 < p \le \infty$ (since, at least in this definition, there is no need to bother about moment conditions for respective atoms and β-quarks). This applies in particular to $p = \infty$ and to all (d, ψ)-sets according to both parts of Definition 22.4 because we have always

$$tr_\Gamma \, B_{pq}^{(\frac{n-d}{p}, \Psi^{\frac{1}{p}})}(\mathbb{R}^n) \subset L_p(\Gamma), \quad 0 < p \le \infty, \quad 0 < q \le \min(1,p). \quad (22.68)$$

We refer to [Mou01b], Proposition 2.14, where $p = \infty$ is covered by (19.21). As for quarkonial characterizations of spaces of type $B_{pq}^{(s, \Psi^a)}(\Gamma)$ we refer to [Bri99] and [Bri01]. There one finds also a detailed discussion of further properties of these spaces, especially in connection with the approach by A. Jonsson and H. Wallin. Further references and comments have been given in 9.34(i) and 9.34(iii).

22.20 Definition

Let Γ be a compact (d, Ψ)-set in $\mathbb{R}^n$ according to (both parts of) Definition 22.4. Let

$$0 < p \le \infty, \quad 0 < q \le \infty, \quad s > 0 \quad \text{and} \quad a \in \mathbb{R}. \quad (22.69)$$

Then

$$B_{pq}^{(s, \Psi^a)}(\Gamma) = tr_\Gamma \, B_{pq}^{(s+\frac{n-d}{p}, \Psi^{\frac{1}{p}+a})}(\mathbb{R}^n) \quad (22.70)$$

equipped with the quasi-norm

$$\left\| f \, | B_{pq}^{(s, \Psi^a)}(\Gamma) \right\| = \inf \left\| g \, | B_{pq}^{(s+\frac{n-d}{p}, \Psi^{\frac{1}{p}+a})}(\mathbb{R}^n) \right\|, \quad (22.71)$$

where the infimum is taken over all $g \in B_{pq}^{(s+\frac{n-d}{p}, \Psi^{\frac{1}{p}+a})}(\mathbb{R}^n)$ with $tr_\Gamma g = f$.

22.21 Remark

The definition is justified by Theorem 22.18, complemented by (22.68) in combination with Definition 22.10(i) and Proposition 22.3(iv). In particular, the spaces on the right-hand side of (22.70) are continuously embedded in the spaces on the left-hand sides of (22.64), (22.68). If $\Psi = 1$ and $0 < d < n$, then we have compact d-sets and the above definition coincides with [Triδ], Definition 20.2, p. 159. There we preferred the letter $\mathbb{B}$ in place of B for reasons explained in [Triδ], 20.3, pp. 160–161. But it seems to be reasonable now to use the same notation as in Definition 9.29 and Theorem 9.33 and to denote the resulting spaces by $B^s_{pq}(\Gamma)$, now for all

$$0 < p \le \infty, \quad 0 < q \le \infty, \quad s > 0. \tag{22.72}$$

It was one of the main aims of [Triδ], Chapter IV, to study compact embeddings between the spaces $B^s_{pq}(\Gamma)$ on compact d-sets in $\mathbb{R}^n$ and to apply these results to the spectral theory of fractal elliptic operators in [Triδ], Chapter V. Now we describe the extension of this theory from d-sets to (d, Ψ)-sets (where we shift the spectral theory to the next section). To avoid awkward formulations we agree on

$$B^{(0, \Psi^a)}_{pq}(\Gamma) = L_p(\Gamma), \quad 0 < p \le \infty, \quad 0 < q \le \infty, \quad a = 0, \tag{22.73}$$

simply as a notation. As always we measure compactness in terms of entropy numbers $e_k(id)$ as introduced in 19.16. Recall $c_+ = \max(c, 0)$ if $c \in \mathbb{R}$.

22.22 Theorem

Let Γ be a compact (d, Ψ)-set in $\mathbb{R}^n$ according to (both parts of) Definition 22.4. Let

$$0 < p_1 \le \infty, \quad 0 < p_2 \le \infty, \quad 0 < q_1 \le \infty, \quad 0 < q_2 \le \infty, \tag{22.74}$$

$s_2 \ge 0$, $a_1 \in \mathbb{R}$, $a_2 \in \mathbb{R}$ (with $a_2 = 0$ if $s_2 = 0$),

$$s_1 > s_2 + d \left(\frac{1}{p_1} - \frac{1}{p_2} \right)_+. \tag{22.75}$$

Then the embedding

$$id : B^{(s_1, \Psi^{a_1})}_{p_1 q_1}(\Gamma) \hookrightarrow B^{(s_2, \Psi^{a_2})}_{p_2 q_2}(\Gamma) \tag{22.76}$$

is compact and for the related entropy numbers,

$$e_k(id) \sim \left(k \, \Psi(k^{-1}) \right)^{-\frac{s_1 - s_2}{d}} \Psi(k^{-1})^{a_2 - a_1}, \quad k \in \mathbb{N}. \tag{22.77}$$

22.23 Remark

By our notational agreement (22.73) if $s_2 = 0$ (and hence $a_2 = 0$) the target space in (22.76) coincides with $L_{p_2}(\Gamma)$. In this version and restricted to $1 < p_1 < \infty$, $1 < p_2 < \infty$, the above theorem coincides with [EdT98], Theorem 5.1, and [EdT99a], Theorem 2.24, where we roughly outlined in the latter paper that the respective proof is similar to that in [Triδ], Section 20. In case of d-sets with $0 < d < n$ and, hence, $\Psi = 1$ the above theorem coincides with [Triδ], Theorem 20.6 on p. 166. Then we have

$$e_k(id) \sim k^{-\frac{s_1-s_2}{d}}, \quad k \in \mathbb{N}. \tag{22.78}$$

The proof in [Triδ] is based on two ingredients: Quarkonial decompositions for the spaces $B^s_{pq}(\mathbb{R}^n)$, which are considered here in the above Sections 2 and 3, on the one hand, and entropy numbers for compact embeddings Id between sequence spaces of type

$$\ell_{\infty,\varrho}\left[\ell_q\left(2^{j\sigma}\ell_p^{E_j}\right)\right] \tag{22.79}$$

as used here in the proof of Theorem 21.3 above, on the other hand. We refer in particular to (21.22)–(21.26). Ignoring technical details, the proofs of (22.78) in [Triδ] on the one hand and of Theorem 21.3 in this book on the other hand and also of the above theorem follow the same scheme. First one needs quarkonial representations. In the above case this is covered by Theorem 22.13 and β-quarks given by (22.46). This reduces, roughly speaking, the embedding (22.76) to an embedding Id of type (21.25), where one has to ask what is meant by E_r and what is the appropriate substitute of $2^{r\varkappa}$. It comes out that one has to choose

$$E_r \sim 2^{rd}\,\Psi^{-1}(2^{-r}), \quad r \in \mathbb{N}_0, \tag{22.80}$$

and $2^{r\varkappa}$ must be replaced in case of $p_2 \geq p_1$ by

$$2^{r\delta}\,\Psi^b(2^{-r}), \quad \delta = s_1 - s_2 + d\left(\frac{1}{p_1} - \frac{1}{p_2}\right), \quad b = a_1 - a_2 + \frac{1}{p_1} - \frac{1}{p_2}. \tag{22.81}$$

This requires an extension of the results obtained in [Triδ], Sections 8 and 9, which we described in (21.44), (21.45). This has been done in [Leo98] and [Leo01] covering also more general cases. A direct proof restricted to (22.80), (22.81) as substitutes in (22.79) has been given in [Mou01b], Proposition 3.9. In this paper one finds also a complete detailed proof of the above theorem.

23 Isotropic fractal drums

23.1 Preliminaries

Let Ω be a bounded C^∞ domain in $\mathbb{R}^n$ where $n \in \mathbb{N}$. Let Γ be a compact set in $\mathbb{R}^n$ with $\Gamma \subset \Omega$. We described in 19.1 and in 19.2 what (in our context) is meant by a fractal drum where the influence of the membrane is characterized by a Radon measure μ with $\Gamma - supp\,\mu$. An adequate mathematical description has been given 19.3 in terms of the operator B in (19.25),

$$B = (-\Delta)^{-1} \circ tr^\Gamma. \tag{23.1}$$

Here $(-\Delta)^{-1}$ has always the same meaning: the inverse of the Dirichlet Laplacian with respect to Ω. In case of d-sets we gave detailed explanations under what circumstances B in (23.1) makes sense and what is meant by tr^Γ. We refer to (19.26) on the one hand and (19.23), (19.24) on the other hand. We now generalize d-sets by (d,Ψ)-sets according to Definition 22.4. Then we have the counterparts (22.64), (22.57) of (19.20), (19.22), respectively. Hence, if $d < n$, then

$$tr^\Gamma \;:\; B_{p,1}^{(\frac{n-d}{p},\Psi^{\frac{1}{p}})}(\Omega) \mapsto B_{p,\infty}^{(-\frac{n-d}{p'},\Psi^{-\frac{1}{p'}})}(\Omega), \quad 1 < p < \infty, \tag{23.2}$$

is the generalization of (19.24). If $p = \infty$ and/or $d = n$, then one can use (22.68) and both parts of Theorem 22.16 in order to extend (23.2) to these cases. In Theorem 19.7 we collected what we know about the properties of B when Γ is a d-set. Now we wish to extend a few assertions to (d,Ψ)-sets, concentrating on the generalization of the distribution of eigenvalues given in case of d-sets by (19.73). This part was taken over from [Triδ] with a reference to [Triδ], Theorems 28.6 and 30.2, pp. 226, 234. However the decisive ingredients for the proofs, entropy numbers for the estimates of the eigenvalues from above and approximation numbers for the estimates from below, have also been used in this book several times, especially in connection with Theorem 19.17. So we refer to these points when only technical adaptions are needed. This applies especially to the estimates of the eigenvalues from below. In the theorem below we clip together Theorem 2.28, Corollary 2.30 (rusty drum), Theorem 2.33 (sintered drum) in [EdT99a] (with [EdT98] as a forerunner). We outline a new and more transparent (so we hope) proof concerning the estimates of the eigenvalues from above. Beside the extension from d-sets to (d,Ψ)-sets, we incorporate also an additional multiplication by a function b on Γ and replace for this purpose tr^Γ in (23.1) and (19.23) by

$$tr_b^\Gamma = id_\Gamma \circ b \circ tr_\Gamma \quad \text{where} \quad b \in L_r(\Gamma). \tag{23.3}$$

Of course, id_Γ and tr_Γ have the previous meaning. Otherwise we use the notation introduced in connection with Theorem 19.7 without further explanations.

23.2 Theorem

Let Ω be a bounded C^∞ domain in $\mathbb{R}^n$ (where $n \in \mathbb{N}$) and let Γ be a compact (d,Ψ)-set according to Definition 22.4 such that $\Gamma \subset \Omega$ and $n - 2 < d \leq n$

(with $0 < d \leq 1$ when $n = 1$). Let b be a positive function on Γ (a.e. with respect to μ) such that

$$b \in L_r(\Gamma) \quad \text{for some } r \text{ with } r > 1 \text{ and} \quad 0 \leq \frac{1}{r} < 1 - \frac{n-2}{d}, \tag{23.4}$$

and for some $c > 0$,

$$b(\gamma) \geq c \quad \text{if} \quad \gamma \in \Gamma_0, \tag{23.5}$$

where Γ_0 is a (d, Ψ)- set with $\Gamma_0 \subset \Gamma$. Then B,

$$B = (-\Delta)^{-1} \circ tr_b^\Gamma, \tag{23.6}$$

with (23.3), is a non-negative compact self-adjoint operator in $\overset{\circ}{H}{}^1(\Omega)$ with null-space

$$N(B) = \overset{\circ}{H}{}^1(\Omega \backslash \Gamma). \tag{23.7}$$

Furthermore, B is generated by the quadratic form

$$(Bf, g)_{\overset{\circ}{H}{}^1(\Omega)} = \int_\Gamma b(\gamma) f(\gamma) \overline{g(\gamma)} \mu(d\gamma) \quad f \in \overset{\circ}{H}{}^1(\Omega), \quad g \in \overset{\circ}{H}{}^1(\Omega), \tag{23.8}$$

with (19.29) as the scalar product in $\overset{\circ}{H}{}^1(\Omega)$. Let ϱ_k be the positive eigenvalues of B, repeated according to multiplicity and ordered by their magnitude,

$$\varrho_1 \geq \varrho_2 \geq \varrho_3 \geq \cdots. \tag{23.9}$$

Then

$$\varrho_k \sim k^{-1} \left(k \Psi(k^{-1}) \right)^{\frac{n-2}{d}}, \quad k \in \mathbb{N}. \tag{23.10}$$

Proof (outline) *Step 1* First we check that B, given by (23.6), is compact in $\overset{\circ}{H}{}^1(\Omega)$ and that

$$e_k(B) \leq c\, k^{-1} \left(k \Psi(k^{-1}) \right)^{\frac{n-2}{d}}, \quad k \in \mathbb{N}, \tag{23.11}$$

where again $e_k(B)$ are the respective entropy numbers according to 19.16. For this purpose we need only (23.4) where b might be even complex. Similarly as before we abbreviate

$$H^{(s,\Psi)} = B_{pv}^{(s,\Psi)} \quad \text{if} \quad p = v = 2. \tag{23.12}$$

By Definition 22.20 we have

$$tr_\Gamma : \overset{\circ}{H}{}^1(\Omega) \mapsto H^{(1-\frac{n-d}{2},\Psi^{-\frac{1}{2}})}(\Gamma). \tag{23.13}$$

By (23.4) we obtain

$$\frac{1}{q} = \frac{1}{2} - \frac{1}{2r} > \frac{n-2}{2d} \quad \text{and} \quad d\left(\frac{1}{2}-\frac{1}{q}\right) < \frac{d+2-n}{2} \tag{23.14}$$

(with $\frac{1}{2} \geq \frac{1}{q} > 0$ if $n = 1$). Hence by Theorem 22.22 with the interpretation (22.73) it follows that

$$id : H^{(1-\frac{n-d}{2},\Psi^{-\frac{1}{2}})}(\Gamma) \mapsto L_q(\Gamma) \tag{23.15}$$

is compact and

$$e_k(id) \sim \left(k\Psi(k^{-1})\right)^{-\frac{d+2-n}{2d}} \Psi(k^{-1})^{\frac{1}{2}} = k^{-\frac{1}{2}}\left(k\Psi(k^{-1})\right)^{\frac{n-2}{2d}}. \tag{23.16}$$

We split b by $b = b_1 \cdot b_2$ with $b_1 \in L_{2r}(\Gamma)$, $b_2 \in L_{2r}(\Gamma)$. Let T^{b_1} be the multiplication operator,

$$T^{b_1} f = b_1 f; \qquad T^{b_1} : H^{(1-\frac{n-d}{2},\Psi^{-\frac{1}{2}})}(\Gamma) \mapsto L_2(\Gamma). \tag{23.17}$$

To justify (23.17) we decompose T^{b_1} by id in (23.15) and the multiplication by b_1 according to Hölder's inequality based on (23.14). In particular, T^{b_1} is compact and by (23.16),

$$e_k(T^{b_1}) \leq c\, k^{-\frac{1}{2}}\left(k\Psi(k^{-1})\right)^{\frac{n-2}{2d}}, \quad k \in \mathbb{N}. \tag{23.18}$$

In 20.5 we introduced spaces $H^s(\Gamma)$ with $s < 0$ in case of d-sets. This can be extended, again by duality, to (d,Ψ)-sets and spaces $H^{(s,\Psi^a)}(\Gamma)$ with $s < 0$ and $a \in \mathbb{R}$. In particular,

$$T^{b_2} f = b_2 f; \qquad T^{b_2} : L_2(\Gamma) \mapsto H^{(-1+\frac{n-d}{2},\Psi^{\frac{1}{2}})}(\Gamma), \tag{23.19}$$

can be obtained as the dual of (23.17) (with $\overline{b}_2$ in place of b_1). In case of mappings between Hilbert spaces the entropy numbers of a compact operator and of its dual coincide. We refer to [ET96], Theorem 1.3.1, p. 9, where one finds also a short proof. As a consequence we have also

$$e_k(T^{b_2}) \leq c\, k^{-\frac{1}{2}}\left(k\Psi(k^{-1})\right)^{\frac{n-2}{2d}}, \quad k \in \mathbb{N}. \tag{23.20}$$

Clipping together (23.17)–(23.20) we get

$$T^b f = b f \, ; \qquad T^b \, : \quad H^{(1-\frac{n-d}{2}, \Psi^{-\frac{1}{2}})}(\Gamma) \mapsto H^{(-1+\frac{n-d}{2}, \Psi^{\frac{1}{2}})}(\Gamma) \, , \qquad (23.21)$$

and

$$e_k(T^b) \le c k^{-1} \left(k \Psi(k^{-1}) \right)^{\frac{n-2}{d}}, \quad k \in \mathbb{N}. \qquad (23.22)$$

The operator B given by (23.6), (23.3), can be factorized as

$$B = (-\Delta)^{-1} \circ id_\Gamma \circ T^b \circ tr_\Gamma \, . \qquad (23.23)$$

By (9.20) the identification operator id_Γ is the dual of tr_Γ. Then we get by (23.13) and the same arguments as in connection with (20.42),

$$id_\Gamma \, : \quad H^{(-1+\frac{n-d}{2}, \Psi^{\frac{1}{2}})}(\Gamma) \mapsto H^{-1}(\Omega) \, . \qquad (23.24)$$

Finally we apply again (20.31). This justifies the factorization of B in (23.23). Now (23.11) follows from (23.22) and the indicated mapping properties of the other factors in (23.23).

Step 2 Now the situation is very much the same as in the relevant parts of Theorem 19.7, covered by Step 1 of its proof with the respective references to [Triδ], Sections 28 and 30. This applies also to (23.7) since we assumed $b(\gamma) > 0$ in Γ a.e. and Proposition 19.5 (appropriately adapted) can be used. In particular, the operator B is non-negative, compact, self-adjoint and generated by the quadratic form in (23.8). As before,

$$\varrho_k \le c k^{-1} \left(k \Psi(k^{-1}) \right)^{\frac{n-2}{d}}, \quad k \in \mathbb{N}, \qquad (23.25)$$

follows now from (23.11) and (19.129). For the estimates of ϱ_k from below one may rely on approximation numbers. One can follow the scheme used in Step 5 of the proof of Theorem 19.17. Furthermore, in [Triδ], Step 4 of the proof of Theorem 28.6 on pp. 228–229, we have given a detailed proof in case of d-sets, resulting in the desired estimate from below in (19.73). This can be taken over, where one has to replace

$$N_j \sim 2^{jd} \quad \text{by} \quad N_j \sim 2^{jd} \Psi^{-1}(2^{-j}) \, , \quad j \in \mathbb{N}_0 \, , \qquad (23.26)$$

according to (22.80). This is the point where one needs (23.5).

23.3 Discussion

The above theorem covers essentially corresponding assertions in [EdT99a] with [EdT98] as a forerunner. The main point of the proof in [EdT99a] and also here is the inequality (23.11). The new argument here is the use of duality resulting in (23.20) and (23.22). In particular in this part of the proof one needs only (23.4), where b might be even complex. Then B remains to be compact but is not necessarily self-adjoint. Then one gets

$$|\varrho_k| \le c\, k^{-1} \left(k\Psi(k^{-1})\right)^{\frac{n-2}{d}}, \quad k \in \mathbb{N}, \qquad (23.27)$$

where now the non-zero eigenvalues are repeated according to algebraic multiplicity and ordered as in (19.128). Otherwise the role of the additional assumptions for b are quite clear by the proof. If $b \ge 0$, but not necessarily positive on Γ, then $N(B)$ might be larger than $\overset{\circ}{H}{}^1(\Omega\setminus\Gamma)$. The assumption (23.5) is needed only for the estimate of the eigenvalues ϱ_k in (23.10) from below by the right-hand side. In [EdT99a] we distinguished between $d < n$, $b = 1$ as the main case and $d = n$, $b = 1$ and $b \in L_r(\Gamma)$ come in as additional cases. Whereas $d < n$, $b = 1$ is largely parallel to the relevant parts of Theorem 19.7 (covered by the quoted assertions form [Triδ]), the two other cases might express some new phenomena. If $d = n$ then we have by (23.10),

$$\varrho_k \sim k^{-\frac{2}{n}} \Psi(k^{-1})^{1-\frac{2}{n}}, \quad k \in \mathbb{N}. \qquad (23.28)$$

Recall that $\varrho_k \sim k^{-\frac{2}{n}}$ is the expected classical Weyl behaviour. Hence (23.28) might be considered as a tiny distortion. Maybe the originally evenly distributed mass of the membrane $\Gamma = \overline{\Omega}$ is crumbling or the membrane becomes rusty but remains otherwise in shape. If this process is getting worse and the membrane is sintering unevenly, then one could try to describe this effect by an additional unevenly distributed function b. However in case of $n = 2$ of direct physical relevance we have always

$$\varrho_k \sim k^{-1}, \quad \text{where} \quad k \in \mathbb{N}, \qquad (23.29)$$

and hence all the above measures are Weyl measures according to Definition 19.13. But one can construct functions b such that the measure $b\mu$ does not satisfy the hypotheses of Theorem 19.17. We refer to 19.18(iii) for a discussion.

Chapter IV

Truncations and Semi-linear Equations

24 Introduction

The aim of this chapter is twofold. We assume that s, p, q are given such that $B_{pq}^s(\mathbb{R}^n)$ is a subspace of $L_1^{loc}(\mathbb{R}^n)$. Let $\mathbb{B}_{pq}^s(\mathbb{R}^n)$ be the real part of $B_{pq}^s(\mathbb{R}^n)$. Then we say that $\mathbb{B}_{pq}^s(\mathbb{R}^n)$ has the truncation property if

$$T^+ : \quad f(x) \mapsto f_+(x) = \max(f(x), 0), \quad x \in \mathbb{R}^n, \tag{24.1}$$

is a bounded map in $\mathbb{B}_{pq}^s(\mathbb{R}^n)$. We call $\left(\frac{1}{p}, s\right)$ a truncation couple if all spaces $\mathbb{B}_{pq}^s(\mathbb{R}^n)$ with $0 < q \leq \infty$ have the truncation property. Let R_n be the collection of all truncation couples. Then

$$R_n = \left\{ \left(\frac{1}{p}, s\right) : 0 \leq \frac{1}{p} < \infty, \; n\left(\frac{1}{p} - 1\right)_+ < s < 1 + \frac{1}{p} \right\}, \tag{24.2}$$

(the shaded region in Fig. 24.1); which might be considered as one of the main results of this chapter. There are similar assertions for the spaces $\mathbb{F}_{pq}^s(\mathbb{R}^n)$ with $\left(\frac{1}{p}, s\right) \in R_n$ (but a curious special case remains open). With T^+ one obtains also the boundedness of operators with the same non-linearity behaviour, for example,

$$f(x) \mapsto \max(f(x), g(x)) = (f - g)_+(x) + g(x), \quad x \in \mathbb{R}^n, \tag{24.3}$$

where, say, $g \in \mathbb{B}_{pq}^s(\mathbb{R}^n)$, and, in particular,

$$T: \quad f(x) \mapsto |f(x)| = 2f_+(x) - f(x), \quad x \in \mathbb{R}^n. \tag{24.4}$$

Problems of this type have attracted some attention. References will be given later. We only mention here the distinguished well-known property,

$$\|Tf \mid W_p^1(\mathbb{R}^n)\| = \|f \mid W_p^1(\mathbb{R}^n)\|, \quad f \in W_p^1(\mathbb{R}^n), \tag{24.5}$$

$1 \leq p < \infty$, which may be found in [GiT77], Lemma 7.6, p. 145, and in [Zie89], 2.1.8, p. 47. We used this remarkable assertion for $p = 2$ in Step 5 of the proof of Theorem 19.7. Besides boundedness we are interested in uniform continuity (or Lipschitz continuity, which in our context is the same) for T^+ and T. The outcome is negative: T^+ (and hence also T) are not Lipschitz continuous in the spaces $\mathbb{B}^s_{pq}(\mathbb{R}^n)$ and $\mathbb{F}^s_{pq}(\mathbb{R}^n)$ with $\left(\frac{1}{p}, s\right) \in R_n$. The lack of Lipschitz continuity is a serious drawback if one wishes to deal with semi-linear equations of type

$$u(x) = \int_{\mathbb{R}^n} K(x-y) \, u_+(y) \, dy + h(x), \quad x \in \mathbb{R}^n, \tag{24.6}$$

or

$$(-\Delta + id) \, u(x) = c \, |u(x)| + h(x), \quad x \in \mathbb{R}^n, \tag{24.7}$$

in spaces having the truncation property. Here, say, $h \in \mathbb{B}^s_{pq}(\mathbb{R}^n)$ with $\left(\frac{1}{p}, s\right) \in R_n$ is given and one asks for the maximal smoothness of solutions $u(x)$. This means $u \in \mathbb{B}^s_{pq}(\mathbb{R}^n)$ for the semi-linear integral equation (24.6). In case of (24.7) one might think of bootstrapping arguments starting from $L_p(\mathbb{R}^n)$ with $1 \leq p \leq \infty$. But this does not work in $\mathbb{R}^n$ if $p < 1$ (but as so often in theories as considered in this book, $p = 1$ is a very popular but nevertheless artificial boundary). Furthermore, as will be indicated at the end of this chapter in 27.10, bootstrapping arguments are of very limited use in connection with the problems treated here. It is the second aim of this chapter to circumvent this difficulty with the help of the Q-method. Let $f \in \mathbb{B}^s_{pq}(\mathbb{R}^n)$ with $\left(\frac{1}{p}, s\right) \in R_n$. Then we have by (2.31) and Corollary 2.12 the quarkonial decomposition

$$f = \sum_{\beta, \nu, m} \lambda^\beta_{\nu m}(f) \, (\beta qu)_{\nu m}(x), \quad x \in \mathbb{R}^n, \tag{24.8}$$

with the optimal coefficients $\lambda^\beta_{\nu m} = \lambda^\beta_{\nu m}(f)$ which depend linearly on f. One may even assume that all functions $(\beta qu)_{\nu m}(x) \geq 0$. Then the operator Q is defined by

$$f(x) \mapsto (Qf)(x) = \sum_{\beta, \nu, m} |\lambda^\beta_{\nu m}(f)| \, (\beta qu)_{\nu m}(x), \quad x \in \mathbb{R}^n. \tag{24.9}$$

It turns out that the operator Q is not only bounded in those spaces $\mathbb{B}^s_{pq}(\mathbb{R}^n)$ and $\mathbb{F}^s_{pq}(\mathbb{R}^n)$ where T^+ (and hence T) are bounded but it is Lipschitz continuous (in sharp contrast to T^+ and T). Hence if one replaces, for example, $u_+(y)$ in (24.6) by $(Qu)(y)$, then one can try to apply Banach's contraction method. One obtains a solution $u^0(x)$ of this modified equation (24.6). If, in addition, the kernel K is non-negative, then $u^0(x)$ is a supersolution of the original equation (24.6). Now one can apply the well-known supersolution technique which, together with the Fatou property of the function spaces involved, paves the way to get solutions of (24.6) with maximal smoothness.

Section 25 deals with the truncation problem. The Q-operator will be treated in Section 26. The Q-method outlined above is considered in Section 27. We are more interested in presenting the method itself than applying it to the most general situations. Hence we restrict ourselves mostly (but not exclusively) to the model equations (24.6), (24.7).

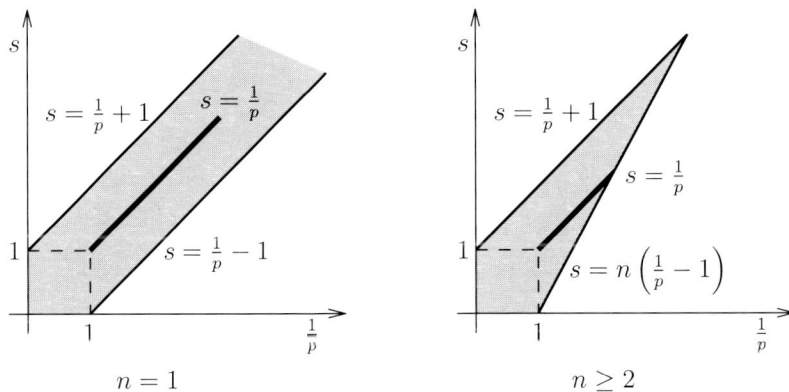

Fig. 24.1

25 Truncations

25.1 Preliminaries

Let $n \in \mathbb{N}$. In this chapter we assume that

$$B^s_{pq}(\mathbb{R}^n) \subset L^{loc}_1(\mathbb{R}^n) \quad \text{and} \quad F^s_{pq}(\mathbb{R}^n) \subset L^{loc}_1(\mathbb{R}^n) \tag{25.1}$$

(if not stated otherwise explicitly). Hence for the F-spaces and the B-spaces, p, s, q are restricted according to Theorem 11.2(i) and (ii), respectively. Let

$$\mathbb{B}^s_{pq}(\mathbb{R}^n) \quad \text{and} \quad \mathbb{F}^s_{pq}(\mathbb{R}^n) \quad \text{be the real parts,} \tag{25.2}$$

consisting of the real valued functions belonging to the corresponding spaces. Obviously, they are quasi-Banach spaces with respect to the same quasi-norms as the original spaces. Sometimes we write $\mathbb{A}_{pq}^s(\mathbb{R}^n)$ if the related context applies both to $B_{pq}^s(\mathbb{R}^n)$ and $F_{pq}^s(\mathbb{R}^n)$ (under the natural restrictions for the parameters s, p, q, which here always means $p < \infty$ for the F-spaces). Then it is clear what is meant by $\mathbb{A}_{pq}^s(\mathbb{R}^n)$. The Sobolev spaces $H_p^s(\mathbb{R}^n)$ and the classical Sobolev spaces $W_p^s(\mathbb{R}^n)$ have the same meaning as in (1.9) and in (1.10), respectively. Let $1 < p < \infty$. Then

$$\mathbb{H}_p^s(\mathbb{R}^n) \text{ with } s \geq 0 \quad \text{and} \quad \mathbb{W}_p^s(\mathbb{R}^n) \text{ with } s \in \mathbb{N}_0 \qquad (25.3)$$

are the corresponding real parts. Again we recall that $a_+ = \max(a, 0)$ if $a \in \mathbb{R}$. We are interested in the mapping properties of the *truncation operators* T and T^+ given by

$$(Tf)(x) = |f|(x) = |f(x)|, \quad x \in \mathbb{R}^n, \qquad (25.4)$$

and

$$(T^+f)(x) = f_+(x) = f(x)_+, \quad x \in \mathbb{R}^n. \qquad (25.5)$$

Here T makes sense for all spaces $A_{pq}^s(\mathbb{R}^n)$ with (25.1), whereas T^+ is restricted to the corresponding real parts $\mathbb{A}_{pq}^s(\mathbb{R}^n)$. In $\mathbb{A}_{pq}^s(\mathbb{R}^n)$ we have (24.4), which can also be written as

$$2T^+ = T + \text{id}. \qquad (25.6)$$

This makes it clear that T and T^+ have in real spaces the same mapping properties (boundedness, continuity). But first we have a look at T in the (complex) spaces $A_{pq}^s(\mathbb{R}^n)$. For given $n \in \mathbb{N}$ we again use our standard abbreviations,

$$\sigma_p = n\left(\frac{1}{p} - 1\right)_+ \quad \text{and} \quad \sigma_{pq} = \left(\frac{1}{\min(p,q)} - 1\right)_+, \qquad (25.7)$$

where $0 < p \leq \infty$ and $0 < q \leq \infty$. As before,

$$(\Delta_h f)(x) = f(x+h) - f(x), \quad x \in \mathbb{R}^n, \ h \in \mathbb{R}^n, \qquad (25.8)$$

are the first differences.

Example 1 Let $n \in \mathbb{N}$,

$$0 < p \leq \infty, \quad 0 < q \leq \infty, \quad \sigma_p < s < 1. \qquad (25.9)$$

By [Triβ], Theorem 2.5.12, p. 110,

$$\|f \,|\, B_{pq}^s(\mathbb{R}^n)\| = \|f \,|\, L_p(\mathbb{R}^n)\| + \left(\int_{|h|\leq 1} |h|^{-sq} \, \|\Delta_h f \,|\, L_p(\mathbb{R}^n)\|^q \, \frac{dh}{|h|^n}\right)^{\frac{1}{q}} \qquad (25.10)$$

(modification if $q = \infty$) is an equivalent quasi-norm in $B^s_{pq}(\mathbb{R}^n)$. We have that

$$\Delta_h|f|(x) = |f(x+h)| - |f(x)| \leq |f(x+h) - f(x)| = |\Delta_h f(x)|. \qquad (25.11)$$

Inserting this in (25.10) one obtains that

$$\|Tf \,|B^s_{pq}(\mathbb{R}^n)\| \leq c \,\|f\,|B^s_{pq}(\mathbb{R}^n)\|, \quad f \in B^s_{pq}(\mathbb{R}^n). \qquad (25.12)$$

There remains a technical point, that of ensuring that $Tf \in B^s_{pq}(\mathbb{R}^n)$. There are two possibilities. First, if for some $f \in L_p(\mathbb{R}^n) \cap S'(\mathbb{R}^n)$ the right-hand side of (25.10) is finite, then $f \in B^s_{pq}(\mathbb{R}^n)$. This assertion is covered by [Triβ], Remark 2.5.12/3, p. 113. This means, applied to our situation, $Tf \in B^s_{pq}(\mathbb{R}^n)$. Secondly, one uses the so-called Fatou property which will be described at the end of this subsection. Hence T is a bounded map in $B^s_{pq}(\mathbb{R}^n)$ with (25.9).

Example 2 Let

$$0 < p < \infty, \quad 0 < q \leq \infty, \quad \sigma_{pq} < s < 1. \qquad (25.13)$$

By [Triβ], Corollary 2.5.11, p. 108 (extended to $q = \infty$),
$\|f\,|F^s_{pq}(\mathbb{R}^n)\|$

$$= \|f\,|L_p(\mathbb{R}^n)\| + \left\| \left[\int_0^1 r^{-sq} \left(\int_{|h|\leq 1} |\Delta_{rh} f(\cdot)|\, dh \right)^q \frac{dr}{r} \right]^{\frac{1}{q}} \bigg| L_p(\mathbb{R}^n) \right\| \qquad (25.14)$$

(modification if $q = \infty$) is an equivalent quasi-norm in $F^s_{pq}(\mathbb{R}^n)$. By (25.11) it follows that

$$\|Tf\,|F^s_{pq}(\mathbb{R}^n)\| \leq c\,\|f\,|F^s_{pq}(\mathbb{R}^n)\|, \quad f \in F^s_{pq}(\mathbb{R}^n). \qquad (25.15)$$

Using again the Fatou property, explained below, we have $Tf \in F^s_{pq}(\mathbb{R}^n)$. Hence T is a bounded map in $F^s_{pq}(\mathbb{R}^n)$ with (25.13).

Example 3 Let $1 < p < \infty$. Recall that $f \in L_p(\mathbb{R}^n)$ belongs to the classical Sobolev space $W^1_p(\mathbb{R}^n)$ if, and only if, the norm in (17.23) is finite (and then it is an equivalent norm). This is covered by the references given there, in particular [Ste70], Proposition 3, p. 139. Then it follows by (25.11) that

$$\|Tf\,|W^1_p(\mathbb{R}^n)\| \leq c\,\|f\,|W^1_p(\mathbb{R}^n)\|, \quad f \in W^1_p(\mathbb{R}^n), \qquad (25.16)$$

and $Tf \subset W^1_p(\mathbb{R}^n)$. Hence T is a bounded map in $W^1_p(\mathbb{R}^n)$.

Example 4 (Real spaces) Let $A^s_{pq}(\mathbb{R}^n)$ be the real part of one of the (complex) spaces $A^s_{pq}(\mathbb{R}^n)$ covered by the Examples 1–3. Then T, and by (25.6) also

T^+, are bounded operators in $\mathbb{A}_{pq}^s(\mathbb{R}^n)$. In case of $\mathbb{W}_p^1(\mathbb{R}^n)$ with $1 < p < \infty$ one has now even the equality (24.5).

Fatou property Again let $A_{pq}^s(\mathbb{R}^n)$ be one of the spaces $B_{pq}^s(\mathbb{R}^n)$ or $F_{pq}^s(\mathbb{R}^n)$ without any restrictions for the parameters, which means

$$0 < p \le \infty, \quad 0 < q \le \infty, \quad s \in \mathbb{R}, \qquad (25.17)$$

(with $p < \infty$ in the F-case). Let $\{g_j\}_{j=1}^\infty$ be a bounded sequence in $A_{pq}^s(\mathbb{R}^n)$ with

$$g_j \to g \quad \text{in} \quad S'(\mathbb{R}^n) \quad \text{if} \quad j \to \infty. \qquad (25.18)$$

Then $g \in A_{pq}^s(\mathbb{R}^n)$ and there is a positive constant c (depending only on the space $A_{pq}^s(\mathbb{R}^n)$ given and the chosen quasi-norm) such that

$$\|g \,|A_{pq}^s(\mathbb{R}^n)\| \le c \sup_j \|g_j \,|A_{pq}^s(\mathbb{R}^n)\|. \qquad (25.19)$$

This follows from the Fourier-analytical characterization of the spaces $A_{pq}^s(\mathbb{R}^n)$ given by (2.37), (2.38), Theorem 3.6, and Fourier multiplier assertions, resulting in the constant c in (25.19). This notation was introduced in [Fra86] in a wider context. We refer also to [RuS96], p. 15.

The examples revisited We apply the Fatou property to the above examples. Let $f \in F_{pq}^s(\mathbb{R}^n)$ with (25.13). Let

$$\varphi^j(x) = \varphi(2^{-j}x), \quad x \in \mathbb{R}^n, \quad j \in \mathbb{N}_0, \qquad (25.20)$$

where φ is given by (2.33). Let

$$f_j(x) = \varphi^j(x) \left(\varphi^j \widehat{f}\right)^{\vee}(x) \quad \text{and} \quad g_j(x) = (Tf_j)(x) = |f_j(x)|, \quad j \in \mathbb{N}_0. \qquad (25.21)$$

Recall that $F_{pq}^s(\mathbb{R}^n) \subset L_r(\mathbb{R}^n)$ for some $1 < r < \infty$. This is well known. We refer to Theorem 11.4 (for a much sharper assertion than needed here) or simply to Fig. 10.1 or Fig. 24.1. Then

$$f_j \to f \quad \text{and} \quad g_j \to g = Tf \quad \text{in } L_r(\mathbb{R}^n) \quad \text{when } j \to \infty. \qquad (25.22)$$

Furthermore $g_j \in \mathcal{C}^\sigma(\mathbb{R}^n)$ for any σ, $0 < \sigma < 1$, where $\mathcal{C}^\sigma(\mathbb{R}^n)$ are the Hölder-Zygmund spaces from (1.11), (1.13). Since every g_j has compact support it follows that $g_j \in A_{pq}^\sigma(\mathbb{R}^n)$, in particular $g_j \in F_{pq}^s(\mathbb{R}^n)$. By (25.15), pointwise and Fourier multipliers, one has that

$$\|g_j \,|F_{pq}^s(\mathbb{R}^n)\| \le c_1 \|f_j \,|F_{pq}^s(\mathbb{R}^n)\| \le c_2 \|f \,|F_{pq}^s(\mathbb{R}^n)\|. \qquad (25.23)$$

By (25.18), (25.19) one obtains $g = Tf \in F_{pq}^s(\mathbb{R}^n)$ and (25.15). This seals the gap left open so far in Example 2 and in a similar way in Example 1.

The main aim of the above examples is the following. Ignoring technical complications (application of the Fatou property), the boundedness of T and T^+ in the indicated spaces with $s < 1$ is a direct consequence of known equivalent quasi-norms. The situation is different if $s \geq 1$. This will be discussed in detail in what follows, where we restrict ourselves to real spaces.

25.2 Definition

Let $\mathbb{A}_{pq}^s(\mathbb{R}^n)$ be the real part of $A_{pq}^s(\mathbb{R}^n)$, where $A_{pq}^s(\mathbb{R}^n)$ stands either for $B_{pq}^s(\mathbb{R}^n)$ or for $F_{pq}^s(\mathbb{R}^n)$, with (25.1).

(i) $\mathbb{A}_{pq}^s(\mathbb{R}^n)$ is said to have the *truncation property* if T^+ is a bounded map in $\mathbb{A}_{pq}^s(\mathbb{R}^n)$.

(ii) T^+ is said to be *uniformly continuous* in $\mathbb{A}_{pq}^s(\mathbb{R}^n)$ if for any ε with $0 < \varepsilon \leq 1$ there is a $\delta = \delta(\varepsilon) > 0$ such that

$$\|f_+ - g_+ \,|\, A_{pq}^s(\mathbb{R}^n)\| \leq \varepsilon \quad \text{when} \quad \|f - g \,|\, A_{pq}^s(\mathbb{R}^n)\| \leq \delta \tag{25.24}$$

for all $f \in \mathbb{A}_{pq}^s(\mathbb{R}^n)$ and $g \in \mathbb{A}_{pq}^s(\mathbb{R}^n)$.

25.3 Remark

Recall that T^+ is given by (25.5). By (25.6) one can replace in the above definition T^+ by T, given by (25.4). This applies also to the following simple observation.

25.4 Proposition

(i) $\mathbb{A}_{pq}^s(\mathbb{R}^n)$ has the truncation property if, and only if, there is a constant $c > 0$ such that

$$\|f_+ \,|\, A_{pq}^s(\mathbb{R}^n)\| \leq c \|f \,|\, A_{pq}^s(\mathbb{R}^n)\| \quad \text{for all} \quad f \in \mathbb{A}_{pq}^s(\mathbb{R}^n). \tag{25.25}$$

(ii) T^+ is uniformly continuous if, and only if, it is Lipschitz continuous: there is a constant $c > 0$ such that

$$\|f_+ - g_+ \,|\, A_{pq}^s(\mathbb{R}^n)\| \leq c \|f - g \,|\, A_{pq}^s(\mathbb{R}^n)\| \tag{25.26}$$

for all $f \in \mathbb{A}_{pq}^s(\mathbb{R}^n)$ and $g \in \mathbb{A}_{pq}^s(\mathbb{R}^n)$.

Proof This follows immediately from Definition 25.2 and the special structure of T^+ (positive 1 homogeneous).

25.5 Proposition

(Necessary condition, $n = 1$)

If $\mathbb{A}_{pq}^s(\mathbb{R})$ has the truncation property, then $s < 1 + \frac{1}{p}$.

Proof *Step 1* If $\mathbb{A}_{pq}^s(\mathbb{R})$ has the truncation property, then

$$\psi(x) = x_+ \varphi(x) \quad \text{with} \quad \varphi \in S(\mathbb{R}), \quad x \in \mathbb{R}, \tag{25.27}$$

belongs to $A_{pq}^s(\mathbb{R})$. Recall that $x_+ = \max(x, 0)$. Taking the first derivative it follows easily that

$$\chi \in A_{pq}^{s-1}(\mathbb{R}), \quad \chi \text{ characteristic function of } [0, 1]. \tag{25.28}$$

It is well known under what conditions χ belongs to $A_{pq}^{s-1}(\mathbb{R})$. By elementary embeddings and [RuS96], p. 53, it follows that $s - 1 < \frac{1}{p}$ in all cases in which $\mathbb{A}_{pq}^s(\mathbb{R})$ has the truncation property with the possible exceptions of $\mathbb{B}_{p\infty}^{1+\frac{1}{p}}(\mathbb{R})$, where $0 < p \leq \infty$, and $\mathbb{B}_{\infty q}^1(\mathbb{R})$ where $0 < q < \infty$. In the latter case one can argue as follows. Taking the second derivative of (25.27), one obtains that $\delta \in B_{\infty q}^{-1}(\mathbb{R})$ for the delta-distribution. Since $q < \infty$, one has a contradiction.

Step 2 It remains to disprove that $\mathbb{B}_{p\infty}^{1+\frac{1}{p}}(\mathbb{R})$ with $0 < p \leq \infty$ has the truncation property. Let $\varphi \in S(\mathbb{R})$ be a real function with $\operatorname{supp} \varphi \subset (-1, 1)$,

$$\varphi(x) \geq 0 \text{ if } x \geq 0, \quad \varphi(-y) = -\varphi(y) \text{ if } y \in \mathbb{R}, \tag{25.29}$$

and $\varphi'(0) > 0$. According to [Triδ], Theorem 13.7 on p. 75,

$$f(x) = \sum_{j=0}^{\infty} 2^{-j} \varphi(2^j x), \quad x \in \mathbb{R}, \tag{25.30}$$

is an atomic decomposition in $B_{p\infty}^{1+\frac{1}{p}}(\mathbb{R})$. In particular,

$$f \in B_{p\infty}^{1+\frac{1}{p}}(\mathbb{R}) \quad \text{for all} \quad 0 < p \leq \infty. \tag{25.31}$$

By (25.30) we have for $k \in \mathbb{N}$ that

$$f(2^{-k}) \sim 2^{-k} k \quad \text{and, hence,} \quad f(x) \sim x |\log x| \text{ if } 0 < x < \frac{1}{2}. \tag{25.32}$$

On the other hand, by Theorem 5.10 and (5.18) it follows that

$$\sup_{x>0} \frac{|g(x)|}{x} \leq c \, \|g \,|\mathcal{C}^1(\mathbb{R})\|, \quad g \in \mathcal{C}^1(\mathbb{R}), \quad \operatorname{supp} g \subset [0, 1]. \tag{25.33}$$

In particular, f_+ does not belong to $\mathcal{C}^1(\mathbb{R})$ and, hence, by embedding, also not to $B_{p\infty}^{1+\frac{1}{p}}(\mathbb{R})$ with $0 < p \leq \infty$. This proves that also $\mathbb{B}_{p\infty}^{1+\frac{1}{p}}(\mathbb{R})$ does not have the truncation property.

25.6 Definition

Let $n \in \mathbb{N}$. Let $0 < p \leq \infty$ and $s \in \mathbb{R}$. Then $\left(\frac{1}{p}, s\right)$ is called a *truncation couple* if for any q, $0 < q \leq \infty$, the space $\mathbb{B}^s_{pq}(\mathbb{R}^n)$ has the truncation property. The collection of all truncation couples is denoted by R_n.

25.7 Proposition

Let $n \in \mathbb{N}$. Then

$$R_n \subset \left\{ \left(\frac{1}{p}, s\right) : \; 0 \leq \frac{1}{p} < \infty, \; n\left(\frac{1}{p}-1\right)_+ < s < 1 + \frac{1}{p} \right\}, \quad (25.34)$$

(the shaded region in Fig. 24.1).

Proof *Step 1* Let $\left(\frac{1}{p}, s\right) \in R_n$. By Definition 25.2, (25.1) and Theorem 11.2(ii) it follows that $s > \sigma_p$, where σ_p is given by (25.7). If $n = 1$, then $s < 1 + \frac{1}{p}$ is a consequence of Proposition 25.5.

Step 2 It remains to prove $s < 1 + \frac{1}{p}$ if $\left(\frac{1}{p}, s\right) \in R_n$ and $n \geq 2$. Since we have in addition $s > \sigma_p$ (we again refer to Fig. 24.1), the spaces $B^s_{pp}(\mathbb{R}^n)$ with $\left(\frac{1}{p}, s\right) \in R_n$ have the Fubini property according to Theorem 4.4. Let $x = (x_1, x')$ with $x' \in \mathbb{R}^{n-1}$ and let

$$\psi(x') \in S(\mathbb{R}^{n-1}), \; \psi(x') \geq 0, \; \psi(0) > 0, \; \psi(y') = 0 \text{ if } |y'| \geq 1, \quad (25.35)$$

be a fixed function. Let $g \in \mathbb{B}^s_{pp}(\mathbb{R})$. Then

$$f(x) = g(x_1)\, \psi(x') \in B^s_{pp}(\mathbb{R}^n). \quad (25.36)$$

This follows (formally) from Theorem 4.4. But it can also be proved directly. One starts with an (optimal) one-dimensional atomic decomposition for $g(x_1)$. An atom at level j is multiplied with $\psi(x')$, which in turn is decomposed by a natural resolution of unity related to the lattice $2^{-j}\mathbb{Z}^n$, resulting for fixed j, after normalization, in $\sim 2^{j(n-1)}$ atoms in $\mathbb{R}^n$. Checking the coefficients one obtains (25.36) and, by Theorem 4.4,

$$\|g\,|B^s_{pp}(\mathbb{R})\| \sim \|f\,|B^s_{pp}(\mathbb{R}^n)\|. \quad (25.37)$$

If $B^s_{pp}(\mathbb{R}^n)$ has the truncation property, then

$$|g(x_1)|\,\psi(x') = |f(x)| \in B^s_{pp}(\mathbb{R}^n), \quad (25.38)$$

and again by Theorem 4.4 we obtain that $Tg = |g| \in B_{pp}^s(\mathbb{R})$ with

$$\|Tg\,|B_{pp}^s(\mathbb{R})\| \le c\,\|g\,|B_{pp}^s(\mathbb{R})\|. \tag{25.39}$$

Hence, $B_{pp}^s(\mathbb{R})$ has the truncation property. Then it follows by Proposition 25.5 that $s < 1 + \frac{1}{p}$.

25.8 Theorem

Let $n \in \mathbb{N}$ and let σ_p and σ_{pq} be given by (25.7).

(i)   Let R_n be the collection of all truncation couples according to Definition 25.6. Then

$$R_n = \left\{ \left(\frac{1}{p}, s\right) \,:\, 0 \le \frac{1}{p} < \infty,\ \sigma_p < s < 1 + \frac{1}{p} \right\} \tag{25.40}$$

(Fig. 24.1).

(ii)   Let either

$$1 < p < \infty,\quad 0 < q \le \infty,\quad \sigma_{pq} < s < 1 + \frac{1}{p}, \tag{25.41}$$

or

$$0 < p \le 1,\quad 0 < q \le \infty,\quad \sigma_{pq} < s < 1 + \frac{1}{p},\quad s \ne \frac{1}{p}. \tag{25.42}$$

Then $F_{pq}^s(\mathbb{R}^n)$ has the truncation property according to Definition 25.2(i), (Fig. 24.1).

Proof   We break the rather long proof into seven steps and shift a technical adaption to 25.9. We give a guide. Steps 1–4 deal with the one-dimensional case $n = 1$, $p = q$, and $s \ne \frac{1}{p}$. We prove that the spaces $\mathbb{B}_p^s(\mathbb{R}) = \mathbb{B}_{pp}^s(\mathbb{R})$ with

$$n = 1,\quad 0 < p \le \infty,\quad \left(\frac{1}{p} - 1\right)_+ < s < 1 + \frac{1}{p},\quad s \ne \frac{1}{p}, \tag{25.43}$$

have the truncation property (Fig. 24.1, $n = 1$). Afterwards we prove part (i) in Step 5. In Step 6 we return to the one-dimensional case and prove that the spaces $\mathbb{F}_{pq}^s(\mathbb{R})$ with

$$n = 1,\quad 0 < p < \infty,\quad 0 < q \le \infty, \tag{25.44}$$

$$\left(\frac{1}{\min(p,q)} - 1\right)_+ < s < 1 + \frac{1}{p},\quad s \ne \frac{1}{p}, \tag{25.45}$$

have the truncation property. Finally Step 7 contains the proof of part (ii).

Step 1 Under the assumption (25.43) we prove in Steps 1–4 that $|f| \in \mathbb{B}_p^s(\mathbb{R})$ if $f \in \mathbb{B}_p^s(\mathbb{R})$ and that there is a constant $c > 0$ with

$$\| \, |f| \, |B_p^s(\mathbb{R})\| \le c \, \|f \, |B_p^s(\mathbb{R})\| \tag{25.46}$$

for all $f \in \mathbb{B}_p^s(\mathbb{R}) = \mathbb{B}_{pp}^s(\mathbb{R})$. By Example 1 in 25.1 we may assume $p < \infty$. We begin with some preparation. Let $\varphi \in S(\mathbb{R})$ be a real even function with

$$supp \, \varphi \subset (-2, 2), \quad \varphi(y) = 1 \quad \text{when} \quad |y| \le 1, \tag{25.47}$$

and let $\varphi_J(x) = \varphi(2^{-J}x)$ where $J \in \mathbb{N}$. Then

$$f_J(x) = (\varphi_J \widehat{f})^\vee(x) \in \mathbb{B}_p^s(\mathbb{R}) \quad \text{if} \quad f \in \mathbb{B}_p^s(\mathbb{R}), \tag{25.48}$$

and, in particular, $f_J(x)$ is a real analytic function. As in (25.22) we have $B_p^s(\mathbb{R}) \subset L_r(\mathbb{R})$ for some r, $1 < r < \infty$. Then

$$f_J \to f \quad \text{and, hence,} \quad |f_J| \to |f| \quad \text{when} \quad J \to \infty, \tag{25.49}$$

in $L_r(\mathbb{R})$. Let us assume that we have (25.46) for all real analytic functions of the above type. Then by the Fourier multiplier properties of $B_p^s(\mathbb{R})$,

$$\| \, |f_J| \, |B_p^s(\mathbb{R})\| \le c_1 \, \|f_J \, |B_p^s(\mathbb{R})\| \le c_2 \, \|f \, |B_p^s(\mathbb{R})\|, \tag{25.50}$$

where c_1 and c_2 are independent of $J \in \mathbb{N}$. Now we can apply the Fatou property described in 25.1. As in (25.23) we obtain $|f| \in B_p^s(\mathbb{R})$ and (25.46). Hence it is sufficient to prove (25.46) for real analytic functions

$$f \in \mathbb{B}_p^s(\mathbb{R}) \quad \text{with} \quad supp \, \widehat{f} \subset \{y \in \mathbb{R} : \quad |y| \le 2^J\} \tag{25.51}$$

for some $J \in \mathbb{N}$.

Step 2 In Step 3, Step 4 and in the technical modification described in 25.9 below, we construct optimal real atomic decompositions for functions f given by (25.51) which respect the zeroes of f. For this purpose we need to prepare by constructing a hierarchy of resolutions of unity on the interval $(0, 1)$. Let $\psi_{0,k}(x)$ with $k \in \mathbb{N}_0$ be non-negative C^∞ functions on $\mathbb{R}_+$ with

$$supp \, \psi_{0,k} \subset (2^{-k-1}, 2^{-k+1}) = Q_k, \quad \sum_{k=0}^\infty \psi_{0,k}(x) = 1 \quad \text{if} \quad x \in (0, 1), \tag{25.52}$$

and

$$|\psi_{0,k}^{(l)}(x)| \le c_l \, 2^{lk}, \quad l \in \mathbb{N}_0, \quad k \in \mathbb{N}_0, \quad x \in \mathbb{R}, \tag{25.53}$$

for some $c_l > 0$ which are independent of x and k. Let χ be a C^∞ function on $\mathbb{R}$ with

$$supp\,\chi \subset (-1,1)\,, \quad \chi(x) \geq 0\,, \quad \sum_{l=-\infty}^{\infty} \chi(x-l) = 1 \quad \text{where} \quad x \in \mathbb{R}\,. \tag{25.54}$$

Put $\chi_{j,l}(x) = \chi(2^j x - l)$ where $j \in \mathbb{N}_0$ and $l \in \mathbb{Z}$. We decompose

$$\psi_{0,0}(x) = \sum_{l} \psi_{0,0}(x)\,\chi_{1,l}(x)\,, \quad x \in \mathbb{R}\,, \tag{25.55}$$

where only, say, $N+1$ functions are different from zero. Put

$$\psi_{1,k}(x) = \psi_{0,k}(x) \quad \text{if} \quad k \geq 1 \quad \text{and} \quad \psi_{1,k}(x) \quad \text{with} \quad k = 0, -1, \ldots, -N\,, \tag{25.56}$$

for the non-vanishing functions on the right-hand side of (25.55). Now we apply this procedure to $\psi_{1,k}(x)$ with $k = 1, 0, -1, \ldots, -N$ with $\chi_{2,l}(x)$ in place of $\chi_{1,l}(x)$, whereas the functions $\psi_{1,k}(x) = \psi_{0,k}(x)$ with $k \geq 2$ remain unchanged. Iteration yields a resolution of unity by non-negative functions $\psi_{j,k}(x)$ where we may assume (after a slight shifting of indices) $k \in \mathbb{Z}$ with $k \geq -2^j$,

$$\sum_{k} \psi_{j,k}(x) = 1 \quad \text{if} \quad x \in (0,1)\,, \tag{25.57}$$

$$\psi_{j,k}(x) = \psi_{0,k}(x) \quad \text{if} \quad k \geq j\,, \tag{25.58}$$

and

$$supp\,\psi_{j,k} \subset \left(2^{-j}(j+1-k) - 2^{-j},\, 2^{-j}(j+1-k) + 2^{-j}\right) \quad \text{if} \quad k < j\,. \tag{25.59}$$

Step 3 We prove (25.46) with (25.43) for

$$f \in \mathbb{B}_p^s(\mathbb{R})\,, \quad supp\,\widehat{f} \subset \{y \in \mathbb{R}\,:\, 2^{J-1} \leq |y| \leq 2^{J+1}\} \tag{25.60}$$

where $J \in \mathbb{N}$. With $J = 0$ the arguments cover also the case

$$supp\,\widehat{f} \subset \{y \in \mathbb{R}\,:\, |y| \leq 2\}\,. \tag{25.61}$$

(In this step we do not need the restriction $s \neq \frac{1}{p}$.) With

$$I_l = \{y \in \mathbb{R}\,:\, |y - 2^{-J}l| \leq c\,2^{-J}\} \tag{25.62}$$

for some suitable $c > 0$, and

$$f_l^* = \sup_{x \in I_l} |f(x)|\,, \quad l \in \mathbb{Z}\,, \tag{25.63}$$

we have that

$$\|f\,|B_p^s(\mathbb{R})\|^p \sim 2^{Jsp}\,\|f\,|L_p(\mathbb{R})\|^p \sim 2^{J(s-\frac{1}{p})p} \sum_l f_l^{*p}. \tag{25.64}$$

(Here and in what follows we assume $p < \infty$. If $p = \infty$ then one has to modify the calculations in the usual way.) The first equivalence in (25.64) comes from (2.37) and (25.60). The second equivalence is covered by the scalar case of [Triβ], Theorem 1.6.2, p. 30 (or [Triβ], Theorem 1.4.1, p. 22, combined with a dilation argument which includes also $p = \infty$). The equivalence constants in (25.64) are independent of $J \in \mathbb{N}_0$. Let $f(0) = 0$ and, say, $f(x) > 0$ in the interval $(0, 2)$. We wish to construct an optimal atomic decomposition of $f(x)$ in the interval $(0, 1)$. We use without further comments atomic decompositions of B_{pq}^s and F_{pq}^s as described in [Triδ], Section 13, especially Theorem 13.8 on p. 75. Let $\psi_{J,k}$ with $k \geq -2^J$ be the resolution of unity in $(0, 1)$ according to (25.57). Let, in analogy to (25.63),

$$f_{J,k}^* = \sup_{y \in \operatorname{supp} \psi_{J,k}} |f(y)|, \quad k \in \mathbb{Z}, \quad k \geq -2^J, \tag{25.65}$$

and

$$f(x) = \sum_{k=-2^J}^{\infty} \varrho_{J,k}\, b_{J,k}(x), \quad x \in (0,1), \tag{25.66}$$

with

$$b_{J,k}(x) = 2^{-(s-\frac{1}{p})\max(k,J)}\, \frac{f(x)}{f_{J,k}^*}\, \psi_{J,k}(x) \tag{25.67}$$

and

$$\varrho_{J,k} = 2^{(s-\frac{1}{p})\max(k,J)}\, f_{J,k}^*. \tag{25.68}$$

We assume that the $b_{J,k}(x)$ are (essentially normalized) atoms in $B_p^s(\mathbb{R})$ according to [Triδ], Theorem 13.8, p. 75 (no moment conditions are necessary). The functions $b_{J,k}(x)$ themselves are correctly normalized. This is not so clear as far as the needed derivatives of $b_{J,k}(x)$ up to order $1+[s]$ are concerned. We shift this technical question to 25.9 below where we modify (25.66) slightly. But this does not influence our reasoning and so we take (25.65)–(25.68) as a model atomic decomposition. We estimate the coefficients $\varrho_{J,k}$ and assume first $k \geq J+1$. Then $\psi_{J,k}(x) = \psi_{0,k}(x)$ by (25.58) with (25.52), (25.53). Since $f(0) = 0$ we have for $x \in \operatorname{supp} \psi_{J,k}$,

$$|f(x)| = \left| \int_0^x f'(y)\, dy \right| \leq c\, 2^{-k} \sup_{0 < y \leq c\, 2^{-k}} |f'(y)|. \tag{25.69}$$

Hence,

$$\sum_{k=J+1}^{\infty} \varrho_{J,k}^{p} \leq c \sup_{0<y\leq c\, 2^{-J}} |f'(y)| \sum_{k=J+1}^{\infty} 2^{k(s-\frac{1}{p}-1)p}$$

$$\leq c'\, 2^{J(s-\frac{1}{p}-1)p} \sup_{0<y\leq c\, 2^{-J}} |f'(y)|^{p}, \qquad (25.70)$$

where we used $s < 1 + \frac{1}{p}$. By (25.64) with f' and $s-1$ in place of f and s, respectively, we have that

$$2^{J(s-1-\frac{1}{p})p} \sum_{l} \sup_{x \in I_l} |f'(x)|^p \leq c\, \|f'\,|B_p^{s-1}(\mathbb{R})\|^p \leq c'\, \|f\,|B_p^{s}(\mathbb{R})\|^p. \qquad (25.71)$$

In particular, the right-hand side of (25.70) is one summand of the left-hand side of (25.71). If $k \leq J$, then $\varrho_{J,k}^{p}$ coincides essentially with related terms on the right-hand side of (25.64). This construction can be made on both sides of any zero of f. Summing up all these (appropriately modified) local representations (25.66) and using (25.64), (25.71) we obtain an optimal atomic decomposition

$$f(x) = \sum_{\nu=0}^{\infty} \sum_{m \in \mathbb{Z}} \lambda_{\nu m}^{J}\, a_{\nu m}^{J}(x) \qquad (25.72)$$

according to [Triδ], Theorem 13.8, p. 75, where the atoms $a_{\nu m}^{J}(x)$ and the real numbers $\lambda_{\nu m}^{J}$ coincide locally with $b_{J,k}(x)$ and $\varrho_{J,k}$, respectively, including the indicated modifications and summations. In particular by (25.64) and (25.71), it follows that

$$\sum_{\nu=0}^{\infty} \sum_{m \in \mathbb{Z}} |\lambda_{\nu m}^{J}|^{p} \leq c\, \|f\,|B_p^{s}(\mathbb{R})\|^{p}, \qquad (25.73)$$

where $c > 0$ is independent of J and f. Since f is real and since the real atoms $a_{\nu m}^{J}(x)$ respect the zeroes of f we have that

$$|f(x)| = \sum_{\nu=0}^{\infty} \sum_{m \in \mathbb{Z}} (sgn\, f(x))\, \lambda_{\nu m}^{J}\, a_{\nu m}^{J}(x), \quad x \in \mathbb{R}. \qquad (25.74)$$

Again by [Triδ], p. 75, and $s > \left(\frac{1}{p} - 1\right)_+$ it follows that (25.74) is an admissible atomic decomposition of $|f(x)|$. In particular,

$$\big\|\, |f|\, |B_p^{s}(\mathbb{R})\big\|^{p} \leq c \sum_{\nu=0}^{\infty} \sum_{m \in \mathbb{Z}} |\lambda_{\nu m}^{J}|^{p} \leq c'\, \|f\,|B_p^{s}(\mathbb{R})\|^{p}, \qquad (25.75)$$

where we used (25.73). Here c and c' are independent of f with (25.60) and $J \in \mathbb{N}_0$.

Step 4 We prove (25.46) with (25.43) for functions f given by (25.51). Again let $p < \infty$; when $p = \infty$ one has to modify the calculations in the usual way. We may assume that

$$f = \sum_{j=0}^{J} f_j \quad \text{with} \quad f_j = \left(\varphi_j \widehat{f}\right)^{\vee}, \quad x \in \mathbb{R}, \tag{25.76}$$

according to (2.37). We ask for the counterpart of (25.66) again under the assumption $f(0) = 0$ and $f(x) > 0$ in the interval $(0, 2)$. Now we use the special structure of the hierarchy of resolutions of unity $\{\psi_{j,k}\}$ as constructed in Step 2. Since

$$\sum_{j=0}^{J} f_j(0) = f(0) = 0 \tag{25.77}$$

we have for $x \in (0, 1)$ that

$$\begin{aligned} f(x) &= \sum_{j=0}^{J} \sum_{k=0}^{\infty} \psi_{0,k}(x) \, (f_j(x) - f_j(0)) \\ &= \sum_{j=0}^{J} \sum_{k=-2^j}^{\infty} \psi_{j,k}(x) \, (f_j(x) - f_j(0)) \\ &= \sum_{j=0}^{J} \sum_{k=-2^j}^{\infty} \varrho_{j,k} \, b_{j,k}(x) - \sum_{j=0}^{J} \sum_{k=-2^j}^{j} f_j(0) \, \psi_{j,k}(x), \end{aligned} \tag{25.78}$$

where $b_{j,k}(x)$ and $\varrho_{j,k}$ have the same meaning as in (25.67), (25.68), respectively, if $k = -2^j, \ldots, j$ (with f_j in place of f). If $k \geq j+1$, then $b_{j,k}(x)$ and $\varrho_{j,k}$ are modified by

$$\varrho_{j,k} = 2^{k(s-\frac{1}{p})} f_{j,k}^{**} \quad \text{with} \quad f_{j,k}^{**} = \sup_{y \in \operatorname{supp} \psi_{j,k}} |f_j(y) - f_j(0)| \tag{25.79}$$

and

$$b_{j,k}(x) = 2^{-k(s-\frac{1}{p})} \frac{f_j(x) - f_j(0)}{f_{j,k}^{**}} \psi_{j,k}(x), \quad x \in (0, 1). \tag{25.80}$$

We have a counterpart of (25.69) with $f_j(x) - f_j(0)$ in place of $f(x)$, which results in a counterpart of (25.70). The terms $\varrho_{j,k} b_{j,k}(x)$ with $k \leq j$ also fit in the above scheme. By the construction of $\{\psi_{j,k}\}$ we have for the remaining terms on the right-hand side of (25.78) that

$$\sum_{j=0}^{J} \sum_{k=-2^j}^{j} f_j(0)\, \psi_{j,k}(x)$$

$$= \sum_{j=0}^{J} f_j(0) \sum_{k=0}^{j} \psi_{0,k}(x)$$

$$= \sum_{k=0}^{J} \psi_{0,k}(x) \sum_{j=k}^{J} f_j(0) \qquad (25.81)$$

$$= \sum_{k=0}^{J} 2^{-k(s-\frac{1}{p})} \psi_{0,k}(x) \left[\sum_{j=k}^{J} 2^{j(s-\frac{1}{p})} f_j(0)\, 2^{-(j-k)(s-\frac{1}{p})} \right]$$

$$= \sum_{k=0}^{J} d_k\, 2^{-k(s-\frac{1}{p})} \psi_{0,k}(x).$$

This is an atomic decomposition in $B_p^s(\mathbb{R})$ with the coefficients d_k. First let $\frac{1}{p} < s < 1 + \frac{1}{p}$. Then

$$\sum_{k=0}^{J} |d_k|^p \le c \sum_{j=0}^{J} 2^{j(s-\frac{1}{p})p} |f_j(0)|^p, \qquad (25.82)$$

which again fits in the above scheme. Secondly, let $\left(\frac{1}{p}-1\right)_+ < s < \frac{1}{p}$. In the middle term in (25.81) one can replace

$$\sum_{j=k}^{J} f_j(0) \quad \text{by} \quad \sum_{j=0}^{k-1} f_j(0). \qquad (25.83)$$

This follows from (25.77). With a similar replacement in d_k one again obtains (25.82). [This does not work if $s = \frac{1}{p}$, excluded for this reason in (25.43).] We have by (25.78) in the interval $(0,1)$, that

$$f(x) = \sum_{j=0}^{J} \sum_{k=j+1}^{\infty} \varrho_{j,k}\, b_{j,k}(x) + f^2(x) = f^1(x) + f^2(x), \qquad (25.84)$$

where $f^2(x)$ collects all the terms in (25.78) with $k \le j$. The atoms in $f^2(x)$ with $k \le j$ resulting from the first sum on the right-hand side of (25.78) belong to different 2^{-j}-levels and, hence, do not interfere with each other in atomic representations. As for the second sum we have (25.81). Hence by the above estimates for the coefficients $\varrho_{j,k}$ and d_k in (25.82) one obtains in a similar way as in Step 3, resulting there in (25.72), (25.73), that

$$f^2(x) = \sum_{\nu=0}^{\infty} \sum_{m \in \mathbb{Z}} \lambda_{\nu m}^2 \, a_{\nu m}^2(x), \quad x \in \mathbb{R}, \tag{25.85}$$

with

$$\sum_{\nu=0}^{\infty} \sum_{m \in \mathbb{Z}} |\lambda_{\nu m}^2|^p \leq c \, \|f \,|B_p^s(\mathbb{R})\|^p, \tag{25.86}$$

where the atoms $a_{\nu m}^2(x)$ respect the zeroes of f. As for f^1 the situation is different. To the interval Q_k in (25.52) the atoms $b_{j,k}(x)$ with, say, $j = 0, \ldots, k$, contribute on the atomic level 2^{-k}. Hence we get a resulting atom

$$b_k(x) = \left(\sum_{j=0}^{k} \varrho_{j,k} \right)^{-1} \sum_{j=0}^{k} \varrho_{j,k} \, b_{j,k}(x) \tag{25.87}$$

with the coefficients

$$\varrho_k = \sum_{j=0}^{k} \varrho_{j,k}, \tag{25.88}$$

where $\varrho_{j,k}$ is given by (25.79). We modify the estimates in (25.69), (25.70) and obtain that

$$\varrho_{j,k} \leq c \, 2^{j(s-\frac{1}{p}-1)} \sup_{0 < y \leq c' \, 2^{-k}} |f'_j(y)| \, 2^{(k-j)(s-\frac{1}{p}-1)} \tag{25.89}$$

and, since $s < 1 + \frac{1}{p}$,

$$\sum_{k=0}^{\infty} \varrho_k^p \leq c \sum_{j=0}^{\infty} 2^{j(s-\frac{1}{p}-1)p} \left[\sup_{0 < y \leq c' \, 2^{-j}} |f'_j(y)| \right]^p. \tag{25.90}$$

Recall that the atoms b_k in (25.87) belong to different 2^{-k} atomic levels and there is no overlapping with the exception of direct neighbours (the last observation is not needed at this moment but of some use in connection with the F_{pq}^s-spaces considered in Step 6 and the modifications described in 25.9 and

25.10 below). Now the counterpart of (25.85), (25.86) is given by the atomic decomposition

$$f^1(x) = \sum_{\nu=0}^{\infty} \sum_{m \in \mathbb{Z}} \lambda_{\nu m}^1 \, a_{\nu m}^1(x), \quad x \in \mathbb{R}, \tag{25.91}$$

with

$$\sum_{\nu=0}^{\infty} \sum_{m \in \mathbb{Z}} |\lambda_{\nu m}^1|^p \leq c \sum_{j=0}^{\infty} \sum_{l \in \mathbb{Z}} 2^{j(s-\frac{1}{p}-1)p} \sup_{|y-2^{-j}l| \leq c\, 2^{-j}} |f_j'(y)|^p$$
$$\leq c' \, \|f' \,|B_p^{s-1}(\mathbb{R})\|^p$$
$$\leq c'' \, \|f \,|B_p^s(\mathbb{R})\|^p, \tag{25.92}$$

where we again used the above-indicated modification of (25.64). The atoms $a_{\nu m}^1(x)$ also respect the zeroes of f. Hence, (25.84), (25.85), (25.91) is the desired atomic decomposition. Now we are in the same position as at the end of Step 3 and arrive finally at (25.46) with (25.43) for functions f given by (25.51). By Step 1 this proves (25.46) with (25.43) for all $f \in \mathbb{B}_p^s(\mathbb{R})$.

Step 5 We prove part (i) of the theorem. Let

$$n \geq 2, \quad 0 < p \leq \infty, \quad s > \sigma_p = n\left(\frac{1}{p} - 1\right)_+. \tag{25.93}$$

According to Theorem 4.4 the spaces $B_p^s(\mathbb{R}^n) = B_{pp}^s(\mathbb{R}^n)$ have the Fubini property and, in particular,

$$\|f \,|B_p^s(\mathbb{R}^n)\| \sim \sum_{j=1}^{n} \left\| \|f^{x^j} \,|B_p^s(\mathbb{R})\| \,|L_p(\mathbb{R}^{n-1}) \right\| \tag{25.94}$$

are equivalent quasi-norms (in the notation used there). We add a technical remark. First one may assume that $f \in \mathbb{B}_p^s(\mathbb{R}^n)$ is so smooth that $|f| \in \mathbb{B}_p^s(\mathbb{R}^n)$ where now s is additionally restricted by $s < 1 + \frac{1}{p}$ and $s \neq \frac{1}{p}$. We add a remark about this technical point in 25.9 below. Then one can extend the considerations afterwards to arbitrary $f \in \mathbb{B}_p^s(\mathbb{R}^n)$ in the same way as at the end of 25.1 using the Fatou property. Hence the question of whether we have

$$\big\| \,|f| \,|B_p^s(\mathbb{R}^n)\big\| \leq c \,\|f \,|B_p^s(\mathbb{R}^n)\|, \quad f \in \mathbb{B}_p^s(\mathbb{R}^n), \tag{25.95}$$

under the restriction (25.93) can be reduced to the one-dimensional case (25.46). This proves (25.95) with

$$n \geq 1, \ 0 < p \leq \infty, \ n\left(\frac{1}{p} - 1\right)_+ = \sigma_p < s < 1 + \frac{1}{p}, \ s \neq \frac{1}{p}. \tag{25.96}$$

It remains to extend (25.95) from $\mathbb{B}_p^s(\mathbb{R}^n)$ to $\mathbb{B}_{pq}^s(\mathbb{R}^n)$ with $0 < q \leq \infty$ and to remove the restriction $s \neq \frac{1}{p}$. We use non linear real interpolation. First we describe the underlying abstract assertions.

Let A_0 and A_1 be two (real or complex) quasi-Banach spaces with $A_1 \subset A_0$ (continuous embedding). Let T be a (non-linear) operator with

$$\|Ta\,|A_1\| \leq c\,\|a\,|A_1\|, \quad a \in A_1, \tag{25.97}$$

(boundedness) and

$$\|Ta^1 - Ta^2\,|A_0\| \leq c\,\|a^1 - a^2\,|A_0\|, \quad a^1 \in A_0, \quad a^2 \in A_0, \tag{25.98}$$

(Lipschitz continuity) for some $c > 0$. Then, whenever $0 < \theta < 1$, $0 < q \leq \infty$,

$$\|Ta\,|(A_0, A_1)_{\theta,q}\| \leq c\,\|a\,|(A_0, A_1)_{\theta,q}\|, \tag{25.99}$$

(boundedness). Here, $(A_0, A_1)_{\theta,q}$ are the usual real interpolation spaces. The general background may be found in [Triα], Chapter 1, or in [BeL76]. The above specific (nonlinear) assertion goes back to [Pee70] and [Tar72a], [Tar72b] (we refer also to the more recent paper [Mal84]). The above formulation coincides essentially with [RuS96], Proposition 1, p. 88.

First let

$$1 \leq p \leq \infty, \quad 0 < \theta < 1, \quad 0 < q \leq \infty, \quad s > 0. \tag{25.100}$$

Then

$$\left(L_p(\mathbb{R}^n), B_p^s(\mathbb{R}^n)\right)_{\theta,q} = B_{pq}^{s\theta}(\mathbb{R}^n) \tag{25.101}$$

with an obvious counterpart if the spaces involved are restricted to their real parts. Of course, $Tf = |f|$ is Lipschitz continuous in $L_p(\mathbb{R}^n)$. Furthermore by (25.95), the operator T is bounded in $\mathbb{B}_p^s(\mathbb{R}^n)$ with (25.96). Application of (25.97)–(25.99) with $A_0 = \mathbb{L}_p(\mathbb{R}^n)$ (the real part of $L_p(\mathbb{R}^n)$) and $A_1 = \mathbb{B}_p^s(\mathbb{R}^n)$ results in

$$\big\|\,|f|\,|B_{pq}^s(\mathbb{R}^n)\big\| \leq c\,\|f\,|B_{pq}^s(\mathbb{R}^n)\|, \quad f \in \mathbb{B}_{pq}^s(\mathbb{R}^n), \tag{25.102}$$

with

$$n \geq 1, \quad 1 \leq p \leq \infty, \quad 0 = \sigma_p < s < 1 + \frac{1}{p}, \quad 0 < q \leq \infty. \tag{25.103}$$

It remains to remove the restriction $p \geq 1$. First we recall that

$$\|f\,|L_p(\mathbb{R}^n)\| + \left(\int_{|h|\leq 1} |h|^{-sq}\,\|\Delta_h^M f\,|L_p(\mathbb{R}^n)\|^q\,\frac{dh}{|h|^n}\right)^{\frac{1}{q}} \tag{25.104}$$

(with the usual modification if $q = \infty$) is an equivalent quasi-norm in the space

$$B^s_{pq}(\mathbb{R}^n) \quad \text{with} \quad 0 < p \le \infty, \quad 0 < q \le \infty, \quad s > \sigma_p.$$

Here $s < M \in \mathbb{N}$ and $\Delta^M_h f$ are the differences according to (1.12). We refer to [Triβ], Theorem 2.5.12, p. 110. Then all these spaces can also be considered as subspaces of $L_p(\mathbb{R}^n)$, even if $0 < p < 1$. Hence they are interpolation couples with the remarkable outcome (25.101), also in its real version, provided that $s\theta > \sigma_p$. This follows from [DeS93], Theorem 6.3, which, in turn, is a special case within a far-reaching non-linear approximation theory for L_p-spaces with $0 < p \le \infty$, based on splines, including interpolation, developed in [DeY91], [DeP88a], [DeP88b], [DeL93]. But then, again by nonlinear interpolation, we have (25.102) now for all

$$n \in \mathbb{N}, \quad 0 < p \le \infty, \quad \sigma_p < s < 1 + \frac{1}{p}, \quad 0 < q \le \infty. \tag{25.105}$$

This covers the right-hand side of (25.40), and completes, together with Proposition 25.7, the proof of part (i) of the theorem.

Step 6 We prove part (ii) of the theorem in two steps. First we modify Step 4 and prove

$$\big\| \, |f| \, | F^s_{pq}(\mathbb{R}) \| \le c \, \| f \, | F^s_{pq}(\mathbb{R}) \| \tag{25.106}$$

with (25.44), (25.45) for real analytic functions

$$f \in \mathbb{F}^s_{pq}(\mathbb{R}), \quad \operatorname{supp} \widehat{f} \subset \{ y \in \mathbb{R} : \ |y| \le 2^J \}, \tag{25.107}$$

for some $J \in \mathbb{N}$. This is the counterpart of (25.46), (25.43), (25.51). We follow the arguments in Step 4, resulting in (25.76)–(25.82) with the splitting (25.84) of f in f^1 and f^2 and the atomic representation (25.85). But now we must replace ℓ_p in (25.86) by the sequence spaces f_{pq} introduced in (2.8), one-dimensional case $n = 1$. We need the counterpart of (25.62)–(25.64), now with f given by (25.107) instead of (25.60). Let

$$I_{j,l} = \{ y \in \mathbb{R} : \ |y - 2^{-j} l| \le c \, 2^{-j} \} \quad \text{where} \quad j \in \mathbb{N}_0 \quad \text{and} \quad l \in \mathbb{Z}, \tag{25.108}$$

(again $c > 0$ is independent of j, l, and, of course, f). Let f be decomposed by (25.76) and

$$f^*_{jl} = \sup_{x \in I_{j,l}} |f_j(x)|, \quad j \in \mathbb{N}_0, \quad l \in \mathbb{Z}. \tag{25.109}$$

As an immediate consequence of [Triγ], 2.3.2, p. 93, we have (in obvious notation) that

$$\left\|\left\{2^{j(s-\frac{1}{p})} f_{jl}^*\right\} | f_{pq}\right\| \sim \|f | F_{pq}^s(\mathbb{R})\|. \tag{25.110}$$

Here the equivalence constants are independent of f and J. Now we refine (25.84) by

$$f(x) = f^1(x) + f^2(x) + f^3(x), \tag{25.111}$$

where $f^1(x)$ has the above meaning, $f^2(x)$ collects all terms with $k \leq j$ resulting by the above procedure from the first sum on the right-hand side of (25.78) and $f^3(x)$ collects all the second sums on the right-hand side of (25.78) resulting by the above procedure. Again we recall that the atoms $a_{\nu m}^1(x)$ of $f^1(x)$ in (25.91) have the property mentioned after (25.90) (controlled overlapping of the respective supports). Let λ^1 be the collection of all $\lambda_{\nu m}^1$ in (25.91) (most of them are zero). The (non-zero) coefficients $\lambda_{\nu m}^1$ originate from (25.88), (25.89), which can be re-written as

$$\varrho_k = \sum_{l=0}^{\infty} \varrho_{k-l,k} \quad \text{(with } \varrho_{m,k} = 0 \text{ if } m < 0\text{)}, \tag{25.112}$$

and

$$\varrho_{k-l,k} \leq c \left[2^{(k-l)(s-\frac{1}{p}-1)} \sup_{0<y\leq c\, 2^{-(k-l)}} |f'_{k-l}(y)| \right] 2^{l(s-\frac{1}{p}-1)}. \tag{25.113}$$

This convolution structure, the influence of index-shifting in the spaces f_{pq} as discussed in (2.103), and the convergence generating factors $2^{l(s-\frac{1}{p}-1)}$ with $s - \frac{1}{p} - 1 < 0$ now prove the counterpart of (25.92),

$$\|\lambda^1 | f_{pq}\| \leq c \left\|\left\{2^{j(s-\frac{1}{p}-1)} f_{jl}^{'*}\right\} | f_{pq}\right\| \sim \|f' | F_{pq}^{s-1}(\mathbb{R})\| \leq c' \|f | F_{pq}^s(\mathbb{R})\|, \tag{25.114}$$

where we used (25.110) with $s - 1$ and f' in place of s and f, respectively. Hence, (25.91) is again an optimal atomic decomposition which respects the zeroes of f. The same argument applies to $f^3(x)$ which originates from (25.81) since the coefficients d_k have a similar convolution structure as in (25.113). Again one has the splitting in

$$\frac{1}{p} < s < 1 + \frac{1}{p} \quad \text{and} \quad \left(\frac{1}{p} - 1\right)_+ < s < \frac{1}{p}.$$

There are no problems as far as f^2 is concerned. This is the usual atomic decomposition. In analogy to (25.85), (25.86) one gets the respective counterparts where one now has to use (25.110). The rest is the same as at the end of Step 4 and Step 3. In particular, since $s > \left(\frac{1}{\min(p,q)} - 1\right)_+$ no moment conditions in the atomic representations for $F^s_{pq}(\mathbb{R})$ are required, [Triδ], Theorem 13.8, p. 75. Then the counterpart of (25.74) is an atomic decomposition for $|f|$. This proves (25.106) with (25.44), (25.45).

Step 7 We prove part (ii) of the theorem. First we remark that (25.106) holds for all $f \in F^s_{pq}(\mathbb{R})$ with (25.44), (25.45). This is again a matter of the Fatou property as indicated in Step 1 with a reference to 25.1, now applied to $F^s_{pq}(\mathbb{R})$. Let

$$n \geq 1, \quad 0 < p < \infty, \quad 0 < q \leq \infty, \quad \sigma_{pq} < s < 1 + \frac{1}{p}, \quad s \neq \frac{1}{p}. \quad (25.115)$$

By Theorem 4.4 the spaces $F^s_{pq}(\mathbb{R}^n)$ with $n \geq 2$ have the Fubini property. Then we can argue as in Step 5. This proves part (ii) with exception of the case $s = \frac{1}{p}$ with $1 < p < \infty$. However this case is covered by Example 2 in 25.1.

25.9 Points left open

We left open two points so far.

(i) We used as a model case that (25.66) with (25.67), (25.68) is an atomic decomposition for f given by (25.60). As mentioned after (25.68) one needs not only a control of the size of $b_{J,k}(x)$ itself but also of its derivatives

$$b^{(m)}_{J,k}(x) \quad \text{with} \quad m = 0, \ldots, 1 + [s], \quad (25.116)$$

according to [Triδ], Theorem 13.8, p. 75 (no moment conditions are required). The shortest way to circumvent this problem is to modify (25.65)–(25.68) as follows. For brevity let $M = 1 + [s]$. Let $f^{(l)*}_{J,k}$ be given by (25.65) with respect to the derivatives $f^{(l)}(y)$, where $l = 0, 1, \ldots, M$. Then we have (25.66) with the modifications of $b_{J,k}(x)$ and $\varrho_{J,k}$ in (25.67), (25.68), respectively, given by

$$b_{J,k}(x) = 2^{-(s-\frac{1}{p})\max(k,J)} \left(\sum_{m=0}^{M} 2^{-m\max(k,J)} f^{(m)*}_{J,k}\right)^{-1} f(x)\,\psi_{J,k}(x) \quad (25.117)$$

and

$$\varrho_{J,k} = 2^{(s-\frac{1}{p})\max(k,J)} \sum_{m=0}^{M} 2^{-m\max(k,J)} f^{(m)*}_{J,k}. \quad (25.118)$$

Now it is clear that all derivatives $b_{J,k}^{(m)}(x)$ with $m = 0, \ldots, M$, are correctly normalized and that, hence, $b_{J,k}(x)$ are atoms. The additional terms in (25.118), compared with (25.68), fit in the above scheme and do not influence the chain of arguments significantly. The right-hand side of (25.70) is now given by

$$c' \sum_{m=1}^{M} 2^{J(s-\frac{1}{p}-m)p} \sup_{0<y\leq c\,2^{-J}} |f^{(m)}(y)|^p. \qquad (25.119)$$

Then the middle term and the last term in (25.71) must be modified by

$$c \sum_{m=1}^{M} \left\| f^{(m)} \,|\, B_p^{s-m}(\mathbb{R}) \right\|^p \leq c' \left\| f \,|\, B_p^s(\mathbb{R}) \right\|^p. \qquad (25.120)$$

The same adaption can be made in Step 4. In particular we have in (25.89) and, using

$$s - m < \frac{1}{p} \quad \text{for} \quad m = 1, \ldots, M,$$

also in (25.90), derivatives up to order M in the same way as in (25.119). This results on the right-hand side of (25.92) in the same terms as in (25.120). Corresponding adaptions can be made in Step 6. We add a comment. In a positivity interval of $f(x)$ there might be small local minima. But this does not influence the above arguments since constant factors in the atoms in (25.67), or (25.117) simply cancel out. In other words, only the behaviour near the zeroes must be considered.

(ii) At the beginning of Step 5 of the above proof it was convenient for us to assume

$$|f| \subset \mathbb{B}_p^s(\mathbb{R}^n) \quad \text{if} \quad f \in \mathbb{B}_p^s(\mathbb{R}^n) \quad \text{is smooth}$$

under the restrictions of s and p indicated there. For justification we recall that the spaces $B_p^s(\mathbb{R}^n)$ in question can be characterized in terms of differences, [Triβ], Remark 3 on p. 113. Together with [Triβ], Theorem 2.5.13, p. 115, and formula (11) on p. 116, combined with the one-dimensional case as treated in Step 4 of the above proof one gets the desired assertion.

25.10 Comments and references

Mapping properties of nonlinear operators in function spaces have attracted a lot of attention, including applications to nonlinear partial differential equations. A far-reaching treatment, especially in connection with (complex and real) spaces of type B_{pq}^s and F_{pq}^s in $\mathbb{R}^n$ and in domains has been given in [RuS96]. It is quite clear that among the large bulk of nonlinear operators worth treating in function spaces, the truncation operators T and T^+, given by (25.4) and (25.5), respectively, are distinguished examples. First discussions of mapping properties of T and of more general operators of power type

$$f \mapsto |f|^\mu, \quad \mu > 0 \quad \text{in} \quad B^s_{pq} \quad \text{and} \quad F^s_{pq} \quad \text{spaces}$$

were given in [Tri84]. Restricted to the case treated here, which means $\mu = 1$, Proposition 1 in [Tri84] coincides essentially with the Examples 1 and 2 in 25.1. If one has in the respective spaces B^s_{pq} and F^s_{pq} equivalent quasi-norms based on first differences $\Delta_h f$, then the boundedness of T is a more or less immediate consequence. Then one has necessarily $s < 1$. In other words, whether T is bounded in B^s_{pq} or F^s_{pq} becomes more complicated if $s \geq 1$ (ignoring the classical Sobolev spaces $W^1_p(\mathbb{R}^n)$ considered in Example 3 in 25.1 and in (24.5)). We refer to [BoM91], [Bod93], [Osw92], and [RuS96], 5.4.1, pp. 350–359. In the last book one finds further references, especially in the Note Section on pp. 391–392, and a description of the state of the art. Roughly speaking, at that time, and covered by the above-mentioned references, the boundedness of T was known in the F-case if $1 < p < \infty$, $1 \leq q \leq \infty$, and in the B-case if $1 \leq p \leq \infty$, $0 < q \leq \infty$ (there are some exceptions and extensions). We followed here [Tri00b].

Of special interest are the Sobolev spaces

$$H^s_p(\mathbb{R}^n) = F^s_{p,2}(\mathbb{R}^n), \quad s \in \mathbb{R}, \quad 0 < p < \infty. \tag{25.121}$$

We extended here the notation introduced in (1.9) to $0 < p < \infty$. Of course, again $\mathbb{H}^s_p(\mathbb{R}^n)$ are the real parts of $H^s_p(\mathbb{R}^n)$ in agreement with (25.3). One might think of $s \geq \sigma_p = n\left(\frac{1}{p} - 1\right)_+$. Then we have (25.1) according to Theorem 11.2(i). Recall that

$$bmo\,(\mathbb{R}^n) = F^0_{\infty,2}(\mathbb{R}^n), \tag{25.122}$$

as considered in 13.7, fits also in the above scheme.

25.11 Corollary

(i) *Let either*

$$1 < p < \infty, \quad 0 \leq s < 1 + \frac{1}{p}, \tag{25.123}$$

or

$$0 < p \leq 1, \quad n\left(\frac{1}{p} - 1\right) < s < 1 + \frac{1}{p}, \quad s \neq \frac{1}{p}. \tag{25.124}$$

Then $\mathbb{H}_p^s(\mathbb{R}^n)$ has the truncation property according to Definition 25.2. Furthermore, the real part of $bmo\,(\mathbb{R}^n)$ has the truncation property.

(ii) Let

$$0 < p < \infty, \quad s \geq 1 + \frac{1}{p} \quad and \quad s > \sigma_p = n\left(\frac{1}{p} - 1\right)_+. \tag{25.125}$$

Then $\mathbb{H}_p^s(\mathbb{R}^n)$ does not have the truncation property.

Proof Of course $L_p(\mathbb{R}^n)$ with $1 < p < \infty$ has the truncation property (real for T and T^+, and complex for T). As for $bmo\,(\mathbb{R}^n)$ one can replace (13.82) by

$$\|f\,|bmo\,(\mathbb{R}^n)\| \sim \sup_{|Q|\leq 1} \left[\inf_{c \in \mathbb{C}} \frac{1}{|Q|} \int_Q |f(x) - c|\,dx \right] + \sup_{|Q|>1} \frac{1}{|Q|} \int_Q |f(x)|\,dx. \tag{25.126}$$

This is well known. It follows also immediately from (13.82). The first term on the right-hand side of (13.82) remains unchanged if one replaces $f(x)$ in each cube by $f(x) - c_Q$ with $c_Q \in \mathbb{C}$, and correspondingly f_Q by $(f - c_Q)_Q = f_Q - c_Q$. Then the left-hand side of (25.126), given by (13.82), can be estimated from above by its right-hand side multiplied with 2. Since the opposite estimate is obvious we obtain (25.126). Now it follows easily that $bmo\,(\mathbb{R}^n)$ has the truncation property (real and complex, the latter restricted to T). The other assertions of part (i) are covered by Theorem 25.8(ii). If $n = 1$ then part (ii) follows from Proposition 25.5. If $n \geq 2$, then the spaces $H_p^s(\mathbb{R}^n)$ have the Fubini property from Theorem 4.4. If one assumes that such a space $\mathbb{H}_p^s(\mathbb{R}^n)$ has the truncation property, then the corresponding space $\mathbb{H}_p^s(\mathbb{R})$ also has the truncation property. This is a contradiction.

25.12 Remark

This corollary clarifies the question of whether or not $\mathbb{H}_p^s(\mathbb{R}^n)$ has the truncation property with exception of the disturbing restriction $s \neq \frac{1}{p}$ in (25.124) and of the limiting case $s = n\left(\frac{1}{p} - 1\right)$ if $p \leq 1$. Similar limiting cases for $\mathbb{B}_{pq}^s(\mathbb{R}^n)$ and $\mathbb{F}_{pq}^s(\mathbb{R}^n)$ with $s = \sigma_p$ and $0 < p \leq \infty$, remain also open ($p < \infty$ for the F-scale).

25.13 Continuity

By [MaM79] any bounded composition operator in $\mathbb{H}_p^1(\mathbb{R}^n)$ with $1 < p < \infty$ is also continuous. This applies in particular to T^+ and T. This assertion was extended in [RuS96], Theorem 3 on p. 377, to

$$\mathbb{B}^s_{pq}(\mathbb{R}^n) \quad \text{with} \quad 1 \le p < \infty, \quad 0 < q < \infty, \quad 0 < s < 1. \tag{25.127}$$

However even for the comparatively simple nonlinear operator T^+ the situation is different if one asks for uniform continuity according to Definition 25.2(ii). By Proposition 25.4(ii) this question is equivalent to the problem of whether T^+ is Lipschitz continuous. By (25.6) in real spaces the nonlinearity behaviour of T and T^+ is the same. Obviously, T, and hence in real spaces T^+, is Lipschitz continuous in $L_p(\mathbb{R}^n)$ with $1 \le p \le \infty$. But otherwise the situation is less favourable.

25.14 Theorem

Let $n \in \mathbb{N}$ and let $\left(\frac{1}{p}, s\right) \in R_n$ be a truncation couple according to (25.40) and Definition 25.6. Let $0 < q \le \infty$. Let $\mathbb{A}^s_{pq}(\mathbb{R}^n)$ be either $\mathbb{B}^s_{pq}(\mathbb{R}^n)$ or $\mathbb{F}^s_{pq}(\mathbb{R}^n)$ (with $p < \infty$ in the F-case). Then T^+ (and hence also T) is not a Lipschitz continuous map in $\mathbb{A}^s_{pq}(\mathbb{R}^n)$.

Proof We break the proof into three steps and justify an interpolation formula used in between, afterwards in 25.15. By (25.6) it is sufficient to prove that T is not Lipschitz continuous.

Step 1 The first (and crucial) step deals with $n = 1$ and $\mathbb{B}^s_p(\mathbb{R}) = \mathbb{B}^s_{pp}(\mathbb{R})$ where

$$0 < p < \infty, \quad \left(\frac{1}{p} - 1\right)_+ < s < 1 + \frac{1}{p}, \quad s \ne \frac{1}{p}. \tag{25.128}$$

Let χ be given by (25.54) and let $j \in \mathbb{N}$. We put

$$f(x) = \sum_{m=0}^{2^{j-1}} \chi(2^j x - 2m) \quad \text{and} \quad g(x) = -\sum_{m=0}^{2^{j-1}} \chi(2^j x - 2m - 1). \tag{25.129}$$

Then $f(x)$, $g(x)$, and

$$|f(x)| - |g(x)| = f(x) + g(x), \quad x \in \mathbb{R}, \tag{25.130}$$

are smoothed zigzag functions with supports in $\left[-2^{-j}, 1 + 2^{-j+1}\right]$, whereas

$$f(x) - g(x) = 1 \quad \text{if} \quad x \in [0, 1]. \tag{25.131}$$

The aim of this step is to prove that

$$\|f - g\,|B^s_p(\mathbb{R})\| \sim \max\left(1, 2^{j(s-\frac{1}{p})}\right) \tag{25.132}$$

and
$$\| |f| - |g| \, |B_p^s(\mathbb{R})\| \sim 2^{js}, \qquad (25.133)$$
where the equivalence constants are independent of $j \in \mathbb{N}$. First we prove (25.132). Let $\psi_{0,k}(x)$ be the same functions as in (25.52), (25.53) and let $\psi_j(x) = \sum_{k=0}^{j} \psi_{0,k}(x)$. Then the ψ_j and $f - g$ are of the same type at least as far as (25.132) is concerned. We have that

$$\begin{aligned}
\|f - g \, |B_p^s(\mathbb{R})\|^p &\sim \|\psi_j \, |B_p^s(\mathbb{R})\|^p \\
&\sim \sum_{k=0}^{j} \|\psi_{0,k} \, |B_p^s(\mathbb{R})\|^p \\
&\sim \sum_{k=0}^{j} 2^{k(s-\frac{1}{p})p} \|\psi_{0,k}(2^{-k}\cdot) \, |B_p^s(\mathbb{R})\|^p \\
&\sim \max\left(1, 2^{j(s-\frac{1}{p})p}\right).
\end{aligned} \qquad (25.134)$$

The first equivalence has been explained above. The second and the third equivalences follow from Theorem 5.14 and Corollary 5.16, respectively. Here one needs $s > \left(\frac{1}{p} - 1\right)_+$ and $B_p^s = F_{pp}^s$. In the last equivalence we used $s \neq \frac{1}{p}$. This proves (25.132). Next we prove (25.133). By (25.129), (25.130) and the localization property from [ET96], Theorem 2.3.2, pp. 35–36, we obtain that

$$\| |f| - |g| \, |B_p^s(\mathbb{R})\| \leq c \, 2^{j(s-\frac{1}{p})} \left(\sum_{m=1}^{2^j} 1\right)^{\frac{1}{p}} = c \, 2^{js}. \qquad (25.135)$$

(Because of the required support properties in [ET96] we do not obtain the desired equivalence immediately.) But (25.135) is also a consequence of atomic decompositions. Hence it remains to prove that there is a constant $c > 0$ such that

$$\| |f| - |g| \, |B_p^s(\mathbb{R})\| \geq c \, 2^{js} \quad \text{for} \quad j \in \mathbb{N}. \qquad (25.136)$$

First let, in addition, $\left(\frac{1}{p} - 1\right)_+ < s < 1$ and let $H(x) = |f(x)| - |g(x)|$. Then we have that
$\|H \, |B_p^s(\mathbb{R})\|^p$

$$\begin{aligned}
&\sim \|H \, |L_p(\mathbb{R})\|^p + \int_0^1 t^{-sp} \sup_{0<\tau<t} \|H(\cdot + \tau) - H(\cdot) \, |L_p(\mathbb{R})\|^p \frac{dt}{t} \\
&\geq c \, 2^{jsp} \sup_{0<\tau \leq 2^{-j}} \|H(\cdot + \tau) - H(\cdot) \, |L_p(\mathbb{R})\|^p \geq c' \, 2^{jsp}, \qquad (25.137)
\end{aligned}$$

where $c > 0$ and $c' > 0$ are independent of $j \in \mathbb{N}$. The equivalent quasi-norm is covered by [Triβ], p. 110. The first estimate is obvious, whereas the last estimate is a consequence of the oscillating nature of H. This proves (25.136) under the indicated restriction for s. Finally, let

$$0 < p < \infty, \quad \left(\frac{1}{p} - 1\right)_+ < s < 1 + \frac{1}{p}. \qquad (25.138)$$

We choose $1 < p_1 < \infty$ and $0 < \theta < 1$ such that

$$s_0 = (1-\theta)s < 1 \quad \text{with} \quad \frac{1}{p_0} = \frac{1-\theta}{p} + \frac{\theta}{p_1}, \quad \frac{1}{q} = \frac{1-\theta}{p} + \frac{\theta}{2}. \qquad (25.139)$$

By the complex interpolation formula in [Triβ], Theorem 2.4.7, p. 69, we have that

$$\left(B_p^s(\mathbb{R}), L_{p_1}(\mathbb{R})\right)_\theta = \left(F_{pp}^s(\mathbb{R}), F_{p_1,2}^0(\mathbb{R})\right)_\theta = F_{p_0,q}^{s_0}(\mathbb{R}) \subset B_{p_0,\infty}^{s_0}(\mathbb{R}). \qquad (25.140)$$

This formula is also covered by the more recent complex interpolation method developed in [MeM00]. By [Triβ], p. 110, there is a counterpart of (25.137) with respect to $B_{p_0,\infty}^{s_0}(\mathbb{R})$. Then there are positive constants $c_1 > 0$, $c_2 > 0$, $c_3 > 0$ such that for all $j \in \mathbb{N}$,

$$\begin{aligned} c_1 \, 2^{j s_0} &\leq \|H \,|\, B_{p_0,\infty}^{s_0}(\mathbb{R})\| \\ &\leq c_2 \, \|H \,|\, B_p^s(\mathbb{R})\|^{1-\theta} \, \|H \,|\, L_{p_1}(\mathbb{R})\|^\theta \\ &\leq c_3 \, \|H \,|\, B_p^s(\mathbb{R})\|^{1-\theta}. \end{aligned} \qquad (25.141)$$

This proves (25.136) for all admitted cases.

Step 2 Let $n \in \mathbb{N}$. We prove the theorem for the spaces $\mathbb{B}_p^s(\mathbb{R}^n)$ under the additional restrictions $p < \infty$ and $s \neq \frac{1}{p}$. With f and g given by (25.129) we put

$$F(x) = f(x_1)\chi(x_2) \cdots \chi(x_n) \quad \text{and} \quad G(x_n) = g(x_1)\chi(x_2) \cdots \chi(x_n). \qquad (25.142)$$

(Of course if $n = 1$ then $F = f$ and $G = g$.) By Theorem 4.4 the spaces $B_p^s(\mathbb{R}^n)$ have the Fubini property. Then it follows from (25.132) and (25.133) that

$$\|F - G \,|\, B_p^s(\mathbb{R}^n)\| \sim \max\left(1, 2^{j(s-\frac{1}{p})}\right) \qquad (25.143)$$

and

$$\| \,|F| - |G| \,|\, B_p^s(\mathbb{R}^n)\| \sim 2^{js}, \qquad (25.144)$$

where again the equivalence constants are independent of $j \in \mathbb{N}$. However (25.143) and (25.144) prove that T is not Lipschitz continuous in $\mathbb{B}_p^s(\mathbb{R}^n)$, provided that, so far, $p < \infty$ and $s \neq \frac{1}{p}$.

Step 3 Now we prove the theorem under the restriction $p < \infty$. We rely on the real interpolation formula (25.101) in the extended version as indicated at the end of Step 5 of the proof of Theorem 25.8. Together with the reiteration theorem of interpolation theory it follows that

$$\left(L_p(\mathbb{R}^n), A_{pq}^s(\mathbb{R}^n)\right)_{\theta,p} = B_p^{s\theta}(\mathbb{R}^n), \tag{25.145}$$

where again A_{pq}^s is either B_{pq}^s or F_{pq}^s with

$$0 < p < \infty, \quad 0 < \theta < 1, \quad \sigma_p < s\theta < s < 1 + \frac{1}{p}, \quad 0 < q \leq \infty. \tag{25.146}$$

Of course, T is Lipschitz continuous in $L_p(\mathbb{R}^n)$. Assume that T is also Lipschitz continuous in $\mathbb{A}_{pq}^s(\mathbb{R}^n)$. Then it follows by real interpolation of Lipschitz operators, [Tar72b], Theorem 4, p. 476, that T is also Lipschitz continuous in $\mathbb{B}_p^{s\theta}(\mathbb{R}^n)$. We may assume that $s\theta \neq \frac{1}{p}$. We get a contradiction to Step 2. This proves the theorem if $p < \infty$. Finally, let $p = \infty$. Let

$$0 < s < 1, \quad 1 < q < \infty, \quad 0 < \theta < 1, \tag{25.147}$$

and

$$\frac{1}{q_0} = \frac{\theta}{q}, \quad s_0 = (1-\theta)s. \tag{25.148}$$

Then we prove in 25.15 below the real interpolation formula

$$B_{q_0}^{s_0}(\mathbb{R}^n) = \left(B_\infty^s(\mathbb{R}^n), L_q(\mathbb{R}^n)\right)_{\theta,q_0}. \tag{25.149}$$

Taking this for granted, we are in the same position as above. Assuming that T is Lipschitz continuous in $\mathbb{B}_\infty^s(\mathbb{R}^n)$ we obtain that T is also Lipschitz continuous in $\mathbb{B}_{q_0}^{s_0}(\mathbb{R}^n)$. But we know that this is not the case. To extend this assertion from $\mathbb{B}_\infty^s(\mathbb{R}^n)$ to $\mathbb{B}_{\infty,q}^s(\mathbb{R}^n)$ with $0 < q \leq \infty$ we follow the above scheme of arguments, now based on the well-known interpolation formula

$$B_\infty^{(1-\theta)s}(\mathbb{R}^n) = \left(B_{\infty,q}^s(\mathbb{R}^n), L_\infty(\mathbb{R}^n)\right)_{\theta,\infty}, \tag{25.150}$$

where $0 < s < 1$ and $0 < \theta < 1$.

25.15 An interpolation formula

We prove (25.149) with (25.148). As usual, $\frac{1}{q} + \frac{1}{q'} = \frac{1}{q_0} + \frac{1}{q_0'} = 1$. Then we have by [Tri$\alpha$], Theorem 2.4.2(a), p. 185 (obviously extended) that

$$B_{q_0'}^{-s_0}(\mathbb{R}^n) = \left(F_{1,1}^{-s}(\mathbb{R}^n), F_{q',2}^0(\mathbb{R}^n)\right)_{\theta, q_0'} \tag{25.151}$$

with

$$1 - \theta + \frac{\theta}{q'} = 1 - \frac{\theta}{q} = \frac{1}{q_0'}. \tag{25.152}$$

By [Triβ], Theorem 2.11.2, p. 178, complemented as in [RuS96], p. 20, the dual of $F_{1,1}^{-s}(\mathbb{R}^n)$ is $B_{\infty}^s(\mathbb{R}^n)$. Hence by the duality theory of real interpolation, [Triα], 1.11.2, p. 69, we have that

$$B_{q_0}^{s_0}(\mathbb{R}^n) = \left(F_{1,1}^{-s}(\mathbb{R}^n), F_{q',2}^0(\mathbb{R}^n)\right)'_{\theta, q_0'} = \left(B_{\infty}^s(\mathbb{R}^n), L_q(\mathbb{R}^n)\right)_{\theta, q_0}. \tag{25.153}$$

This proves (25.149).

25.16 Hölder continuity

The bad continuity properties of T^+ and T as described in Theorem 25.14 can also be demonstrated by looking for best Hölder exponents, when these operators are considered as mappings between different source and target spaces. We restrict ourselves to an example. Again we put $B_p^\sigma = B_{pp}^\sigma$. By (25.6) the operators T and T^+ always have the same continuity properties. Hence we may restrict the formulation to, say, T.

25.17 Corollary

Let

$$1 \le p \le \infty, \quad 0 < s < 1 + \frac{1}{p} \quad \text{and} \quad 0 < \theta < 1. \tag{25.154}$$

Let for some α with $0 \le \alpha \le 1$ and some $c > 0$,

$\|Tf - Tg \,|\, B_p^{s(1-\theta)}(\mathbb{R}^n)\|$

$$\le c \left(\|f \,|\, B_p^s(\mathbb{R}^n)\| + \|g \,|\, B_p^s(\mathbb{R}^n)\|\right)^{1-\alpha} \|f - g \,|\, L_p(\mathbb{R}^n)\|^\alpha \tag{25.155}$$

for all $f \in B_p^s(\mathbb{R}^n)$ and all $g \in B_p^s(\mathbb{R}^n)$. Then $0 \le \alpha \le \theta$.

Proof *Step 1* First we remark that (25.155) with $\alpha = \theta$ follows from Theorem 25.8 and the well-known interpolation formula

$$B_p^{s(1-\theta)}(\mathbb{R}^n) = \left(B_p^s(\mathbb{R}^n), L_p(\mathbb{R}^n)\right)_{\theta,p}. \qquad (25.156)$$

This proves (25.155) for $0 \leq \alpha \leq \theta$.

Step 2 It remains to prove that $\alpha = \theta$ is the best possible exponent in (25.155). We use the functions F and G in (25.142). Then we have (25.144) and by the same arguments as there that

$$\|F \,|B_p^s(\mathbb{R}^n)\| \sim \|G \,|B_p^s(\mathbb{R}^n)\| \sim 2^{js}, \quad j \in \mathbb{N}. \qquad (25.157)$$

This follows from Step 1 and Step 2 of the proof of Theorem 25.14, where the restriction $s \neq \frac{1}{p}$ is not needed for this purpose. Inserting these equivalences, and (25.144) with $s(1-\theta)$ in place of s, in (25.155) we obtain that

$$\begin{aligned}
2^{js(1-\theta)} &\leq c_1 \,\|TF - TG \,|B_p^{s(1-\theta)}(\mathbb{R}^n)\| \\
&\leq c_2 \left(\|F\,|B_p^s(\mathbb{R}^n)\| + \|G\,|B_p^s(\mathbb{R}^n)\|\right)^{1-\alpha} \|F - G\,|L_p(\mathbb{R}^n)\|^\alpha \\
&\leq c_3 \, 2^{js(1-\alpha)}, \quad j \in \mathbb{N},
\end{aligned} \qquad (25.158)$$

where the positive constants c_1, c_2, c_3 are independent of $j \in \mathbb{N}$. Hence, $\alpha \leq \theta$.

26 The Q-operator

26.1 Preliminaries

As described in Section 24 it is our intention to develop a smoothness theory for some semi-linear integral equations and differential equations with prototypes (24.6), (24.7). It is natural to interpret the right hand side of (24.6) as an operator and to ask for a fixed point in a suitable (real) function space. As a minimal requirement, the operators T and T^+, given by (25.4) and (25.5), should be bounded in such function spaces. By Theorem 25.8 we have a rather satisfactory answer. Let, for example, $h \in B_{pq}^s(\mathbb{R}^n)$ in (24.6) with $\left(\frac{1}{p}, s\right) \in R_n$ according to (25.40). Then it is natural to ask for solutions $u \in B_{pq}^s(\mathbb{R}^n)$ with (24.6). However at least at first glance there are no fixed point theorems which can be applied. Since we are in $\mathbb{R}^n$ there is no compactness such that Schauder or Leray-Schauder arguments can be used. Banach's fixed point theorem (contraction mappings) in $B_{pq}^s(\mathbb{R}^n)$ does not work if $\left(\frac{1}{p}, s\right) \in R_n$ as above, since by Theorem 25.14 the operators T^+ and T have poor continuity properties. This is the point where the Q-operator, roughly described so far in (24.8), (24.9), comes in. It has much better properties than T and T^+. Under some positivity assumptions for the kernels K it produces supersolutions of (24.6) and its generalizations. Together with the supersolution technique which will be described in Section 27 one gets finally solutions of (24.6), and also of (24.7), with optimal smoothness properties. In the present section we study the operator Q. Based on quarkonial decompositions as proved in Section 2 one obtains rather easily the described assertions.

First we collect what we need. Let $n \in \mathbb{N}$ and let

$$(\mathbb{R}_+)^n = \{y \in \mathbb{R}^n : y = (y_1, \ldots, y_n) \text{ with } y_j > 0\}. \tag{26.1}$$

In specification of Definition 2.4 we assume that ψ is a non-negative C^∞ function in $\mathbb{R}^n$ with

$$supp\, \psi \subset \{y \in (\mathbb{R}_+)^n : |y| < 2^r\} \tag{26.2}$$

for some $r > 0$ and with (2.15). Otherwise the $\psi^\beta(x) = x^\beta \psi(x)$ have the same meaning as there, where $\beta \in \mathbb{N}_0^n$. By (26.2) we have now that

$$\psi^\beta(x) \geq 0 \quad \text{for all} \quad x \in \mathbb{R}^n \quad \text{and} \quad \beta \in \mathbb{N}_0^n. \tag{26.3}$$

The related $(s,p) - \beta$-quarks $(\beta qu)_{\nu m}(x)$ were introduced in Definition 2.4. In particular they are also non-negative. Let $B_{pq}^s(\mathbb{R}^n)$ and $F_{pq}^s(\mathbb{R}^n)$ be the spaces according to Definition 2.6. We are not interested in the most general cases. So we concentrate mainly on the spaces $B_{pq}^s(\mathbb{R}^n)$ and complement the results obtained by a look at the Sobolev spaces

$$H_p^s(\mathbb{R}^n) = F_{p,2}^s(\mathbb{R}^n). \tag{26.4}$$

In other words, in what follows we always assume that $n \in \mathbb{N}$ and that

$$0 < p \leq \infty, \quad 0 < q \leq \infty, \quad s > \sigma_p = n\left(\frac{1}{p} - 1\right)_+, \tag{26.5}$$

with $p < \infty$ in case of the Sobolev spaces in (26.4). By the discussion in 2.7 we have in all these spaces the quarkonial decomposition

$$f(x) = \sum_{\beta,\nu,m} \lambda_{\nu m}^\beta (\beta qu)_{\nu m}(x), \quad x \in \mathbb{R}^n, \tag{26.6}$$

which converges absolutely and (hence) unconditionally in $L_{\overline{p}}(\mathbb{R}^n)$ with $\overline{p} = \max(p,1)$. By Theorem 2.9 these spaces coincide with the usual ones. We refer also to 2.11–2.13. In particular by Corollary 2.12 there are optimal coefficients

$$\lambda_{\nu m}^\beta = \lambda_{\nu m}^\beta(f), \tag{26.7}$$

which depend linearly on f. Furthermore we may assume that the generating functions $\Psi_{\nu m}^{\beta,\varrho}$ in Corollary 2.12 are real. We justify this remark. Let φ, and hence φ_k, in 2.8 and $\varkappa_k$ in (2.58) be even (with respect to the origin). Then the inverse Fourier transforms of these functions are real, and, by (2.86) the functions $\Psi_{\nu m}^{\beta,\varrho}$ are also real. In particular, if f is a real function then its coefficients $\lambda_{\nu m}^\beta(f)$ in (26.6), (26.7) are also real. We summarize our point of view.

The even functions φ_k and $\varkappa_k$ are fixed according to 2.8 and (2.58); and the (real) functions $\Psi_{\nu m}^{\beta,\varrho}$ are calculated as in Corollary 2.12, in particular by (2.86).

Let p, q, s be given by (26.5), with $p < \infty$ in case of the Sobolev spaces (26.4). Then we have (25.1). As in (25.2) the real parts of $B_{pq}^s(\mathbb{R}^n)$ and $H_p^s(\mathbb{R}^n)$ are denoted by

$$\mathbb{B}_{pq}^s(\mathbb{R}^n) \quad \text{and} \quad \mathbb{H}_p^s(\mathbb{R}^n), \quad \text{respectively}. \tag{26.8}$$

We have optimal quarkonial decompositions (26.6) in all these spaces $B_{pq}^s(\mathbb{R}^n)$ and $H_p^s(\mathbb{R}^n)$. If, in addition, f belongs to the real parts according to (26.8), then the representation (26.6) is also real. Recall what is meant by *optimal*. Let b_{pq} be the sequence spaces introduced in (2.7) and let $\|\lambda\,|b_{pq}\|_\varrho$ be given by (2.23). Let the coefficients $\lambda_{\nu m}^\beta(f)$ be calculated as indicated above. Then

$$\|f\,|B_{pq}^s(\mathbb{R}^n)\| \sim \|\lambda(f)\,|b_{pq}\|_\varrho, \quad f \subset B_{pq}^\varepsilon(\mathbb{R}^n), \tag{26.9}$$

where the equivalence constants are independent of f. We use the abbreviations introduced in Definition 2.6; in particular $\lambda(f)$ refers to (2.22) with (26.7), and $\varrho > r$ has the same meaning as there. In case of $H_p^s(\mathbb{R}^n)$ one can argue in a similar way, where b_{pq} must be replaced by $f_{p,2}$. Hence *optimal coefficients* means that we have (26.9).

We introduce the operator Q for the full spaces $B_{pq}^s(\mathbb{R}^n)$ and $H_p^s(\mathbb{R}^n)$.

26.2 Definition

Let p, q, s be given by (26.5) (with $p < \infty$ in case of the H-spaces). Let $f \in B_{pq}^s(\mathbb{R}^n)$ or $f \in H_p^s(\mathbb{R}^n)$ and let

$$f(x) = \sum_{\beta,\nu,m} \lambda_{\nu m}^\beta(f)\,(\beta qu)_{\nu m}(x), \quad x \in \mathbb{R}^n, \tag{26.10}$$

be the above distinguished quarkonial decomposition (26.6), (26.7), (26.9), where $(\beta qu)_{\nu m}$ are $(s,p) - \beta$-quarks according to Definition 2.4, now based on (26.2). Then the operator Q is given by

$$(Qf)(x) = \sum_{\beta,\nu,m} |\lambda_{\nu m}^\beta(f)|\,(\beta qu)_{\nu m}(x), \quad x \in \mathbb{R}^n \tag{26.11}$$

26.3 Remark

First we remark that

$$Qf \in B_{pq}^s(\mathbb{R}^n) \quad \text{if} \quad f \in B_{pq}^s(\mathbb{R}^n), \qquad (26.12)$$

(similarly for $H_p^s(\mathbb{R}^n)$). This is covered by Definition 2.6, the discussion in 2.7 and the detailed arguments in Step 1 of the proof of Theorem 2.9, in particular, on the role of $\varrho > r$. Hence, the right-hand side of (26.11) converges absolutely in $L_{\bar{p}}(\mathbb{R}^n)$ with $\bar{p} = \max(p, 1)$ and, hence, unconditionally in $S'(\mathbb{R}^n)$. By (26.9) it follows that there is a constant $c > 0$ such that

$$\|Qf \,|B_{pq}^s(\mathbb{R}^n)\| \le c \,\|f \,|B_{pq}^s(\mathbb{R}^n)\| \quad \text{for all} \quad f \in B_{pq}^s(\mathbb{R}^n), \qquad (26.13)$$

(and an obvious counterpart with $H_p^s(\mathbb{R}^n)$ in place of $B_{pq}^s(\mathbb{R}^n)$). Hence, Q is a bounded operator in all these spaces. But it has much better continuity properties than the operators T and (in real spaces) T^+ considered in the previous section. Obviously, we have that

$$|f(x)| \le (Qf)(x), \quad x \in \mathbb{R}^n, \quad f \in B_{pq}^s(\mathbb{R}^n). \qquad (26.14)$$

26.4 Theorem

Let $n \in \mathbb{N}$ and let p, q, s be given by (26.5). Let Q be the operator according to Definition 26.2.

(i) *There is a constant $c > 0$ such that*

$$\|Qf - Qg \,|B_{pq}^s(\mathbb{R}^n)\| \le c \,\|f - g \,|B_{pq}^s(\mathbb{R}^n)\| \qquad (26.15)$$

for all $f \in B_{pq}^s(\mathbb{R}^n)$ and all $g \in B_{pq}^s(\mathbb{R}^n)$ (Lipschitz continuity).

(ii) *Let, in addition, $p < \infty$. Then there is a constant $c > 0$ such that*

$$\|Qf - Qg \,|H_p^s(\mathbb{R}^n)\| \le c \,\|f - g \,|H_p^s(\mathbb{R}^n)\| \qquad (26.16)$$

for all $f \in H_p^s(\mathbb{R}^n)$ and $g \in H_p^s(\mathbb{R}^n)$.

Proof After the above preparations the proof is simple. We prove part (i). By (26.10), (26.11), and the obvious counterparts with g in place of f, we have that

$$Qf - Qg = \sum_{\beta, \nu, m} \left(|\lambda_{\nu m}^\beta(f)| - |\lambda_{\nu m}^\beta(g)| \right) (\beta q u)_{\nu m}(x). \qquad (26.17)$$

This is an admitted quarkonial decomposition of $Qf - Qg$ in $B^s_{pq}(\mathbb{R}^n)$ according to Definition 2.6 and the above considerations. Using the linearity of the distinguished coefficients $\lambda^\beta_{\nu m}(f)$ we have that

$$\left| |\lambda^\beta_{\nu m}(f)| - |\lambda^\beta_{\nu m}(g)| \right| \leq \left| \lambda^\beta_{\nu m}(f) - \lambda^\beta_{\nu m}(g) \right| = \left| \lambda^\beta_{\nu m}(f-g) \right|. \tag{26.18}$$

By (26.9) and the discussion there the sequence $\lambda(f-g)$ is optimal. Then it follows that

$$\|Qf - Qg \,|B^s_{pq}(\mathbb{R}^n)\| \leq c \, \|f - g \,|B^s_{pq}(\mathbb{R}^n)\| \tag{26.19}$$

for some $c > 0$ and all $f \in B^s_{pq}(\mathbb{R}^n)$ and $g \in B^s_{pq}(\mathbb{R}^n)$. This proves (26.15). The proof of (26.16) is the same.

26.5 Remark

Let $\left(\frac{1}{p}, s\right) \in R_n$ be a truncation couple according to Definition 25.6 and (25.40). Let $0 < q \leq \infty$. Then (26.14) can be complemented by

$$(T^+ f)(x) \leq (Tf)(x) \leq (Qf)(x), \quad x \in \mathbb{R}^n, \quad f \in B^s_{pq}(\mathbb{R}^n). \tag{26.20}$$

By the above theorem and Theorem 25.14 the continuity properties of Q are much better than the continuity properties of T and T^+. This will be the decisive observation in the applications given in the following section. We followed [Tri01].

27 Semi-linear equations; the Q-method

27.1 Semi-linear integral equations

Let $K \in L_1(\mathbb{R}^n)$ be real and let $h \in L_p(\mathbb{R}^n)$ be real, where $1 \leq p \leq \infty$. Then one can apply Banach's contraction theorem to find a real (unique) solution $u \in L_p(\mathbb{R}^n)$ of

$$u(x) = \int_{\mathbb{R}^n} K(y) \, u_+(x-y) \, dy + h(x), \quad x \in \mathbb{R}^n, \tag{27.1}$$

provided that $\|K \,|L_1(\mathbb{R}^n)\|$ is sufficiently small. This follows from the obvious fact that the truncation operator T^+, given by (25.5), is Lipschitz continuous in $L_p(\mathbb{R}^n)$. But what can be said if h belongs to some real spaces as, for example, in (26.8)? Since T^+ is involved the assumption that $\left(\frac{1}{p}, s\right) \in R_n$ is a truncation couple according to Definition 25.6 and Theorem 25.8 seems to be natural, or at least reasonable. If one has no further information for the kernel K than $K \in L_1(\mathbb{R}^n)$, then one must restrict, in addition, the considerations to Banach spaces. What remains is the shaded region in Fig. 27.1. By Theorem 25.8 the operator T^+ is bounded in, say, $\mathbb{B}^s_{pq}(\mathbb{R}^n)$ if $\left(\frac{1}{p}, s\right)$ belongs to the shaded region. But by Theorem 25.14 the operator T^+ is not Lipschitz continuous. Hence there is no direct way, based on Banach's contraction mapping theorem, to deal with equations of type (27.1) in some spaces $\mathbb{B}^s_{pq}(\mathbb{R}^n)$. It is the main aim of this section to describe the Q-method which circumvents these difficulties. Then the kernels involved must be, in addition, non-negative. Otherwise one combines the Q-operator considered in Section 26 with supersolution techniques. Although we are more interested in the method and not in the most general applications, we wish to make clear that the convolution structure of (27.1) is not needed. We deal with the more general semi-linear integral equation

$$u(x) = \int_{\mathbb{R}^n} K(x,y)\, u_+(x-y)\, dy + h(x), \quad x \in \mathbb{R}^n, \qquad (27.2)$$

where the non-negative kernels $K(x,y)$ belong to some hybrid spaces of type $L_1(\mathcal{C}^\sigma)(\mathbb{R}^{2n})$. Recall that

$$\mathcal{C}^\sigma(\mathbb{R}^n) = B^\sigma_\infty(\mathbb{R}^n) = B^\sigma_{\infty,\infty}(\mathbb{R}^n), \quad \sigma > 0, \qquad (27.3)$$

are the classical Hölder-Zygmund spaces.

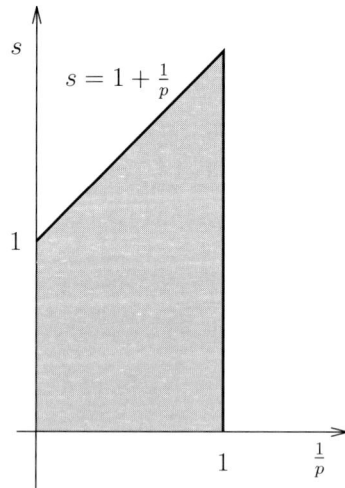

Fig. 27.1

27.2 Definition

Let $\sigma > 0$. Then $L_1(\mathcal{C}^\sigma)(\mathbb{R}^{2n})$ is the collection of all (complex-valued) functions $f(x, y)$ with $x \in \mathbb{R}^n$ and $y \in \mathbb{R}^n$ such that

$$\|f \,|\, L_1(\mathcal{C}^\sigma)(\mathbb{R}^{2n})\| = \int_{\mathbb{R}^n} \|f(\cdot, y) \,|\, \mathcal{C}^\sigma(\mathbb{R}^n)\| \, dy < \infty. \tag{27.4}$$

27.3 Remark

Of course, $L_1(\mathcal{C}^\sigma)(\mathbb{R}^{2n})$ is a Banach space. For any fixed $y \in \mathbb{R}^n$ the $\mathcal{C}^\sigma$-norm in (27.4) is applied to $x \mapsto f(x, y)$. If $K(y) \in L_1(\mathbb{R}^n)$ then we put $K(x, y) = K(y)$ and have

$$K \in L_1(\mathcal{C}^\sigma)(\mathbb{R}^{2n}) \quad \text{for any} \quad \sigma > 0.$$

Recall that the real spaces $\mathbb{B}_{pq}^s(\mathbb{R}^n)$ and $\mathbb{H}_p^s(\mathbb{R}^n)$ in (26.8) are specifications of (25.1), (25.2).

27.4 Theorem

Let $n \in \mathbb{N}$,

$$1 \leq p \leq \infty, \quad 1 \leq q \leq \infty, \quad 0 < s < 1 + \frac{1}{p}, \tag{27.5}$$

(the shaded region in Fig. 27.1). Let $\sigma > s$ and let

$$K(x, y) \geq 0 \quad \text{in} \quad \mathbb{R}^n \times \mathbb{R}^n \quad \text{with} \quad K \in L_1(\mathcal{C}^\sigma)(\mathbb{R}^{2n}).$$

Then there is a positive number ε with the following property.

(i) If $\|K \,|\, L_1(\mathcal{C}^\sigma)(\mathbb{R}^{2n})\| \leq \varepsilon$, then for any $h \in \mathbb{B}_{pq}^s(\mathbb{R}^n)$ the equation (27.2) has a uniquely determined solution $u \in \mathbb{B}_{pq}^s(\mathbb{R}^n)$.

(ii) Let, in addition, $p < \infty$ and $\left(\frac{1}{p}, s\right) \neq (1, 1)$. If $\|K \,|\, L_1(\mathcal{C}^\sigma)(\mathbb{R}^{2n})\| \leq \varepsilon$, then for any $h \in \mathbb{H}_p^s(\mathbb{R}^n)$ the equation (27.2) has a uniquely determined solution $u \in \mathbb{H}_p^s(\mathbb{R}^n)$.

Proof In the first (and main) step we prove part (i) for the model equation (27.1). Here we present what we wish to call the Q-method. In Steps 2 and 3 we extend this assertion from B-spaces to H-spaces and from (27.1) to (27.2).

Step 1 We prove part (i) for the model equation (27.1). Let $h \in \mathbb{B}_{pq}^s(\mathbb{R}^n)$ with (27.5). By the interpretation given in 27.3 we must prove that (27.1) has a unique solution $u \in \mathbb{B}_{pq}^s(\mathbb{R}^n)$ under the assumptions that $\|K \,|\, L_1(\mathbb{R}^n)\|$ is small and $K(y) \geq 0$. Let T^+ and Q be the operators introduced in (25.5) and in Definition 26.2, respectively. Let B and B^Q be given by

$$(Bu)(x) = \int_{\mathbb{R}^n} K(y)\,(T^+u)(x-y)\,dy + h(x)\,, \quad x \in \mathbb{R}^n\,, \tag{27.6}$$

and

$$(B^Q u)(x) = \int_{\mathbb{R}^n} K(y)\,(Qu)(x-y)\,dy + h(x)\,, \quad x \in \mathbb{R}^n\,. \tag{27.7}$$

By (27.5) we can apply Theorem 25.8(i) to T^+ and Theorem 26.4 to Q. Using the triangle inequality for Banach spaces we obtain that

$$\|Bu\,|B^s_{pq}(\mathbb{R}^n)\| \leq c_1\,\|K\,|L_1(\mathbb{R}^n)\|\,\|u\,|B^s_{pq}(\mathbb{R}^n)\| + \|h\,|B^s_{pq}(\mathbb{R}^n)\|\,, \tag{27.8}$$

$$\|B^Q u\,|B^s_{pq}(\mathbb{R}^n)\| \leq c_2\,\|K\,|L_1(\mathbb{R}^n)\|\,\|u\,|B^s_{pq}(\mathbb{R}^n)\| + \|h\,|B^s_{pq}(\mathbb{R}^n)\|\,, \tag{27.9}$$

and

$$\|B^Q u - B^Q v\,|B^s_{pq}(\mathbb{R}^n)\| \leq c_3\,\|K\,|L_1(\mathbb{R}^n)\|\,\|u-v\,|B^s_{pq}(\mathbb{R}^n)\|\,, \tag{27.10}$$

where c_1, c_2, c_3 are positive constants. Let

$$c_3\,\|K\,|L_1(\mathbb{R}^n)\| < 1 \tag{27.11}$$

in (27.10). Then B^Q is a contraction in $B^s_{pq}(\mathbb{R}^n)$ and hence also in $\mathbb{B}^s_{pq}(\mathbb{R}^n)$. We use Banach's contraction mapping theorem. It goes back to S. Banach, [Ban22], and may be found in any book on non-linear functional analysis or fixed-point theorems. We refer, for example, to [Sma74], p. 2. Hence, under the assumption (27.11) the equation (27.7) has for given $h \in \mathbb{B}^s_{pq}(\mathbb{R}^n)$ a uniquely determined solution $u^0(x) \in \mathbb{B}^s_{pq}(\mathbb{R}^n)$,

$$u^0(x) = \int_{\mathbb{R}^n} K(y)\,(Qu^0)(x-y)\,dy + h(x)\,, \quad x \in \mathbb{R}^n\,. \tag{27.12}$$

Let $j \in \mathbb{N}$ and let, by iteration,

$$u^j(x) = (Bu^{j-1})(x) = \int_{\mathbb{R}^n} K(y)\,(T^+ u^{j-1})(x-y)\,dy + h(x)\,. \tag{27.13}$$

Again by Theorem 25.8 we have $u^j(x) \in \mathbb{B}^s_{pq}(\mathbb{R}^n)$. Since $K(y) \geq 0$ it follows by (26.20) that $u^0(x)$ is a *supersolution* for B,

$$u^1(x) = \int_{\mathbb{R}^n} K(y)\,(T^+ u^0)(x-y)\,dy + h(x) \leq u^0(x)\,. \tag{27.14}$$

27. Semi-linear equations; the Q-method

We obtain by iteration a monotonically decreasing sequence,

$$h(x) \le u^{j+1}(x) \le u^j(x) \le \cdots \le u^0(x), \quad x \in \mathbb{R}^n, \quad j \subset \mathbb{N}. \qquad (27.15)$$

Since both $h \in L_p(\mathbb{R}^n)$ and $u^0 \in L_p(\mathbb{R}^n)$ it follows by (27.15) and Lebesgue's bounded convergence theorem that

$$u^j(x) \to u(x) \quad \text{in} \quad L_p(\mathbb{R}^n). \qquad (27.16)$$

We wish to prove that $u \in \mathbb{B}^s_{pq}(\mathbb{R}^n)$ and that u is a solution of (27.1). By (27.8), (27.9) and (27.12), (27.13) we have for sufficiently small $\|K\,|L_1(\mathbb{R}^n)\|$ that

$$\|u^0\,|B^s_{pq}(\mathbb{R}^n)\| \le 2\,\|h\,|B^s_{pq}(\mathbb{R}^n)\|, \qquad (27.17)$$

$$\|u^j\,|B^s_{pq}(\mathbb{R}^n)\| \le \frac{1}{2}\|u^{j-1}\,|B^s_{pq}(\mathbb{R}^n)\| + \|h\,|B^s_{pq}(\mathbb{R}^n)\|, \quad j \in \mathbb{N}, \qquad (27.18)$$

and by iteration

$$\|u^j\,|B^s_{pq}(\mathbb{R}^n)\| \le 2\,\|h\,|B^s_{pq}(\mathbb{R}^n)\|, \quad j \in \mathbb{N}. \qquad (27.19)$$

Since $\{u^j\}$ is uniformly bounded in $B^s_{pq}(\mathbb{R}^n)$ and $u^j \to u$ in $S'(\mathbb{R}^n)$ we can apply the Fatou property of $B^s_{pq}(\mathbb{R}^n)$ as described in 25.1. Hence, $u \in B^s_{pq}(\mathbb{R}^n)$. Obviously, u is real. Then it follows that $u \in \mathbb{B}^s_{pq}(\mathbb{R}^n)$. By (27.15) and the monotone pointwise convergence in (27.13) (which can be assumed) we obtain that

$$u(x) = \int_{\mathbb{R}^n} K(y)\,(T^+u)(x-y)\,dy + h(x), \quad x \in \mathbb{R}^n, \qquad (27.20)$$

is the solution we are looking for. The uniqueness in $B^s_{pq}(\mathbb{R}^n)$ follows from the uniqueness in $L_p(\mathbb{R}^n)$, where Banach's contraction mapping theorem is available.

Step 2 The proof of part (ii) for the model equation (27.1) is the same. Now we rely on Theorem 25.8(ii) and on Theorem 26.4. This explains also the curious restriction $\left(\frac{1}{p}, s\right) \ne (1,1)$ which we could not remove.

Step 3 Now we prove the theorem for the general equation (27.2). We replace $K(y)$ in (27.6), (27.7) by $K(x,y)$. Then we obtain that

$$\|Bu\,|B^s_{pq}(\mathbb{R}^n)\| \le \int_{\mathbb{R}^n} \|K(\cdot,y)\,(T^+u)(\cdot-y)\,|B^s_{pq}(\mathbb{R}^n)\|\,dy + \|h\,|B^s_{pq}(\mathbb{R}^n)\|. \qquad (27.21)$$

For fixed $y \in \mathbb{R}^n$ the function $x \mapsto K(x,y)$ is a pointwise multiplier in $B^s_{pq}(\mathbb{R}^n)$ and we have by [Triγ], Corollary on p. 205, that

$$\begin{aligned}\|K(\cdot,y)\,(T^+u)(\cdot-y)\,|B^s_{pq}(\mathbb{R}^n)\| &\\
&\leq c\,\|K(\cdot,y)\,|\mathcal{C}^\sigma(\mathbb{R}^n)\|\,\|T^+u\,|B^s_{pq}(\mathbb{R}^n)\| \qquad (27.22)\\
&\leq c'\,\|K(\cdot,y)\,|\mathcal{C}^\sigma(\mathbb{R}^n)\|\,\|u\,|B^s_{pq}(\mathbb{R}^n)\|\,.\end{aligned}$$

Inserting this estimate in (27.21) we obtain the counterpart of (27.8),

$$\|Bu\,|B^s_{pq}(\mathbb{R}^n)\| \leq c\,\|K\,|L_1(\mathcal{C}^\sigma)(\mathbb{R}^{2n})\|\,\|u\,|B^s_{pq}(\mathbb{R}^n)\| + \|h\,|B^s_{pq}(\mathbb{R}^n)\|\,. \quad (27.23)$$

Obviously there are similar counterparts of (27.9), (27.10). Afterwards one can follow the arguments given in Step 1 and Step 2.

27.5 Semi-linear differential equations

Again let $-\Delta = -\sum_{j=1}^n \frac{\partial^2}{\partial x_j^2}$ be the Laplacian in $\mathbb{R}^n$ and let id be the identity. Let $\varepsilon > 0$ and, say, $h \in \mathbb{B}^s_{pq}(\mathbb{R}^n)$. We ask for solutions $u(x)$ of

$$(-\Delta + id)\,u(x) = \varepsilon\,u_+(x) + h(x)\,, \qquad x \in \mathbb{R}^n\,. \quad (27.24)$$

Now we prefer (27.24) instead of (24.7). But this is unimportant. By (24.4) one can reduce (24.7) essentially to (27.24). Then we have $-\Delta + \lambda\,id$ with some $\lambda > 0$ on the left-hand side of (27.24). But this does not influence the arguments. Let $G(x)$ be the Green's function of $-\Delta + id$. Then (27.24) can be reformulated as

$$u(x) = \varepsilon \int_{\mathbb{R}^n} G(y)\,u_+(x-y)\,dy + H(x)\,, \qquad x \in \mathbb{R}^n\,, \quad (27.25)$$

where

$$H(x) = (-\Delta + id)^{-1}h \in \mathbb{B}^{s+2}_{pq}(\mathbb{R}^n)\,. \quad (27.26)$$

By the well-known properties of $G(y)$ which may be found in 17.1, we have that

$$G(y) > 0 \quad \text{with} \quad y \in \mathbb{R}^n \quad \text{and} \quad G \in L_1(\mathbb{R}^n)\,. \quad (27.27)$$

Hence (27.25) is a special case of (24.6) and one can apply Theorem 27.4. But the situation is now much better. In particular under the restrictions (27.5) there is no need to apply the Q-method. To discuss this point we assume temporarily that

$$1 < p < \infty\,, \quad 0 < s < 1 + \frac{1}{p}\,, \quad \varepsilon > 0 \quad \text{small}\,, \quad (27.28)$$

27. Semi-linear equations; the Q-method

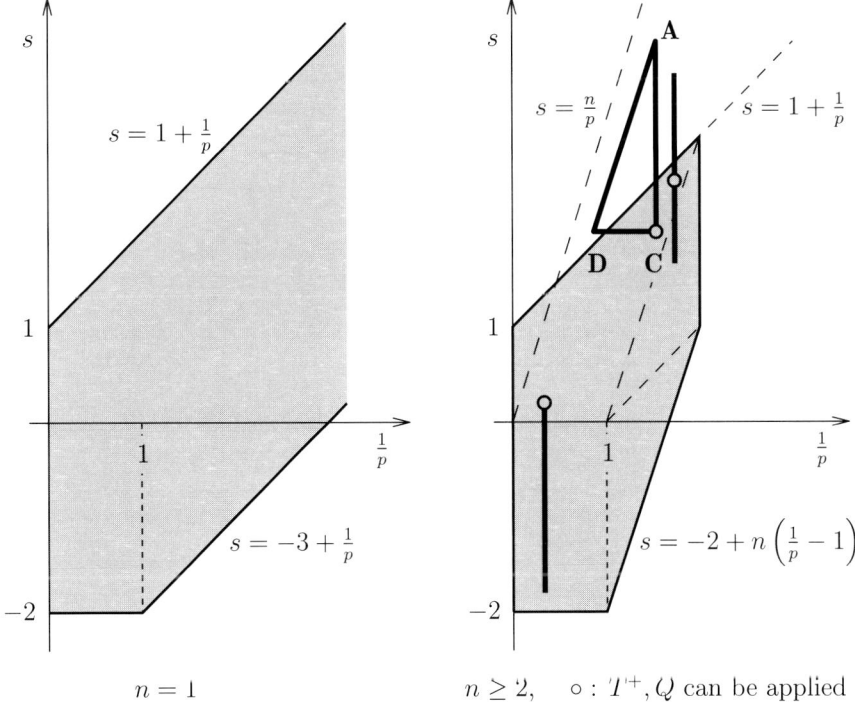

Fig. 27.2

in (27.24) and in (27.25). Then one can apply Banach's contraction mapping theorem in $L_p(\mathbb{R}^n)$. By the well-known lifting properties of $(-\Delta + id)^{-1}$ one obtains $u \in W_p^2(\mathbb{R}^n)$ for the uniquely determined solution of (27.24) in $L_p(\mathbb{R}^n)$. By (27.28) and Theorem 25.8 we have $u_+ \in \mathbb{B}_{pq}^s(\mathbb{R}^n)$ and, hence, by (27.25), (27.26) that

$$u \in \mathbb{B}_{pq}^{s+2}(\mathbb{R}^n). \tag{27.29}$$

This is the maximal smoothness which can be expected. One can extend this argument to $p = 1$ and $p = \infty$, but not to $p < 1$. There is also a difference between the indicated problems in $\mathbb{R}^n$ and corresponding (boundary value) problems of type (27.24) in bounded (smooth) domains Ω. In the latter case one has not only bootstrapping arguments at hand but one can also step from spaces, say, $\mathbb{B}_{p_1,q}^s(\Omega)$ with $p_1 > 1$ to spaces $\mathbb{B}_{p_2,q}^s(\Omega)$ with $p_2 < 1$, since these spaces are monotonically included. Hence for bounded domains any point $\left(\frac{1}{p}, s\right) \subset R_n$ according to (25.40) can be reached in this way starting from

$L_r(\Omega)$ with $1 < r \le \infty$. The situation in $\mathbb{R}^n$ is different. By Theorem 25.8 and Fig. 24.1 we have the natural region R_n where T^+ and, by Theorem 26.4, also Q can be applied. Furthermore there is the indicated lifting property of $(-\Delta + id)^{-1}$ improving the smoothness by 2. Hence we get the natural

$$\text{region of admissibility} \quad R_n(T^+, -\Delta + id)$$

in Fig. 27.2 for the problem (27.24) given by

$$R_n(T^+, -\Delta + id) = \bigcup_{0 \le \lambda \le 2} \left\{ \left(\frac{1}{p}, s\right) : \left(\frac{1}{p}, s+\lambda\right) \in R_n \right\}. \tag{27.30}$$

To avoid a wrong impression we add the following remark. If one replaces $-\Delta + id$ in (27.24) by a somewhat more complicated operator, for example, $a(x)(-\Delta + id)$, where $a(x)$ is a positive function with some singularities, then bootstrapping arguments do not work any longer. We return to this point in the later considerations and refer in particular to 27.10. In other words, for semi-linear differential equations of slightly more complicated type than (27.24) bootstrapping is of no use.

27.6 Theorem

Let $n \in \mathbb{N}$ and

$$0 < p \le \infty, \quad 0 < q \le \infty, \quad n\left(\frac{1}{p} - 1\right)_+ < s + \lambda < 1 + \frac{1}{p}, \tag{27.31}$$

for some $\lambda \in [0, 2]$ (the shaded region in Fig. 27.2). Then there is a number $\varepsilon_0 > 0$ with the following property.

(i) If $0 < \varepsilon \le \varepsilon_0$, then for any $h \in \mathbb{B}_{pq}^s(\mathbb{R}^n)$ the equation (27.24) has a uniquely determined solution $u \in \mathbb{B}_{pq}^{s+2}(\mathbb{R}^n)$.

(ii) Let, in addition, $p < \infty$. If $0 < \varepsilon \le \varepsilon_0$, then for any $h \in \mathbb{H}_p^s(\mathbb{R}^n)$ the equation (27.24) has a uniquely determined solution $\mathbb{H}_p^{s+2}(\mathbb{R}^n)$.

Proof *Step 1* We prove part (i). We modify Step 1 of the proof of Theorem 27.4. The functions h and H have the same meaning as in (27.24)–(27.26). The counterparts of B and B^Q in (27.6) and (27.7) are given by

$$(Bu)(x) = \varepsilon(-\Delta + id)^{-1} \circ T^+ u(x) + H(x) \tag{27.32}$$

$$= \varepsilon \int_{\mathbb{R}^n} G(y)(T^+ u)(x-y)\, dy + H(x), \quad x \in \mathbb{R}^n,$$

and

$$(B^Q u)(x) - \varepsilon(-\Delta + id)^{-1} \circ Qu(x) + H(x) \qquad (27.33)$$
$$= \varepsilon \int_{\mathbb{R}^n} G(y)\,(Qu)(x-y)\,dy + H(x), \quad x \in \mathbb{R}^n.$$

Recall that $G(y) > 0$ if $y \in \mathbb{R}^n$ and $G \in L_1(\mathbb{R}^n)$. Let $\lambda \in [0,2]$ be chosen such that (27.31) is satisfied. This corresponds to ○ on the line segment in Fig. 27.2. Then both T^+ and Q can be applied in $\mathbb{B}_{pq}^{s+\lambda}(\mathbb{R}^n)$. We consider B and B^Q in $\mathbb{B}_{pq}^{s+\lambda}(\mathbb{R}^n)$ and obtain the counterpart of (27.10),

$$\|B^Q u - B^Q v\,|B_{pq}^{s+\lambda}(\mathbb{R}^n)\| \le c\varepsilon\,\|u-v\,|B_{pq}^{s+\lambda}(\mathbb{R}^n)\|, \qquad (27.34)$$

and corresponding counterparts of (27.8), (27.9) with H in place of h. Now one can proceed as in Step 1 of the proof of Theorem 27.4 with the following modifications. First we remark that $\mathbb{B}_{pq}^{s+\lambda}(\mathbb{R}^n)$ is a complete metric space with the metric ϱ,

$$\varrho(f,g) = \|f - g\,|B_{pq}^{s+\lambda}(\mathbb{R}^n)\|^{\varkappa}, \quad \varkappa = \min(1,p,q). \qquad (27.35)$$

Hence Banach's contraction mapping theorem can be applied, [Sma74], p. 2. Secondly, using well-known embedding theorems, one must modify (27.16) in case of $p < 1$ by

$$u^j(x) \to u(x) \quad \text{in} \quad L_r(\mathbb{R}^n) \quad \text{with} \quad 1 \le r < \infty, \quad s+\lambda - \frac{n}{p} > -\frac{n}{r}. \qquad (27.36)$$

With the modifications indicated one can follow Step 1 of the proof of Theorem 27.4. Hence there is a (uniquely determined) solution $u \in \mathbb{B}_{pq}^{s+\lambda}(\mathbb{R}^n)$ of

$$u(x) = \varepsilon(-\Delta + id)^{-1} \circ T^+ u(x) + H(x). \qquad (27.37)$$

The first term on the right-hand side of (27.37) belongs even to $\mathbb{B}_{pq}^{s+\lambda+2}(\mathbb{R}^n)$. Since $H \in \mathbb{B}_{pq}^{s+2}(\mathbb{R}^n)$ we get finally $u \in \mathbb{B}_{pq}^{s+2}(\mathbb{R}^n)$.

Step 2 The proof of part (ii) is the same. Now we can choose $\lambda \in [0,2]$ in such a way that $s + \lambda \ne \frac{1}{p}$ avoiding the exceptional cases in Theorem 25.8 concerning the boundedness of T^+.

27.7 Two comments

We are more interested in presenting the Q-method than in the most general applications. This applies both to Theorem 27.4 and, to an even larger extent, to Theorem 27.6. In connection with (27.24) we relied on the properties (27.27) for its Green's function. Here $G \in L_1(\mathbb{R}^n)$ follows from the exponential decay of $G(y)$ if $|y| \to \infty$. But these properties of Green's functions, positivity and exponential decay, are well known for a large class of second order elliptic differential operators in $\mathbb{R}^n$. All that one needs may be found in [Dav89]. This applies also to some functions of these operators, in particular to fractional powers. In other words there is a good chance to replace $-\Delta + id$ in (27.24) by some classes of elliptic pseudodifferential operators.

Secondly, we add a comment on the supersolution technique. It goes back to O. Perron, [Per23], in connection with boundary value problems for the Laplacian in bounded domains in $\mathbb{R}^n$. This method was used later on, especially in the 1970s and 1980s, in connection with nonlinear equations of type

$$Lu = f(x, u) \quad \text{and} \quad Lu = f(x, u, \nabla u), \qquad (27.38)$$

where L is a second-order linear elliptic differential operator. The underlying function spaces are preferably Hölder spaces, but occasionally also classical Sobolev spaces. In contrast to our very special nonlinearities $f(u) = |u|$ and $f(u) = u_+$, the admitted nonlinearities in (27.38) are described in qualitative terms, by smoothness and growth conditions. Details of the use of the supersolution and subsolution technique in the just-outlined context may be found in [Ama76], [AmC78], [KaK78], [FKY87], to mention only a few typical papers. At this moment it is not so clear whether the techniques developed in these and related papers can be combined with the methods presented in this chapter. But it might be worth consideration. As an indispensable ingredient one needs sharp mapping properties of composition operators of type $u \mapsto f(x, u)$ in function spaces of type B^s_{pq} and F^s_{pq}. We refer to [RuS96] where one finds far-reaching assertions in this direction.

27.8 Local singularities

As just said there might be many possibilities to apply the Q-method. We have a closer look at how to incorporate local singularities in the right-hand side of (27.24), which then looks like

$$(-\Delta + id)\, u(x) = \varepsilon\, (1 + b(x))\, u_+(x) + h(x), \quad x \in \mathbb{R}^n, \qquad (27.39)$$

where $b(x) \geq 0$. To find out which type of singularities can be admitted one may consult Fig. 27.2. In the proof of Theorem 27.6 we used suitable points on the vertical line segments of length 2 characterized by ○, say the point **C** representing for example $H^s_p(\mathbb{R}^n)$. However, instead of going down from **A** to **C** on the vertical line, one may use sharp embeddings from **A** to **D** and afterward Hölder inequalities at level s to step from **D** to **C**. This is the point where b

27. Semi-linear equations; the Q-method

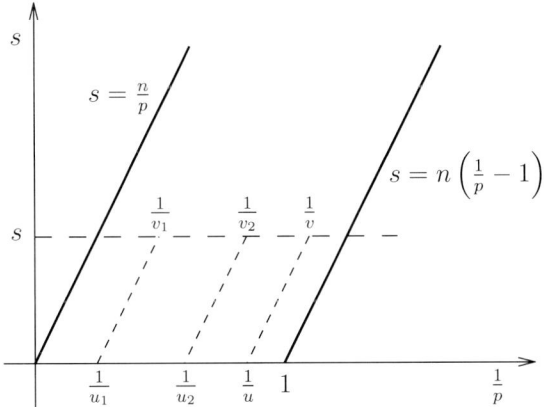

Fig. 27.3

comes in. The underlying theory has been developed in [SiT95] and may also be found in [ET96], 2.4 and [RuS96], 4.8.2, pp. 238–239. We describe a special case given in [ET96], p. 54, formula (17). Let

$$1 < u_1 < \infty, \quad 1 < u_2 < \infty, \quad \frac{1}{u} = \frac{1}{u_1} + \frac{1}{u_2} < 1, \quad s > 0, \qquad (27.40)$$

and

$$\frac{1}{v_1} = \frac{1}{u_1} + \frac{s}{n}, \quad \frac{1}{v_2} = \frac{1}{u_2} + \frac{s}{n}, \quad \frac{1}{v} = \frac{1}{u} + \frac{s}{n}, \qquad (27.41)$$

as indicated in Fig. 27.3. Then we have that

$$H^s_{v_1}(\mathbb{R}^n) \cdot H^s_{v_2}(\mathbb{R}^n) \subset H^s_v(\mathbb{R}^n). \qquad (27.42)$$

Hence, the classical Hölder inequality

$$L_{u_1}(\mathbb{R}^n) \cdot L_{u_2}(\mathbb{R}^n) \subset L_u(\mathbb{R}^n)$$

is shifted along lines of slope n at the level s. In our case we have the triangle **ADC** in Fig. 27.2 with the base line **DC** of length $\frac{2}{n}$. Let $n \geq 3$. Then $\frac{2}{n} < 1$ and it makes sense to assume that the triangle **ADC** is located as seen in Fig. 27.2 between the two lines $s = \frac{n}{p}$ and $s = n\left(\frac{1}{p} - 1\right)$, which appear also in Fig. 27.3. Let

$$\mathbf{C} \sim H^s_p(\mathbb{R}^n), \quad \text{then} \quad \mathbf{A} \sim H^{s+2}_p(\mathbb{R}^n) \quad \text{and} \quad \mathbf{D} \sim H^s_r(\mathbb{R}^n) \qquad (27.43)$$

with $\frac{1}{r} = \frac{1}{p} - \frac{2}{n}$. To be sure that not only **C** but also **A** and **D** are points within the distinguished strip, we must assume in addition that

$$\frac{1}{q} = \frac{2}{n} + \frac{s}{n} = \frac{s+2}{n} < \frac{1}{p}. \tag{27.44}$$

By (27.42) and (27.44) we have that

$$H_r^s(\mathbb{R}^n) \cdot H_q^s(\mathbb{R}^n) \subset H_p^s(\mathbb{R}^n). \tag{27.45}$$

Hence,

$$b \in H_q^s(\mathbb{R}^n) \quad \text{with} \quad \frac{1}{q} = \frac{2}{n} + \frac{s}{n} \tag{27.46}$$

is a reasonable condition for b we are looking for. To get a feeling for what type of singularities can be included we remark that

$$H_q^s(\mathbb{R}^n) \subset L_{\frac{n}{2}}(\mathbb{R}^n). \tag{27.47}$$

This is well known and also covered by Theorem 11.4. Let ψ be a non-negative C^∞ function in $\mathbb{R}^n$ with compact support and $\psi(0) > 0$. Then we have that

$$b(x) = |x|^{-2} \, |\log|x||^{-\varkappa} \, \psi(x) \in H_q^s(\mathbb{R}^n), \quad 0 < s < \infty, \quad \frac{1}{q} = \frac{2+s}{n}, \tag{27.48}$$

if, and only if, $\varkappa > \frac{1}{q}$. We give a short proof. Using (27.44) we have by Theorem 15.2 for the corresponding growth envelope that

$$\mathfrak{E}_G H_q^s = \left(t^{-\frac{2}{n}}, q\right), \tag{27.49}$$

and, hence by (15.9),

$$\left(\int_0^\varepsilon \left(t^{\frac{2}{n}} b^*(t)\right)^q \frac{dt}{t}\right)^{\frac{1}{q}} \leq c \, \|b \,|\, H_q^s(\mathbb{R}^n)\|. \tag{27.50}$$

One may assume that $\psi(x)$ in (27.48) is identically 1 near the origin. Then it follows that

$$b^*(t) \sim t^{-\frac{2}{n}} |\log t|^{-\varkappa}, \quad 0 < t < \varepsilon. \tag{27.51}$$

If $\varkappa \leq \frac{1}{q}$ then the integral on the left-hand side of (27.50) diverges and b does not belong to $H_q^s(\mathbb{R}^n)$. Let $\varkappa > \frac{1}{q}$. Let $B_{j,l}$ with $j = J, J+1, \ldots$, and $l = 1, \ldots, L$, be balls of radius $c2^{-j}$ for some $c > 0$ and

$$B_{j,l} \subset \{y : \ 2^{-j-1} \leq |y| \leq 2^{-j+1}\}. \tag{27.52}$$

If properly located and if $\{\psi_{jl}\}$ is a suitable subordinated resolution of unity with the usual properties, then

$$b(x) = \sum_{j=J}^{\infty} \sum_{l=1}^{L} j^{-\varkappa} \left[j^{\varkappa} |x|^{-2} |\log |x||^{-\varkappa} \psi_{j,l}(x) \right], \quad |x| < \varepsilon, \qquad (27.53)$$

is an atomic decomposition in $H_q^s(\mathbb{R}^n)$ of $b(x)$ near the origin. This follows from $s - \frac{n}{q} = -2$, Theorem 13.8 in [Triδ], p. 75, and, with $\lambda_{jl} = j^{-\varkappa}$,

$$\|\lambda\,|f_{q,2}\| \sim \left(\sum_{j=J}^{\infty} j^{-\varkappa q} \right)^{\frac{1}{q}} < \infty. \qquad (27.54)$$

This makes clear which type of singularities of $b(x)$ in (27.39) can be expected.

27.9 Corollary

Let $n \geq 3$,

$$0 < p < \infty, \quad n\left(\frac{1}{p} - 1\right)_+ < s < 1 + \frac{1}{p}, \quad p < \frac{n}{s+2} = q, \qquad (27.55)$$

and $s \neq \frac{1}{p}$ if $p \leq 1$. Let

$$b(x) \geq 0 \quad \text{and} \quad b \in H_q^s(\mathbb{R}^n). \qquad (27.56)$$

Then there is a number $\varepsilon_0 > 0$ with the following property. If $0 < \varepsilon \leq \varepsilon_0$, then for any $h \in \mathbb{H}_p^s(\mathbb{R}^n)$ the equation (27.39) has a uniquely determined solution $u \in \mathbb{H}_p^{s+2}(\mathbb{R}^n)$.

Proof Since $b(x) \geq 0$, equation (27.39) can be rewritten as

$$(-\Delta + id)\,u(x) = \varepsilon \left(T^+(1+b)u \right)(x) + h(x), \quad x \in \mathbb{R}^n. \qquad (27.57)$$

Furthermore by the discussion in 27.8, in particular by (27.43), (27.44), the triangle **ADC** is located as indicated in Fig 27.2 and as discussed there. The counterpart of the operator B in (27.32) is now given by

$$(Bu)(x) = (B_1 u)(x) + (B_b u)(x) + H(x), \quad x \in \mathbb{R}^n, \qquad (27.58)$$

with

$$\begin{aligned}(B_1 u)(x) &= \varepsilon\,(-\Delta + id)^{-1} \circ T^+ u(x), & (27.59)\\ (B_b u)(x) &= \varepsilon\,(-\Delta + id)^{-1} \circ (T^+ b) u(x). & (27.60)\end{aligned}$$

One has similar counterparts for the operator B^Q in (27.33), where one must replace T^+ in (27.59), (27.60) by Q. The operators B_1 and B_b (and their counterparts B_1^Q and B_b^Q) are treated differently. They are decomposed by

$$B_1 = \varepsilon\,(-\Delta + id)^{-1} \circ T^+ \circ id_1 \quad \text{and} \quad B_b = \varepsilon\,(-\Delta + id)^{-1} \circ T^+ \circ b \circ id_b \tag{27.61}$$

with

$$id_1 \;:\; \mathbb{H}_p^{s+2}(\mathbb{R}^n) \mapsto \mathbb{H}_p^s(\mathbb{R}^n)\,, \tag{27.62}$$

$$T^+ \;:\; \mathbb{H}_p^s(\mathbb{R}^n) \mapsto \mathbb{H}_p^s(\mathbb{R}^n)\,, \tag{27.63}$$

$$\varepsilon\,(-\Delta + id)^{-1} \;:\; \mathbb{H}_p^s(\mathbb{R}^n) \mapsto \mathbb{H}_p^{s+2}(\mathbb{R}^n)\,, \tag{27.64}$$

and

$$id_b \;:\; \mathbb{H}_p^{s+2}(\mathbb{R}^n) \mapsto \mathbb{H}_r^s(\mathbb{R}^n)\,, \tag{27.65}$$

$$b \;:\; \mathbb{H}_r^s(\mathbb{R}^n) \mapsto \mathbb{H}_p^s(\mathbb{R}^n)\,, \tag{27.66}$$

$$T^+ \;:\; \mathbb{H}_p^s(\mathbb{R}^n) \mapsto \mathbb{H}_p^s(\mathbb{R}^n)\,, \tag{27.67}$$

$$\varepsilon\,(-\Delta + id)^{-1} \;:\; \mathbb{H}_p^s(\mathbb{R}^n) \mapsto \mathbb{H}_p^{s+2}(\mathbb{R}^n)\,. \tag{27.68}$$

The necessary explanations have been given in 27.8. Recall that (27.65) is a well-known embedding with constant differential dimension. We refer to [ET96], p. 45, formula (8). Otherwise (27.62)–(27.64) correspond to the line segment **AC** and (27.65)–(27.68) to the triangle **ADC**. In particular, (27.66) is covered by (27.45). One deals in a similar way with the operator B^Q. Then one obtains the counterparts of (27.8)–(27.10) now with respect to the space $\mathbb{H}_p^{s+2}(\mathbb{R}^n)$. Afterwards the rest is the same as before.

27.10 Comments; the Q-method revisited

We followed [Tri01]. The aim to present the above corollary is twofold. First we wanted to discuss which type of (local and global) singularities, expressed by the function b, can be admitted. Secondly, the corollary sheds new light on the usefulness of the Q-method. Let p and s be given by (27.28). We discussed in 27.5 that under these restrictions the Q-method is not needed to get maximal regularity for solutions u of (27.24). In this particular case one can use bootstrapping arguments. But the latter method does not work if one replaces (27.24) by (27.39). The mapping properties of

$$(-\Delta + id)^{-1} \quad \text{and} \quad (-\Delta + id)^{-1} \circ b$$

are different. In other words, in contrast to the Q-method bootstrapping arguments are restricted to rather special situations.

References

[Ada71] Adams, D. R., Traces of potentials arising from translation invariant operators. Ann. Sc. Norm. Super. Pisa **25** (1971), 203–217

[Ada73] Adams, D. R., A trace inequality for generalized potentials. Stud. Math. **48** (1973), 99–105

[Ada88] Adams, D. R., A sharp inequality of J. Moser for higher order derivatives. Annals Math. **128** (1988), 385–98

[Ada89] Adams, D. R., The classification problem for the capacities associated with the Besov and Triebel-Lizorkin spaces. In: Approximation and Function Spaces, Banach Center Publ. **22**, 9–24, Warszawa, PWN Polish Scientific Publishers, 1989

[AdH96] Adams, D. R. and Hedberg, L. I., Function spaces and potential theory. Berlin, Springer, 1996

[Adm75] Adams, R. A., Sobolev spaces. New York, Academic Press, 1975

[AlK00] Albeverio, S. and Kurasov, P., Singular perturbations of differential operators. Cambridge Univ. Press, 2000

[Ama76] Amann, H., Fixed point equations and nonlinear eigenvalue problems in ordered Banach spaces. SIAM Review **18** (1976), 620–709

[AmC78] Amann, H. and Crandall, M. G., On some existence theorems for semi-linear elliptic equations. Indiana Univ. Math. J. **27** (1978), 779–790

[Ban22] Banach, S., Sur les opérations dans les ensembles abstraits et leur application aux équations intégrales. Fund. Math. **3** (1922), 133–181

[BeL76] Bergh, L. and Löfström, J., Interpolation spaces, an introduction. Berlin, Springer, 1976

[Ber98] Berger, G., Eigenvalue distribution of elliptic operators of second order with Neumann boundary conditions in a snowflake domain. Math. Nachr. **220** (2000), 11–32

[BeR80] Bennett, C. and Rudnik, K., On Lorentz-Zygmund spaces. Dissertationes Math. **175** (1980), 1–72

[BeS88] Bennett, C. and Sharpley, R., Interpolation of operators. Boston, Academic Press, 1988

[BIN75] Besov, O. V., Il'in, V. P. and Nikol'skij, S. M., Integral representations of functions and embedding theorems (Russian). Moskva, Nauka, 1975 (there are two English translations)

[BiS74] Birman, M. S. and Solomyak, M. Z., Quantitative analysis in Sobolev embedding theorems and applications to spectral theory. In: Tenth Math. Summer School Katsiveli/Nalchik 1972. Inst. Mat. Akad. Nauk Ukrain. SSR, Kiev, 1974, 5–189 (Russian). (English translation: Amer. Math. Soc. Transl. **114** (1980), 1–132)

[BKLN88] Besov, O. V., Kudrjavcev, L. D., Lizorkin, P. I. and Nikol'skij, S. M., Studies on the theory of spaces of differentiable functions of several variables (Russian). Trudy Mat. Inst. Steklov **182** (1988), 68–127. (English translation, Proc. Steklov Inst. Math. **182** (1990), 73–140)

[Bod93] Bourdaud, G., The functional calculus in Sobolev spaces. In: Proc. Conf. Function spaces, differential operators and nonlinear analysis. Teubner-Texte Math. **133**, Teubner, Stuttgart, 1993, 127–142

[BoL00] Bourdaud, G. and Lanza de Cristoforis, M., Functional calculus on Hölder-Zygmund spaces. Preprint, Paris, 2000

[BoM91] Bourdaud, G. and Meyer, Y., Fonctions qui opèrent sur les espaces de Sobolev. J. Funct. Anal. **97** (1991), 351–360

[Bor70] Borzov, V. V., Quantitative characteristics of singular measures. Probl. Mat. Fiz. **4**, Leningrad, 1970, 42–47 (Russian). (English translation: Topics Math. Phys. **4**, New York, Plenum Press, 1971)

[Bou56] Bourbaki, N., Élements de Mathématique. XXI, Livre VI, Ch. 5, Intégration des mesures. Paris, Hermann, 1956

[Bri99] Bricchi, M., On some properties of (d, Ψ)-sets and related Besov spaces. Jenaer Schriften Math/Inf/99/31, Jena, 1999

[Bri00] Bricchi, M., On the relationship between Besov spaces $B_{pq}^{(s,\Psi)}(\mathbb{R}^n)$ and L_p-spaces defined on a (d, Ψ)-set.
Jenaer Schriften Math/Inf/00/13, Jena, 2000

[Bri01] Bricchi, M., PhD Thesis, Jena, 2001

[Bru72] Brudnyi, Y. A., On the rearrangement of a smooth function. Uspechi Mat. Nauk **27** (1972), 165–166 (Russian)

[Bru76] Brudnyi, Y. A., On the scale of spaces $L_p^{\lambda,\Theta}$ and precise embedding theorems. In: Embedding theorems and applications. Alma-Ata, Nauka, 1976, 23–27 (Russian)

[BrW80] Brézis, H. and Wainger, S., A note on limiting cases of Sobolev embeddings and convolution inequalities. Comm. Part. Diff. Equations **5** (1980), 773–789

[Byl94] Bylund, P., Besov spaces and measures on arbitrary sets. PhD Theses, Umeå, 1994

[Cae99] Caetano, A. M., Approximation by functions of compact support in Besov-Triebel-Lizorkin spaces on irregular domains. Studia Math. **142** (2000), 47–63

[Cae00] Caetano, A. M., On fractals which are not so terrible. Preprint, CM00/I-14, Univ. Aveiro, 2000

[Carl81] Carl, B., Entropy numbers, s-numbers and eigenvalue problems. J. Funct. Analysis **41** (1981), 290–306

[CaT80] Carl, B. and Triebel, H., Inequalities between eigenvalues, entropy numbers, and related quantities of compact operators in Banach spaces. Math. Ann. **251** (1980), 129–133

[CDS99] Chang, D.-C., Dafni, G. and Stein, E. M., Hardy spaces, BMO, and boundary value problems for the Laplacian on a smooth domain. Trans. Amer. Math. Soc. **351** (1999), 1605–1661

[CKS92] Chang, D.-C., Krantz, S. G. and Stein, E. M., Hardy spaces and elliptic boundary-value problems. In: Proc. Madison Symp. Complex Analysis. Contemporary Math. **137**, AMS, 1992, 119–131

[CKS93] Chang, D.-C., Krantz, S. G. and Stein, E. M., H^p theory on a smooth domain in $\mathbb{R}^N$ and elliptic boundary value problems. J. Funct. Anal. **144** (1993), 286–347

[CoH24] Courant, H. and Hilbert, D., Methoden der mathematischen Physik. Berlin, Springer, 1993 (4. Auflage), (1. Auflage, 1924)

[CoK00] Cobos, F. and Kühn, T., Entropy numbers of Besov spaces in generalized Lipschitz spaces. Preprint, Leipzig, 2000

[COV99] Cascante, C., Ortega, J. M. and Verbitsky, I. E., Trace inequalities of Sobolev type in the upper triangle case. Proc. London Math. Soc. **80** (2000), 391–414

[CoW71] Coifman, R. and Weiss, G., Analyse harmonique noncommutative sur certains espaces homogènes. Lect. Notes Math. **242**. Berlin, Springer, 1971

[CwP98] Cwikel, M. and Pustylnik, E., Sobolev type embeddings in the limiting case. Journ. Fourier Anal. Applications **4** (1998), 433–446

[Dav89] Davies, E. B., Heat kernels and spectral theory. Cambridge Univ. Press, 1989

[Dav95] Davies, E. B., Spectral theory and differential operators. Cambridge Univ. Press, 1995

[DeL93] DeVore, R. A. and Lorentz, G. G., Constructive approximation. Berlin, Springer, 1993

[DeP88a] DeVore, R. A. and Popov, V. A., Interpolation of Besov spaces. Trans. Amer. Math. Soc. **305** (1988), 397–414

[DeP88b] DeVore, R. A. and Popov, V. A., Interpolation spaces and nonlinear approximation. In: Proc. Conf. Function spaces and applications, Lect. Notes Math. **1302**, Berlin, Springer, 1988, 191–205

[DeS93] DeVore, R. A. and Sharpley, R. C., Besov spaces on domains in $\mathbb{R}^d$. Trans. Amer. Math. Soc. **335** (1993), 843–864

[DeY91] DeVore, R. A. and Yu, X. M., K-functionals for Besov spaces. J. Approximation Theory **67** (1991), 38–50

[Die75] Dieudonné, J., Grundzüge der modernen Analysis 2. Berlin, VEB Deutscher Verl. Wissenschaften, 1975

[Din95] Dintelmann, P., Classes of Fourier multipliers and Besov-Nikol'skij spaces. Math. Nachr. **173** (1995), 115–130

[EdE87] Edmunds, D. E. and Evans, W. D., Spectral theory and differential operators. Oxford, Clarendon Press, 1987

[EdH99] Edmunds, D. E. and Haroske, D. D., Spaces of Lipschitz type, embeddings and entropy numbers. Dissertationes Math. **380** (1999), 1–43

[EdH00] Edmunds, D. E. and Haroske, D. D., Embeddings in spaces of Lipschitz type, entropy and approximation numbers, and applications. J. Approximation Theory **104** (2000), 226–271

[EdK95] Edmunds, D. E. and Krbec, M., Two limiting cases of Sobolev embeddings. Houston J. Math. **21** (1995), 119–128

[EdT95] Edmunds, D. E. and Triebel, H., Logarithmic Sobolev spaces and their applications to spectral theory. Proc. London Math. Soc. **71** (1995), 333–371

[EdT98] Edmunds, D. and Triebel, H., Spectral theory for isotropic fractal drums. C. R. Acad. Sci. Paris **326**, Sér. I, (1998), 1269–1274

[EdT99a] Edmunds, D. and Triebel, H., Eigenfrequencies of isotropic fractal drums. Operator Theory: Advances Appl. **110** (1999), 81–102

[EdT99b] Edmunds, D. E. and Triebel, H., Sharp Sobolev embeddings and related Hardy inequalities: The critical case. Math. Nachr. **207** (1999), 79–92

[EgK90] Egorov, Ju. B. and Kondratev, B. A., On the negative spectrum of elliptic operators. Mat. Sbornik **181** (1990), 147–166 (Russian)

[EKP00] Edmunds, D. E., Kerman, R. and Pick, L., Optimal Sobolev embeddings involving rearrangement-invariant quasi-norms. J. Funct. Anal. **170** (2000), 307–355

[ET96] Edmunds, D. E. and Triebel, H., Function spaces, entropy numbers, differential operators. Cambridge Univ. Press, 1996

[EvH93] Evans, W. D. and Harris, D. J., Fractals, trees and the Neumann Laplacian. Math. Ann. **296** (1993), 493–527

[FaJ00] Farkas, W. and Jacob, N., Sobolev spaces on non-smooth domains and Dirichlet forms related to subordinate reflecting diffusions. Math. Nachr. (to appear)

[Fal85] Falconer, K. J., The geometry of fractal sets. Cambridge Univ. Press, 1985

[Fal90] Falconer, K. J., Fractal geometry. Chichester, Wiley, 1990

[Fal97] Falconer, K. J., Techniques in fractal geometry. Chichester, Wiley, 1997

[Far98] Farkas, W., Eigenvalue distributions of some fractal semi-elliptic differential operators. Math. Z. **236** (2001), 291–320

[Far99] Farkas, W., The behaviour of the eigenvalues for a class of operators related to some self-affine fractals in R^2. Z. Analysis Anwendungen **18** (1999), 875–893

[Far00] Farkas, W., Atomic and subatomic decompositions in anisotropic function spaces. Math. Nachr. **209** (2000), 83–113

[FaT99] Farkas, W. and Triebel, H., The distribution of eigenfrequencies of anisotropic fractal drums. J. London Math. Soc. **60** (1999), 224–236

[FJS00] Farkas, W., Johnsen, J. and Sickel, W., Traces of anisotropic Besov-Lizorkin-Triebel spaces – a complete treatment of borderline cases. Math. Bohemica **125** (2000), 1–37

[FJW91] Frazier, M., Jawerth, B. and Weiss, G., Littlewood-Paley theory and the study of function spaces. CBMS-AMS Regional Conf. Ser. **79**, Providence, AMS, 1991

[FKY87] Fukagai, N., Kusano, T. and Yoshida, K., Some remarks on the supersolution-subsolution method for superlinear elliptic equations. J. Math. Analysis Applications **123** (1987), 131–141

[FLS96] Fusco, N., Lions, P.-L. and Sbordone, C., Sobolev embedding theorems in borderline case. Proc. Amer. Math. Soc. **124** (1996), 562–565

[FMM98] Fabes, E., Mendez, O. and Mitrea, M., Boundary layers on Sobolev-Besov spaces and Poisson's equation for the Laplacian in Lipschitz domains. J. Funct. Analysis **159** (1998), 323–368

[Fra86] Franke, J., On the spaces F_{pq}^s of Triebel-Lizorkin type: Pointwise multipliers and spaces on domains. Math. Nachr. **125** (1986), 29–68

[FrJ85] Frazier, M. and Jawerth, B., Decomposition of Besov spaces. Indiana Univ. Math. J. **34** (1985), 777–799

[FrJ90]   Frazier, M. and Jawerth, B., A discrete transform and decompositions of distribution spaces. J. Funct. Analysis **93** (1990), 34–170
[FrR95]   Franke, J. and Runst, T., Regular elliptic boundary value problems in Besov-Triebel-Lizorkin spaces. Math. Nachr. **174** (1995), 113–149
[GiT77]   Gilbarg, D. and Trudinger, N. S., Elliptic partial differential equations of second order. Berlin, Springer, 1977
[Gol85]   Gold'man, M. L., Embedding of generalized Nikol'skij-Besov spaces in Lorentz spaces. Trudy Mat. Inst. Steklov **172** (1985), 128–139 (Russian)
[Gol86]   Gold'man, M. L., On embedding of constructive and structural Lipschitz spaces in symmetric spaces. Trudy Mat. Inst. Steklov **173** (1986), 90–112 (Russian). (English translation: Proc. Steklov Inst. Math. **4** (1987), 93–118)
[Had28]   Hardy, G. H., Notes on some points in the integral calculus (LXIV). Messenger of Math. **57** (1928), 12–16
[HaL99]   Han, Y.-S. and Lin, C.-C., Embedding theorem on spaces of homogeneous type. Preprint, Auburn, 1999
[Ham00]   Hambly, B. M., Heat kernels and spectral asymptotics for some random Sierpinski gaskets. In: Fractal geometry and stochastics II. Basel, Birkhäuser, 2000, 239–267
[Han94]   Han, Y.-S., Triebel-Lizorkin spaces on homogeneous spaces. Studia Math. **108** (1994), 247–273
[Han98]   Han, Y.-S., Plancherel-Polya type inequality on spaces of homogeneous type and its applications. Proc. Amer. Math. Soc. **126** (1998), 3315–3327
[Har95]   Haroske, D. D., Approximation numbers in some weighted function spaces. J. Approximation Theory **83** (1995), 104–136
[Har97]   Haroske, D. D., Embeddings of some weighted function spaces on $\mathbb{R}^n$; entropy and approximation numbers. An. Univ. Craiova, Ser. Mat. Inform. **24** (1997), 1–44
[Har98]   Haroske, D. D., Some logarithmic function spaces, entropy numbers, applications to spectral theory. Dissertationes Math. **373** (1998), 1–59
[Har00a] Haroske, D. D., On more general Lipschitz spaces. Zeitschr. Analysis Anwendungen **19** (2000), 781–799
[Har00b] Haroske, D. D., Embeddings in spaces of Lipschitz type, entropy and approximation numbers. In: Function Spaces, Differential Operators and Nonlin. Analysis. Prague, Math. Inst. Czech Acad. Sci., 2000, 99–112
[Har00c] Haroske, D. D., Logarithmic Sobolev spaces on $\mathbb{R}^n$; entropy numbers, and some applications. Forum Math. **12** (2000), 257–313

[Har01]   Haroske, D. D., Habilitationsschrift, Jena 2001
[Has79]   Hansson, K., Embedding theorems of Sobolev type in potential theory. Math. Scand. **45** (1979), 77–102
[HaS94]   Han, Y.-S. and Sawyer, E. T., Littlewood-Paley theory on spaces of homogeneous type and the classical function spaces. Memoirs Amer. Math. Soc. **530**. Providence, Amer. Math. Soc., 1994
[HaT94a] Haroske, D. and Triebel, H., Entropy numbers in weighted function spaces and eigenvalue distributions of some degenerate pseudodifferential operators I. Math. Nachr. **167** (1994), 131–156
[HaT94b] Haroske, D. and Triebel, H., Entropy numbers in weighted function spaces and eigenvalue distributions of some degenerate pseudodifferential operators II. Math. Nachr. **168** (1994), 109–137
[Hed84]   Hedberg, L. I., Spectral synthesis in Sobolev spaces. Linear and complex analysis problem book. Lect. Notes Math. **1043**, Berlin, Springer, 1984, 435–437
[HeL97]   He, C. Q. and Lapidus, M. L., Generalized Minkowski content, spectrum of fractal drums, fractal strings and the Riemann zeta-function. Memoirs Amer. Math. Soc. **608** (1997)
[Her68]   Herz, C., Lipschitz spaces and Bernstein's theorem on absolutely convergent Fourier transforms. J. Math. Mech. **18** (1968), 283–324
[HeW83]   Hedberg, L. I. and Wolff, T., Thin sets in nonlinear potential theory. Ann. Inst. Fourier (Grenoble) **33**:4 (1983), 161–187
[HLP52]   Hardy, G. H., Littlewood, J. E. and Polya, G., Inequalities. Sec. ed., Cambridge Univ. Press, 1952 (first ed., 1934)
[HLY99a] Han, Y.-S., Lu, S.-Z. and Yang, D., Inhomogeneous Triebel-Lizorkin spaces on spaces of homogeneous type. Math. Sci. Res. Hot-Line **3** (9) (1999), 1–29
[HLY99b] Han, Y.-S., Lu, S.-Z. and Yang, D., Inhomogeneous Besov and Triebel-Lizorkin spaces on spaces of homogeneous type. Approx. Theory its Appl. **15**:3 (1999), 37–65
[Hut81]   Hutchinson, J. E., Fractals and self similarity. Indiana Univ. Math. J. **30** (1981), 713–747
[Jaw77]   Jawerth, B., Some observations on Besov and Lizorkin-Triebel spaces. Math. Scand. **40** (1977), 94–104
[JeK95]   Jerison, D. and Kenig, C. E., The inhomogeneous Dirichlet problem in Lipschitz domains. J. Funct. Analysis **130** (1995), 161–219
[Joh00]   Johnsen, J., Traces of Besov spaces revisited. Zeitschr. Analysis Anwendungen **19** (2000), 763–779

[Jon93a] Jonsson, A., Besov spaces on closed sets by means of atomic decompositions. Umeå Report **7**, 1993

[Jon93b] Jonsson, A., Atomic decomposition of Besov spaces on closed sets. In: Function spaces, differential operators and non-linear analysis. Teubner-Text Math. **133**, Leipzig, Teubner, 1993, 285–289

[Jon94] Jonsson, A., Besov spaces on closed subsets of $\mathbb{R}^n$. Trans. Amer. Math. Soc. **341** (1994), 355–370

[Jon96] Jonsson, A., Brownian motion on fractals and function spaces. Math. Z. **222** (1996), 495–504

[Jon98a] Jonsson, A., Wavelets on fractals and Besov spaces. J. Fourier Analysis Appl. **4** (1998), 329–340

[Jon98b] Jonsson, A., Haar wavelets of higher order on fractals and regularity of functions. Research Report, Umeå, 1998

[JoW84] Jonsson, A. and Wallin, H., Function spaces on subsets of $\mathbb{R}^n$. Math. reports **2**, 1, London, Harwood acad. publ., 1984

[JoW95] Jonsson, A. and Wallin, H., The dual of Besov spaces on fractals. Studia Math. **112** (1995), 285–300

[JoW97] Jonsson, A. and Wallin, H., Boundary value problems and Brownian motion on fractals. Chaos, Solitons & Fractals **8** (1997), 191–205

[JPW90] Jawerth, B., Pérez, C. and Welland, G., The positive cone in Triebel-Lizorkin spaces and the relation among potential and maximal operators. In: Harmonic Analysis and Partial Differential Equations (Boca Raton, FL 1988), 71–91, Contemp. Math. **107**. Providence, Amer. Math. Soc., 1990

[KaK78] Kazdan, J. L. and Kramer, R. J., Invariant criteria for existence of solutions to second order quasilinear elliptic equations. Comm. Pure Appl. Math. **31** (1978), 619–645

[Kal80] Kaljabin, G. A., The description of functions of classes of Besov-Lizorkin-Triebel type (Russian). Trudy Mat. Inst. Steklov **156** (1980), 82–109

[Kal85] Kaljabin, G. A., Theorems on extensions, multipliers and diffeomorphisms for generalized Sobolev-Liouville classes in domains with Lipschitz boundary (Russian). Trudy Mat. Inst. Steklov **172** (1985), 173–186

[KaL87] Kaljabin, G. A. and Lizorkin, P. I., Spaces of functions of generalized smoothness. Math. Nachr. **133** (1987), 7–32

[Ken94] Kenig, C. E., Harmonic analysis techniques for second order elliptic boundary value problems. CBMS, Regional Conf. Series **83**, Providence, Amer. Math. Soc., 1994

[Kol89] Kolyada, V. I., Rearrangement of functions and embedding theorems. Uspechi Mat. Nauk **44** (1989), 61–98 (Russian)

[Kol98] Kolyada, V. I., Rearrangement of functions and embeddings of anisotropic spaces of Sobolev type. East Journal Approximations **4** (1998), 111–199

[KrS96] Krbec, M. and Schmeisser, H.-J., Limiting embeddings. The case of missing derivatives. Richerche Math. **45** (1996), 423–447

[KrS98] Krbec, M. and Schmeisser, H.-J., Embeddings of Brézis-Wainger type. The case of missing derivatives. Proc. Royal Soc. Edinb. (to appear)

[KuN88] Kudrjavcev, L. D. and Nikol'skij, S.M., Spaces of differentiable functions of several variables and embedding theorems (Russian). In: Itogi nauki i techniki, Ser. Sovremennye problemj mat. **26**. Moskva, Akad. Nauk SSSR, 1988, 5–157. (English translation: In: Analysis III, Spaces of differentiable functions, Encyclopaedia of Math. Sciences **26**. Heidelberg, Springer, 1990, 4–140)

[KuS00] Kühn, T. and Schonbek, T. P., Entropy numbers of diagonal operators between vector-valued sequence spaces. Preprint, Leipzig, 2000

[Lan93] Lang, S., Real and functional analysis, 3rd ed., New York, Springer, 1993

[Lap91] Lapidus, M. L., Fractal drums, inverse spectral problems for elliptic operators and a partial resolution of the Weyl-Berry conjecture. Trans. Amer. Math. Soc. **325** (1991), 465–529

[Leo98] Leopold, H.-G., Limiting embeddings and entropy numbers. Preprint, Jena, 1998

[Leo00a] Leopold, H.-G., Embeddings and entropy numbers in Besov spaces of generalized smoothness. Proc. Conf. Function Spaces V, Poznan 1998 (to appear)

[Leo00b] Leopold, H.-G., Embeddings for general weighted sequence spaces and entropy numbers. Proc. Conf. function spaces, differential operators, nonlin. analysis, Syöte (Finland), 1999. Praha, Math. Institute Acad. Sci. Czech Rep., 2000, 170–186

[Leo01] Leopold, H.-G., Embeddings and entropy numbers for general weighted sequence spaces: The non-limiting case. Georgian Math. Journ. **17** (2000), 731–743

[Lio98] Lions, P.-L., Mathematical topics in fluid mechanics, Vol. 2. Oxford, Clarendon Press, 1998

[Liz86] Lizorkin, P. I., Spaces of generalized smoothness. Appendix to the Russian ed. of [Triβ]. Moscow, Mir, 1986, 381–415 (Russian)

[Mal84] Maligranda, L., Interpolation of locally Hölder operators. Studia Math. **78** (1984), 289–296

[Mall95] Malliavin, P., Integration and probability. New York, Springer, 1995

[MaM79] Marcus, M. and Mizel, V. J., Every superposition operator mapping one Sobolev space into another is continuous. J. Funct. Anal. **33** (1979), 217–229

[MaN95] Maz'ya, V. G. and Netrusov, Yu. V., Some counterexamples for the theory of Sobolev spaces on bad domains. Potential Analysis **4** (1995), 47–65

[Mar87a] Marschall, J., The trace of Sobolev-Slobodeckij spaces on Lipschitz domains. Manuscripta Math. **58** (1987), 47–65

[Mar87b] Marschall, J., Some remarks on Triebel spaces. Studia Math. **87** (1987), 79–92

[Mar95] Marschall, J., On the boundedness and compactness of nonregular pseudo-differential operators. Math. Nachr. **175** (1995), 231–262

[Mat95] Mattila, P., Geometry of sets and measures in euclidean spaces. Cambridge Univ. Press, 1995

[MaV95] Maz'ya, V. G. and Verbitsky, I. E., Capacitary inequalities for fractional integrals with applications to partial differential equations and Sobolev multipliers. Ark. mat. **33** (1995), 81–115

[Maz85] Maz'ya (Maz'ja), V. G., Sobolev spaces. Berlin, Springer, 1985

[MeM00] Mendez, O. and Mitrea, M., The Banach envelopes of Besov and Triebel-Lizorkin spaces and applications to partial differential equations. Journ. Fourier Anal. Applications **6** (2000), 503–531

[Mey92] Meyer, Y., Wavelets and operators. Cambridge Univ. Press, 1992

[Miy90] Miyachi, A., H^p spaces over open subsets of $\mathbb{R}^n$. Studia Math. **95** (1990), 205–228

[Mos71] Moser, J., A sharp form of an inequality by N. Trudinger. Indiana Univ. Math. J. **20** (1971), 1077–1092

[Mou99] Moura, S. D. de, Some properties of the spaces $F_{pq}^{(s,\Psi)}(\mathbb{R}^n)$ and $B_{pq}^{(s,\Psi)}(\mathbb{R}^n)$. Report, Coimbra, 1999

[Mou01a] Moura, S. D. de, PhD Thesis, Coimbra, 2001

[Mou01b] Moura, S. D. de, Function spaces of generalized smoothness. Dissertationes Math. (to appear)

[NaS94] Naimark, K. and Solomyak, M., On the eigenvalue behaviour for a class of operators related to self-similar measures on $\mathbb{R}^d$. C. R. Acad. Sci., Paris **319** (I) (1994), 827–842

[NaS95] Naimark, K. and Solomyak, M., The eigenvalue behaviour for the boundary value problems related to self-similar measures on $\mathbb{R}^d$. Math. Research Letters **2** (1995), 279–298

[NaS00] Naimark, K. and Solomyak, M., Eigenvalue distributions of some fractal semi-elliptic differential operators: combinatorial approach. Integr. Equat. Oper. Th. (to appear)

[Net87a] Netrusov, Ju. V., Embedding theorems of Besov spaces into ideal spaces (Russian). Zap. Naučn. Sem. Leningrad. Otdel. Mat. Inst. Steklov (LOMI) **159** (1987), 69–82. (English translation: J. Soviet Math. **47** (1989), 2871–2881)

[Net87b] Netrusov, Ju. V., Embedding theorems for Lizorkin-Triebel spaces (Russian). Zap. Naučn. Sem. Leningrad. Otdel. Mat. Inst. Steklov (LOMI) **159** (1987), 103–112. (English translation: J. Soviet Math. **47** (1989), 2896–2903)

[Net89a] Netrusov, Ju. V., Sets of singularities of functions of spaces of Besov and Lizorkin-Triebel type (Russian). Trudy Mat. Inst. Steklov **187** (1989), 162–177. (English translation: Proc. Steklov Inst. Math. **187** (1990), 185–203)

[Net89b] Netrusov, Ju. V., Metric estimates of the capacities of sets in Besov spaces (Russian). Trudy Mat. Inst. Steklov **190** (1989), 159–185. (English translation: Proc. Steklov Inst. Math. **190** (1992), 167–192)

[Net92] Netrusov, Ju. V., Spectral synthesis in spaces of smooth functions. Dokl. akad. nauk **325** (1992), 923–925 (Russian). (English translation: Russian Acad. Sci. Dokl. Math. **46** (1993), 135–137)

[Nik77] Nikol'skij, S. M., Approximation of functions of several variables and embedding theorems (Russian). Sec. ed., Moskva, Nauka, 1977. (First ed., Moskva, Nauka, 1969; English translation, Berlin, Springer, 1975)

[OpK90] Opic, B. and Kufner, A., Hardy-type inequalities. Pitman Research Notes Math. Series. Harlow, Longman Scientific & Technical, 1990

[Orw51] Orwell, G., Animal farm. Harmondsworth, Penguin books Ltd., 1951

[Osw92] Oswald, P., On the boundedness of the mapping $f \mapsto |f|$ in Besov spaces. Comment. Univ. Carolina **33** (1992), 57–66

[Pee66] Peetre, J., Espaces d'interpolation et théorème de Soboleff. Ann. Inst. Fourier **16** (1966), 279–317

[Pee70] Peetre, J., Interpolation of Lipschitz operators and metric spaces. Mathematica **12** (1970), 325–334

[Pee76] Peetre, J., New thoughts on Besov spaces. Duke Univ. Math. Series. Durham, Univ., 1976

[Per23] Perron, O., Eine neue Behandlung der Randwertaufgabe $\Delta u = 0$. Math. Zeitschr. **18** (1923), 42–54

[Pic99] Pick, L., Optimal Sobolev embeddings. In: Nonlinear Analysis, Function Spaces, Applications **6**. Prague, Math. Inst. Czech. Acad. Sci., 1999, 156–199

[Poh65] Pohozaev, S., On eigenfunctions of the equation $\Delta u + \lambda f(u) = 0$. Dokl. Akad. Nauk SSSR **165** (1965), 36–39 (Russian)

[ReS78] Reed, M. and Simon, B., Methods of modern mathematical physics, IV. Analysis of operators. New York, Academic Press, 1978

[Rie'54] Riemann, B., Ueber die Darstellbarkeit einer Function durch eine trigonometrische Reihe. Habilitationsschrift, Univ. Göttingen, 1854. (In: Bernhard Riemann, Gesammelte math. Werke, wissenschaftl. Nachlass und Nachträge. Leipzig, Teubner-Verlag, 1990, 227–264)

[RuS96] Runst, T. and Sickel, W., Sobolev spaces of fractional order, Nemytskij operators, and nonlinear partial differential equations. Berlin, W. de Gruyter, 1996

[Scho98a] Schott, T., Function spaces with exponential weights I. Math. Nachr. **189** (1998), 221–242

[Scho98b] Schott, T., Function spaces with exponential weights II. Math. Nachr. **196** (1998), 231–250

[She99] Shevchik, V., Spectral properties of some semi-elliptic operators in L_p spaces. Math. Nachr. **202** (1999), 151–162

[Shu92] Shubin, M. A., Spectral theory of elliptic operators on non-compact manifolds. Astérisque **207** (1992), 35–108

[Sic99] Sickel, W., On pointwise multipliers for $F_{pq}^s(\mathbb{R}^n)$ in case $\sigma_{p,q} < s < \frac{n}{p}$. Annali Mat. pura applicata **176** (1999), 209–250

[SiT95] Sickel, W. and Triebel, H., Hölder inequalities and sharp embeddings in function spaces of B_{pq}^s and F_{pq}^s type. Z. Analysis Anwendungen **14** (1995), 105–140

[Skr98] Skrzypczak, L., Mapping properties of pseudodifferential operators on manifolds with bounded geometry. J. London Math. Soc. **57** (1998), 721–738

[Sma74] Smart, D. R., Fixed point theorems. Cambridge Univ. Press, 1974

[Sob50] Sobolev, S. L., Einige Anwendungen der Funktionalanalysis auf Gleichungen der mathematischen Physik. Berlin, Akademie Verlag, 1964. (Translation from Russian, Leningrad, 1950)

[Sol94] Solomyak, M., A remark on Hardy inequalities. Integral Equ. Operator Theory **19** (1994), 120–124

[ST87] Schmeisser, H.-J. and Triebel, H., Topics in Fourier analysis and function spaces. Chichester, Wiley, 1987

[Ste70] Stein, E. M., Singular integrals and differentiability properties of functions. Princeton Univ. Press, 1970

[Str67] Strichartz, R. S., Multipliers on fractional Sobolev spaces. J. Math. Mech. **16** (1967), 1031–1060

[Str68] Strichartz, R. S., Fubini-type theorems. Annali Scuola Norm. Sup. Pisa **22** (1968), 399–408

[Str72] Strichartz, R. S., A note on Trudinger's extension of Sobolev's inequality. Indiana Univ. Math. J. **21** (1972), 841–842

[Str83] Strichartz, R. S., Analysis of the Laplacian on the complete Riemannian manifold. J. Funct. Analysis **52** (1983), 48–79

[Tai96] Taira, K., Bifurcation for nonlinear elliptic boundary value problems, I. Collect. Math. **47** (1996), 207–229

[Tar72a] Tartar, L., Interpolation non linéaire. Bull. Soc. Math. France, Mémoire **31–32** (1972), 375–380

[Tar72b] Tartar, L., Interpolation non linéaire et régularité. J. Funct. Anal. **9** (1972), 469–489

[Tay96] Taylor, M. E., Partial differential equations, I. New York, Springer, 1996

[Tay00] Taylor, M. E., Tools for PDE: Pseudodifferential operators, paradifferential operators, and layer potentials. Providence, Amer. Math. Soc., 2000

[Tor91] Torres, R. H., Boundedness results for operators with singular kernels on distribution spaces. Memoirs Amer. Math. Soc. **442**. Providence, Amer. Math. Soc., 1991

[Triα] Triebel, H., Interpolation theory, function spaces, differential operators. Amsterdam, North-Holland, 1978 (Sec. ed. Heidelberg, Barth, 1995)

[Triβ] Triebel, H., Theory of function spaces. Basel, Birkhäuser, 1983

[Triγ] Triebel, H., Theory of function spaces II. Basel, Birkhäuser, 1992

[Triδ] Triebel, H., Fractals and spectra. Basel, Birkhäuser, 1997

[Tri78] Triebel, H., Spaces of Besov-Hardy-Sobolev type. Leipzig, Teubner, 1978

[Tri84] Triebel, H., Mapping properties of non-linear operators generated by $\Phi(u) = |u|^\varrho$ and by holomorphic $\Phi(u)$ in function spaces of Besov-Hardy-Sobolev type. Boundary value problems for elliptic differential equations of type $\Delta u = f(x) + \Phi(u)$. Math. Nachr. **117** (1984), 193–213

[Tri88] Triebel, H., On a class of weighted function spaces and related pseudodifferential operators. Studia Math. **90** (1988), 37–68

[Tri92] Triebel, H., Higher analysis. Leipzig, Barth, 1992

[Tri93] Triebel, H., Approximation numbers and entropy numbers of embeddings of fractional Besov-Sobolev spaces in Orlicz spaces. Proc. London Math. Soc. **66** (1993), 589–618

[Tri99a] Triebel, H., Decompositions of function spaces. Progress Nonlin. Diff. Equations Applications **35** (1999), 691–730

[Tri99b] Triebel, H., Hardy inequalities in function spaces. Mathematica Bohemica **124** (1999), 123–130

[Tri99c] Triebel, H., Function spaces and spectra of elliptic operators on a class of hyperbolic manifolds. Studia Math. **134** (1999), 179–202

[Tri99d] Triebel, H., Sharp Sobolev embeddings and related Hardy inequalities: The sub-critical case. Math. Nachr. **208** (1999), 167–178

[Tri00a] Triebel, H., Taylor expansions of distributions. Numer. Funct. Anal. and Optimiz. **21** (2000), 307–317

[Tri00b] Triebel, H., Truncations of functions. Forum Math. **12** (2000), 731–756

[Tri01] Triebel, H., Regularity theory for some semi-linear equations: the Q-method. Forum Math. **13** (2001), 1–19

[Tru67] Trudinger, N., On embeddings into Orlicz spaces and some applications. J. Math. Mech. **17** (1967), 473–483

[TrW96] Triebel, H. and Winkelvoss, H., Intrinsic atomic characterizations of function spaces on domains. Math. Z. **221** (1996), 647–673

[Ver99] Verbitsky, I. E., Superlinear equations, potential theory and weighted norm inequalities. In: Nonlinear Analysis, Function Spaces Applications **6**. Prague, Math. Institute Czech. Acad. Sci., 1999, 223–269

[Vis98] Vishik, M., Hydrodynamics in Besov spaces. Arch. Rational Mech. Anal. **145** (1998), 197–214

[Wal91] Wallin, H., The trace to the boundary of Sobolev spaces on a snowflake. Manuscripta Math. **73** (1991), 117–125

[Wig88] Wingren, P., Lipschitz spaces and interpolating polynomials on subsets of euclidean space. In: Function spaces and applications, Lecture Notes Math. **1302**, Berlin, Springer, 1988, 424–435

[Win95] Winkelvoss, H., Function spaces related to fractals. Intrinsic atomic characterizations of function spaces on domains. PhD-Thesis, Jena, 1995

[Yud61] Yudovich, V. I., Some estimates connected with integral operators and with solutions of elliptic equations. Soviet Math. Doklady **2** (1961), 746–749

[Zah00] Zähle, M., Harmonic calculus on fractals – a measure geometric approach II. Preprint, Jena, 2000

[Zan00] Zanger, D. Z., The inhomogeneous Neumann problem in Lipschitz domains. Comm. Part. Diff. Equations **25** (2000), 1771–1808

[Zie89] Ziemer, W. P., Weakly differentiable functions. New York, Springer, 1989

[Zyg45] Zygmund, A., Smooth functions. Duke Math. Journ. **12** (1945), 47–76

[Zyg77] Zygmund, A., Trigonometric series, sec. ed. Cambridge Univ. Press, 1977

Symbols

Sets
$B(x,c)$, 25, 120, 144, 153
$\mathbb{C}$, 10
$IR(n)$, 155
K^-, 51
K_t^+, 54
$MR(n)$, 155
$\mathbb{N}$, 10
$\mathbb{N}_0$, 10
$\mathbb{N}_0^n$, 10
(M,g), 84
rQ, 10, 125
$Q_{\nu m}$, 10, 125
$\mathbb{R}$, 10
$\mathbb{R}^n$, 2, 10
$\mathbb{R}_+^n$, 51
$(\mathbb{R}_+)^n$, 386
$\mathbb{R}_-^n$, 54
U_B, 278
$\mathbb{Z}$, 10
$\mathbb{Z}^n$, 10

Spaces
A_{pq}^s, 44
$A_{pq}^s(\mathbb{R}^n)$, 35, 44, 358, 361
$\mathbb{A}_{pq}^\sigma(\mathbb{R}^n)$, 359, 361
$A_{pq}^s(\Omega)$, 44

$\mathring{A}_{pq}^s(\Omega)$, 44
$\widetilde{A}_{pq}^s(\Omega)$, 44

$B_{pq}^s(\mathbb{R}^n)$, 13, 16, 29
$B_p^s(\mathbb{R}^n)$, 102, 113
$B_{pq}^s(\mathbb{R}^n, w(\cdot))$, 119
$B_p^s(\mathbb{R}^n, \langle x \rangle^\alpha)$, 102
$tr_\Gamma B_{pq}^s(\mathbb{R}^n)$, 147, 151
$B_{pq}^{s,\Gamma}(\mathbb{R}^n)$, 158
$B_{pq}^{(1,-b)}(\mathbb{R}^n)$, 247
$B_{pq,\Gamma}^\sigma(\mathbb{R}^n)$, 159, 260
$B_{pq}^s(\Gamma)$, 147, 151
$B_{pq}^s(\Gamma)_\varrho$, 149
$B_{pq}^{(s,\Psi)}(\mathbb{R}^n)$, 333, 340
$B_{pq}^{(s,\Psi),\Gamma}(\mathbb{R}^n)$, 343
$B_{pq}^{(s,\Psi^a)}(\Gamma)$, 346
$\mathbb{B}_{pq}^s(\mathbb{R}^n)$, 355, 357, 387
$\mathbb{B}_p^s(\mathbb{R})$, 364
$B_{pq}^s(\Omega)$, 41
$B_{pq}^s(\overline{\Omega})$, 155
$B_p^s(\Omega)$, 113
$tr_\Gamma B_{pq}^s(\Omega)$, 156
$\widetilde{B}_{pq}^s(\Omega)$, 44, 52, 53
$\widetilde{B}_p^s(\Omega)$, 113
$\widetilde{B}_{pq}^{s,\Gamma}(\mathbb{R}^n)$, 158

$\overset{\circ}{B}{}^s_{pq}(\Omega)$, 44

$\overset{\circ}{B}{}^s_{p}(\Omega)$, 113

b_p, 106

b_{pq}, 11

$\|\lambda\,|b_{pq}\|_\varrho$, 13

$\|\lambda\,|b_p\|_\varrho$, 106

$b^s_p(\mathbb{R}^n)$, 113

$b^s_p(\Omega)$, 113

$\widetilde{b}^s_p(\Omega)$, 113

$\|\lambda|b^\Omega_p\|_\varrho$, 115

b^Γ_{pq}, 148

$\|\lambda|b^\Gamma_{pq}\|_\varrho$, 149

$bmo(\mathbb{R}^n)$, 4, 216, 379

$C(\mathbb{R}^n)$, 168

$C(\Gamma)$, 309

$C^1(\mathbb{R}^n)$, 168

$\mathcal{C}^s(\mathbb{R}^n)$, 4, 390

$\mathcal{C}^s(\Omega)$, 254

$C^\infty(\overline{\Omega})$, 46, 156, 227

$D(\Omega)$, 41, 43, 112

$D'(\Omega)$, 43, 112

$D_\Omega(\mathbb{R}^n)$, 47

$D(\Gamma)$, 298

$D'(\Gamma)$, 298

$F^s_{pq}(\mathbb{R}^n)$, 13, 16, 29

$F^s_{pq}(\mathbb{R}^n, w)$, 85

$F^s_{pq}(\mathbb{R}^n, w(\cdot))$, 119

$tr_\Gamma F^s_{pq}(\mathbb{R}^n)$, 128, 132

$\overset{+}{F}{}^\sigma_{uv}(\mathbb{R}^n)$, 126

$F^s_{pq}(M)$, 88

$F^s_{pq}(M, g^\varkappa)$, 89

$F^s_{pq}(\Omega)$, 41

$F^s_{pq}(\overline{\Omega})$, 155

$F^{(s,\Psi)}_{pq}(\mathbb{R}^n)$, 333, 340

$\mathbb{F}^s_{pq}(\mathbb{R}^n)$, 357

$\widetilde{F}^s_{pq}(\Omega)$, 42, 44

$\widetilde{F}^s_{pq}(\overline{\Omega})$, 44

$\overset{\circ}{F}{}^s_{pq}(\Omega)$, 42

$tr_\Gamma F^s_{pq}(\Omega)$, 156

f_{pq}, 11

$\|\lambda|f_{pq}\|_\varrho$, 13

f^Ω_{pq}, 73, 98

$\|\lambda|f^\Omega_{pq}\|_\varrho$, 73

$H^s(\mathbb{R}^n)$, 296

$H^s_p(\mathbb{R}^n)$, 3, 378, 386

$H^s(M)$, 312

$H^s_p(M)$, 81

$H^s(M, g^\varkappa)$, 312

$H^s_p(M, g^\varkappa)$, 90

$H^s(\Gamma)$, 157, 296, 298

$H^s(\Omega)$, 254, 296

$H^s_p(\Omega)$, 254

$H^{(s,\Psi)}$, 350

$H^s_\Gamma(\mathbb{R}^n)$, 157

$\mathbb{H}^s_p(\mathbb{R}^n)$, 358, 379, 387

$H^\sigma_{p,\Gamma}(\mathbb{R}^n)$, 260

$H^s_p(\log H)_a(\Omega)$, 249

$\overset{\circ}{H}{}^1(\Omega)$, 255, 264, 296

$(\cdot,\cdot)_{\overset{\circ}{H}{}^1(\Omega)}$, 257

$h_p(\mathbb{R}^n)$, 4

$L(A,B), L(A)$, 278

$L_p(\mathbb{R}^n)$, 2, 10

$L_1^{loc}(\mathbb{R}^n)$, 161, 167
$L_r(\Gamma)$, 123
$L_1(\Omega, g^\tau)$, 99
$L_p(\Omega, d^\sigma)$, 43
$L_{ru}(I_\varepsilon)$, 174
$L_r(\log L)_a(I_\varepsilon)$, 174
$L_p(\log L)_a(\Omega)$, 249
$L_{ru}(\log L)_a(I_\varepsilon)$, 175
$L_1(\mathcal{C}^\sigma)(\mathbb{R}^{2n})$, 391
$Lip(\mathbb{R}^n)$, 164
$Lip^{(1,-\alpha)}(\Omega)$, 227
$Lip_{pq}^{(1,-\alpha)}(\mathbb{R}^n)$, 248
$Lip(\Gamma)$, 309
ℓ_p, 7
$\|\lambda\,|\ell_p\|_\varrho$, 7
$\ell_q\left(2^{j\sigma}\ell_p^{E_j}\right)$, 314
$\ell_{\infty,\varkappa}\left[\ell_q\left(2^{k\sigma}\ell_\mu^{L_k}\right)\right]$, 284, 314

$S(\mathbb{R}^n)$, 3, 10
$S'(\mathbb{R}^n)$, 3, 10
$S(\Omega)$, 112
$S'(\Omega)$, 113

$W_p^s(\mathbb{R}^n)$, 3
$\mathbb{W}_p^s(\mathbb{R}^n)$, 358
$W_p^s(M)$, 82
$W_p^s(\Omega)$, 41
$\widetilde{W}_p^s(\Omega)$, 41
$\mathring{W}_p^s(\Omega)$, 41

Operators

B, 257, 258, 299
$\widehat{\varphi}, F\varphi$, 14
$\varphi^\vee, F^{-1}\varphi$, 15

H_β, 311, 322
H^β, 311, 329
H_β^b, 328
I_σ, 3, 27
I_s, 121
Δ, 27
Δ_g, 84, 92, 312
∇, 198
$(-\Delta)^{-1}$, 254, 255
id_Γ, 123
$ext(f|\Omega)$, 48
tr_Γ, 121, 123
tr^Γ, 252, 256
tr_b^Γ, 349
$tr_{\partial\Omega}$, 50
T, 355, 358
T^+, 355, 358
Q, 356, 387

Functions, numbers, relations

a_+, 7, 12, 212, 329
$[a]$, 212
$a_k(T)$, 279
$(\beta qu)_{\nu m}$, 12, 72, 98, 147, 341
$(\beta qu)_{\nu m}^L$, 28, 79, 115
$(\beta qu)_{\nu m}^{*L}$, 105
$C_{\alpha,p}(K)$, 260
$d_{t,u}^N f$, 37
$D_{t,u}^N g$, 54
$d(x), D(x), dist(x, \partial\Omega)$, 41, 43, 83, 235, 240
$\dim_H \Gamma$, 335
$e_k(T)$, 279
$f_{\nu m}$, 125

$f^*(t)$, 161, 181
$f^{**}(t)$, 161
$\mu_f(t)$, 161, 181

$\mathcal{E}_G|A_{pq}^s(t)$, 162, 190
$\mathcal{E}_G A_{pq}^s, [\mathcal{E}_G A_{pq}^s]$, 192
$\mathfrak{E}_G(A_{pq}^s)$, 164
$\mathfrak{E}_G A_{pq}^s$, 193
$\mathfrak{E}_{G,\Omega} A_{pq}^s$, 216
$\mathcal{E}_C|A_{pq}^{1+\frac{n}{p}}(t)$, 165, 197
$\mathcal{E}_C A_{pq}^{1+\frac{n}{p}}, [\mathcal{E}_C A_{pq}^{1+\frac{n}{p}}]$, 198
$\mathfrak{E}_C(A_{pq}^{1+\frac{n}{p}})$, 165
$\mathfrak{E}_C A_{pq}^{1+\frac{n}{p}}$, 198
$Ess\,Spec$, 322

$\mathcal{H}^d$, 133, 135, 256, 296

M_a, 56
$M_{\mu,s}f$, 128
$M_\mu^\alpha f$, 136
μ_{uv}^t, 145
μ_f, 161

N_β, 311, 322
N^β, 329
N_β^b, 328
R_n, 355, 363
$R_n(T^+, -\Delta + id)$, 396

$\omega(f,t)$, 164, 196
$\widetilde{\omega}(f,t)$, 164, 196
$\overline{\omega}(f,t)$, 196
$\omega_m(f,t)_p$, 247

$Spec$, 322
σ_p, 12, 35

σ_{pq}, 12, 35
$\widetilde{W}_{s,p}\mu$, 134
x^β, 10
$\langle\xi\rangle$, 3, 102
Δ_h, Δ_h^l, 4
$\chi_{\nu m}^{(p)}$, 10

$\sum_{\beta,\nu,m}$, 21, 30, 76, 149

$\sim$, 83, 256
$\#A$, 322

Index

Animal Farm, 186
approach, Weierstrassian, 6, 9, 24, 32, 33
atom, 80, 81, 205
atom, hydrogen, 325
atom, $(s, p, \Psi)_{K,L}$, 341

Banach, contraction mapping theorem, 392, 397
Big Bang, 329

capacity, (α, p), 261
capacity, $(1,2)$, 309
case, borderline, 225
case, critical, 162, 170, 173, 177, 202, 237
case, degenerate, 291
case, sub-critical, 162, 170, 173, 176, 198, 229, 238
case, super-critical, 164, 170, 173, 177, 218
coefficient, Cauchy, 24, 33
coefficient, optimal, 22, 23, 27, 32, 76, 107, 386
condition, ball, 138
condition, doubling, 139, 277
condition, interior ball, 156
condition, moment, 28
condition, outer ball, 310
condition, strong separation, 337
constant, Planck's, 311, 325
continuity, Lipschitz, 373, 388
continuous, Lipschitz, 361

continuous, uniformly, 361
convergence, absolute, 21
convergence, unconditional, 8, 21, 30, 74, 76
covering, locally finite, 86
couple, truncation, 355, 363
criterion, D. R. Adams, 132
criterion, Wiener, 309

d-domain, 86
d-domain, thorny starlike, 88, 312
d-set, 120, 135, 153, 251, 295, 330
(d, Ψ)-set, 120, 153, 253, 334
decomposition, Weyl, 157
differences, 4
dimension, differential, 172
dimension, Hausdorff, 335
dimension, interior Minkowski, 87
Dirichlet Laplacian, 254, 255
distance, 43, 235
distribution function, 161, 163, 181, 183
distribution, regular, 167
distribution, tempered, 10, 104
distribution, tempered, in domains, 113, 117
domain, minimally regular, 155
domain, interior regular, 155
drum, fractal, 254, 274

embedding, 93, 94
embedding, continuous, 167
embedding, limiting, 173

envelope, continuity, 165, 198
envelope, growth, 164, 193, 216
equation, semi-linear integral, 389, 390
equation, semi-linear differential, 394
euclidean n-space, 2
extension, 48

formula, Cauchy, 5, 9, 24
function, admissible, 333
function, continuity envelope, 165, 198
function, Green's, 244
function, growth envelope, 163, 192, 194, 216
function, maximal, 128, 136
function, monotone, 333
function, smooth, 226
function system, pseudo iterated, 337
function, tame, 289

geometry, bounded, 81, 83

Habilitationsschrift, Riemann, 226
Hamiltonian, 325

IFS, 289, 337
ψIFS, 337
inequality, Carl, 279
inequality, Hardy, 42, 52, 53, 236
inequality, Hardy-Littlewood maximal, 64
injectivity radius, 81, 83
interpolation, non-linear, 373
interpolation, real, 178

K-functional, 262

lattice, 10
lattice, approximate, 25, 144
lattice, irregular, 71

Lebesgue point, 260
limit, semi-classical, 325
Lipschitz function, 87
localization, 42, 59
localization, refined, 68

manifold, Riemannian, 81, 84, 312
means, ball, 37, 54, 66
means, cone, 54
means, local, 125
measure, diffuse, 139, 276
measure, directionally strongly diffuse, 278
measure, Hausdorff, 133, 256, 296
measure, multifractal, 289
measure, Radon, 122
measure, strongly diffuse, 277
measure, Weyl, 274, 288, 335
membrane, rusty, 353
method, Q, 357, 391, 402
modulus of continuity, 164, 196, 247
modulus of continuity, divided, 164, 196
monomial, 10
multi-index, 10
multiplier, pointwise, 67
music, of the ferns, 288

number, approximation, 279
number, entropy, 279
Nullstellenfreiheit, 271

operator, hydrogen-like, 326, 329
operator, identification, 124
operator, Laplace-Beltrami, 84, 92, 312
operator, pseudodifferential, 321
operator, Q, 356, 387
operator, semi-elliptic, 259
operator, trace, 120, 123, 349
operator, truncation, 358

potential, Bessel, 121
potential, Hedberg-Wolff, 134
potential, Riesz, 121, 308
potential, single layer, 294, 300, 301
primus inter pares, 186
principle, Birman-Schwinger, 323
principle, Max-Min, 323
procedure, constructive, 22
property, Fatou, 360
property, Fubini, 34, 35, 36
property, Markov, 158
property, truncation, 355, 361

quark, generalized, 25, 33, 72
quark, $(s,p) - \beta$, 12, 72, 147
quark, $(s,p)^L - \beta$, 28, 115
quark, $(s,p,\varkappa) - \beta$, 98
quark, $(s,p,\alpha)^{*L} - \beta$, 105
quark, $(s,p,\Psi) - \beta$, 341
quasi-everywhere, (α, p), 261
quasi metric, 159

rearrangement, non-increasing, 161, 181
region of admissibility, 396
regular, interior, 155
regular, minimally, 155
resolution of unity, 5, 6, 15, 25, 88, 145
resolution of unity, approximate, 67, 97
resolution of unity, approximate, family, 72, 97

set, pseudo self-similar, 337, 338
solution, classical, 309
space, B_{pq}^s, 13, 29, 30, 102
space, F_{pq}^s, 13, 29, 31, 88
space, bounded mean oscillation, 4, 169, 216, 379
space, classical Besov, 4

space, classical Sobolev, 3, 34, 41, 82
space, classical target, 168
space, Hardy, 4
space, Hölder-Zygmund, 4, 254
space, homogeneous type, 159
space, Lebesgue, 2, 10
space, Lipschitz type, 248
space, logarithmic Sobolev, 250
space, Lorentz, 174
space, Lorentz-Zygmund, 175
space, quasi-metric, 159
space, Schwartz, 10, 103
space, sequence, 11, 73, 98, 148
space, sequence, modified, 26
space, Sobolev, 3, 82, 379, 387
space, trace, 128, 131, 136, 138, 151, 152
space, very classical Sobolev, 255
space, Zygmund, 174
spectral synthesis, 159, 264
spectrum, essential, 322
spectrum, negative, 311, 322, 328
supersolution, 392, 398

Taylor expansion, 101, 111, 117
transform, Fourier, 15
transform, Fourier, inverse, 15

weight, 85